Future of Sustainable Agriculture in Saline Environments

Future of Sustainable Agriculture in Saline Environments

Edited by

Katarzyna Negacz, Pier Vellinga, Edward Barrett-Lennard,
Redouane Choukr-Allah, and Theo Elzenga

CRC Press
Taylor & Francis Group
Boca Raton London New York

CRC Press is an imprint of the
Taylor & Francis Group, an **informa** business

First edition published 2022
by CRC Press
6000 Broken Sound Parkway NW, Suite 300, Boca Raton, FL 33487-2742

and by CRC Press
2 Park Square, Milton Park, Abingdon, Oxon, OX14 4RN

CRC Press is an imprint of Taylor & Francis Group, LLC

ISBN: 978-0-367-62146-9 (hbk)
ISBN: 978-0-367-63147-5 (pbk)
ISBN: 978-1-003-11232-7 (ebk)

DOI: 10.1201/9781003112327

Typeset in Times
by KnowledgeWorks Global Ltd.

Contents

Preface

Climate Change triggers the imagination. What will the coastal fringes of the world look like with a rise in sea level of one meter or more? How can food production survive in the arid and semi-arid zones under higher temperatures and more persistent droughts? Emission control should limit global temperature rise to 2°C. But even then, droughts will become more persistent and sea level will keep on rising for many hundreds of years. Agriculture can help to capture carbon from the atmosphere but there is more to it: agriculture will have to adapt. Increasing the efficiency of water use is a number one priority, but in many regions of the world, this will not prevent an increase in the salinity levels in water and soils.

For these reasons, an increasing number of researchers and practitioners are exploring ways to produce food under saline soil and water conditions. Fortunately, they can learn from earlier practices and experience as many regions have a long tradition of struggling with salinization. Mutual learning is the major reason why some 200 experts and practitioners participated in the International Saline Futures Conference held in Leeuwarden, the Netherlands in September 2019.

The presentations and discussions at the conference revealed a strong sense of urgency. Better use of degraded or potentially degraded lands due to salinization will contribute to important global sustainable development goals (SDGs) such as reducing poverty, conservation of land and water resources, food security and economic growth, and the preservation of livelihoods in rural areas.

The conference called for action to better promote the need for increasing capacities and opportunities. But more is needed. The network of practitioners needs to be expanded and strengthened by building capacity. There is a need to support existing regional centers for research and set up new centers. And experiments and pilot projects require significant investments and participation by government and private sector actors.

This book presents a snapshot of current 'the state of the art' in saline agriculture including strategies for the future. The first section (Chapters 1 – 11) provides an overview of the situation and strategies in Australia, Middle East and North Africa, Eurasian countries, Bangladesh, and the North Sea region. It also describes the opportunities and barriers including articles on the economic aspects. The second section (Chapters 12-22) is focused on salination arising in coastal and river deltas and small islands as a result of climate change and sea level rise. It presents in several chapters how salinization differs by region, dependent on the hydro-geologic conditions. It discusses the strategies of creating new value chains based on the production and consumption of saline grown food products. The third section (Chapters 23-33) presents the progress in scientific understanding of the plant physiological aspects of salt sensitivity and salt tolerance. Finally, we would like to draw your attention to the first chapter: "Call to Action".

As editors of this book, we have organized an independent review of all articles. The content remains the full responsibility of the authors. We want to thank the Wadden Academy and the SalFar project participants for their participation and financial support.

Katarzyna Negacz
Pier Vellinga
Edward Barrett-Lennard
Redouane Choukr-Allah
Theo Elzenga

Acknowledgments

We would like to extend our thanks to reviewers of chapters included in this book:

Abdelaziz Sbai, Institute of Agronomy and Veterinary Hassan II, Rabta Morocco

Abdulrasoul M., Alomran King Saud University Riyadh, Saudi Arabia

Ana Delaunay, University of Lisbon, Portugal

Andrew Noble, Agriculture Research for Development Advisor

Anne Mette Teigen Asselin de Williencourt, Norwegian University of Life Sciences (NMBU)

Asgeir Almas, Norwegian University of Life Sciences (NMBU)

Atiq Rahman, Bangladesh Centre for Advanced Sciences Dhaka, Bangladesh

Bas Bruning, The Salt Doctors

Detlaf Steng, Ökowerk Emden, Germany

Dionysia-Angeliki Lyra, ICBA, UAE

Erick Verbruggen, University of Antwerp, Belgium

Erik Meijles, Faculty of Spatial Sciences University of Groningen, The Netherlands

Hanna Dijkstra, Institute for Environmental Studies, Vrije Universiteit Amsterdam, The Netherlands

Iain Gould, University of Lincoln Lincoln, UK

Jeroen De Waegemaeker, ILVO, Belgium

Joca Jansen, Wetterskip Fryslân, Leeuwarden, The Netherlands

Jouke Velstra, Acacia Water, Gouda, The Netherlands

Laurids Christiansen, Horsekaer, Denmark

Margot Faber, Province of Groningen, The Netherlands

Maria Konyuskova, Eurasian Center for Food Security, Russia

Marlise Vroom, Foodboosters, The Netherlands

Mohamed Hachicha National research Institute of rural engineering, Water and Forests (INRGREF), Tunis, Tunisia

Onno Kuik, Institute for Environmental Studies, Vrije Universiteit Amsterdam, The Netherlands

Oscar Widerberg, Institute for Environmental Studies, Vrije Universiteit Amsterdam, The Netherlands

Ragab Ragab Centre for Ecology and Hydrology, CEH, Natural Environment Research Council, NERC, Wallingford, UK

Richard W. Bell, Murdoch University, Australia

Richard George, The University of Western Australia, Department of Agriculture and Food, Australia

Sarah Garre, ILVO, Belgium

Tine te Winkel, Acacia Water, Gouda, The Netherlands

Editors

Katarzyna Negacz is a postdoctoral researcher at the Vrije Universiteit Amsterdam and cooperates with Wadden Academy. For more than 12 years she has been involved in research and practice related to sustainable development. After completing her studies in economics and law, she earned a doctoral degree in environmental economics at the Warsaw School of Economics for her research on the evolution of green consumption in Taiwan. She conducted research in Switzerland, Poland, Spain, Taiwan, Germany, and the Netherlands. Her current research focuses on the potential of saline degraded lands for sustainable food production and transnational biodiversity governance.

Pier Vellinga earned a PhD in coastal protection at Delft Technical University. He has a chair on climate change at the Vrije Universiteit Amsterdam since 1990. His teaching, research, and publications (about 200) focus on the implications of climate change regarding water, energy, and food. He joined Wageningen University in 2007 as a professor in climate change. Over the years he has fulfilled many different board positions in NGOs, research programs, and UN, EU, and governmental committees and financial institutions. For 30 years he has been an advisor to the Venice Water Authorities on the protection of Venice and its lagoon, a work successfully completed in 2020.

Edward Barrett-Lennard works in the Department of Primary Industries and Regional Development (DPIRD) of Western Australia, Murdoch University, and The University of Western Australia. For more than 35 years Professor Barrett-Lennard has been a passionate researcher and advocate of the need to develop saline agricultural farming systems in response to landscape salinization and climate change. His interests lie at the intersection between practical agriculture, agronomy, soil science, and ecophysiology. He is the author/editor of four books, more than 70 papers, and numerous other publications. Professor Barrett-Lennard has worked in Australia (mostly), Pakistan, Bangladesh, India, Iraq, and Vietnam.

Redouane Choukr-Allah is a horticultural, soil, and water environmental expert with more than 35 years of experience in the use of saline water and the use of pretreated sewage in Horticulture. He earned a PhD in environment horticulture at the University of Minnesota, St. Paul, United States. He also served as a technical coordinator of a 12 million project, financed by USAID on the water resources sustainability in Morocco. He served as head of the Horticulture Department from 1983 to 1996 and as head of the salinity and plant nutrition laboratory since 1996. He served at ICBA as a senior fellow scientist in horticulture and a Section Head of Crop Diversification and Genetics. He has produced numerous publications, including edited books, research reports, articles in peer-reviewed international journals, and books in the field of nonconventional water.

Theo Elzenga earned an MSc in biology at the University of Amsterdam and a PhD at Groningen on nutrient and CO_2 acquisition by plants. After working as a postdoctoral student at Wageningen University and at the University of Washington in Seattle, he returned to Groningen, where he has held a chair in ecophysiology of plants since 2000. His teaching focuses on the adaptation and acclimation of plants to adverse conditions. He was Director of the Centre of Ecological and Evolutionary Studies, Director of the Graduate School of Ecology and Evolution, and Director of the Undergraduate School of the Faculty of Science and Engineering. He is on advisory panels on agricultural development and the safety of genetically engineered organisms.

Contributors

Hayatullah Ahmadzai
International Center for Biosaline Agriculture
Dubai, United Arab Emirates

Basem Al Khawaldeh
Khalifa Fund for Enterprise Development
Dubai, United Arab Emirates
Khawla Mohammed Al Marzouqi
Abu Dhabi Agriculture and Food Safety Authority
Abu Dhabi, United Arab Emirates

Ohod Saleh Al Masjedi
Abu Dhabi Agriculture and Food Safety Authority
Abu Dhabi, United Arab Emirates

Mohamed Al Muhairi
Abu Dhabi Agriculture and Food Safety Authority
Abu Dhabi, United Arab Emirates

Mansoor Khamees Al Tamimi
Environmental Agency
Abu Dhabi, United Arab Emirates

Henrik Aronsson
Department of Biological & Environmental Sciences
University of Gothenburg
Gothenburg, Sweden

Giulia Atzori
Department of Agriculture, Food, Environment and Forestry
University of Florence
Florence, Italy

Maryam Bahrami
Water Engineering Department and Drought Research Center
Shiraz University
Shiraz, Iran

Edward G. Barrett-Lennard
Land Management Group, Agriculture Discipline
College of Science, Health, Engineering and Education
Murdoch University
Department of Primary Industries and Regional Development
South Perth, Australia
and
School of Agriculture and Environment
The University of Western Australia
Nedlands, Australia

Isa Camara Beauchampet
Vrije Universiteit Amsterdam
Amsterdam, The Netherlands

Richard W. Bell
Land Management Group, Agriculture Discipline
College of Science, Health, Engineering and Education
Murdoch University
Perth, Australia

Gary Bosworth
Department Newcastle Business School
Northumbria University
Newcastle upon Tyne, UK
and
Lincoln International Business School
University of Lincoln
Lincoln, UK

Bas Bruning
The Salt Doctors
Den Burg, The Netherlands

Abdelghani Chakhchar
Laboratory of Biotechnology and Plant Physiology
Mohammed V University
Rabat, Morocco

Tanmay Chaturvedi
Department of Energy Technology
Aalborg University
Esbjerg, Denmark

Redouane Choukr-Allah
Salinity and Plant Nutrition Laboratory
Institute of Agronomy and Veterinary Hassan II
Ait-Melloul, Morocco

Laurids Siig Christensen
Smagen af Danmark
Copenhagen, Denmark

Aslak H. C. Christiansen
Department of Plant and Environmental Science
University of Copenhagen
Frederiksberg, Denmark

Luca Cobre
Global Food Industries – Healthy Farm
Sharjah, United Arab Emirates

William K. Cornwell
School of Biological, Earth and Environmental Sciences
University of New South Wales
Kensington, Australia

Wasel Abdelwahid Abou Dahr
Environmental Agency
Abu Dhabi, United Arab Emirates

Mohamed Abdel Hamyd Dawoud
Environmental Agency
Abu Dhabi, United Arab Emirates

Arjen De Vos
The Salt Doctors
Den Burg, The Netherlands

Mindert de Vries
Van Hall University of Applied Sciences
Leeuwarden, The Netherlands
and
Deltares
Delft, The Netherlands

Jeroen De Waegemaeker
Flanders Research Institute for Agriculture, Fisheries and Food (ILVO)
Melle, Belgium

Mare Anne de Wit
Vrije Universiteit Amsterdam
Amsterdam, The Netherlands
Stichting De Zilte Smaak
Hoorn, Terschelling

Susanne Eich-Greatorex
Faculty of Environmental Sciences and Natural Resource Management (MINA)
Norwegian University of Life Sciences
NMBU
As, Norway

Theo Elzenga
Department of Ecophysiology of Plants
University of Groningen
Groningen, The Netherlands

Gualbert Oude Essink
Department of Groundwater Management
Utrecht University
Utrecht, The Netherlands

Iwona Gołębiewska
Bioscience Engineering and Earth and Life Institute
Université Catholique de Louvain
Louvain-la-Neuve, Belgium

Andrés Parra González
Department of Irrigation
Centro de Edafología y Biología Aplicada del Segura (CSIC)
University of Murcia, Espinardo
Espinardo, Spain

Iain Gould
Lincoln Institute for Agri-Food Technology
University of Lincoln
Lincoln, UK

Zeinab Hazbavi
Department of Natural Resources
Faculty of Agriculture and Natural Resources
and
Water Management Research Center
University of Mohaghegh Ardabili
Ardabil, Iran

Abdelaziz Hirich
African Sustainable Agriculture Research Institute (ASARI)
Mohammed VI Polytechnic University
Laayoune, Morocco
and
International Center for Biosaline Agriculture
Dubai, United Arab Emirate

Md. Iqbal Hossain
Tala
Satkhira, Bangladesh

Md. Sahadat Hossain
Department of Environmental Science
Stamford University, Bangladesh
Dhaka, Bangladesh

Joca Jansen
Wetterskip Fryslân
Leeuwarden, The Netherlands

Živko Jovanović
Faculty of Biology
University of Belgrade
Belgrade, Serbia

Enamul Kabir
Agrotechnology Discipline
Khulna University
Khulna, Bangladesh

Leena Karrasch
Department für Wirtschafts- und Rechtswissenschaften
University of Oldenburg
Oldenburg, Germany

Angelica Kaus
Province of Groningen
Groningen, The Netherlands

Gulchekhra Khasankhanova
Research and Development Institute "UZGIP"
Ministry of Water Resources
Tashkent, Uzbekistan

Maria Konyushkova
Eurasian Center for Food Security
Moscow, Russia

Alexander Krenke
Institute of Geography
Russian Academy of Sciences
Petersburg, Russia

Klaas Laansma
De Wikel
Groningen, The Netherlands

Efstathios Lampakis
Aquaculture and Aquaponics Specialist
Dubai, United Arab Emirates

Dionysia-Angeliki Lyra
International Center for Biosaline Agriculture
Dubai, United Arab Emirates

Nizamatdin Mamutov
Ecology Department GIS Center
Berdakh Karakalpak State University
Nukus, Uzbekistan

Johan Medenblik
Provincie Fryslân
Leeuwarden, The Netherlands

Afrin Jahan Mila
Land Management Group, Agriculture Discipline
College of Science, Health, Engineering and Education
Murdoch University
Perth, Australia
and
Irrigation and Water Management Division
Bangladesh Agricultural Research Institute
Gazipur, Bangladesh

Meis Moukayed
School of Arts and Sciences
American University in Dubai
Dubai, United Arab Emirates

Sasirekha Munikumar
Department of Ecophysiology of Plants
University of Groningen
Groningen, The Netherlands

Karaba N. Nataraja
Department of Crop Physiology
University of Agricultural Sciences
Bengaluru, Karnataka

Katarzyna Negacz
IVM
Vrije Universiteit Amsterdam, Wadden Academy
Amsterdam, The Netherlands

Hayley C. Norman
CSIRO Agriculture and Food
Wembley, Australia

Titian Oterdoom
Programma naar een Rijke Waddenzee
Leeuwarden, The Netherlands

Yevgenia Pankova
Dokuchaev Soil Science Institute
Moscow, Russia

Simon Pearson
Lincoln Institute for Agri-Food Technology
University of Lincoln
Lincoln, UK

Jacek Plewa
Global Food Industries – Healthy Farm
Sharjah, United Arab Emirates

Svetlana Radović
Faculty of Biology
University of Belgrade
Belgrade, Serbia

Muhammad Abdur Rahaman
Climate Change Adaptation Mitigation Experiment and Training (CAMET) Park
Noakhali, Bangladesh

Atiq Rahman
Bangladesh Center for Advanced Studies
Dhaka, Bangladesh

Marc van Rijsselberghe
Salt Farm Foundation
Den Burg, The Netherlands

Aaltje Rispens
Stichting Proefboerderijen Noordelijke Akkerbouw
Munnekezijl, The Netherlands

Elke Rogge
Flanders Research Institute for Agriculture, Fisheries and Food (ILVO)
Melle, Belgium

Jelte Rozema
Systems Ecology
Vrije Universiteit Amsterdam
Amsterdam, The Netherlands

Eric Ruto
Lincoln Business School
University of Lincoln
Lincoln, UK

Ali Reza Sepaskhah
Water Engineering Department
and
Drought Research Center
Shiraz University
Shiraz, Iran

Mohammad Shahid
International Center for Biosaline Agriculture
Dubai, United Arab Emirates

Mostafa Zabihi Silabi
Department of Watershed Management Engineering
Faculty of Natural Resources
Tarbiat Modares University
Tehran, Iran

Linda Smit
Stichting Proefboerderijen Noordelijke Akkerbouw
Munnekezijl, The Netherlands

Victor Statov
GIS Center
Berdakh Karakalpak State University
Nukus, Uzbekistan

Rezvan Talebnejad
Water Engineering Department
and
Drought Research Center
Shiraz University
Shiraz, Iran

Fatima Mohammed Bin Tarsh
Abu Dhabi Agriculture and Food Safety Authority
Abu Dhabi, United Arab Emirates

Mette H. Thomsen
Department of Energy Technology
Aalborg University
Esbjerg, Denmark

Sumitha Thushar
International Center for Biosaline Agriculture
Dubai, United Arab Emirates

Domna Tzemi
School of Geography
University of Lincoln
Lincoln, UK
and
Agricultural Production and Resource Economics
Technical University of Munich
Munich, Germany

Md. Nasir Uddin
Department/Institute
Bangladesh Centre for Advance Studies (BCAS)
Dhaka, Bangladesh

Pier Vellinga
Wadden Academy/IVM
Vrije Universiteit Amsterdam
Amsterdam, The Netherlands

Jouke Velstra
Acacia Water B.V.
Gouda, The Netherlands

Liping Wang
Department of Ecophysiology of Plants
University of Groningen
Groningen, The Netherlands

Jacqueline Wijbenga
Stichting De Zilte Smaak
Hoorn, Terschelling

Tine te Winkel
Acacia Water B.V.
Gouda, The Netherlands

Barbara Wolthuis
Bangladesh Center for Advanced Studies
Dhaka, Bangladesh

Isobel Wright
EPSRC Centre for Doctoral Training in Agri-Food Robotics
University of Lincoln
Lincoln, UK

Junjie Yi
Department of Ecophysiology of Plants
University of Groningen
Groningen, The Netherlands

Section I

Saline Agriculture: Global State of the Art and Strategies

1 Saline Agriculture
A Call to Action

Pier Vellinga, Atiq Rahman, Barbara Wolthuis, Edward G. Barrett-Lennard, Redouane Choukr-Allah, Theo Elzenga, Angelica Kaus, and Katarzyna Negacz

CONTENTS

1.1 INTRODUCTION: SALINE AGRICULTURE FROM THE CRADLE OF CIVILIZATION TO THE PRESENT AND FUTURE

The association between human civilizations and salinity has existed for thousands of years. The primary causes of the decline of the ancient Mesopotamian civilizations (located in modern Iraq) were three major salinization events: the first and most severe was from 2400 BC to 1700 BC, the second was between 1300 and 900 BC and the third occurred after 1200 AD (Jacobsen and Adams 1958). The reasons for these failures are familiar to irrigated agriculturalists today: the over-irrigation of land led to a rising water-table and consequent salinity and waterlogging, and there was major silting of water-courses and canals (Gelburd 1985; Shahid et al. 2018). In a similar manner to these ancient civilizations, most agricultural production today is still largely based on the use of freshwater resources and salinity remains a major threat (Figure 1.1).

The extent of salinization is difficult to estimate. The total area of saline and sodic lands is likely to be ~10% of arable land worldwide (Shahid et al. 2018). Ghassemi et al. (1995) estimated that ~20% of irrigated land is salt-affected; however, remote sensing studies show that in some countries up to 50% of irrigated land is salt-affected (Metternicht and Zinck 2003). The desertification of irrigated lands amounts to ~1.5 Mha around the world (Sentis 1996). The total surface area which

DOI: 10.1201/9781003112327-1

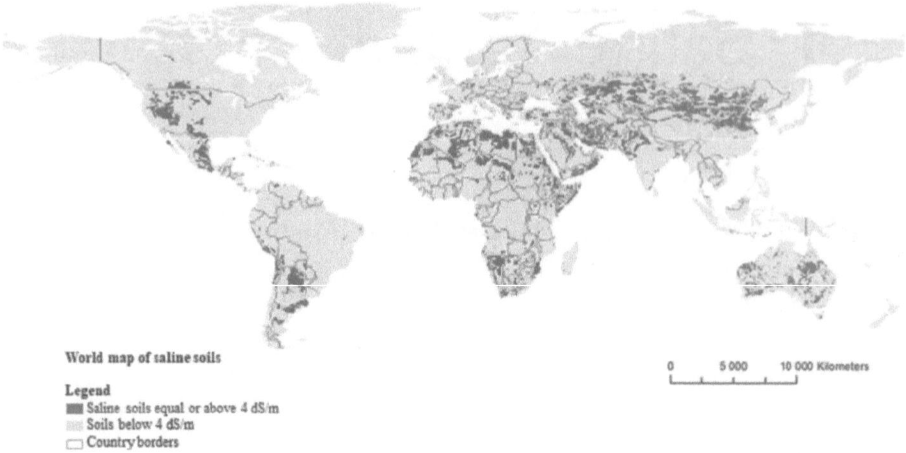

FIGURE 1.1 World map representing countries with salinity problems based on Negacz et al. (2019)

can be used for saline agriculture depends on several conditions such as water availability and soil fertility (Negacz et al. 2019).

More efficient use of freshwater and irrigation systems can play a major role in restoring freshwater agriculture in degraded soils. However, many countries with salinity problems are experiencing major water scarcity problems. Hoekstra and Mekkonen (2016) have developed a global map indicating the number of months in which water demand exceeds supply (Figure 1.2).

Many of the factors leading to soil salinization are being exacerbated by climate change; projections indicate more persistent droughts, an acceleration of sea-level rise and more extreme weather events projected by the Intergovernmental Panel on Climate Change (Pörtner et al. 2019). Salinization problems will become increasingly manifest in many coastal areas and wetlands, deltas of major rivers and small

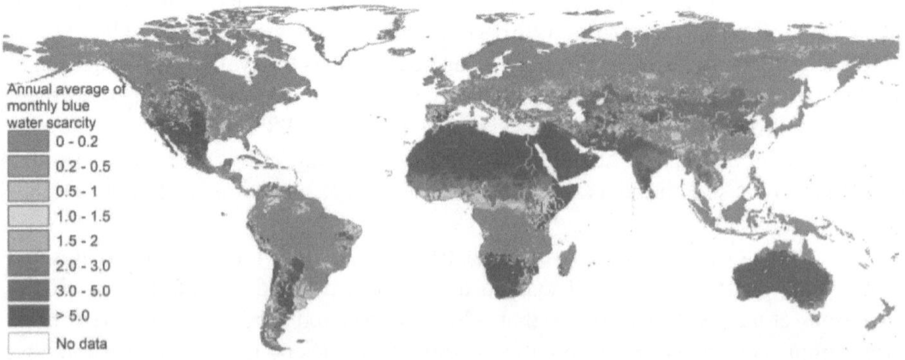

FIGURE 1.2 The number of months per year in which blue water scarcity exceeds 1.0 at 30 × 30 arc min resolution for period: 1996–2005. (From Mekonnen and Hoekstra, 2016.)

islands, not only through sea-level rise and more severe floods associated with storms, but also through groundwater intrusion by saline waters (Tiggeloven et al. 2020). In regions with semi-arid to arid climates, the availability of good quality water will continue to decrease so irrigation will continue using lower quality groundwater (FAO 2019b).

1.2 INITIATIVES TO GROW CROPS UNDER SALINE CONDITIONS

In the last few decades, there have been several initiatives in exploring the feasibility of growing food under saline conditions. The U.S. Salinity Laboratory was one of the earliest initiatives launched in 1954. More recently, the International Centre for Biosaline Agriculture (ICBA), initiated in 2000 in the United Arabic Emirates, has led the way in conducting research on problems and solutions for agricultural productivity under saline conditions (ICBA 2019, 20). Similar initiatives, usually at a smaller scale, have been undertaken in other countries like Australia, the Netherlands, Russia, China, Morocco and Egypt. Indigenous food production practices illustrate the wide variety of saline tolerant crops and agricultural practices that used to occur. Several international initiatives have been taken to explore and re-introduce indigenous knowledge and practices for food production in saline soils; examples are the International Partnership for the Satoyama Initiative.

Food production with saline soil and water is currently a marginal business and there is continuing momentum in maintaining the status quo. Often public money is used to ensure the continuity of freshwater availability (Chapter 13 of this book). Furthermore, there is an important element of cultural heritage and tradition in freshwater agriculture that keeps farmers and agricultural policymakers on the traditional track of continuity in freshwater provision to farmers at whatever the economic and environmental cost. For example, in Pakistan, farmers have the right to continue receiving freshwater through the canal command system even though there is increasing evidence that this distribution is becoming unsustainable now and the deficits will be even greater in the future (Zawahri 2011).

The explanation for this may be that under the present market conditions, with agricultural subsidies and the exclusion of environmental costs and externalities, saline agriculture cannot compete with freshwater agriculture. This market distortion makes it generally more financially attractive to overdraw on freshwater and land resources than to invest in food production with saline soils and water. Furthermore, under present international market conditions, it is seen to be "cheaper" to deforest virgin areas or drain freshwater lakes and wetlands, than to revitalize saline lands. However, markets do not reflect the real costs of the use of land. Compared to trees, grasses absorb only a fraction of CO_2. Deforestation to create more land for cattle grazing therefore means a massive loss of compensation of CO_2 and is a high driver of climate change. So far, the FAO life cycle assessment does not take that into account, whereas research has clearly shown the relation between deforestation, meat production, water scarcity and CO_2 implications on land use (Schmidinger and Stehfest 2012). Therefore, revitalization and regeneration of soils is a viable option in many situations. It does require specific skills in water and soil management, and it takes several years of investment before production levels are at their potential.

The pressure to repair the market failure is increasing from three directions: (1) growing food demand and subsequent calls for more land (FAO 2009), (2) the urgency to reduce greenhouse gas emissions and subsequent calls to stop deforestation and drainage of wetlands, and (3) the growing opportunities to produce food on salinized lands (Chapter 7 of this book).

Recently the FAO (2019a) has initiated a thematic working group with the aim "to explore the opportunities offered by saline environments (water and soil) for agriculture". The European Commission in its EIP-AGRI program (EIP-AGRI 2019) has initiated a focus group on saline agriculture around the question: "How to maintain agricultural productivity by preventing, reducing or adapting to soil salinity".

Recent initiatives cover not only crops, soil and water but also circular saline farming (farming that includes the use of all residual biological products in the production chain). Multi-functional solutions are being explored such as combinations of food production and flood protection through activities like mangrove-based agroforestry. In addition, a number of traditional/indigenous crops with salt-tolerant varieties are being identified. Nature-based solutions are being explored such as wetland farming and marine farming next to hydroponics (floating agriculture) and integral farming (ICBA 2015b).

However, as saline farming is about new food products and new markets, it is difficult to develop specific food chains of significant volumes. It is recognized among experts for example that quinoa (ICBA 2015a) and potatoes (van Straten et al. 2019) have salt-tolerant varieties with some 80% yield and excellent quality when produced under moderately saline soil and water conditions. Still, this potential is not used even though there are millions of hectares of underutilized moderately saline land areas. An additional challenge is the production of dedicated seeds as seed rights can have a positive as well as a negative effect on the use of salt-tolerant varieties/species. In conclusion, there are a number of groups and regions experimenting in saline agriculture. However, the field is fragmented and the development of saline farming appears to be an uphill battle.

1.3 FOUR REASONS TO INVEST IN PRODUCING FOOD UNDER SALINE CONDITIONS

During the 2019 International Saline Futures Conference at Leeuwarden in the Netherlands, researchers and practitioners presented a range of reasons for boosting investments in the exploration and further development of food production under saline conditions. The arguments can be summarized in four lines:

1. *There is a need to address growing freshwater scarcity.* Freshwater scarcity is growing. The area of salinized land is growing rapidly. Public funding of freshwater for agriculture is likely to reach its limits. Chapter 4 of this book describes the situation regarding salinization for the North African countries. Chapter 3 of this book provides an overview for the situation in Australia. Increasing the efficiency of freshwater use is one line of action; a parallel one is the introduction of salt tolerant species in combination with specific water and soil management practices. These two lines have synergistic benefits.

2. *There is a need to stop the loss of biodiversity while meeting growing food demand.* Population growth inducing the global food demand exerts pressure on land-use change. The smart use of saline lands can prevent the progressing destruction of unique ecosystems and biodiversity hotspots, such as the remaining forests and coastal wetlands.
3. *There is a need to adapt to climate change.* Climate change is bringing more severe weather events such as droughts, erratic rainfall patterns, more severe cyclones and accelerated sea-level rise. More frequent and severe coastal storm surges will cause salinity intrusions into river-deltas and low-lying coastal areas and small islands, resulting in increasing salt stress and yield losses. The inundation of fertile low lands is likely to become an existential threat to agricultural livelihoods in many places around the world, leading to the internal displacement or migration of local inhabitants as described by Rahman and Uddin (2021). Food production under saline conditions and innovation in this field could help to create an economic and social perspective for the regions and populations affected.
4. *There is a need to increase opportunities and capacities to produce food under saline soil and water conditions.* A wide range of experiments around the world is showing a large economic potential for innovative saline agriculture systems. Chapter 29 of this book reports a series of successes in growing quinoa in regions of North Africa. Chapter 11 of this book report on innovations in the field of integral saline farming where aquaculture and crop production are being combined. Chapter 21 of this book report on successful pilots on potato yields under saline conditions in Bangladesh. Chapter 30 of this book explores the potential of edible halophytes as new crops in saline agriculture using the example of the ice plant.

1.4 TWO LINES OF ACTION

Based on the reasoning presented above, we propose to focus on two lines of action:

1. *Building the international community of saline agricultural science and practice* by organizing meetings for sharing knowledge; by setting up specialized journals, social media platforms, conferences and webinars; by developing a network of cutting-edge science, innovations and policy solutions in agriculture and water management.
2. *Enhancing investment* in research, experimental centers and pilots for food production under saline conditions. Agendas of climate change adaptation and food production under saline conditions provide opportunities for new initiatives by national and international organizations such as the FAO, World Food Program (WFP) and multilateral financing organizations like the Green Climate Fund, the World Bank, the Development Finance Institutions (DFI's) and private impact investors. A major bottleneck is the availability of "bankable" projects. To meet this challenge, it is important to invest in capacity building, set-up local pilot projects and establish regional centers of excellence for saline agricultural research and development, and engage with investors at an early stage to allow for economic upswing.

1.5 IN SUPPORT OF THE UN SUSTAINABLE DEVELOPMENT GOALS

Increasing investments in saline agriculture are fully in line with the majority of the UN Sustainable Development Goals. Chapter 2 of this book shows that in particular "Zero hunger" (SDG 2), "Decent work and economic growth" (SDG 8) and "Addressing freshwater scarcity" (SDG 6) are supported. In addition, "Partnerships for the goals" (SDG 17) in the context of saline agriculture is most relevant. Saline agriculture and the underlying hydro-geological conditions vary greatly by geographic area. It is clear that the situation and the relevant processes differ enormously by region and within regions, they differ by location. However, experts and practitioners can unite in the quest to grow ecologically and economically viable crops in under conditions of freshwater scarcity and moderately to highly saline soils.

Enhancing investment in saline agriculture requires the development of new partnerships. Partnerships between research and practitioners in the fields of agriculture, water, economics and food are all equally important: partnerships should also include experts in rural sociology, economics and finance. In fact, a transdisciplinary community is required to move forward. New investments are necessary because the international and national communities are relatively small and fragmented at present.

Relatively new is the concern of farmers in delta regions with abundant rainfall, such as in South East Asia as described by Chapter 8 of this book and the situation around the North Sea as described by Chapter 5 of this book. Sea level rise, more frequent flooding, saline groundwater seepage and more persistent droughts are increasingly threatening the yields in these fertile lands. The prospect of sea-level rise by 1–2 meters over the next one hundred years is creating a sense of urgency.

1.6 TOWARD AN INNOVATIVE AGENDA

We argue that any agenda on food production under saline conditions should be transdisciplinary and multinational covering field experiments as well as socio-economic research and policy evaluation. A local focus with stakeholder participation is crucial, as is (inter)national knowledge sharing and dissemination. Financial support and new investments are also necessary as the international and national communities are at present relatively small and fragmented. Figure 1.3 illustrates how different stakeholders and different scientific disciplines could work together.

The 2019 International Saline Futures Conference presented the following fields for investment in capacity building, and research and development:

1. identifying and improving salt tolerant crop varieties
2. innovation in farming practices: exploring regenerative techniques and practices enhancing the carbon content of soils, integral farming including aquaculture and crop farming, hydroponics
3. evaluation and innovation considering the full value chain including product and market development and promotion
4. field testing and large-scale pilot projects

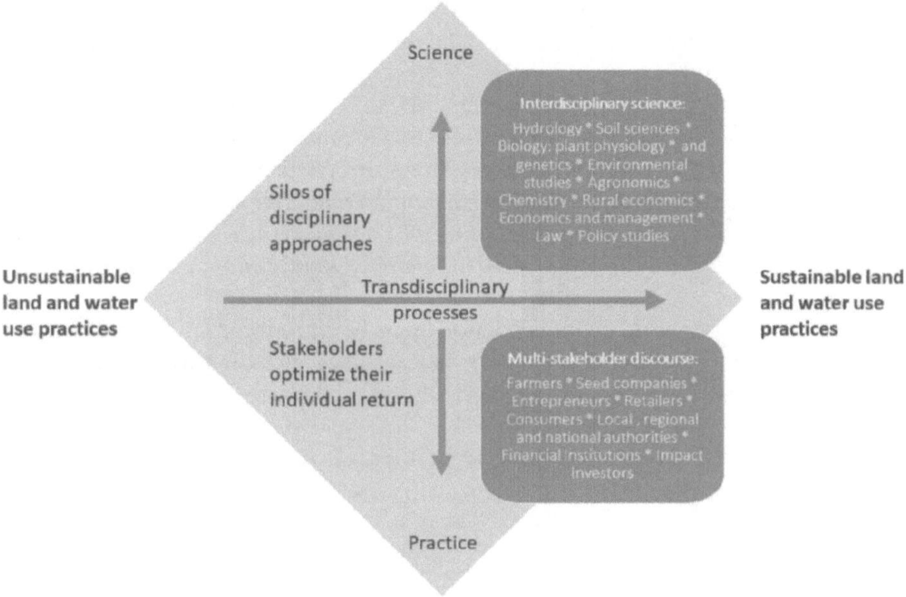

FIGURE 1.3 Illustration of a transdisciplinary approach toward food production under saline conditions. (Inspired by R.W. Scholz, G. Steiner, Sustainability Science, August 2015, doi: 10.1007/s11625-015-0326-4.)

5. training and capacity building, including the establishment of an international scientific journal and on-line networks
6. development of regional centers for research and large-scale pilots
7. development of supportive land and water use policies
8. creation and implementation of investment opportunities

We argue that existing national salinity centers such as in the USA, Australia, Morocco, Egypt, Jordan and India together with international centers such as ICBA should be reinforced in terms of scope and budget. But new centers also need to be established. The fertile soils of the world's deltas threatened by climate change and related salt water flooding and seepage require urgent attention. Innovations in coastal agriculture, land use and water and soil management are of crucial importance to ensure the longer-term food production in deltas. Innovation in land use in these areas is equally required to avoid loss of economic welfare and subsequent migration away from these climate-vulnerable areas (Hassani et al. 2020).

1.7 CONCLUSION

As participants of the 2019 International Saline Futures Conference, we conclude that the opportunities to produce food, fuel, forage and fiber under saline soil and water conditions deserve much more attention at an international and national level for four reasons: (1) *The need to address the growing freshwater scarcity*, (2) *the need to stop the loss of biodiversity while meeting growing food demand*, (3) *the need to adapt to climate change*, and (4) *the increasing capacities to produce food under saline soil and water conditions*. We propose a transdisciplinary approach in developing and implementing a "Research to Development Continuum" to ensure a continuous interaction between farmers, researchers, marketeers and investors. Two lines of investment are recommended to address this challenge: (A) *capacity building and strengthening the international community including its national and local stakeholders*, and (B) *strengthening existing and setting up new expertise centers in particular through a combination of two agenda's: climate change adaptation and saline agriculture*.

These conclusions and recommendations will be introduced in the discussions with national and international business and agencies including agriculture, food and seed companies, waterboards and agricultural ministries, the FAO, World Food Program (WFP), World Bank, UNEP, IsDB, Development Finance Institutions (DFI's), the Green Climate Fund and Private Impact Investors.

REFERENCES

Atzori, Giulia. 2021. "The Potential of Edible Halophytes as New Crops in Saline Agriculture: The Ice Plant (*Mesembryanthemum crystallinum* L.) Case Study." In *Future of Sustainable Agriculture in Saline Environments*. Routledge, Taylor & Francis Group, London.

Barrett-Lennard, Edward G., and Hayley Norman. 2021. "Agriculture in Salinizing Landscapes in Southern Australia: Selected Research 'Snapshots.'" In *Future of Sustainable Agriculture in Saline Environments*. Routledge, Taylor & Francis Group, London.

Beauchampet, Isa Camara. 2021. "Stakeholder Perspectives on the Issue of Salinization in Agriculture in The Netherlands." In *Future of Sustainable Agriculture in Saline Environments*. Routledge, Taylor & Francis Group, London.

Choukr-Allah, Redouane. 2021. "Use and Management of Saline Water for Irrigation in the Near East and North Africa (NENA) Region." In *Future of Sustainable Agriculture in Saline Environments*. Routledge, Taylor & Francis Group, London.

EIP-AGRI. 2019. "European Innovation Partnership 'Agricultural Productivity and Sustainability' – EIP-AGRI – European Commission." *EIP-AGRI*./eip/agriculture/en/european-innovation-partnership-agricultural.

FAO. 2009. "High Level Expert Forum - How to Feed the World in 2050." FAO. http://www.fao.org/fileadmin/templates/wsfs/docs/Issues_papers/HLEF2050_Global_Agriculture.pdf.

———. 2019a. "Saline Agriculture | Land & Water | Food and Agriculture Organization of the United Nations | Land & Water | Food and Agriculture Organization of the United Nations." http://www.fao.org/index.php?id=99230.

———. 2019b. "The Multi-Faced Role of Soil in the Near East and North Africa." Policy brief. Global Soil Partnership. FAO. http://www.fao.org/3/ca7123en/CA7123EN.pdf.

Gelburd, Diane E. 1985. "Managing Salinity Lessons from the Past." *Journal of Soil and Water Conservation* 40 (4). Soil and Water Conservation Society: 329–31.

Ghassemi, Fereidoun, Anthony J. Jakeman, and Henry A. Nix. 1995. *Salinisation of Land and Water Resources: Human Causes, Extent, Management and Case Studies*. CAB International.

Gould, Iain, Jeroen De Waegemaeker, Domna Tzemi, Isobel Wright, Simon Pearson, Eric Ruto, Leena Karrasch, et al. 2021. "Salinization Threats to Agriculture across the North

Sea Region." In *Future of Sustainable Agriculture in Saline Environments.* Routledge, Taylor & Francis Group, London.

Hassani, A., Azapagic, A., & Shokri, N. (2020). Predicting long-term dynamics of soil salinity and sodicity on a global scale. *Proceedings of the National Academy of Sciences*, 117(52), 33017–33027.

ICBA. 2015a. "Quinoa for Marginal Environments, Project brief International Centre for Biosaline Agriculture." Project Brief. ICBA. https://www.biosaline.org/sites/default/files/Projectbrieffiles/Quinoa-Project_Brief-Final-2.pdf.

———. 2015b. "Integrated Aqua-Agriculture for Enhanced Food and Water Security." International Center for Biosaline Agriculture. April 28. https://www.biosaline.org/projects/integrated-aqua-agriculture-enhanced-food-and-water-security.

———. 2019. "ICBA at 20: A Look at History." International Center for Biosaline Agriculture. December 19. https://www.biosaline.org/corporate-publications/icba-20-look-history.

Jacobsen, Thorkild, and Robert M. Adams. 1958. "Salt and Silt in Ancient Mesopotamian Agriculture." *Science* 128 (3334). JSTOR: 1251–58.

Lyra, Dionysia Angeliki, Efstathios Lampakis, Mohamed Al Muhairi, Fatima Mohammed Bin Tarsh, Mohamed Abdel Hamyd Dawoud, Basem Al Khawaldeh, Meis Moukayed, et al. 2021. "From Desert Farm to Fork: Value Chain Development for Innovative Salicornia-Based Food Products." In *Future of Sustainable Agriculture in Saline Environments.* Routledge, Taylor & Francis Group, London.

Mekonnen, Mesfin M., and Arjen Y. Hoekstra. 2016. "Four Billion People Facing Severe Water Scarcity." *Science Advances* 2 (2). American Association for the Advancement of Science: e1500323. doi:10.1126/sciadv.1500323.

Metternicht G.I. and Zinck J.A. 2003. Sensing of Soil Salinity: Potentials and Constraints." *Remote Sensing of Environment* 85 (1). Elsevier: 1–20.

Negacz, Katarzyna, Bas Bruning, and Pier Vellinga. 2021. "Achieving Multiple Sustainable Development Goals through Saline Agriculture." In *Future of Sustainable Agriculture in Saline Environments.* Routledge, Taylor & Francis Group, London.

Negacz, Katarzyna, and Pier Vellinga. 2021. "Cost or Benefit? Estimating the Global Economic Potential of Saline Agriculture." In *Future of Sustainable Agriculture in Saline Environments.* Routledge, Taylor & Francis Group, London.

Negacz, Katarzyna, Pier Vellinga, and Arjen de Vos. 2019. "Saline Degraded Lands Worldwide: Identifying Most Promising Areas for Saline Agriculture." *Paper Presented at the Saline Futures Conference.*

Pörtner, H. O., D. C. Roberts, V. Masson-Delmotte, P. Zhai, M. Tignor, E. Poloczanska, K. Mintenbeck, M. Nicolai, A. Okem, and J. Petzold. 2019. "IPCC, 2019: Summary for Policymakers." *IPCC Special Report on the Ocean and Cryosphere in a Changing Climate.*

Rahman, Atiq, and Md. Nasir Uddin. 2021. "Challenges and Opportunities for Saline Agriculture in Coastal Bangladesh." In *Future of Sustainable Agriculture in Saline Environments.* Routledge, Taylor & Francis Group, London.

Schmidinger, Kurt, and Elke Stehfest. 2012. "Including CO_2 Implications of Land Occupation in LCAs—Method and Example for Livestock Products." *The International Journal of Life Cycle Assessment* 17 (8): 962–72. doi:10.1007/s11367-012-0434-7.

Sentis, I. 1996. "Soil Salinization and Land Desertification." *Soil Degradation and Desertification in Mediterranean Environments.* Logroño, Spain, Geoforma Ediciones, 105–29.

Shahid, Mohammad. 2021. "Response of Quinoa to High Salinity under Arid Conditions." In *Future of Sustainable Agriculture in Saline Environments.* Routledge, Taylor & Francis Group, London.

Shahid, Shabbir A., Mohammad Zaman, and Lee Heng. 2018. "Soil Salinity: Historical Perspectives and a World Overview of the Problem." In *Guideline for Salinity*

Assessment, Mitigation and Adaptation Using Nuclear and Related Techniques, edited by Mohammad Zaman, Shabbir A. Shahid, and Lee Heng, 43–53. Cham: Springer International Publishing. doi:10.1007/978-3-319-96190-3_2.

Tiggeloven, Timothy, Hans de Moel, Hessel C. Winsemius, Dirk Eilander, Gilles Erkens, Eskedar Gebremedhin, Andres Diaz Loaiza, et al. 2020. "Global-Scale Benefit–Cost Analysis of Coastal Flood Adaptation to Different Flood Risk Drivers Using Structural Measures." *Natural Hazards and Earth System Sciences* 20 (4). Copernicus GmbH: 1025–44. https://doi.org/10.5194/nhess-20-1025-2020.

van Straten, G., A. C. de Vos, J. Rozema, B. Bruning, and P. M. van Bodegom. 2019. "An Improved Methodology to Evaluate Crop Salt Tolerance from Field Trials." *Agricultural Water Management* 213. Elsevier: 375–87.

Vos, Arjen de, Andres Parra González, and Bas Bruning. 2021. "Case Study: Putting Saline Agriculture into Practice – A Case Study from Bangladesh." In *Future of Sustainable Agriculture in Saline Environments*. Routledge, Taylor & Francis Group, London.

World Data Lab. 2020. "Water Scarcity Clock." https://worldwater.io/?utm_source=google&utm_medium=search&utm_campaign=WaterscarcityData&campaignid=6444167483&adgroupid=77198318295&adid=376808482557&gclid=Cj0KCQjw2or8BRCNARIsAC_ppyYcIxTPIIjWJt26sSt1BVQPWWhiuZ6cODYw0JAQ4BD0DrwLaS2zQf8aAnl3EALw_wcB.

Zawahri, Neda A. 2011. "Using Freshwater Resources to Rehabilitate Refugees and Build Transboundary Cooperation." *Water International* 36 (2): 167–77. doi:10.1080/02508060.2011.557994.

2 Achieving Multiple Sustainable Development Goals through Saline Agriculture

Katarzyna Negacz, Bas Bruning, and Pier Vellinga

CONTENTS

2.1 INTRODUCTION

Recent research shows that saline agriculture is gaining popularity as a management technique for saline soils (Dagar et al. 2016, 2019; De Waegemaeker 2019). This form of revitalisation is an integrated approach addressing multiple sectors at the same time. There is a need to better understand the impact of saline agriculture on society, the economy and the environment as well as to uncover potential synergies and trade-offs within saline agriculture. The United Nations Sustainable Development Goals (SDGs) provide a systematic and reliable framework to address nexus topics (Stoorvogel et al. 2017; Hülsmann & Ardakanian 2018; Liu et al. 2018;

DOI: 10.1201/9781003112327-2

van Noordwijk et al. 2018). These goals were set by the United Nations in 2015 as a way to achieve a more sustainable future. Table 2.1 presents the list of the goals with their main focus. We argue saline agriculture can best be viewed as a multi-sectorial topic, as it acts across multiple sectors and touches upon multiple SDGs.

Soil salinization is one of the reasons for soil degradation and has an impact on land use, water supply, soil fertility, and plant (and animal) community composition. It is defined as the accumulation of water-soluble salts in the soil to a level that impacts agricultural production, environmental health, and economic welfare (FAO 2011). Salinization is a worldwide problem occurring on more than 400 million ha (more than twice the total area of European farmland) and the salt-affected land area is likely to increase rapidly as a result of climate change and sea-level rise (Joe-Wong et al. 2019). It occurs in more than 75 countries on 20% of the global irrigated land (Ghassemi et al. 1995). It creates both land and water issues, having a major impact on land productivity and crop production (Datta & Jong 2002). Salinity has an adverse effect on crops because a low osmotic pressure hampers the absorption of water and because soluble salts can accumulate to toxic levels in plant tissues (Munns and Tester 2008).

The negative effects of salinisation on crop growth and the increasing land surface area suffering from it have an effect on food security and sustainability. Food security relates to all people having at all times, "physical, social and economic access to sufficient, safe and nutritious food to meet their dietary needs and food preferences for an active and healthy life" (FAO 2009). Food sustainability includes economic, social and environmental issues, representing the three classical dimensions

TABLE 2.1
17 Sustainable Development Goals

Goal Number	Theme
SDG1	No Poverty
SDG2	Zero Hunger
SDG3	Good Health and Well-being
SDG4	Quality Education
SDG5	Gender Equality
SDG6	Clean Water and Sanitation
SDG7	Affordable and Clean Energy
SDG8	Decent Work and Economic Growth
SDG9	Industry, Innovation and Infrastructure
SDG10	Reduced Inequality
SDG11	Sustainable Cities and Communities
SDG12	Responsible Consumption and Production
SDG13	Climate Action
SDG14	Life Below Water
SDG15	Life on Land
SDG16	Peace and Justice Strong Institutions
SDG17	Partnerships to Achieve the Goals

of sustainable development. Current farming practices exploit considerable amounts of natural resources, i.e. major shares of all ice-free land (33%), freshwater (70%) and energy production (20%) (Smil 2001; Aiking 2014). Due to the continuing pressure on resources and land, as well as population growth leading to increased demands, food prices are expected to rise by 70–90% by 2030 (KPMG International et al. 2012). As a result, new innovative solutions need to be studied to increase food production through higher yields on degraded lands and to minimise pressure on the environment.

One of these options is saline agriculture. It involves irrigation solutions, different soil and water management and different crop species and variety choices. Thanks to these actions, despite degradation, saline lands can be further used for agricultural purposes. The choice of the methods to be applied on a selected area will depend on multiple factors such as the geomorphological and environmental aspects of the site, the socio-economic environment, the capacity of services and operational and maintenance factors (FAO 2018). The implementation of saline agriculture does not come without a cost. The management techniques usually require an initial investment in the irrigation system, equipment and/or seeds. There is also the risk of off-site effects. For example, saline irrigation could result in the pollution of groundwater or cause salinization of adjacent good quality land. These can also be seen as costs. At the same time, it brings benefits of enhanced global cooperation, the inclusion of private partners and civic society, as well as inspirational value for countries around the world.

Food production lies at the centre of saline agriculture. Rockström & Sukhdev (2016) argue that all SDGs are linked to sustainable and healthy diets. They highlight that economies and societies are embedded in the environment connecting all related SDGs. In particular, they relate food to eradicating poverty (SDG 1) and famine (SDG 2), implementing gender equality (SDG 5), providing decent jobs (SDG 8) and reducing inequality (SDG 10). This approach suggests that the SDGs should be examined not separately but as a system of direct and indirect interconnections. These interconnections can occur at various stages of food production.

This paper addresses the research question of which SGDs are directly and indirectly related to the revitalisation of saline soils through saline agriculture. Our hypothesis is that saline agriculture supports the SDGs of food security (SDG 2), the use of freshwater resources (SDG 6), adaptation to climate change (SDG 13) and sustainable livelihoods (SDG 8). If not managed properly, it has the potential to have adverse effects on the marine (SDG 14) and terrestrial (SDG 15) biodiversity.

2.2 METHODS

To answer the research question, we applied a two-step research process. First, we constructed a simplified Drivers-Pressures-State-Impacts-Response (DPSIR) scheme to investigate the relationships between causes and consequences of salinization, and their links to SDGs. Second, we conducted semi-structured interviews with experts to discuss constraints and opportunities for saline agriculture and examine which SDGs' areas appear most often.

2.2.1 DPSIR Framework

Building on previous studies on salinization, we designed a simplified DPSIR scheme (Cooper 2013; Patrício et al. 2016). It was used to present connections between natural and social sciences related to the topic, and show the flow between actions and possible solutions for policymakers.

The DPSIR framework is a tool used to structure and understand various environmental and socio-economic activities better. Drivers can be described as "the social, demographic and economic developments in societies and the corresponding changes in lifestyles, overall levels of consumption and production patterns" (van Teeffelen 2017). These are the activities that are undertaken to enhance human well-being and welfare, often defined as the sectors that satisfy human needs (e.g. agriculture, industry, transport). Further, the pressure is a means by which the driver causes a change in the state. Then, states are changes in the properties of the natural environment. Consequently, an impact is an effect on welfare caused by the change in the state. Finally, responses are actions taken in reply to the changes in states and impacts (van Teeffelen 2017, pp. 43–55). The DPSIR framework is well fitted to analyse anthropocentric trade-offs in environmental decision-making, e.g. through cost-benefit analysis or input-output models (Cooper 2013).

The DPSIR scheme presented here was created in the research process which can be divided into two stages:

- *Creating a database:* A database of 72 documents, including scientific literature, conference proceedings, official publications and reports, related to the potential of saline agriculture was created using the following keywords in a Google Scholar search: saline agriculture, saline agriculture potential, saline agriculture benefits, saline agriculture challenges. Additional literature was added based on the recommendations from five experts in the field. The documents in the database were reviewed in order to make sure they addressed saline agriculture.
- *Quantitative content analysis:* Then the database was automatically searched with several keywords per SDG in the Atlas.ti software to score the number of times these SDG terms were mentioned. Keywords, such as desalination, agricultur* or climate change, associated with the 17 SDGs, can be found in the Appendix. Further, we quantified the number of SDGs mentioned in our database by counting the keywords associated to the SDGs and expressing them relative to the total number or presence of SDG keywords. Then, we summarised this information by designing a simplified DPSIR graph. For example, for SDG 6, we selected "Desalination" as one of the keywords because it is described in the targets. Then we scanned 72 publications related to saline agriculture for quotations with this keyword using Atlas.ti software. Further, we counted the number of quotations in which "Desalination" appears. We summed quotations for all the keywords for SDG 6. After analysing all the SDGs in this way, we converted the sum for each SDG into a percentage of total quotations.

The findings were compared with findings from the semi-structured interviews.

2.2.2 Semi-Structured Interviews

Interviews with experts were conducted to understand various underlying conditions for saline agriculture. First, based on a literature review, we developed a questionnaire. Second, a pilot interview was conducted which allowed us to adjust questions and restructure the questionnaire. All experts were asked a similar set of questions, which was modified in certain cases to better fit their field of expertise. Third, experts for interviews were selected based on their publications and work in the field, as well as through the snowball sampling method (Christopoulos 2009). Another factor for respondents' selection was the geographical area of their expertise. Maximum variation sampling was used to provide a full picture of global potential. Eight of the experts consulted worked for large research centres, two for universities, two for consulting and training companies, and two for governmental institutions. Their areas of expertise included ecology and agriculture (three), land restoration (four), economics (three), policy (three), climate modelling (three), and soil research (two). Most scientists researched more than one field. Their expertise is concentrated in Australia, Bangladesh, Central Asia countries (e.g. Uzbekistan), Kenya, Middle East countries (e.g. Kuwait, United Arab Emirates, Qatar, Oman), Netherlands, Niger, Morocco, Pakistan, Russia, Sweden, Spain and the United Kingdom. Interviewees were coded from El to E11. Each interview was transcribed and summarised.

We conducted quantitative and qualitative content analysis. The summaries were coded automatically using selected keywords matching SDGs (see Appendix) to count their presence (direct quotations) and coded manually for expressions matching SDGs (indirect quotations). We defined a direct quotation as one including keywords assigned to an SDG. An indirect quotation was understood to be an expression which relates to an SDG but did not use selected keywords. This two-fold approach allowed for more precise analysis of the interviews. Further, we examined the summaries and selected four overarching SDG topics emerging from the experts' interviews.

2.3 RESULTS

This section presents the results of the DPSIR analysis and interviews with the experts. To facilitate comparison between the two approaches, we express the results in a comparative manner.

2.3.1 From Drivers to Responses: SDGs in the Saline Agriculture

Figure 2.1 shows the main causes and consequences of the salinization process. Due to the complexity of salinization and location-specific issues, only main phenomena were included in the graph which allowed us to track connections to SDGs. The DPSIR categories presented in Figure 2.1 are derived from the documents in our database.

Each category of the DPSIR framework relates to several SDGs which we outline in Table 2.2. For example, the drivers of salinization are mostly related to SDG 13 ("Climate action") and SDG 2 ("Zero hunger") and their targets.

Further, we investigated which SDGs were the most related to saline agriculture following the procedure described in Section 2.2. We scored the number of times each keyword related to SDG was mentioned in our database (Figure 2.2).

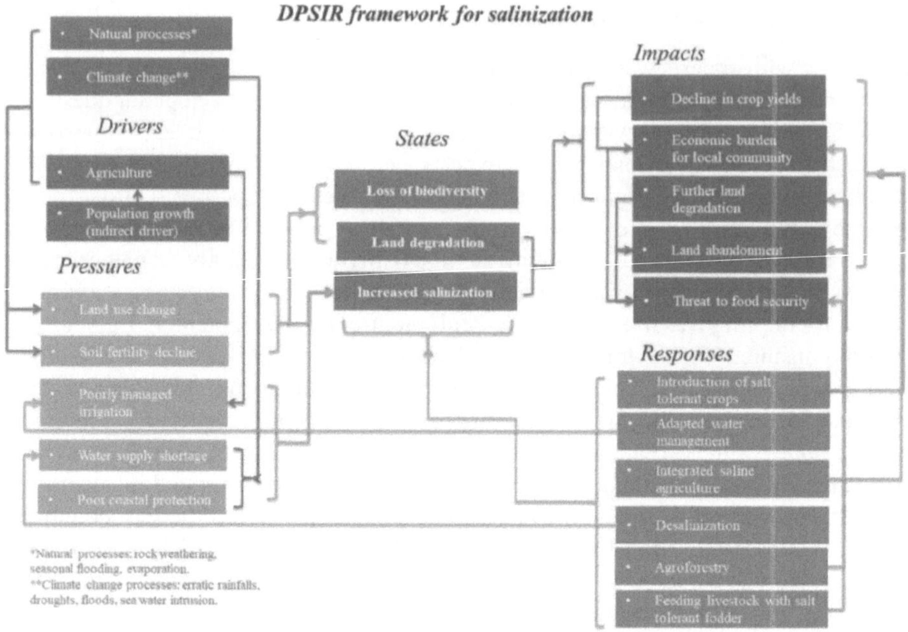

FIGURE 2.1 DPSIR research framework for salinization on a global, regional and local level. The two categories in the upper left corner of the figure, natural processes and climate change, are formally not part of the DPSIR's driver category but strongly affect direct and indirect drivers related to the process. The arrows represent casual relations between elements.

TABLE 2.2
DPSIR-Related SDGs and SDG Targets

DPSIR Category	SDG Goal	SDG Target
Drivers	SDG 13 ("Climate action")	13.1., 13.2.
	SDG 2 ("Zero hunger")	2.3, 2.6
Pressures	SDG 15 ("Life on land")	15.3, 15.5, 15.9
	SDG 6 ("Clear water and sanitation")	6.1, 6.3, 6.4, 6.6., 6.7
	SDG 14 ("Life below water")	14.2
States	SDG 15 ("Life on land")	15.3, 15.5
	SDG 6 ("Clear water and sanitation")	6.3, 6.4, 6.5, 6.6, 6.7
Impacts	SDG 8 ("Decent work and economic growth")	8.2, 8.3, 8.5
	SDG 2 ("Zero hunger")	2.1, 2.3, 2.4,2.5, 2.6, 2.7
	SDG 15 ("Life on land")	15.3, 15.5
Responses	SDG 9 ("Industry, innovation and infrastructure")	9.1, 9.4, 9.5, 9.7
	SDG 2 ("Zero hunger")	2.1, 2.3, 2.4,2.5, 2.6, 2.7
	SDG 6 ("Clear water and sanitation")	6.3, 6.4, 6.5, 6.6, 6.7

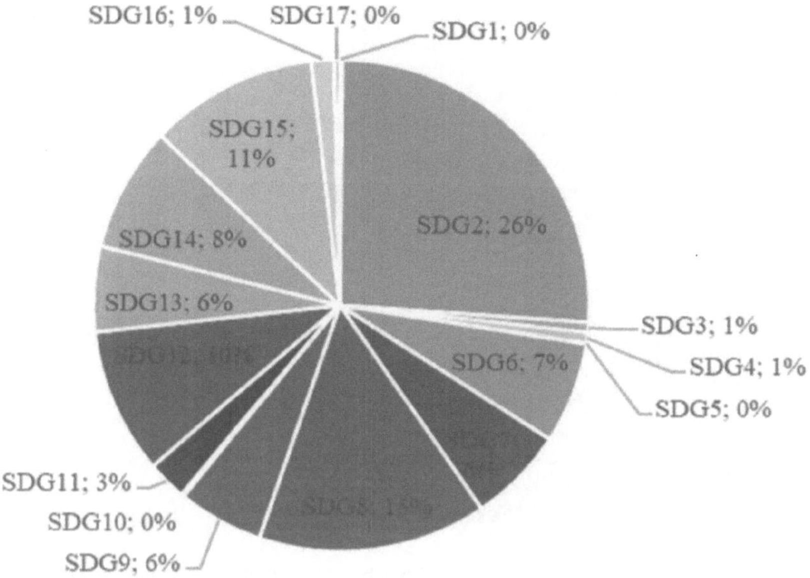

FIGURE 2.2 SDGs appearing in the articles by percentage of total quotations coded.

SDG 2 ("Zero hunger") was the most common (26%), followed by SDG 8 ("Decent work and economic growth") with 15% and SDG 15 ("Life on land", 11%). The least mentioned goals were SDG 1 ("No poverty"), SDG 5 ("Gender equality"), SDG10 ("Reduced inequality"), SDG 17 ("Partnerships"), scoring less than 1%.

2.3.2 SDGs according to Experts

We analysed the number of SDGs which appeared in the interviews (Figure 2.3). All but one SDG were mentioned directly or indirectly. SDG 2 "Zero hunger" accounted for 25% of total quotations. The second most mentioned was SDG 8 "Decent work and economic growth" related to sustainable livelihood (15%). The third most common was SDG 6 "Clean water and sanitation" (13%) (Figure 2.3).

We also examined which SDGs were mentioned directly and indirectly in the interviews (Figure 2.4). For the direct references, SDG 2 and SDG 8 appeared most commonly, followed by SDG 14. For the indirect references, SDG 6 and SDG 8 were referred to most often, followed by SDG 2 and SDG 9.

The interviews revealed a number of overarching themes from the perspective of reaching SDGs. First, there were certain conditions for the introduction of salt-tolerant crops which could be related to SDG 2 on food production. The second emerging theme was in impacts on economic development and possible scalability linked to SDG 8 economic development and SDG 12 on sustainable production and consumption. Third, interviewees mentioned some possible trade-offs of the revitalisation of saline degraded lands in the context of SDG 4 on education and SDG 15 on life on land. Fourth, they pointed to the impact of climate change on saline agriculture practices in the context of SDG 13 on climate.

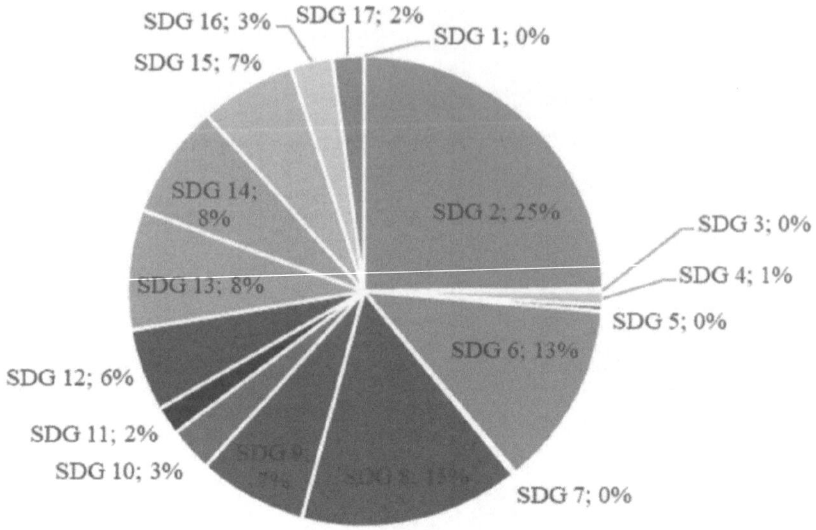

FIGURE 2.3 SDGs appearing in the interviews by percentage of total quotations coded.

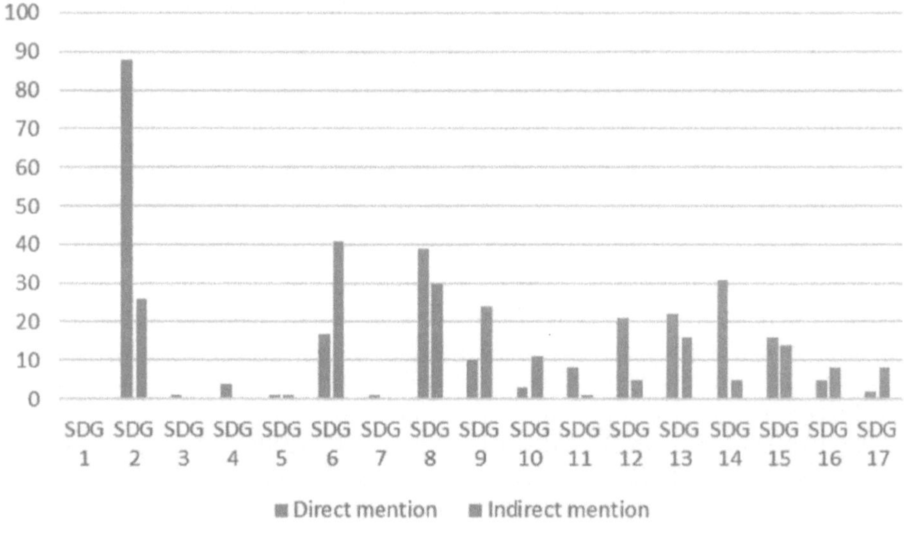

FIGURE 2.4 SDGs mentioned directly and indirectly in interviews.

2.3.2.1 SDG 2: Conditions for the Introduction of Salt-Tolerant Crops

This section outlines several conditions for growing salt-tolerant crops, including biophysical inputs, economic incentives, past habits, water management and a tailored approach for each case.

Experts named multiple conditions for the optimal use of saline soils and introduction of salt-tolerant crops. These conditions were summarised by two researchers: "[…] the future agricultural models should have the following characteristics: 1) require low use of chemical inputs, 2) ensure maximal water and nutrient recycling, 3) be cheap, cost-effective, 4) be easy to operate, 5) be nutrient-dense. Nutrient-sensitive agriculture (nutritionally rich foods, dietary diversity, and food fortification) is really important as integrating different components gives a more thorough spectrum of elements, nutrients and vitamins" (E5[1]). One respondent suggested the importance of the measurement of salinity level, the water system management and presence of markets for the products (E10).

An important factor is taking into consideration the perspective of farming communities, including the opportunities for income diversification by use of integrated systems including crop cultivation and aquaculture (E5). E1 and E3 highlighted the importance of their access to resources, like fertilizers, seeds and additives, and a need for financial assistance, in the form of subsidies. The alternative solution mentioned by E3 and E2 is conservation or ecological agriculture. It is often an option for larger farms which can withstand yield fluctuations.

Current habits and practices, as well as historical conditions, constitute an important factor because farmers are unlikely to switch to a new type of crop (E1) and may traditionally prefer livestock farming, e.g. on saline pastures in Russia (E9).

Another key factor is proper water management, including leaching, drainage and monitoring (E9, E3, E2), which allows farmers to lower or keep the salinity level stable. Poor management may lead to increased salinization or high economic and environmental costs, e.g. when water is flushed through the fields without limits in Uzbekistan (E9). On the other hand, improving water harvesting practices, e.g. by establishing rainwater collecting points and underground storages, may help to maintain stable water supplies in regions with a changeable climate like Bangladesh (E1).

Proper research, including soil and water sampling, assessing the impact on the environment, seems to be the key to success. No one technique would fit all the areas: "The first step is always to identify the causes of salinization in the area supplemented with scientific diagnostics of the problems. Once these problems (salinity levels, sodicity level, hardpan, water table etc.) are diagnosed and based on what problems are existing at the site of interest, a combination of best management practices (integrated soil reclamation-irrigation & drainage; physical & chemical methods, biological-salt tolerant crops) is to be used which are site-specific. For example, if a hardpan is existing it must be broken down (subsoiling, chiselling etc.), high water table (drainage system to be installed), soil sodicity can be corrected using chemical amendment like gypsum etc. Salinity can be managed by using salt-tolerant crops and subsequent leaching fraction to maintain the root zone salinity below the crop threshold. Then, the crops are chosen and there is a large variety as it comes to salt tolerance level" (E10).

2.3.2.2 SDG 8 and SDG 12: Impact on Economic Development and Scalability

This section presents economic impacts of salinization including portfolio diversification, market development and the feasibility of upscaling.

Although economic aspects are widely discussed, little research has been done to fully estimate the economic aspects of saline agriculture for farmers with an assessment of scalability beyond a few hectares. For many farmers, portfolio diversification is fundamental as it makes them more resistant to sudden changes or less productive years. It helps not only the individuals but is also important for national food security and independence in this regard (E4, E5, E1). Saline agriculture has the potential to increase various farmers' income from diverse backgrounds, although in different ways. For farmers from developing countries, like Bangladesh, it is a way to avoid poverty and provide better living conditions. In a developed country, e.g. Australia, it is a source of competitive advantage (E3, E2).

Economic potential depends on the target country and the focus, like nutrition and food security, building market potential or commercialisation. The market for saline products is a factor often mentioned by experts. It is clear that to develop supply, the demand must exist. It would largely differ among countries. For example, developing countries struggling with food security are more likely to accept salt-tolerant varieties of conventional crops. At the opposite extreme, in developed countries, there is a high-end market for halophytes as delicacies (e.g. Salicornia has a very good selling price in the United Kingdom reaching 12 GBP/kg). Building the market relates also to creating a supply chain for necessary materials, such as seeds or fertilizers. The markets currently exist in the Netherlands, the United Kingdom, United States of America and Belgium (E5). Another perspective is the pharmaceutical properties of these plants (E5). However, in some countries, there is no present economic imperative to cultivate saline degraded lands. This is especially valid for countries with a large amount of only slightly salt-affected surface area and rich organic soils, like Russia. It also depends on the key sectors in the country Gross Domestic Product (GDP), which for Uzbekistan is agriculture, thus being a point of focus. For others, like Iran, the political factors, e.g. relative isolation, may be a barrier to the flow of know-how and access to markets.

Finally, the most important factor for scaling up is the salinity level. Each salinity level has its economics which should not be compared with conventional agriculture. "The scaling up must be done where similar soil, water and environmental conditions may be existing, similar to where technology is going to be transferred. There must be a reason for scaling up based on market demand. If such information is not available then pilot-scale testing must be done before scaling up" (E10). Creating trust and establishing a business model is a way to scale up through establishing model farms and assigning the leading farmers who can later train their community (E1).

2.3.2.3 SDG 4 and SDG 15: Possible Trade-Offs

This section presents possible problems involved in application of saline agriculture such as improper water and soil management or threat for biodiversity.

Despite many benefits of saline agriculture, there are certain downsides of increasing human activities on saline degraded lands. The first one is the lack of knowledge because saline agriculture projects are generally multidisciplinary and require cooperation with an irrigation engineer, soil, horticultural and halophyte experts. The farmers become inspired by their neighbours, but do not seek expert advice and instead try to implement schemes themselves, which can lead to failure (E5). Salinization is often treated as a complex process with various phases. The management techniques may worsen the situation for a certain time, e.g. through dissolving the salt in the soil by leaching and pushing it to the upper layer. Eventually, the salinity level decreases, but the process may take many years (E9). Following this perspective, experts highlight the pressure on water systems and potential pollution from fertilizers, insecticides and pesticides (E10), but also returning extreme brine to the sea after desalination.

Increased agricultural production may affect biodiversity, especially if the lands are considered degraded, but are just semi-arid (E3). However, degradation also includes the low provision of ecosystem services, so agricultural restoration can be beneficial for ecosystem services linked to agro-biodiversity (E11).

2.3.2.4 SDG 13: Impact of Climate Change

This section presents effects of climate change on the process of salinization according to experts.

The impact of climate change on the process of salinization depends on the type of salinity and the level at which the salinity occurs. In some cases, if the soils are saline for geomorphic reasons and groundwater is very deep (more than 5–10 m), desertification may have no effect (E9). Climate change has a geographic pattern. For the Aral and Caspian Sea region, the main problem is sea-level change, which can be caused not only by climate but also by geology or through human-induced processes.

In some places, like the Netherlands, where the salinity issue is not visible now, climate change may lead to shifting climatic conditions in the future and a requirement to apply adaptive measures. Again, proper management is mentioned as one of the ways to avoid the negative consequences of climate change (E10).

Finally, the assessment of the climate change effect will depend greatly on the time frame. It may be minor within the next 20 years but have major implications in 50 years (E11). The experts perceive saline agriculture as a backup option for national food security and autonomy in case of unexpected or long-term climatic changes (E5).

2.4 DISCUSSION

Our findings show that almost all SDGs are related to the revitalisation of saline soils through saline agriculture. In particular, saline agriculture is highly related to SDG 2 and SDG 8 according to both methods used in this paper. In this way, it addresses food and water security through agricultural activities and water management (see Table 2.3).

There is surprising agreement between the two methods in the top- and lowest-ranked SDGs. The numbers differ (albeit slightly) for the SDGs moderately related to saline agriculture.

SDGs related to inequality, gender equality and partnerships for the goals are among the least scored in both the methods. However, saline agriculture creates an opportunity to address those goals, which should be more highlighted in future projects. Similarly, even though many current projects are focused on educating farmers, this is not reflected in the DPSIR analysis and the interviews. The reason for this may be a disparity between science and practice. In particular, the researchers often focus

TABLE 2.3
Comparison of Number of Quotations between Our Two Methods

SDG		DPSIR (%)	Interviews (%)
SDG1	"No poverty"	0	0
SDG2	"Zero hunger"	26	25
SDG3	"Good health and well-being"	1	0
SDG4	"Quality education"	1	1
SDG5	"Gender equality"	0	0
SDG6	"Clean water and sanitation"	7	13
SDG7	"Affordable and clean energy"	6	0
SDG8	"Decent work and economic growth"	15	15
SDG9	"Industry, innovation and infrastructure"	6	7
SDG10	"Reduced inequalities"	0	3
SDG11	"Sustainable cities and communities"	3	2
SDG12	"Responsible consumption and production"	10	6
SDG13	"Climate action"	6	8
SDG14	"Life below water"	8	8
SDG15	"Life on land"	11	7
SDG16	"Peace, justice and strong institutions"	1	3
SDG17	"Partnerships for the goals"	0	2

on innovations in plant physiology rather than on how to make the results of their studies accessible to a wider audience.

Based on the results of this study, future research could explore how saline agriculture fits in the DPSIR framework related to food production as a valuable addition to conventional agriculture as it comes to addressing food security. Our study of saline agriculture confirms Rockström & Sukhdev's (2016) argument that all SDGs are related to food. These food security and sustainability issues are present in developing countries (see, e.g. Ladeiro 2012; Chapter 21 of this book). The nexus nature of saline agriculture can be a sustainable land management practice in countries confronting salt-induced land degradation (Qadir et al. 2014).

The DPSIR analysis in this paper partially overlaps with the results obtained by Ruto et al. (2018). However, the causes, impacts and state of the salinization described in their report focus more on the physical and economic aspects of the process.

Our results partly align with SDGs reported in the study of Chapter 21 in this book. Its findings include SDG 2 and SDG 13 which score high in this research, but also SDG 1, SDG 3, SDG 5 and SDG 17 which are among the least addressed SDGs in our study. The reason for this difference may be that the study of Chapter 21 focused on a community-oriented project.

What is surprising in our study is the relatively weak link to SDG 5 "gender equality", which is often mentioned by other studies and programs, e.g. network for the Arab Women Leaders in Agriculture (AWLA) fellowship program (ICBA, 2020). Also, SDG 1 "No poverty" was not covered in the interviews, which may come from the fact that farmers involved in the community live above the poverty line. Finally, the partnerships (SDG 17) were mentioned less often than expected, while it seems that the community of practice and science is emerging. A reason for this result may be a choice of the keywords for this SDG. Future studies could explore it more in the context of funding possibilities and development aid.

Additionally, the link to climate is also weaker than expected. Saline agriculture can be more pronounced as a countermeasure for economic and climate migration. For example, in Bangladesh, increasingly severe floods will push salt water more inward across land (Chen & Mueller 2018). This will affect the livelihoods of the coastal and delta-area people (SDG 11) as locally there will be less productive agricultural activity (SDG 8). As a result, people are likely to migrate away from the Delta with many negative social and political consequences. Subsequently, SDG 8 ensuring decent and economic growth in coastal areas is a major goal supporting food production under saline soil and water conditions.

Finally, implementation of saline agriculture may involve some trade-offs, as Bailis & Yu (2012) suggest. These trade-offs could happen in places of ecological or cultural significance. Potential synergies and barriers among SDGs in relation to saline agriculture should be further investigated.

2.5 CONCLUSION

Our study analyses in a systematic way SDGs connected to salinization by employing the DPSIR framework and the analysis of semi-structured expert interviews. We conclude that both methods consequently point to SDG 2 ("Zero hunger") and SDG 8 ("Decent work and economic growth") as being strongly related to saline agriculture. These results are partially in line with our hypothesis pointing towards SDG 2, SDG 6, SDG 8, SDG 13, SDG 14 and SDG 15.

Based on our findings, we formulate the following recommendations:

- The revitalisation of saline soils through saline agriculture can foster achieving SDG 2 and SDG 8, especially in salt-affected regions struggling with food and water security.
- The management of saline soils can create workplaces for local farmers, increase income through higher yields than with conventional crops, and prevent or reduce economic and climate migrations (SDG 8).
- Revitalisation projects should focus more on the education (SDG 4) of underprivileged target groups (SDG 10) and woman (SDG 5) to allow for synergy effects. It could be achieved by partnerships (SDG17) among various stakeholders.

Our analysis shows that saline agriculture is a nexus area linking food production, water management and sustainable economic development. Investments in this method of revitalisation can address multiple SDGs at the same time. Policymakers and other actors should take these findings into account as saline agriculture is an effective way to meet SDGs and combat the challenges of deteriorating food security and increasing climate change.

APPENDIX – CODE BOOK

Sustainable Development Goals SDG	Directly Mentioned (Interviews)	Directly Mentioned (DPSIR)
GOAL 1: No Poverty	Poverty	Poverty
GOAL 2: Zero Hunger	Food Agricultur* Genetic Seeds	Food Agricultur* Genetic Seeds
GOAL 3: Good Health and Well-being	Health Well-being	Health Well-being
GOAL 4: Quality Education	Education Training Teach	Education Training Teach
GOAL 5: Gender Equality	Gender Woman	Gender Woman
GOAL 6: Clean Water and Sanitation	Clean Desalination Wastewater Reuse Groundwater	Clean Desalination Wastewater Reuse Groundwater
GOAL 7: Affordable and Clean Energy	Energy Biomass	Energy Biomass
GOAL 8: Decent Work and Economic Growth	Employment Economy Economic Growth	Employment Economy Economic Growth
GOAL 9: Industry, Innovation and Infrastructure	Industry Innovat* Technolog*	Industry Innovat* Technolog*
GOAL 10: Reduced Inequality	Equality Migration	Equality Migration
GOAL 11: Sustainable Cities and Communities	Cities Urban Community	Cities Urban Community
GOAL 12: Responsible Consumption and Production	Production Consumption Consumer	Production Consumption Consumer
GOAL 13: Climate Action	Climate change	Climate change
GOAL 14: Life Below Water	Water (mannually cross-checked) Marine Coastal Fish	(Water excluded) Marine Coastal Fish
GOAL 15: Life on Land	Land degradation Soil degradation Biodiversity Desert Ecosystem	Land degradation Soil degradation Biodiversity Desert Ecosystem

(*Continued*)

Sustainable Development Goals SDG	Directly Mentioned (Interviews)	Directly Mentioned (DPSIR)
GOAL 16: Peace and Justice Strong Institutions	Institution Law Regulation	Institution Law Regulation
GOAL 17: Partnerships to achieve the Goal	Partnership Cooperation	Partnership Cooperation

ENDNOTE

1. Interviewees were coded from E1 to E11.

ACKNOWLEDGMENTS

We would like to express our gratitude to all the experts participating in the interviews for their time, patience and sharing their knowledge.

REFERENCES

Aiking, H. (2014). Protein production: Planet, profit, plus people?–. *The American Journal of Clinical Nutrition*, *100*(suppl_1), 483S–489S.

Bailis, R., & Yu, E. (2012). Environmental and social implications of integrated seawater agriculture systems producing Salicornia bigelovii for biofuel. *Biofuels*, *3*(5), 555–574.

Chen, J., & Mueller, V. (2018). Coastal climate change, soil salinity and human migration in Bangladesh. *Nature Climate Change*, *8*(11), 981–985.

Christopoulos, D. (2009). Peer Esteem Snowballing: A methodology for expert surveys. *Eurostat Conference for New Techniques and Technologies for Statistics*, 171–179.

Cooper, P. (2013). Socio-ecological accounting: DPSWR, a modified DPSIR framework, and its application to marine ecosystems. *Ecological Economics*, *94*, 106–115. https://doi.org/10.1016/j.ecolecon.2013.07.010

Dagar, J. C., Sharma, P. C., Sharma, D. K., & Singh, A. K. (2016). *Innovative saline agriculture*. Springer.

Dagar, J. C., Yadav, R. K., & Sharma, P. C. (2019). *Research developments in saline agriculture*. Springer.

Datta, K. K., & Jong, C. de. (2002). Adverse effect of waterlogging and soil salinity on crop and land productivity in northwest region of Haryana, India. *Agricultural Water Management*, *57*(3), 223–238. https://doi.org/10.1016/S0378-3774(02)00058-6

De Waegemaeker, J. (2019). *SalFar framework on salinization processes*. A report by ILVO for the Interreg VB North Sea Region project Saline Farming (SalFar). http://scholar.googleusercontent.com/scholar?q=cache:sh0r_1_FkcsJ:scholar.google.com/+De+Waegemaeker,+J.+(2019).+SalFar+framework+on+salinization+processes.+&hl=pl&as_sdt=0,5

FAO. (2009). *World summit on food security. Declaration of the World Summit on food security*. FAO. http://ftp.fao.org/docrep/ fao/Meeting/018/k6050e.pdf

FAO. (2011). *The state of the world's land and water resources for food and agriculture: managing systems at risk*. Food and Agriculture Organization of the United Nations.

FAO. (2018). *More information on Salt-affected soils | FAO SOILS PORTAL | Food and Agriculture Organization of the United Nations*. http://www.fao.org/soils-portal/soil-management/management-of-some-problem-soils/salt-affected-soils/more-information-on-salt-affected-soils/en/

Ghassemi, F., Jakeman, A. J., & Nix, H. A. (1995). *Salinisation of land and water resources: Human causes, extent, management and case studies.* CAB International.

Hülsmann, S., & Ardakanian, R. (Eds.). (2018). *Managing water, soil and waste resources to achieve sustainable development goals.* Springer International Publishing. https://doi.org/10.1007/978-3-319-75163-4

ICBA. (2020, May 4). *ICBA launches virtual alumnae network to support Arab women researchers | Awla.* https://www.awlafellowships.org/news/icba-launches-virtual-alumnae-network-support-arab-women-researchers

Joe-Wong, C., Schlesinger, D., Chow, A., & Myneni, S. C. B. (2019). Sea level rise produces abundant organobromines in salt-affected coastal wetlands. *Geochemical Perspectives Letters, 10,* 31–35.

KPMG International, De Boer, Y., & van Bergen, B. (2012). *Expect the unexpected: Building business value in a changing world.* KPMG International.

Ladeiro, B. (2012). Saline agriculture in the 21st century: Using salt contaminated resources to cope food requirements. *Journal of Botany,* 2012, 1–7. https://doi.org/10.1155/2012/310705

Liu, J., Hull, V., Godfray, H. C. J., Tilman, D., Gleick, P., Hoff, H., Pahl-Wostl, C., Xu, Z., Chung, M. G., Sun, J., & Li, S. (2018). Nexus approaches to global sustainable development. *Nature Sustainability, 1*(9), 466–476. https://doi.org/10.1038/s41893-018-0135-8

Munns, R., & Tester, M. (2008). Mechanisms of salinity tolerance. *Annu. Rev. Plant Biol., 59,* 651–681.

Patrício, J., Elliott, M., Mazik, K., Papadopoulou, K.-N., & Smith, C. J. (2016). DPSIR—two decades of trying to develop a unifying framework for marine environmental management? *Frontiers in Marine Science, 3.* https://doi.org/10.3389/fmars.2016.00177

Qadir, M., Quillérou, E., Nangia, V., Murtaza, G., Singh, M., Thomas, R. J., Drechsel, P., & Noble, A. d. (2014). Economics of salt-induced land degradation and restoration. *Natural Resources Forum, 38*(4), 282–295. https://doi.org/10.1111/1477-8947.12054

Rockström, J., & Sukhdev, P. (2016, June 14). *How food connects all the SDGs—Stockholm Resilience Centre* [Text]. https://www.stockholmresilience.org/research/research-news/2016-06-14-how-food-connects-all-the-sdgs.html

Ruto, E., Gould, I., Bosworth, G., & Wright, I. (2018). *The state, causes and impact of soil salinization: A global overview* (pp. 1–13). SalFar Work Package 3 Baseline January 2018 Report.

Smil, V. (2001). *Feeding the world: A challenge for the twenty-first century.* MIT press.

Stoorvogel, J. J., Bakkenes, M., & Brink, B. J. E. (2017). To what extent did we change our soils? A global comparison of natural and current conditions. *Land degradation & development, 28*(7), 1982–1991.

van Noordwijk, M., Duguma, L. A., Dewi, S., Leimona, B., Catacutan, D. C., Lusiana, B., Öborn, I., Hairiah, K., & Minang, P. A. (2018). SDG synergy between agriculture and forestry in the food, energy, water and income nexus: Reinventing agroforestry? *Current Opinion in Environmental Sustainability, 34,* 33–42. https://doi.org/10.1016/j.cosust.2018.09.003

van Teeffelen, A. (2017, September 4). *Causes and consequences of environmental change.* ERM, Amsterdam.

Vos, Arjen de, Andres Parra González, and Bas Bruning. 2021. "Case Study: Putting Saline Agriculture into Practice – A Case Study from Bangladesh." In *Future of Sustainable Agriculture in Saline Environments.* Routledge, Taylor & Francis Group, London.

3 Agriculture in Salinising Landscapes in Southern Australia
Selected Research 'Snapshots'

Edward G. Barrett-Lennard and Hayley C. Norman

CONTENTS

3.1 INTRODUCTION

Annual crops and pastures are widely grown in the 250–600 mm rainfall zone of Australia under non-irrigated conditions. This non-irrigated land can be affected by two salinity problems: salinity induced by the presence of a shallow water-table caused by the removal of the original forest (often called 'dryland' salinity) and that caused by soil dispersion due to sodicity and soil alkalinity (often called 'transient' salinity) (Rengasamy 2006; Barrett-Lennard et al. 2016). The former stress is caused by the formation of semi-permanent shallow water-tables and affects the growth of agricultural crops in all years. By contrast, the latter stress affects crop growth particularly in dry years, when the salt concentration in the soil increases, the water content of the soil decreases and the salinity of the soil solution is therefore elevated. The levels of salinity associated with these stresses can be very different. Soils affected by transient salinity typically fall into the slightly to moderately saline

DOI: 10.1201/9781003112327-3

TABLE 3.1
Categorisation of Salinity in Australia in Terms of the EC_e Ranges That Fall into the following $EC_{1:5}$ Ranges (by Soil Texture)

		$EC_{1:5}$ Range for Different Soil Textures		
Term	**EC_e Range (dS/m)**	**For Sands**	**For Loams**	**For Clays**
Non-saline	0 – 2	0 – 0.14	0 – 0.18	0 – 0.25
Slightly saline	2 – 4	0.15 – 0.28	0.19 – 0.36	0.26 – 0.50
Moderately saline	4 – 8	0.29 – 0.57	0.37 – 0.72	0.51 – 1.00
Highly saline	8 – 16	0.58 – 1.14	0 73 – 1.45	1.01 – 2.00
Severely saline	16 – 32	1.15 – 2.28	1.46 – 2.90	2.01 – 4.00
Extremely saline	≥32	≥2.29	≥2.91	≥4.01

Source: After Barrett-Lennard et al. 2008.

range (Table 3.1), so these soils are generally sown to annual crops such as wheat, barley and canola. By contrast, those soils affected by dryland salinity often have salinities in the highly, severely and extremely saline ranges (Table 3.1), so these soils are generally reserved for the growth of saltland pastures based around the use of halophytes or are otherwise abandoned.

One issue that immediately confronts the reader wanting to obtain an overview of salinity in Australia – its scale, the kinds of research that have been conducted and the kinds of adaptations that communities have made – is that salinity is a State rather than a Federal issue, so the answers to these questions generally reside within State Government agencies, and integrated answers to questions at the national scale may be difficult to find.[1] Nevertheless, there have been federal initiatives (such as the National Dryland Salinity Program, National Action Plan on Salinity and Water Quality and several Cooperative Research Centres) and networks of interstate collaboration (such as the National Program on the Utilisation and Rehabilitation of Saline Land – PURSL) that have enabled national collaborations and syntheses to be developed around particular issues.

The total area of land used for the growth of grain crops in Australia is ~112 Mha (Rengasamy 2002). In 2002, the Australian Bureau of Statistics surveyed farmers at the national level about the extent of dryland salinity on agricultural land (Table 3.2). This showed that dryland salinity affected nearly 20,000 farms around Australia, nearly 2 Mha showed signs of salinity, 0.8 Mha were so severely affected that the land was not able to be productively used and that WA was the most severely affected State.

The areas of land associated with transient salinity (i.e. associated with dispersive soils) are not precisely known but might be substantially greater (Rengasamy, 2002). In regions where the salinity problem is greatest, many farmers manage mixed cropping and livestock systems with annual crops and sheep for meat and wool (Norman et al. 2016a).

TABLE 3.2
Results of National Australian Bureau of Statistics Survey of Extent of Dryland Salinity in 2002

State	Number of Farms with Salinity	Land Showing Signs of Salinity ('000 ha)	Salinised Land Unable to Be Used for Production ('000 ha)
NSW	3108	124	44
Vic.	4834	139	60
Qld	993	107	40
SA	3328	350	105
WA	6918	1241	567
Tas.	390	6	2
NT	8	2	2
Total Australia	19,579	1,969	821

Source: ABS 2002.

3.2 CAUSES OF SALINITY

Salinity was recognised in Australia as an important constraint in catchments used to collect water for towns and cities from the early 1920s (Wood 1924) and in agricultural landscapes from the 1930s (Teakle and Burvill 1938). Beginning with Burvill (1956), a series of surveys of farmers in Western Australia conducted by the Australian Bureau of Statistics every 4–7 years showed that there was a continuing deterioration in the areas of previously arable land that became too saline to grow conventional crops and pastures. The effect of these surveys (1955, 1962, 1979, 1984, 1989, 2002) was to induce a strong level of community concern about the apparent inexorable increase in salinity with time.

The reputed causes of salinity in rainfed environments in Australia were initially confusing: today two causes are recognised. The most severe form of salinity, induced by the presence of a shallow water-table, is caused by the clearing of the original native forests, shrublands and perennial grasslands (which used virtually all the rainfall) and their replacement with annual crops and pastures (which used less than all the rain) (Wood 1924). Water-tables at the time of European settlement 100–150 years ago were ~5–50 m below ground level. The net percolation of water deep into the soil caused a rise in water-table, bringing salt stored in the profile to the soil surface. When the water-table reached a critical depth, the land became too saline for the growth of annual crops and pastures. Nulsen (1981) provided a set of plausible, critical depths to the water-table for plants of different salt tolerance in the wheatbelt of Western Australia: ~2.2 m for the growth of wheat crops, ~1.8 m depth for the growth of barley crops and ~1.5 m depth for the growth of salt-tolerant annual barley grass. The problems of shallow water tables in dryland (non-irrigated landscapes) are not confined to southern Australia alone: Ghassemi et al. (1995) note that have been reports of similar effects in the Great Plains region of North America, and in South Africa, Turkey, Thailand, India and Argentina.

There were two important later elaborations were to this hydrological explanation of salinity. Firstly, it was recognised that groundwater flow, and therefore the expression of salinity, could be strongly influenced by geomorphic structures in the landscape (e.g. weathered dolerite dykes, bedrock highs, fractures in rock aquifers and the presence of semi-confining sedimentary layers; George et al. 1997). Secondly, it was recognised that the surface hydrology of soils was also important. In landscapes cleared of their original native vegetation, runoff from hillsides accumulated at low points in the landscape exacerbating groundwater rise (reviewed by Barrett-Lennard et al. 2005). However, the complication remained that in semi-arid areas salinity was not necessarily associated with the presence of shallow water-tables (Teakle and Burvill 1938). It later became clear that many Australian soils, particularly sodic alkaline clays of the semi-arid environment were naturally dispersive and could accumulate over hundreds of years the small amounts of salt that fell in the rain (Hingston and Gailitis 1975) to levels that impacted on crop growth in relatively dry years (Rengasamy 2006; Barrett-Lennard et al. 2016).

3.3 WATERLOGGING-SALINITY INTERACTIONS

A further level of complexity became apparent when it was realised that another by-product of shallow water-tables was seasonal waterlogging, and waterlogging interacted with salinity to further constrain crop growth. Empirical observations in the 1970s by an influential farmer named Harry Whittington (who later established a farming group of over 1000 members, WISALTS) suggested that the symptoms of salinity could be abated by surface water management by installation of throughflow interceptor banks (Conacher et al. 1983). Confusion therefore ensued: Whittington's supporters claimed that the implementation of the interceptor banks prevented salinity; however, the truth is that his 'interceptor banks' probably changed shallow soil hydrology, decreasing waterlogging. Research that began in the 1970s showed that in saline landscapes, waterlogging increased the uptake of Na^+ and Cl^-, and decreased the uptake of K^+ by crop plants (reviewed by Barrett-Lennard 1986, 2003; Barrett-Lennard and Shabala 2013) and this knowledge helped settle the confusion. More recently, Bennett et al. (2009) have published a matrix suggesting that the ideal location for different crop and forage plants in salt-affected land can be determined by relating the general level of soil salinity and waterlogging in the landscape to the known tolerances of the plants to these two stresses. Today, the amelioration of waterlogging with shallow (and some deep) drainage, and the planting of appropriate crops and pastures with a combination of tolerances to salinity and waterlogging are regarded as the optimal solution for farming saline landscapes affected by shallow water-tables. The message to farmers that emerged from this was to 'put the right plant in the right location'.

3.4 USE OF TREES TO CONTROL SHALLOW WATER-TABLES

Agricultural systems in the rainfed landscapes of southern Australia are generally of low productivity due to the low rainfall and low fertility of these soils; expensive drainage options to lower water-tables are therefore unlikely to be adopted without substantial publicly funded subsidies. For this reason, there became a strong research focus on the use of perennial plants acting as forms of 'biological pumps' to lower water-tables. The first systematic attempt at this was through the planting of trees. The thinking was that if salinity was caused by the development of shallow water-tables as a consequence of the clearing of deep-rooted native forests, then perhaps it could also be reversed by re-establishing trees back into the landscape. In a series of 'ventilated-chamber' experiments, Greenwood and colleagues were able to show that stands of trees were able to use water at rates in excess of local rainfall: this could only be possible if the plants were accessing local groundwater and therefore decreasing the risk of salinity in the landscape.

In one of these experiments (Greenwood et al. 1985), trees were established ~6 years earlier in two dense plantations in an area receiving ~770 mm annual average rainfall; one plantation was near the top of a hill (~8 m above the water-table) and the other was located mid-slope (~5 m above the water-table). Evapotranspiration from the trees was measured for 24 h every 4 weeks for a year; this was used to estimate the annual evapotranspiration (summarised in Figure 3.1). Evapotranspiration was measured similarly from an adjacent annual pasture while it was green. It was found that annual evapotranspiration from the pasture accounted for ~57% of annual rainfall over the year; the balance of the rainfall was clearly infiltrating into the deeper soil increasing the risk of salinity. By contrast, evapotranspiration from the trees accounted for 240–390% of annual rainfall; the water used in excess of rainfall must have come from soils adjacent to the trees (Figure 3.1). The conclusion was that the trees were capable of using water in excess of rainfall and therefore decreasing salinity risk in the landscape.

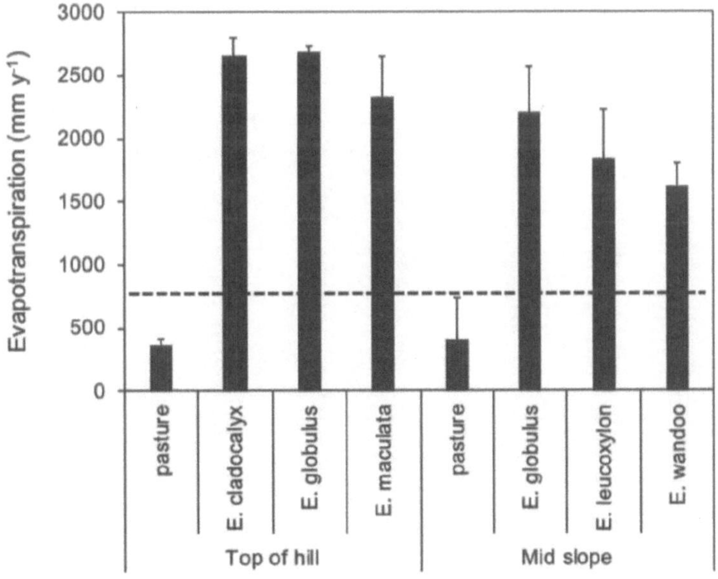

FIGURE 3.1 Annual evapotranspiration by annual pasture and stands of eucalyptus trees growing in a 770 mm annual average rainfall zone (dotted line) at two locations in Western Australia (After Greenwood et al. 1985).

One problem with the use of relatively small areas of trees for short periods to estimate rates of tree water use is that as trees draw-down local water-tables there will generally be a mass flow of water from outside the plantation into the root-zone of the trees; this ensures high rates of water use in the short term. In general, the experiments of Greenwood et al. measured water use by small plots of trees over short intervals of time (1–2 years after establishment); their experiments, therefore, exaggerated the effects that trees might have on groundwater at catchment scales. To make the case more strongly what was needed were experiments that examined the effects of trees on depth to water-table (the key driver of salinity) planted over larger areas and measured over longer periods of time. A later paper (Bennett and George 2008) showed that in 24 investigations in catchments receiving 450–820 mm of average annual rainfall and where trees had been planted at a catchment scale for 10–21 years, there was a significant ($P < 0.001$) effect on the rate of water-table drawdown of the percent of a catchment with trees (i.e. remnant vegetation and revegetation) (Figure 3.2). On average to achieve a rate of water-table drawdown of 0.1 m per year, catchments needed to have ~30% of their area covered by remnant vegetation and revegetation.

How large does a tree need to be to be able to use enough soil water to avert a salinity problem? 10 m? What about 10 cm? One of the interesting themes of research in Australia is that perennial pasture plants may be able to act as 'functional mimics' of trees and use enough water to dry soils, lower water-tables and prevent salinity (Hatton and Nulsen 1999). The critical factor here is not the height of the transpiring perennial plant but the depth of its root system. This has led to the notion of 'phase farming' (Latta et al. 2001, 2002), in which perennial pasture plants like lucerne (alfalfa) are grown on a field for 1–2 years to deplete stored soil water and these are

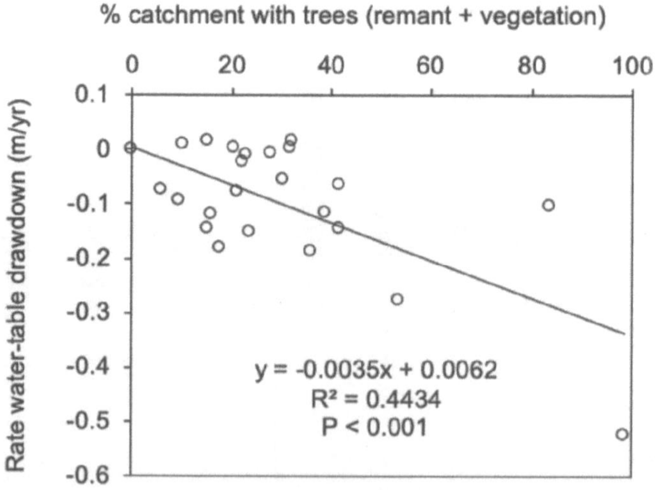

FIGURE 3.2 Effect of tree vegetation (remnant vegetation and revegetation) on the rate of water-table drawdown (m/yr) for catchment with a shallow water-table in Western Australia (After Bennett and George 2008). Rates of water-table drawdown were measured over periods of 10–21 years.

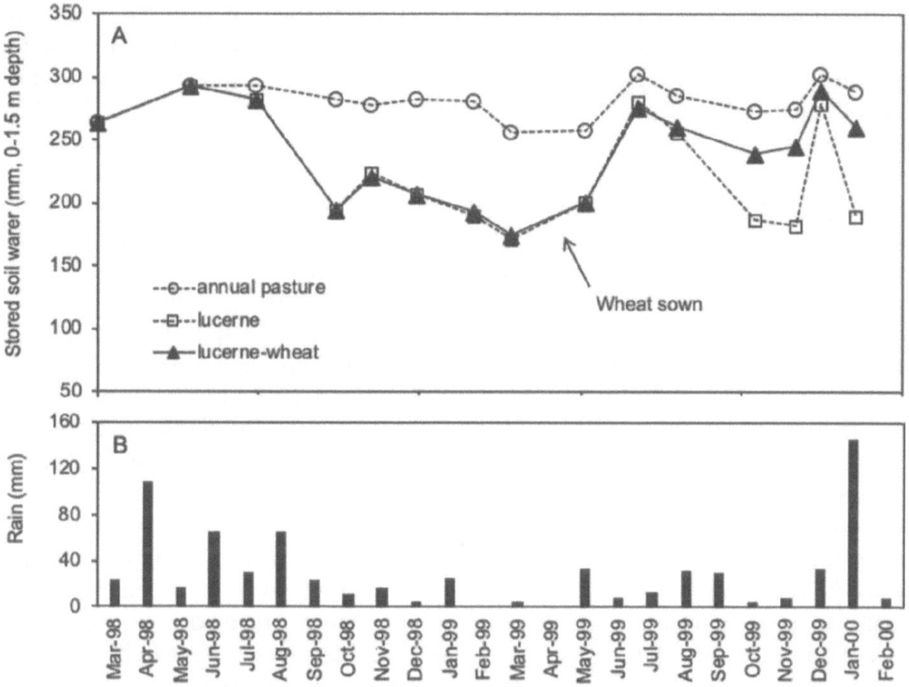

FIGURE 3.3 Results of phase farming trial at Newdegate (After Latta et al. 2002): (A) Amounts of water stored in the upper 1.5 m of the soil profile at a Newdegate field site when sown to an annual clover-based pasture, to lucerne, or to a lucerne-wheat rotation; and (B) Monthly rainfall at site (Bureau of Meteorology). Each treatment was replicated three times in a fully randomised block design. Plots were 12 × 25 m in area. Wheat was sown at point indicated by arrow.

then replaced by more profitable annual grain crops for several years during which time the soil profile wets up again.

The concept can be illustrated using data from a typical field trial comparison of the soil water content in the upper 1.5 m of the soil profile beneath an annual pasture, a perennial lucerne pasture and a lucerne-wheat rotation (Latta et al. 2002). Figure 3.3 shows the variation over a 2-year period in the amount of water stored in the upper 1.5 m of the soil profile when planted to these three options and the monthly rainfall. It can be seen that irrespective of monthly rainfall (Figure 3.3b) the soils planted to the annual pasture option contained ~250–300 mm of water in the upper 1.5 m of the soil profile (Figure 3.3a). Presumably, this water content was close to the soil's drained upper limit, and rainfall in excess of that required to bring the soil to the drained upper limit percolated deeper into the soil profile. By contrast, once the lucerne pasture had established, the soils planted to this perennial had 20–100 mm less stored soil water in the soil profile; there would have been a less deep percolation of rainfall into the soil profile with this treatment. Finally, where the lucerne was replaced by wheat, there was an increase in soil water content; this option would have leaked water into the deeper soil profile. These observations led the researchers to conclude that the growth of lucerne alternating with wheat might be both a more sustainable and financially robust solution, with the wheat crops providing the bulk of the financial gain to the farmer, and the lucerne ensuring that the soils were periodically dried down, decreasing the long-term salinity outcomes in the landscape.

While plot- and catchment-scale research could show that perennial plants were capable of using groundwater, lowering water-tables and decreasing salinity, different (modelling) approaches were needed to determine the likely benefits of plant interventions at a landscape scale. One model used to do this was 'Flowtube' (Dawes et al. 2000). This is a simple two-dimensional model that can predict the likely impacts of any hydrological intervention (such as the implementation of perennial vegetation) in a 'slice' through a catchment (from the top of a hill to the bottom of a valley) on the depth to groundwater along that slice. In one study, models were built at three locations (in the 330, 400 and 600 mm annual rainfall zones of WA) to examine the expected long-term impacts of a series of plausible hydrological interventions (low, medium or high changes to the water balance) on groundwater depths along the flowtube (George et al. 2001). Simulations began with the present-day water-table depth profile in each catchment (as determined from piezometers) and continued at daily intervals for 100 years into the future. The model was calibrated by comparing the computed rates of groundwater rise along the flowtube with the historic measured rates of groundwater rise in piezometers in each catchment. What became clear from the simulations was that irrespective of the location, with the do-nothing scenario most (72–94%) of each catchment was at risk of salinisation (i.e. a water-table less than 1 m in depth would occur within 100 years). Furthermore, even with low, moderate or high levels of hydrological intervention (defined in specific terms for each catchment), the outcomes in terms of land salinity remained challenging with 46–79%, 41–52% and 34–48% respectively of the three catchments being affected (Table 3.3). These predictions of salinity outcomes for non-irrigated landscapes, even with the implementation of major remedial farming practice change, remain as one of the most confronting commentaries on the sustainability of agriculture in southern Australia.

TABLE 3.3
Percent of Flowtube with Water-Table Less than 1 m after 100 Years

	Hydrological Intervention			
Rainfall (Location)	**None**	**Low**	**Moderate**	**High**
330 mm (North Baandee)	89	79	52	34
400 mm (Toolibin)	94	62	61	48
600 mm (Date Creek)	72	46	41	38

Source: After George et al. 2001.

3.5 SALTLAND PASTURES

If salinity cannot be cheaply avoided by revegetating farmland with perennial vegetation (trees and pastures) then can it be lived with? One of the obvious ways to obtain production from saline land is to plant it to salt-tolerant plants (halophytes) that have value as forages for ruminant livestock (such as sheep and cattle). Australian researchers and farmers have been working collaboratively to incorporate salt-tolerant forages into their farming systems for more than 80 years. Some of the key dates and events associated with the development of herbaceous species and halophytic shrubs for saltland pastures are summarised in Tables 3.4 and 3.5. For each stream of activity, there were relatively similar research and developmental processes. The research started with the identification and dissemination of suitable species; later research focused on how to make these an economically viable part of the farming system. However, in the remaining space that we have here, we want to focus mostly on nutritive value, i.e. the suite of factors that make halophytes suited to be a source of nutrients for grazing animals.

TABLE 3.4
Highlights in the Development of Herbaceous Species for Saltland in Southern Australia

Year	Activity
1945–1951	Farmers attest to value of tall wheatgrass (*Thinopyrum ponticum*) and salt-water couch (*Paspalum vaginatum*) in WA. Researchers confirm their results (Teakle and Burvill 1945; Burvill and Marshall 1951).
1954–1959	Screening of herbaceous germplasm (69 grasses and forbs) at Kojonup WA. Puccinellia (*Puccinellia ciliata*) and tall wheat grass are best adapted (Rogers and Bailey 1963).
Mid 1960s–1980s	Adoption of puccinellia in WA, NSW and SA. Tall wheatgrass strongly adopted in NSW (Malcolm and Smith 1965; Fleck 1967; Hamilton 1972; Lay 1990).
1992–1996	Growth of puccinellia in SA is found to respond strongly to N fertiliser (McCarthy 1992; Herrmann 1996)
1985–1995	The annual legume balansa clover (*Trifolium michelianum*) is released by the SA Department of Agriculture in 1985; its value for waterlogged/marginally saline soils is established in VIC and WA (Rogers and Noble 1991; Rogers and West 1993; Evans 1995).
1996–1999	Screening of perennial grasses (30 genotypes) occurs at six NSW sites. Tall wheatgrass, puccinellia, salt-water couch and kikuyu (*Pennisetum clandestinum*) are best adapted (Semple et al. 1998; 2003).
1995–2013	A patented clone of *Distichlis spicata* is introduced from the US and tested in SA, WA and VIC (Leake et al. 2001; Sargeant et al. 2001). However, *D. spicata* that was irrigated with saline effluent had relatively poor feeding value for sheep in metabolism crates (Lymbery et al. 2013).
2000–2011	The annual legume messina (*Melilotus seculis*) is introduced from Spain and tested in SA and WA (Nichols et al. 2008). The species fails to reproduce under field conditions until suitable rhizobial symbionts are selected (Bonython et al. 2011). Messina and its rhizobium are commercially released.

TABLE 3.5
Highlights in the Development of Halophytic Shrubs for Saltland in Southern Australia

Year	Activity
Late 1890s	Value of saltbushes (*Atriplex* species) noted for pastoral locations (Turner 1897).
1959	Planting of small-leaf bluebush (*Maireana brevifolia*), old man saltbush (*Atriplex nummularia*) and creeping saltbush (*Atriplex semibaccata*) advocated for saltland in WA wheatbelt (Smith and Malcolm 1959).
1966–1990	WA Department of Agriculture builds halophyte collection (accessions listed in Malcolm and Clarke 1971; Malcolm et al. 1984). The collection peaked at more than 1000 accessions.
1968–1982	120 halophytic shrubs from five genera screened at three WA sites (Malcolm and Clarke 1971). 25 genotypes selected for further work at 14 WA sites (Malcolm and Swaan 1989). Outstanding genotypes are small-leaf bluebush, river saltbush (*Atriplex amnicola*), samphire (*Tecticornia pergranulata*), wavy-leaf saltbush (*Atriplex undulata*) and quailbrush (*Atriplex lentiformis*).
1976–1982	First 'niche' seeder built for direct seeding of saltbush. Programme of direct seeding trials commence with river saltbush, wavy-leaf saltbush, quailbrush and small leaf bluebush (Malcolm and Allen 1981; Malcolm et al. 1980, 1982, 2003; Malcolm and Swaan 1985; see also summary by Barrett-Lennard et al. 2016).
1985–1994	Halophytic shrubs from WA programme trialled in SA, VIC and NSW with mixed success (Lay 1990; West 1990; Barson 1994).
1991	Saltbush establishment improved by planting commercially raised seedlings with tree planters (Barrett-Lennard et al.1991).
1994	Saltbush criticised on the basis that the plants have low nutritive value and sheep lose condition (Warren and Casson 1994).
1998	Importance of understorey beneath shrubs emphasised as part of the feed on offer (Barrett-Lennard and Ewing 1998).
2001–2002	CSIRO reviews the opportunity for animal production from saline land (Masters et al. 2001) and conducts benchmarking studies identifying halophytes and understorey species growing on saline farmland in three casestudies (Norman et al. 2002).
2002–2009	Grazing experiments show that some saltbush species have higher nutritional value than others, and there are differences in relative palatability within species (Norman et al. 2004; Tiong et al. 2004; Norman et al. 2008). A carbon isotope method is calibrated for the prediction of saltbush in the diet of sheep (Norman et al. 2009a). Sheep select between 10 and 40% saltbush in diet; dietary selection is related to the nutritional value of the non-salty alternative (Norman et al. 2009b; 2010a). Large producer demonstration sites are established (Thomas et al. 2009).
2002–2012	'Sustainable Grazing on Saline Lands' project compares productivity, environmental health and hydrological outcomes of saline land that has been revegetated with saltbush. Saltbush with understorey was shown to quadruple grazing value, half water runoff and reduce salt export by 90% (Bennett et al. 2012).
2002–2008	'Salty Diets' project in the CRC Salinity explores impacts of high salt diets on ruminant health and productivity. Voluntary salt intake levels for sheep documented (Masters et al. 2005b). Student projects explore various forms of exposure (*in utero* and with mother) to improve utilisation of saltland pastures (Blache et al. 2007; Thomas et al. 2007; Chadwick et al. 2009; Digby et al. 2010; Norman et al. 2016b).

(Continued)

TABLE 3.5 (*Continued*)
Highlights in the Development of Halophytic Shrubs for Saltland in Southern Australia

Year	Activity
2005	First review on importance of nutritive value in saltland pastures published; subsequently updated in 2015 (Masters et al. 2005a; Masters 2015).
2005–2010	*In vivo* organic matter digestion of native Australian shrubs by sheep compared with *in vitro* and *in sacco* predictions (Norman et al. 2010b). Near-infrared spectroscopy methods redeveloped to allow for rapid and inexpensive prediction of nutritional value of novel shrubs (Norman and Masters 2010c).
2006–2010	Whole farm economic modelling of saltland pastures shows that profits are increased more by improvement in nutritive value than by improvement in biomass (O'Connell et al. 2006). Improved nutritive value becomes a plant selection goal (Monjardino et al. 2010).
2009–2014	CRC Dryland Salinity commences old man saltbush improvement programme; 60,000 plants, two sub-species, three sites. First commercial high nutritive value old man saltbush clone ('Anameka™') released in 2014 (Norman et al. 2016a).
2004-2019	The role of saltbush as an antioxidant to improve sheep health and meat quality is explored (Pearce et al. 2005; Pearce et al. 2010; Fancote et al. 2013; Norman et al. 2019).
2013	Interactions between plants and environment, and impact on feeding value reviewed (Norman et al. 2013). Oxalate found to accumulate in leaves of old man saltbush plants fed nitrate but not ammonium (Al Daini et al. 2013).
2015–2018	Ability of halophytic shrubs to reduce methane emissions from sheep investigated (Li et al. 2016; 2018).
2019–2022	CSIRO's 'No More Gaps' project develops agronomic packages for management of high nutritive value old man saltbushes and seed coatings for direct seeding.

Halophytes have been defined as plants that can complete their life cycles at salt concentrations in the soil solution greater than 200 millimoles per litre (i.e. EC_w values > 20 dS m^{-1}) (Flowers and Colmer, 2015); they have strengths and weaknesses as forages (Masters et al. 2005a; Norman et al. 2013). Many species particularly from the family Amaranthaceae are able to grow in saline environments because they use salt for the osmotic adjustment of cell vacuoles (Flowers and Colmer, 2015). This can be a problem for the use of these plants as forages because the accumulated salt has no nutritive value for ruminants and must be excreted (Masters et al. 2005a; Norman et al. 2013). Sheep grazing halophytes need to be supplied with large amounts of fresh water (they can drink up to 12 L of water per day), and they will decrease their feed intake as the concentration of the salt in the diet increases (Warren et al. 1990; Masters et al. 2005b). Other disadvantages of halophytes are: (a) they can have low digestibility of the organic matter (which means that they have low energy concentrations for grazing animals), (b) they can be unpalatable, and (c) they can accumulate high (>6% DM) concentrations of potentially toxic compounds like oxalate in their leaves (Malcolm et al. 1988; Norman et al. 2004; Masters et al. 2007; Masters 2015). In old man saltbush (*Atriplex nummularia*), accumulation of oxalate was shown to be associated with the uptake of nitrate, as opposed to ammonium, by the plants (Al Daini et al. 2013).

On the other hand, halophytes can have nutritional benefits for animals. Many halophytic species osmotically adjust the cellular cytoplasm through the accumulation of small molecular weight 'compatible solutes' such as glycine-betaine (Storey et al. 1977). These compounds are rich in N, which can be converted into protein in the rumen if the animals have sufficient energy, often from other feed sources, in the diet (Masters et al. 2005a). Finally, saltland pastures may assist ruminants to assist with oxidative damage associated with heat stress. Vitamin E deficiency and associated muscular myopathy is widespread in sheep grazing dry pastures and crop residues during the hot summer and autumn seasons in Western Australia (White and Rewell 2007). Research in the 2000s found that leaves of halophytes like saltbush (*Atriplex* species) are an excellent source of vitamin E and sheep grazing saltbush are not susceptible to muscular myopathy (Pearce et al. 2005). Furthermore, access to saltbush leaves for several weeks can provide enough vitamin E for lambs to maintain health through autumn feedlotting and improve meat quality (Pearce et al. 2010; Fancote et al. 2013). In addition to vitamin E, saltbushes are rich in the minerals associated with antioxidant pathways (Norman et al. 2019).

Our modern understanding is that halophytic shrubs like saltbushes (*Atriplex* spp.) can be an important part of the fodder supply on a farm. However, best animal and economic performance will occur if farmers follow a number of principles. These include:

1. Graze the saltland pasture during the most feed-limited time of year – in southern Australia, this is often in summer and autumn when senesced crops and annual pastures provide the most of the feedbase (O'Connell et al. 2006).
2. Establish pastures using palatable high nutritional value species on moderately to highly saline landscapes, not severely-extremely saline landscapes. Feed intake is limited by high salt in the diet, so the most productive systems often contain a mixed range of species such as rows of perennial halophytes accompanied by an understorey of more salt-sensitive annual pasture plants (e.g. Norman et al. 2010a). The system is not suited to soils of extreme salinity: the euhalophytes (e.g. *Tecticornia pergranulata*) that grow in these environments accumulate too much salt in their foliage (often accounting for ~40% of shoot DM) and these soils will not support the required low-salt understorey. Naturally occurring annual pasture species can be used as indicators of the better soils (Bennett and Barrett-Lennard 2013).
3. Train the animals to use the feed; they may not recognise that novel shrubs are food. Training is often achieved by introducing inexperienced lambs to the feed while still joined to their more experienced mothers (Norman et al. 2016b). *In utero* exposure of lambs provides another opportunity to improve the intake and utilisation of saltbush (Chadwick et al. 2009).
4. Use it or lose it. The growth of long-lived, woody perennial shrubs slows as the plants reach a critical leaf area and move into their reproductive phase. They will drop excess leaves when water-stressed. A complete annual defoliation improves plant growth rates and produces new leaves of higher nutritional value (Wilmot and Norman 2006).
5. Supply animals with large amounts of cool fresh water.

3.6 DEVELOPMENT OF ANAMEKA™ SALTBUSH

Whole-farm economic modelling in 2006 suggested that increasing the metabolisable energy value of saltland pastures was the key to improving the profitability of systems (O'Connell et al. 2006). Germplasm of *Atriplex nummularia* (old man saltbush) was collected from across the native range in Australia and genotypes from 27 populations and two subspecies were compared at three research sites. At the same time, methods were developed to allow for the rapid and inexpensive screening of the energy value of the shrubs to grazing animals (Norman et al. 2010b,c). Sheep were used to identify plants that were consistently preferred; these plants tended to have higher energy values and lower sulphur (Norman et al. 2009; 2015). Over 8 years in a sequence of on-farm experiments, the team identified old man saltbush genotypes with 20% higher organic matter digestibility, greater acceptability to sheep and up to eight times more edible biomass production than the mean of the collection (Norman et al. 2016). In 2015, the first clonal cultivar was commercialised by CSIRO (cv. Anameka™). Whole-farm economic analysis suggests that Anameka, planted on soils that are marginal for crop production, can double the profitability of saltbush plantations on farms (Monjardino et al. 2010). Over two million Anameka™ shrubs, sold as clonal cuttings, have now been planted by over 200 Australian farmers. The team are now developing elite seed lines and methods to aid saltbush establishment from seeds.

3.7 ELECTROMAGNETIC INDUCTION AS A TOOL IN CONDUCTING FIELD TRIALS

Any student of the saline agricultural literature will soon find that many plant experiments are conducted under controlled conditions (e.g. salt flushed lysimeters, nutrient solution cultures, etc.) but fewer are conducted under general field conditions. This occurs despite the fact that such research is generally intended to produce better outcomes in the field. One of the reasons for this is that soil salinity is often spatially and temporally heterogenous and it is, therefore, difficult to layout and block the plots making up field trials in a manner that gives appropriate statistical rigour. For researchers in Australia, this changed in the 1980s because of the availability of instruments like the EM38 (Geonics Ltd) and the DualEM that use the principle of electromagnetic induction to measure variation in the apparent electrical conductivity of the soil (EC_a). Electromagnetic induction has been used at a range of scales to survey spatial variation in salinity for a number of decades (Spies and Woodgate 2005). The recent innovation has been to recognise that at the plot scale measurements of EC_a can be correlated with measures of soil $EC_{1:5}$ and/or the salinity of the soil solution and that these measures can then be used statistically to correct grain yields for variation in soil salinity across a site (e.g. Setter et al. 2016; Asif et al. 2018).

An example of the use of the EM38 for this purpose can be seen in the work of Setter et al. (2016). These researchers tested the variation in grain yield of 90 wheat and 27 barley cultivars in separate experiments in a saline field in the wheatbelt of Western Australia in two growing seasons 2009 and 2011. Experiments of this kind

TABLE 3.6
Significance (*P*-Value) of the Terms Used in Statistical Models to Account for Variation in Grain Yield in the Field

Crop	Year	EM38 (June)	Row	Column	Genotype
Wheat	2009	< 0.001	0.0418	< 0.001	< 0.001
	2011	Ns	< 0.001	0.0141	< 0.001
Barley	2009	< 0.001	0.0062	0.0036	< 0.001
	2011	< 0.001	Ns	Ns	< 0.001

Source: After Setter et al. 2016.

are normally conducted using a rectangular array of plots indexed by rows and columns, with replication along the rows and columns. This enables a statistical analysis of grain yield that takes account of linear variation in the direction of increasing rows and columns. However, in this work, the researchers also measured EC_a values with an EM38 in June (about a month after crop establishment). The significance of the linear effects of EM38, row and column effects on the variation, and the final impact of these covariates on the significance of cultivar effects are summarised in Table 3.6. It can be seen that there were highly significant effects of EM38 reading ($P < 0.001$) for three of the four experiments, but the linear effects of row and column were less significant (only one of the four effects in each case had $P < 0.001$). Thankfully however, the combination of all covariates was sufficient for the effects of genotype to be significant at $P < 0.001$ in all four experiments, and the research team was able to conclude that the variables associated with high grain yield in this environment were genetic; successful cultivars were adapted to local conditions, had high salt-tolerance and early flowering (Setter et al. 2016).

3.8 SALINE FUTURES AND CLIMATE CHANGE

Will climate change make salinity outcomes better or worse in southern Australia? The answer to this question will depend on the primary cause of salinity. For soils affected by the presence of shallow water-tables, the risk of salinity may well be ameliorated overall by the drier conditions that are likely to occur. Indeed, there is already some evidence for this. Since 1975, annual rainfall has decreased by 10–25% over much of the southwest of Australia (McFarlane et al. 2020). About 700 bores across the southwest of Australia have a sufficiently long record of measurement to estimate groundwater levels trends before and after 2000. Two-thirds of all monitored bores had rising groundwater levels prior to 2000, but only 40% rose between 2000 and 2012 (McFarlane et al. 2020). The implication is clear: climate change may help ameliorate dryland salinity.

However, the effects of climate change on cropped soils affected by transient salinity in southern Australia, is expected to be different. For these dispersive alkaline soils, salt has accumulated over hundreds of years from rainfall; there is no expectation that climate change will decrease the concentration of salt in these soils, and if the soils become drier, then the salinity of the soil solution will increase. This will affect the growth of annual crops and pastures, but particularly those that are most salt sensitive. The adverse effects can be expected to be greatest for salt sensitive crops like field peas, will be lower for wheat, and least for more tolerant crops like barley and canola.

We conclude on a challenging note. The species developed for saltland revegetation across southern Australia may have a broader application than saltland alone. Many of these species are endemic to arid and well as saline environments. With climate change, we expect that the growth of traditional crops and pastures will become riskier in the driest agricultural areas of southern Australia (annual average rainfall less than 280 mm). What could this land be used for? Modelling suggests that one of the most effective strategies that Australian farmers use to manage risk is to diversify their farms across crop and livestock enterprises (Ghahramani et al. 2019). We believe that much of this land could be suited to sheep and cattle breeding systems based on the growth of elite perennial halophytic shrubs augmented by perennial grasses and short-season annual legumes. The eventual area of these permanent shrub-based pasture farming systems could extend over millions of hectares. Researchers in partnership with farmers need to use the skills and expertise that have been honed in the building of saltland pastures to explore this exciting (but also daunting) new opportunity.

ACKNOWLEDGEMENTS

We acknowledge support and contributions from many co-funders and collaborators across many projects. This includes participants in the CRC Plant Based Management of Dryland Salinity and the CRC Future Farm Industries. Various projects have received support from the Australian Government Department of Agriculture, Water and the Environment, Meat and Livestock Australia, Australian Wool Innovation, the Grains Research and Development Corporation and the Australian Centre for International Agricultural Research. Many Australian agencies have contributed to this research, including teams from CSIRO, WA Department of Primary Industries and Regional Development, New South Wales Department of Primary Industries, WA Department of Biodiversity, Conservation and Attractions, Department of Water Land and Biodiversity Conservation in South Australia and the South Australian Research and Development Institute, University of Western Australia, Curtin University and Murdoch University. We are very grateful to all the Australian farmers who have hosted and support our research and the team at Chatfield's Tree Nursery. We are grateful for the helpful comments of Dr Richard George on this manuscript.

ENDNOTE

1. The names of Australian States are abbreviated as follows: New South Wales (NSW), Victoria (VIC), Queensland (QLD), South Australia (SA), Western Australia (WA) and Tasmania (TAS).

REFERENCES

ABS (2002) Salinity on Australian farms. Australian Bureau of Statistics, Canberra (www.abs.gov.au/AUSSTATS/abs@.nsf/Lookup/4615.0Main+Features12002)

Al Daini, H, Norman, HC, Young P and Barrett-Lennard, EG (2013) The source of nitrogen (NH_4^+ or NO_3^-) affects the concentration of oxalate in the shoots and the growth of *Atriplex nummularia* (oldman saltbush). Functional Plant Biology 40, 1057–1064.

Asif MA, Schilling RK, Tilbrook J, Brien C, Dowling K, Rabie H, Short L, Trittermann C, Garcia A, Barrett-Lennard EG, Berger B, Mather DE, Gilliham M, Tester M, Fleury D, Roy SJ and Pearson AS (2018) Identification of novel salt tolerance QTL in the Excalibur × Kukri doubled haploid (DH) wheat population. Theoretical and Applied Genetics 131, 2179–2196.

Barrett-Lennard EG (1986) Effects of waterlogging on the growth and NaCl uptake by vascular plants under saline conditions. Reclamation and Revegetation Research 5, 245–261.

Barrett-Lennard EG (2003) The interaction between waterlogging and salinity in higher plants: causes, consequences and implications. Plant and Soil 253, 35–54.

Barrett-Lennard E and Ewing M (1998) Saltland Pastures? They are feasible and sustainable – we need a new design. In: Proceedings of the 5th National Conference on Productive Use and Rehabilitation of Saline Lands, 10–12 March, Tamworth, pp. 160–161.

Barrett-Lennard E, Frost F, Vlahos S and Richards N (1991) Revegetating salt-affected land with shrubs. Journal of Agriculture of Western Australia 32, 124–129.

Barrett-Lennard EG, George RJ, Hamilton G, Norman HC and Masters DG (2005) Multidisciplinary approaches suggest profitable and sustainable farming systems for valley floors at risk of salinity. Australian Journal of Experimental Agriculture 45, 1415–1424.

Barrett-Lennard EG and Shabala SN (2013) The waterlogging/salinity interaction in higher plants revisited – focusing on the hypoxia-induced disturbance to K^+ homeostasis. Functional Plant Biology 40, 872–882.

Barrett-Lennard EG, Bennett SJ and Colmer TD (2008) Standardising terminology for describing the level of salinity in soils. 2nd International Salinity Forum, Adelaide, 31 March – 3 April, 4 pp.

Barrett-Lennard EG, Norman HC and Dixon K (2016) Improving saltland revegetation through understanding the 'recruitment niche': potential lessons for ecological restoration in extreme environments. Restoration Ecology 24, S91–S97.

Bennett D and George R (2008) Long term monitoring of groundwater levels at 24 sites in Western Australia shows that integrated farm forestry systems have little impact on salinity. Proceedings of the 2nd International Salinity Forum, 31 March – 3 April, Adelaide, Australia.

Bennett SJ and Barrett-Lennard EG (2013) Predictions of watertable depth and soil salinity levels for land capability assessment using site indicator species. Crop and Pasture Science 64, 285–294.

Bennett SJ, Barrett-Lennard EG and Colmer TD (2009) Salinity and waterlogging as constraints to saltland pasture production: a review. Agriculture, Ecosystems and Environment 129, 349–360.

Blache D, Grandison MJ, Masters DG, Dynes RA, Blackberry MA and Martin GB (2007). Relationships between metabolic endocrine systems and voluntary feed intake in Merino sheep fed a high salt diet. Australian Journal of Experimental Agriculture 47, 544–550.

Bonython AL, Ballard RA, Charman N, Nichols PGH and Craig AD (2011) New strains of rhizobia that nodulate regenerating messina (*Melilotus siculus*) plants in saline soils. Crop and Pasture Science 62, 427–436.

Burvill GH (1956) Salt land survey, 1955. Journal of Agriculture of Western Australia 5, 113–120.

Burvill GH and Marshall AJ (1951) *Paspalum vaginatum* or seashore paspalum. Journal of Agriculture of Western Australia, 28, 191–194.
Chadwick MA, Vercoe PE, Williams IH and Revell DK (2009) Dietary exposure of pregnant ewes to salt dictates how their offspring respond to salt. Physiology and Behavior 97, 437–445.
Conacher AJ, Combes PL, Smith PA and McLellan RC (1983) Evaluation of throughflow interceptors for controlling secondary soil and water salinity in dryland agricultural areas of southwestern Australia: I. Questionnaire surveys. Applied Geography 3, 29–44.
Dawes WR, Stauffacher M and Walker GR (2000) Calibration and Modelling of Groundwater Processes in The Liverpool Plains. CSIRO Land and Water Technical Report 5/00, Canberra, Australia.
Digby SN, Masters DG, Blache D, Hynd PI and Revell DK (2010) Offspring born to ewes fed high salt during pregnancy have altered responses to oral salt loads. Animal 4, 81–88.
Evans P (1995) Developing pasture legumes for waterlogged/salt affected areas in medium rainfall zones of Western Australia. Final Report, Project DAW 164, International Wool Secretariat, 20 pp.
Fancote CR, Vercoe PE, Pearce KL, Williams IH and Norman HC (2013) Backgrounding lambs on saltbush provides an effective source of Vitamin E that can prevent Vitamin E deficiency and reduce the incidence of subclinical nutritional myopathy during summer and autumn. Animal Production Science 53, 247–255.
Fleck BC (1967) A note on the performance of *Agropyron elongatum* (Host.) Beauv. and *Puccinellia* (Parl.) sp. in revegetation of saline areas. Journal of the Soil Conservation Service of NSW 23, 261–269.
Flowers TJ and Colmer TD (2015) Plant salt tolerance: adaptations in halophytes. Annals of Botany 115, 327–331.
George R, McFarlane D and Nulsen B (1997) Salinity threatens the viability of agriculture and ecosystems in Western Australia. Hydrogeology Journal 5, 6–21.
George RJ, Clarke CJ and Hatton T (2001) Computer-modelled groundwater response to recharge management for dryland salinity control in Western Australia. Advances in Environmental Monitoring and Modelling 2, 3–35.
Ghahramani A, Howden, MS, del Prado A, Thomas DT, Moore AD, Ji B (2019) Climate change impact, adaptation and mitigation in temperate grazing systems: a review. Sustainability 11, 7224.
Ghassemi F, Jakeman AJ and Nix HA (1995) Salinisation of Land and Water Resources: Human Causes, Extent, Management and Case Studies. University of New South Wales Press Ltd, Sydney, pp. 526.
Greenwood EAN, Klein L, Beresford JD and Watson GD (1985) Differences in annual evaporation between grazed pasture and *Eucalyptus* species in plantations on a saline farm catchment. Journal of Hydrology 78, 261–278.
Hamilton GJ (1972) Investigations into reclamation of dryland saline soils. Journal of the Soil Conservation Service of NSW 28, 191–211.
Hatton TJ and Nulsen RA (1999) Towards achieving functional ecosystem mimicry with respect to water cycling in southern Australian agriculture. Agroforestry Systems 45, 203–214.
Herrmann T (1996) Puccinellia – maximum productivity from saltland. In: Proceedings of the 4th National Conference on Productive Use and Rehabilitation of Saline Lands, 25–30 March, Albany. Promaco Conventions Pty Ltd, Canning Bridge, pp. 297–302.
Hingston FJ and Gailitis V (1975) The geographic variation of salt precipitated over Western Australia. Australian Journal of Soil Research 14, 319–335.
Latta RA, Blacklow LJ and Cocks PS (2001) Comparative soil water, pasture production, and crop yields in phase farming systems with lucerne and annual pasture in Western Australia. Australian Journal of Agricultural Research 52, 295–303.

Latta RA, Cocks PS and Matthews C (2002) Lucerne pastures to sustain agricultural production in southwestern Australia. Agricultural Water Management 53, 99–109.

Lay B (1990) Salt land revegetation: a South Australian overview. In: Revegetation of Saline Land (edited by BA Myers and DW West), Institute for Irrigation and Salinity Research, Tatura, Victoria, pp. 15–20.

Leake J, Barrett-Lennard E, Yensen N and Prefumo J (2001) NyPa Distichlis cultivars: rehabilitation of highly saline areas for forage, turf and grain. Final Report Project RAS98-74. Rural Industries Research and Development Corporation, Canberra.

Li X, Young P and Norman HC (2016) Nitrate and sulphate accumulating saltbush reduces methane emissions in the ANKOM fermentation system. In: '6th Greenhouse Gas and Animal Agriculture Conference, Melbourne'. p. 102.

Li X, Norman HC, Hendry JK, Hulm E, Young P, Speijers J and Wilmot MG (2018) The impact of supplementation with *Rhagodia preissii* and *Atriplex nummularia* on wool production, mineral balance and enteric methane emissions of Merino sheep. Grass and Forage Science 73, 381–391.

Lymbery AJ, Kay GD, Doupe RG, Partridge GJ and Norman HC (2013) The potential of a salt-tolerant plant (*Distichlis spicata* cv. NyPa Forage) to treat effluent from inland saline aquaculture and provide livestock feed on salt-affected farmland. Science of the Total Environment 445, 192–201.

Malcolm CV and Allen RJ (1981) The Mallen niche seeder for plant establishment on difficult sites. Australian Rangeland Journal 3, 106–109.

Malcolm CV and Clarke AJ (1971) Progress Report No. 1. Collection and testing of forage plants for saline and arid areas. Technical Bulletin No. 8, Department of Agriculture of Western Australia, South Perth, 39 pp.

Malcolm CV and Smith ST (1965) Puccinellia – outstanding saltland grass. Journal of Agriculture of Western Australia 6, 153–156.

Malcolm CV and Swaan TC (1985) Soil mulches and sprayed coatings and seed washing to aid chenopod establishment on saline soil. Australian Rangeland Journal 7, 22–28.

Malcolm CV and Swaan TC (1989) Screening shrubs for establishment and survival on salt-affected soils in south-western Australia. Technical Bulletin 81, Department of Agriculture of Western Australia, South Perth, 35 pp.

Malcolm CV, Swaan TC and Ridings HI (1980) Niche seeding for broad scale forage shrub establishment on saline soils. In: International Symposium on Salt Affected Soils, Symposium Papers. Central Soil Salinity Research Institute, Karnal, India, pp. 539–544.

Malcolm CV, Clarke AJ and Swaan TC (1984) Plant collections for saltland revegetation and soil conservation. Technical Bulletin 65, Department of Agriculture of Western Australia, South Perth.

Malcolm CV, Hillman BJ, Swaan TC, Denby C, Carlson D and D'Antuono M (1982) Black paint soil amendment and mulch effects on chenopod establishment in a saline soil. Journal of Arid Environments 5, 179–189.

Malcolm CV, Lindley VA, O'Leary JW, Runciman HV and Barrett-Lennard EG (2003) Halophyte and glycophyte salt tolerance at germination and the establishment of halophyte shrubs in saline environments. Plant and Soil 253, 171–185.

Malcolm CV, Clarke AJ, D'Antuono MF and Swaan TC (1988) Effects of plant spacing and soil conditions on the growth of five *Atriplex* species. Agriculture, Ecosystems and Environment 21, 265–279.

Masters DG (2015). Assessing the feeding value of halophytes. In: Halophytic and Salt tolerant Feedstuffs: Impacts on Nutrition, Physiology and Reproduction of Livestock (edited by HM El Shaer and VR Squires). CRC Press. pp. 89–105.

Masters DG, Norman HC and Dynes RA (2001) Opportunities and limitations for animal production from saline land. Asian-Australasian Journal of Animal Sciences 14, 199–211.

Masters DG, Norman HC and Barrett-Lennard EG (2005a) Agricultural systems for saline soil: the potential role of livestock. Asian-Australasian Journal of Animal Science 18, 296–300.

Masters DG, Rintoul AJ, Dynes RA, Pearce KL and Norman HC (2005b) Feed intake and production in sheep fed diets high in sodium and potassium. Australian Journal of Agricultural Research 56, 427–434.

Masters DG, Benes SE, and Norman HC (2007) Biosaline agriculture for forage and livestock production. Agriculture, Ecosystems and Environment 119, 234–248.

McCarthy DG (1992) Salt tolerant grasses – mediterranean environment. In: National Workshop on Productive Use of Saline Land (edited by TN Herrmann), South Australian Department of Agriculture, pp. 28–35.

McFarlane D, George R, Ruprecht J, Charles S and Hodgson G (2020) Runoff and groundwater responses to climate change in south west Australia. Journal of the Royal Society of Western Australia 103, 9–27.

Monjardino M, Revell DK and Pannell DJ (2010) The potential contribution of forage shrubs to economic returns and environmental management in Australian dryland agricultural systems. Agricultural Systems 103, 187–197.

Nichols PGH, Rogers ME, Craig AD, Albertsen TO, Miller SM, McClements DR, Hughes SJ, D'Antuono MF and Dear BS (2008) Production and persistence of temperate perennial grasses and legumes at five saline sites in southern Australia. Australian Journal of Experimental Agriculture 48, 536–552.

Norman HC, Dynes RA and Masters DG (2002) An integrated approach to the planning, training and management for the revegetation of saline land. Final Report to State Salinity Council. CSIRO Livestock Industries (in collaboration with the Saltland Pastures Association), Wembley WA, 17 pp.

Norman HC, Freind C, Masters DG, Rintoul AJ, Dynes RA and Williams IH (2004) Variation within and between two saltbush species in plant composition and subsequent selection by sheep. Australian Journal of Agricultural Research 55, 999–1007.

Norman HC, Masters DG, Wilmot MG and Rintoul AJ (2008) Effect of supplementation with grain, hay or straw on the performance of weaner Merino sheep grazing old man (*Atriplex nummularia*) or river (*Atriplex amnicola*) saltbush. Grass and Forage Science 63, 179–192.

Norman HC, Wilmot MG, Thomas DT, Masters DG and Revell DK (2009a) Stable carbon isotopes accurately predict diet selection by sheep fed mixtures of C-3 annual pastures and saltbush or C-4 perennial grasses. Livestock Science 121, 162–172.

Norman HC, Revell DK and Masters DG (2009b) Sheep avoid eating saltbushes with high sulphur concentrations. Proceedings of the XIth International Symposium on Ruminant Physiology (ISRP), Clermont-Ferrand, France, September 6 to 9.

Norman HC, Wilmot MG, Thomas DT, Barrett-Lennard EG and Masters DG (2010a) Sheep production, plant growth and nutritive value of a saltbush-based pasture system subject to rotational grazing or set stocking. Small Ruminant Research 91, 103–109.

Norman HC, Revell DK, Mayberry DE, Rintoul AJ, Wilmot MG and Masters DG (2010b) Comparison of in vivo organic matter digestion of native Australian shrubs by sheep to *in vitro* and *in sacco* predictions. Small Ruminant Research 91, 69–80.

Norman HC and Masters DG (2010c) Predicting the nutritive value of saltbushes (*Atriplex* spp) with near infrared reflectance spectroscopy. In Proceedings of the International Conference on Management of Soil and Groundwater Salinization in Arid Regions, 11–14 January 2010, Muscat Oman. Edited by M. Ahmed and S. Al-Rawahy, SQU Press, Muscat, Oman, pp. 51–57.

Norman HC, Masters DG and Barrett-Lennard EG (2013) Halophytes as forages in saline landscapes: interactions between plant genotype and environment change their feeding value to ruminants. Environmental and Experimental Botany 92, 96–109.

Norman HC, Hulm E and Wilmot MG (2016a) Improving the feeding value of old man saltbush for saline production systems in Australia. In: Halophytic and Salt tolerant Feedstuffs: Impacts on Nutrition, Physiology and Reproduction of Livestock (edited by HM El Shaer and VR Squires) CRC Press, pp. 79–86.

Norman HC, Thomas D, Wilmot MG and Revell DK (2016b) Programming lambs to improve utilisation of novel forages. Australian Society of Animal Production Biannual Conference, Adelaide, Australia 4–8 July 2016.

Norman HC, Duncan EG and Masters DG (2019) Halophytic shrubs accumulate minerals associated with antioxidant pathways. Grass and Forage Science 74, 345–355.

Nulsen RA (1981) Critical depth to saline groundwater in non-irrigated situations. Australian Journal of Soil Research 19, 83–86.

O'Connell M, Young J and Kingwell R (2006) The economic value of saltland pastures in a mixed farming system in Western Australia. Agricultural Systems 89, 371–389.

Pearce KL, Masters DG, Jacob RH, Smith G and Pethick DW (2005) Plasma and tissue a-tocopherol concentrations and meat colour stability in sheep grazing saltbush (*Atriplex* spp.). Australian Journal of Agricultural Research 56, 663–672.

Pearce KL, Norman HC and Hopkins DL (2010) The role of saltbush-based pasture systems for the production of high quality sheep and goat meat. Small Ruminant Research 91, 29–38.

Rengasamy P (2002) Transient salinity and subsoil constraints to dryland farming in Australian sodic soils: an overview. Australian Journal of Experimental Agriculture 42, 351–361.

Rengasamy P (2006). World salinization with emphasis on Australia. Journal of Experimental Botany 57, 1017–1023.

Rogers AL and Bailey ET (1963) Salt tolerance trials with forage plants in south-western Australia. Australian Journal of Experimental Agriculture and Animal Husbandry 3, 125–130.

Rogers ME and Noble CL (1991) The effect of NaCl on the establishment and growth of balansa clover (*Trifolium michelianum* Savi Var. balansae Boiss.). Australian Journal of Agricultural Research 42, 847–857.

Rogers ME and West DW (1993) The effects of rootzone salinity and hypoxia on shoot and root growth in *Trifolium* species. Annals of Botany 72, 503–509.

Sargeant MR, Rogers ME, White RE and Batey T (2001) *Distichlis spicata* – a grass for highly saline areas. In: Wanted – Sustainable Futures for Saline Lands – Proceedings of the 7th National Conference on Productive Use and Rehabilitation of Saline Lands, Launceston, TAS, 20–33 March, pp. 212–213.

Semple W, Beale G, Cole I, Gardiner T, Glasson A, Koen T, Parker B, Phillips B, Reynolds K, Thearle L and the Windellama Landcare Group (1998). Evaluation of perennial grasses for saline site revegetation. In: Managing Saltland into the 21st Century: Dollars and Sense from Salt – Proceedings of 5th National Conference on the Productive Use and Rehabilitation of Saline Lands (edited by NE Marcar and AKM Afzal Hossain), PNM Editorial Publications, Canberra, pp. 157–159.

Semple WS, Cole IA and Koen TB (2003). Performance of some perennial grasses on severely salinised sites of the inland slopes of New South Wales. Australian Journal of Experimental Agriculture 43, 357–371.

Setter TL, Waters I, Stefanova K, Munns R and Barrett-Lennard EG (2016) Salt tolerance, date of flowering and rain affect the productivity of wheat and barley on rainfed saline land. Field Crops Research 194, 31–42.

Smith ST and Malcolm CV (1959) Bringing wheatbelt saltland back into production. Journal of Agriculture of Western Australia 8, 263–267.

Spies B and Woodgate P (2005) Salinity Mapping Methods in the Australian Context. Report for the Programs Committee of Natural Resource Management Ministerial Council, through Land and Water Australia and the National Dryland Salinity Program. Department of the Environment and Heritage; and Agriculture, Fisheries and Forestry, Canberra, 47 pp.

Storey R, Ahmad N and Wyn Jones RG (1977) Taxonomic and ecological aspects of the distribution of glycinebetaine and related compounds in plants. Oecologia 27, 319–332.

Teakle LJH and Burvill GH (1938) The movement of soluble salts in soils under light rainfall conditions. Journal of Agriculture of Western Australia 15, 218–245.

Teakle LJH and Burvill GH (1945) The management of salt lands in Western Australia. Journal of Agriculture of Western Australia 22, 87–93.

Thomas DT, Rintoul AJ and Masters DG (2007) Sheep select combinations of high and low sodium chloride, energy and crude protein feed that improve their diet. Applied Animal and Behaviour Science 105, 140–153.

Thomas DT, White CL, Hardy J, Collins JP, Ryder A and Norman HC (2009) An on-farm evaluation of the capability of saline land for livestock production in southern Australia. Animal Production Science 49, 79–83.

Tiong MK, Masters DG, Norman HC, Milton JTB and Rintoul AJ (2004) Variation within and between four halophytic shrub species collected from five saline environments. Animal Production in Australia 25, 327.

Turner F (1897) West Australian saltbushes. In: The West Australian Settler's Guide and Farmer's Handbook, Part III, Chapter II (edited by L Lindley-Cowen), E.S. Wigg and Son, Printers, pp. 418–431.

Warren BE, Bunny CJ and Bryant ER (1990) A preliminary examination of the nutritive value of four saltbush (*Atriplex*) species. Proceedings of the Australian Society of Animal Production 18, 424–427.

Warren BE and Casson T (1994) Sheep and saltbush – are they compatible? Proceedings of the 3rd national workshop on Productive Use of Saline Land, Echuca 15-17 March, Murray Darling Basin Commission and Salt Action, pp. 125–129.

West DW (1990) Revegetation of saline lands – perspectives from Victoria. In: Revegetation of Saline Land (edited by BA Myers and DW West), Institute for Irrigation and Salinity Research, Tatura, Victoria, pp. 5–11.

White CL and Rewell L (2007) Vitamin E and selenium status of sheep during autumn in Western Australia and its relationship to the incidence of apparent white muscle disease. Australian Journal of Experimental Agriculture 47, 535–543.

Wilmot MG and Norman HC (2006 July). Saltbush biomass in a saline grazing system - use it or lose it. Proceedings of the Australian Society of Animal Production 26th Biennial Conference Science and Industry - Hand in glove. Perth, Australia. 10–14.

Wood WE (1924) Increase of salt in soil and streams following the destruction of native vegetation. Journal of the Royal Society of Western Australia 10, 35–47.

4 Use and Management of Saline Water for Irrigation in the Near East and North Africa (NENA) Region

Redouane Choukr-Allah

CONTENTS

4.1 WATER SCARCITY IN THE NENA REGION

The Near East and North Africa (NENA) region covers 19 countries where water scarcity (FAO, 2017) is one of the major challenges (Figure 4.1). This constraint contributes to the degradation of important irrigated areas, affecting the increase of food production needed to address the increase in population. NENA regions account for about 6% of the world's population with only 1% of the world's renewable water resources (Mahmoud, 2013; Abu-Zeid, 2013). Most countries of the NENA region, are forced to use non-conventional water resources, including saline water. Saline water is loosely

DOI: 10.1201/9781003112327-4

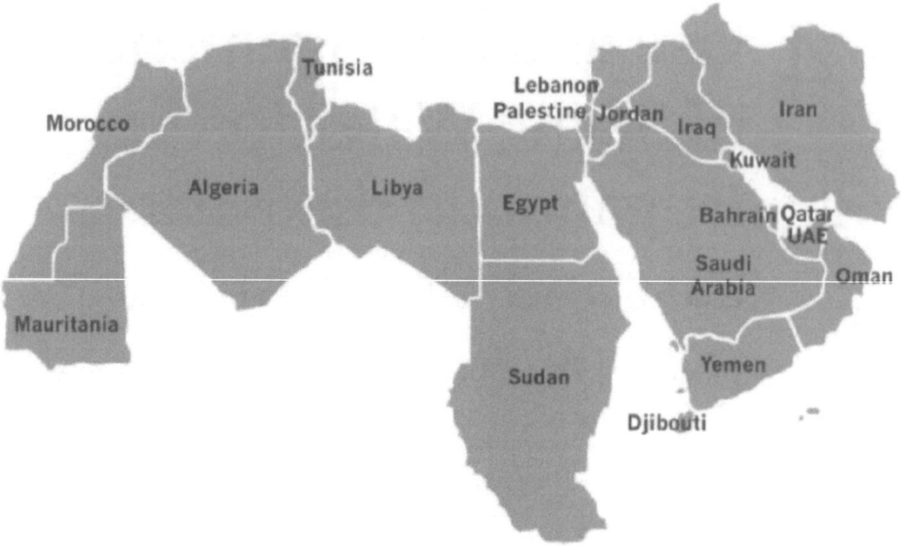

FIGURE 4.1 The Near East and North Africa (NENA) region covers 19 countries from Mauritania to Iran.

defined as water that is more saline than fresh water, but not as saline as seawater. It covers a large range of salinity (FAO, 2018) levels varying from 500 to 35,000 total dissolved solids (TDS) (mg/L). Despite the presence of large amounts of such saline water (e.g. 2 billion m^3 of groundwater in Morocco, 13.7 billion m^3 in Iran), it is only used in limited amounts for irrigation, even though this water has the potential to be used to grow a number of crops if appropriate management practices are followed. But the successful, long-term use of saline water requires background knowledge of scientific principles combined with proper management in order to minimize the negative impact of salinity on the soil, the crop and the environment.

The NENA region is one of the most water scarce regions in the world (Table 4.1). The average annual precipitation is estimated at ~150 mm. At the same time, the renewable water resources per capita is one of the lowest in the world due

TABLE 4.1
Total Renewable Water Resources per Capita in NENA Countries

Country	Total Renewable Water Resources per Capita (m^3)
Iran, Iraq, Lebanon, Mauretania, Sudan	>1000
Egypt, Morocco	500–1000
Algeria, Jordan, Libya, WBG, Oman, Syria, Tunisia	100–500
Bahrain, Kuwait, Qatar, KSA, UAE, Yemen	<100

Source: FAO AQUASTAT Survey – 2008.

to population growth (690 m^3 per capita/year in the region vs. 6400 m^3 per capita/year in the world) (Abu-Zeid, 2013).

There is growing concern over the declining availability of freshwater, and the ever-increasing demands on low quality water.

In general, in arid and semi-arid areas, two realities are recognized. The first is that for all practical purposes fresh water resources are finite and most of the economically viable development of these resources has already been implemented; thus the potential to expand this resource base is marginal. The second is that water quality degradation resulting from urban industrial and agriculture wastewater pollution is reducing the volume of freshwater. Moreover, this is exacerbated with climate change, increasing population, rapid urbanization and the associated expansion of economic activities, all of which require more water, putting tremendous strain on the already limited and fragile resource.

There is no doubt that the water requirements in arid and semi-arid regions will continue to increase significantly during the next decades. The traditional response of increasing water supply to meet higher demands will no longer be adequate in the future. This implies the use, euse and recycling of the non-conventional water resources as an additional water source, particularly in the irrigation sector.

In the majority of the arid and semi-arid regions, the slow progress in agricultural development as well as the decline in food production is not only limited by water shortage, but equally, by rapid soil salinization, bringing nearly 30% of arable lands (FAO, 2008) out of production (Table 4.2).

TABLE 4.2
Appropriate Management Practices

Crop Selection and Management	- Selection of crops tolerant to salinity and specific ions - Identification of critical growing stage affected by salinity - Intercropping: irrigating the least salt-tolerant crop first, then using drainage water to irrigate another crop which is relatively more salt tolerant.
Water Management	- Types of irrigation practice (e.g. trickle, furrow, flood, etc.) - Application system, method, schedule - Monitoring of irrigation water quality - Leaching requirements - Land drainage - using saline drainage water to grow fish, algae and shrimp
Land Management	- Levelling, tillage, ploughing, mulching - tillage: i.e. the mechanical operation for seedbed preparation, to break up surface crust, increase organic matter and nutrient availability - ploughing: beneficial on stratified soils having impermeable layers laying between permeable layers - mulching: reduces soil evaporation and temperature
Soil Improvement	- Application of chemical or organic amendments (e.g. $CaSO_4$, organic matter) to neutralize soil reaction and replace exchangeable sodium by calcium - Mixing with sands to increase the permeability of a fine-textured surface soil - Application of adequate fertilizers, type of fertilizers (preferably acid) - Timing and placement of mineral fertilizers - Regular monitoring of soil salinity

Issues related to salinization include the concentration of total salts (salinity) and the concentration of sodium relative to calcium and magnesium (sodicity). Salinity has direct effects on the growth and development of plants. Sodic conditions may cause an important deterioration of the soil physical properties, indirectly affecting crop growth through increased surface crusting, poor water infiltration and reduced root zone aeration.

Salt-affected soils vary in extent by country from 10 to 15% in Algeria to over 50% of arable land in Iraq (FAO, Status of the World's Soil Resources: Main Report, 2015). In Iraq and Syria, about 50% of reclaimed lands in the Euphrates plain are seriously affected by salinization and waterlogging (CAMRE/UNEP/ACSAD, 1996), and in Yemen, approximately 60% of the 0.5 Mha of irrigated land is slightly to moderately saline, and another 40% has levels of salinity that prevent farming (FAO, Status of World's Soil Resources: Main Report, 2015). However, a comprehensive assessment on the extent of salt-affected soils globally and in particular the NENA region is lacking.

The successful use of saline water for irrigation requires a basic understanding of the scientific principles affecting the interactions between climate, the applied water, the soil and the crop. Equally important is the application of suitable technology and management practices (Table 4.2) that will facilitate the optional use of this poor-quality water. A higher level of management is needed to successfully use saline water and the adoption of new irrigation management practices will likely be necessary. Since climate, water quality, soil type and crop tolerance to salinity varies from location to location, site-specific and appropriate on-farm management practices need to be developed to attenuate the negative impact of salinity on soil, plant and the environment.

Different types of saline water reuse exist (agricultural drainage water, groundwater and treated wastewater) and are widely used (FAO, AWC, 2018). However, the potential negative impacts of such waters, that include increases in soil salinity, yield reductions, deterioration of soil quality and costs associated with these negative impacts should be considered. Saline water can be used for irrigation directly, mixed or blended with good quality water, used in a cyclic manner (i.e. fresh water followed by saline water) or desalinated prior to irrigation. But to do so effectively, proper management (choice of the adapted crops, irrigation system, leaching, drainage) are needed and care must be exercised to monitor water, soil and crop to ensure long-term deterioration is not taking place.

4.2 USE OF SALINE WATER RESOURCES FOR IRRIGATION IN NENA REGION

In the agriculture sector, the use of saline water resources as an additional source for irrigation is highly recommended (Abu-Zeid and Hamdy, 2008), especially in water stressed regions such as the NENA region, to satisfy increasing water demand for irrigation, subsequently expanding the irrigated areas and reducing existing gaps in food and fiber production.

There is ample evidence to illustrate the widespread availability of saline waters and a wide range of experience exists around the world with respect to using them for irrigation under different conditions (Naeimi and Zehtabian, 2011). This evidence and experience demonstrate that water of much higher salinities than those commonly used in irrigation can be used effectively to produce selected crops under appropriate field management. However, the use and reuse of such non-conventional water resources for crop production is complex as it is inter-linked with different aspects of the environment, health, industry, agriculture and water resources.

Recognizing these complex inter-linkages, efforts are being directed to the development and use of saline water resources, notably artesian, drainage and saline groundwater water for irrigation. This will certainly result in generating greater amounts of water for irrigation.

It is vital to assess the suitability of such water for irrigation and the subsequent effect on the composition of the soil water and crop (Rhoades, 1972).

4.3 STATUS AND PRACTICES OF SALINE WATER USE IN SELECTED COUNTRIES OF THE NENA REGION

The NENA region is faced with a wide range of salinity problems and each country in this region has a unique set of rules and regulations to protect water quality (Abou-Hadid, 2003). Therefore, it is not surprising that there are also examples of the successful use of saline water in this region. The agricultural practices in the NENA region countries are a matter of experience gained by the farmers depending on water availability and prevailing agricultural conditions and economic factors. Each country has its own experience in producing crops that is specific to its local conditions (FAO and AWC, 2018). Also, each country has its own crop varieties which are a result of its research work and farmers' experiences. Many research and published papers and reports present case studies of the use of saline water in agriculture particularly under conditions of water scarcity in NENA region.

The following is a brief review on saline water use and practices in different NENA countries:

4.3.1 Algeria

In Algeria, desertification is putting extreme stress on irrigated agriculture due to the fast rate of soil and water salinization, resulting in the drastic reduction of arable land with productive agricultural potential. This phenomenon is most notable in the western part of the country where major irrigation schemes are located. Out of a total area of 140,000 ha in this part of Algeria, 30% consists of very saline soils (EC_e >8 dS/m). About 90% of Algeria is in the Sahara Desert where rainfall is rare but at the same time, this arid region has large underground reserves of saline water (Daddi Bouhoun et al., 2013). In the Sahara region, saline groundwater is used for irrigation to grow date palm and alfalfa.

Lands irrigated with saline water are exhibiting salinity problems that differ under different bioclimatic conditions. Meanwhile soil degradation advances, due to the combined action of water salinity, and the insufficiency and/or lack of drainage systems. An assessment of irrigated areas reveals the existence and extent of the salinity level of much of the agricultural land in the West and South of the country (Lahouati and Halim, 2012).

In the Ouargla region most of palm groves have poor drainage, a shallow water table (1.6 m) and high salinity (EC_w 34 dS/m). The drains are the open type and their maintenance is not regular. This situation of managing irrigation-drainage promotes waterlogging in soils. Soil salinity (ECe) ranges between 4.6 and 9 dS/m. This accumulation of salts is due to the dynamic ascending and descending of salts respectively under the effect of capillarity from the shallow groundwater and leaching by irrigation.

4.3.2 Egypt

Egypt is a country with about 5000 years of experience in irrigation. Nevertheless, the country's economy suffers from severe salinity problems due to irrigation with low quality water and poor drainage systems. About 33% of the cultivated land is already salinized (Abo Soliman and Halim, 2012).

Salinity problems are widespread. Almost 30–40% of irrigated farmlands are salt affected. It is estimated that 60% of the Northern cultivated land and 20% of the Middle and Southern Delta regions have salt-affected soils. In the Nile Valley, i.e. Upper Egypt, salt-affected soils account for about 25% of the cultivated areas (Figure 4.2). In addition, many areas of the reclaimed desert land adjacent to the

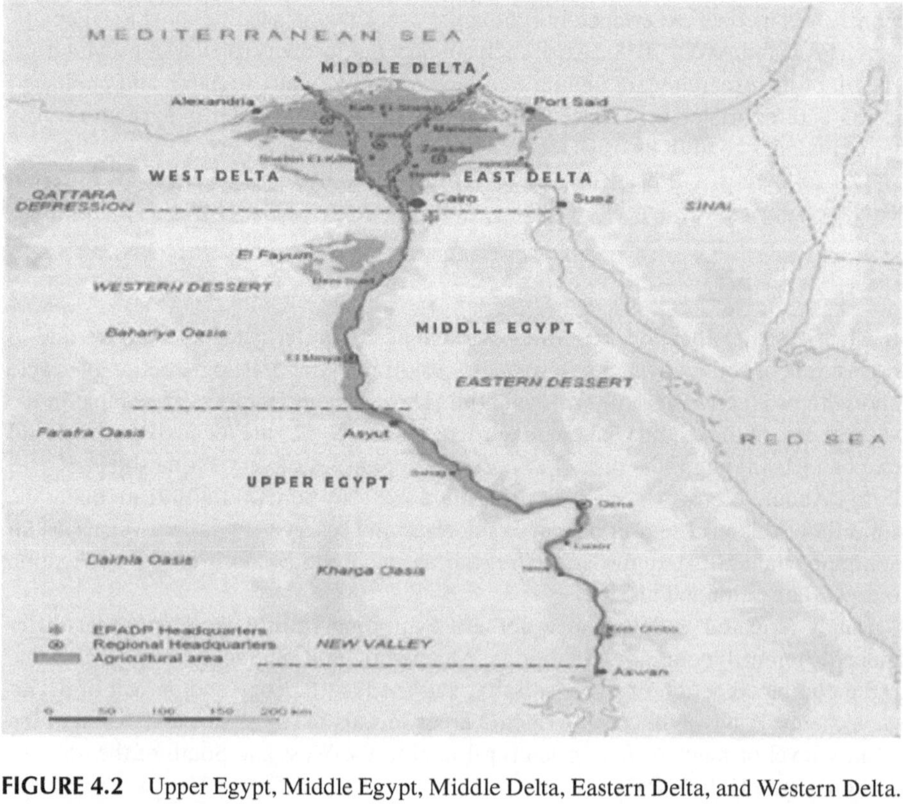

FIGURE 4.2 Upper Egypt, Middle Egypt, Middle Delta, Eastern Delta, and Western Delta.

Nile Valley and Delta, as well as Sinai and Oases, suffer from waterlogging and high salinity (Saad El-Dein and Galal, 2017).

The needed increase in food production to support the acceleration of population growth (2.7%), compels the country to use all sources of water (i.e. drainage water, groundwater and treated sewage water) for the expansion of irrigated agriculture. The drainage water presently used for irrigation amounts to 7 billion m^3per annum and it is likely to increase in the future.

The policy of the Egyptian Government is to use drainage water (up to EC_w 4.5 dS/m) after it is blended with fresh Nile water. Another emerging strategy of alternating different types of water quality has been introduced lately. Research has shown that it is possible to irrigate sensitive crops (maize, pepper, onion, alfalfa, etc.) directly with drainage water in rotation with fresh Nile water, and salt tolerant crops (wheat, cotton, sugar beet, etc.) and moderately sensitive crops (tomato, lettuce, potato, sunflower, etc.) can be irrigated with drainage water but after seedling establishment with fresh Nile water. Based on these results, the Governorate is planning to reclaim 4000 ha using drainage water (Abo Soliman and Halim, 2012).

Crops now grown are mostly forage, cereals and vegetables. In the Delta, saline waters of EC_w 2.5 to 4 dS/m have been used successfully to grow vegetables under greenhouse conditions. In the New Valley (Oases, Siwa, Bahariya, Farafra, Dakhla and Kharga) there is potential to irrigate about 60,000 ha using groundwater (salinity ranging from EC_w 0.5 dS/m to 6.0 dS/m), of which 17,000 ha are already under cultivation. Siwa Oasis has the largest naturally flowing springs in the New Valley. Siwa once contained a thousand springs, of salinity ranging from EC_w 2 to 4 dS/m, which were used successfully to irrigate olive and date-palm orchards, with some scattered forage areas.

The Ministry of Water Resources and Irrigation (MWRI) manages the operation of the pumping stations and the reuse volume is therefore well monitored and recorded. The drainage water salinity ranges between EC_w 1.3 and 4.0 dS/m except in the most Northern part of the Delta near the Mediterranean Sea coast, where drainage water salinity exceeds EC_w 5 dS/m in some locations. Growers in the Beheira, Kafr-El-Sheikh, Damietta and Dakhlia Governorates have used drainage water directly to irrigate barley, berseem clover, cotton, rice, sugar beet and wheat, although yields are not optimal. However, with good management and crop selection practices, growers have successfully used drainage water with EC_w 2–2.5 dS/m without adverse effects (Rhoades et al., 1992).

Historically, the natural flooding from the Nile not only supplied a continual source of nutrients but also provided a natural flushing of salts to the Mediterranean Sea. Ultimately, the difference in the long-term sustainability of irrigated agriculture in both areas was attributed to salinity control via the leaching of salts.

In the Edkawy region in the delta, tomato plants are cultivated in a special way to use the maximum available rain and natural resources (Abou-Hadid, 2003). After filling the waterlogged soil with sand and arranging the irrigation and drainage systems, the drains and irrigation systems are constructed, and the land is ready for cultivation. The nursery starts in August while the temperature is warm and the growth of the seedlings is quick and uniform. The seedlings stay in the nursery for about 45–60 days and are ready to be transplanted at the end of September or early

October. Irrigation of the seedlings is kept under control to allow for initial good growth and later appropriate hardening before transplanting. The fields are prepared for cultivation by digging furrows that are short (10–15 m) and deep enough to avoid the capillary riser

4.3.3 Iraq

The water resources of Iraq depend largely on the surface water of the Tigris and Euphrates rivers and most of the natural renewable water resources of Iraq come from outside the country (Rahi and Halihan, 2018). Both the Tigris and the Euphrates are transnational rivers, originating in Turkey. Between 75 and 85% of the cropped area is generally planted to grains (mostly wheat and barley). Two-thirds of Iraq's cereal production occurs within the irrigated zone that runs along and between the Tigris and Euphrates rivers. The salinity of the Euphrates in Iraq has increased due to: (1) the decrease in quantity and the increase in salinity of the flow that is entering the country due to the Turkish South-eastern Anatolia Project (GAP), (2) the recharge to the river from Al Tharthar Lake and (3) drainage return flows from irrigated fields within Iraq (Rahi and Halihan 2018). The salinity at the lower regions of the river has increased to a point at which the river water is no longer useful for most municipal or agricultural purposes. Half of the irrigated areas in central and southern Iraq were found to be degraded due to waterlogging and salinity (Abdul Halim and Halim 2012). The absence of drainage facilities and, to a lesser extent, the irrigation practices (flooding) were the major causes of these problems. By 1989 a total of 700,000 ha had been reclaimed at a cost of around US$2000/ha. According to more recent estimates 4% of the irrigated areas were severely saline, 50% moderately saline and 20% slightly saline. Irrigation with highly saline waters (more than 1500 ppm) has been practiced for date palm trees since 1977. The use of brackish groundwater is also reported for tomato irrigation in the south of the country (Rahi and Halihan, 2010).

The main option available to mitigate the salinity of the river and to restore the ecosystem is to maintain a minimum instream flow (MIF) (also referred to as environmental flow requirements) (Partow, 2001).

4.3.4 Iran

Salt-affected area in Iran has increased from 15.5 Mha in 1960 to more than 25 Mha in 2008 (Qadir et al., 2008). The volume of marginal water resources is about 12% of the potential renewable surface water resources of the country. The total area with saline groundwater resources is 350,222 km^2 with an annual abstraction volume of 13.7 km^3 (Qadir et al., 2008). The use of saline water for crop production has a long history in Iran. Management practices employed by the farmers in using these waters are similar to those practiced with the use of non-saline waters (Cheragi and Halim, 2012). In general, crop production is based on using high inputs of seeds, fertilizer and water. Agronomic practices such as land preparation, irrigation methods and crop rotation are suboptimal. More information on Iran can be found in Chapter 8 (Hazbavi and Silabi, 2021) in this book.

4.3.5 Jordan

Over 60% of Jordan's agricultural produce is grown in the Jordan Valley. Here, 4.5% of the water resources have a salinity over 2000 ppm and 46% of cultivated soils are moderately to strongly saline (EC_e 4.5–14.1 dS/m) due to the lack of natural flooding to flush the irrigated lands and leach salts and also, due to the low rainfall and high evaporation.

Brackish water for direct use or after desalination appears to offer the highest potential for augmenting the country's water resources. Brackish springs (67) have been identified in various parts of the country with a total average discharge estimated to be approximately 46 million m^3/year. As such, when referring to statistics about brackish water, the quality, quantity and location of this resource needs to be carefully studied in order to assess its potential for use (Ammari et al., 2013).

Modern desalination technologies applied to brackish water (salinities between 2000 and 8000 ppm pumped from wells at depths between 100 and 150 m) offer effective alternatives in a variety of circumstances. In 2015, there were 52 private desalination plants operated by farmers to desalinate brackish water for irrigation purposes and desalinate about 10 million m^3 annually. This irrigation water is used particularly for bananas, a crop of high market value.

The progressive increase in soil salinity in the Jordan Valley is attributed to unsustainable agricultural practices and inputs, a deteriorating quality of irrigation water, the lack of advanced irrigation technologies and efficient drainage systems and improper land management. Also, fertilization and irrigation practices are not based upon sound recommendations that consider the pedoclimatic conditions and crop demands (Al-Rifaee, 2013).

Several trials have been conducted at the research stations of Al-Karamah (Jordan valley) and in Al-Khalidiyah (upper land) to select salt tolerant crop species adapted to the local conditions. These include barley, triticale, wheat, pearl millet, sunflower, sesbania and elephant grass.

The Lower Jordan River defines the international border between Israel and the West Bank on the west and the Hashemite Kingdom of Jordan on the east. Decades of diversion of upstream good-quality water and the direct dumping of saline water and wastewater have severely damaged the river's ecological system. The salinity of the Lower Jordan River has risen significantly to ~5400 mg Cl/L in summer (Al-Rifaee, 2013), endangering its capability to supply water, even to saline-resistant crops such as palms, which are one of the main agricultural products of the Jordan Valley (Figure 4.3).

The Jordan Valley Authority partnered with the German Technical Cooperation (GTC) to evaluate crop production in the middle and southern portions of the Jordan Valley with saline water ranging between EC_w 2 and 7 dS/m over a four year period (GTC, 2003). The goal was to try and develop guidelines for growers in this region that were related to local conditions and practices

4.3.6 Tunisia

Tunisia is among the semi-arid countries faced with serious problems of salinization and water scarcity. Approximately 1.5 Mha, or roughly 10% of the country's area is

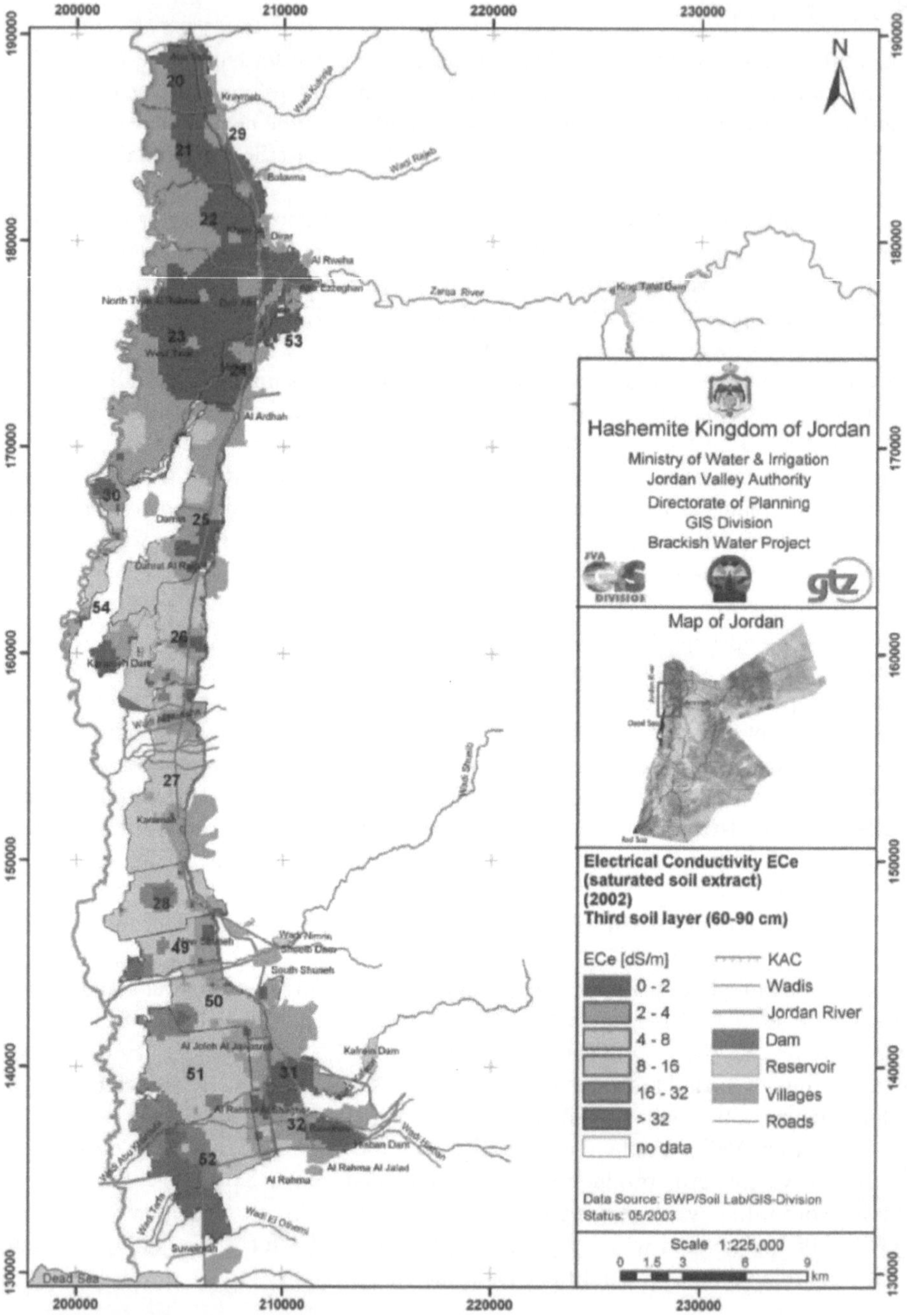

FIGURE 4.3 Soil Salinity map in the Jordan Valley, Jordan (Ammari et al., 2013).

affected by different degrees of salinity and about 25% of water resources have a salt concentration exceeding 3g/L (Hachicha, 2007).

As a result, about 50% of the total irrigated area is considered to be at high risk of salinization (Achour and Halim, 2012). Several initiatives have been implemented since the 1960s by the National Institute of Rural Engineering Water and Forests to evaluate the performance of different management practices including irrigation scheduling, improvements in drainage efficiency, the development of appropriate irrigation systems, physical and chemical techniques (i.e. soil levelling, ploughing, better planting techniques, amino-acid application, etc.), soil amendments and the introduction of salt tolerant crop varieties and new species that could be used to sustainably utilize the irrigated areas that are affected by salinity.

Although the use of saline water for irrigation is a strategy to mitigate water shortage, poor management of saline water for irrigation has resulted in secondary salinization and a series of environmental problems (Kumar et al., 2015; Lei, 2015). These problems will become worse under climate change, in areas of unfavorable soil, with over exploitation of groundwater, with improper cropping patterns and with sea-water intrusion (Heydari, 2019).

The impact of the use of saline water on agricultural production has negatively impacted the environment and the socio-economy of farmers' communities. The main constraint to the use of saline water for agricultural production is primarily the absence of efficient drainage systems in several irrigation command areas. Irrigation with saline water and agricultural development are possible through proper techniques and management of the irrigation water, leaching of salts, adapted farming techniques and choice of salt-tolerant plant varieties.

Based on the importance of the salinity problem in Tunisia, research projects covering some Tunisian regions have been conducted to evaluate the adoption and performance of different management strategies to improve crop production under salt and drought conditions. Research studies related to soil salinity control include: (1) the cultivation of alternative and tolerant-salt varieties such, new cultivars of olive tree, quinoa, jatropha, sesbania and aloe vera, (2) irrigation water management using drip irrigation and sub-surface drip irrigation and (3) improvement of crop tolerance to salinity by application of exogenous proline. The results have shown beneficial effects of different management strategies on the growth and yield of crops, on soil and water properties and the tolerance of the majority of alternative crops to salinity and drought conditions, which has confirmed the possibility of using low quality water for agriculture. However, further research is required in the development of new tools for salinity assessment and the application of biotechnology to improve crop adaptation to limit salt stress effects.

4.3.7 Morocco

The sustainability of irrigated agriculture is threatened by the salinization of land and water resources in Morocco. These problems are the result of seepage from unlined canals, inadequate provision of surface and subsurface drainage, poor water management, inappropriate cultural practices and use of saline water for irrigation. Approximately, 30% of the irrigated area is salt-affected and average yield losses may

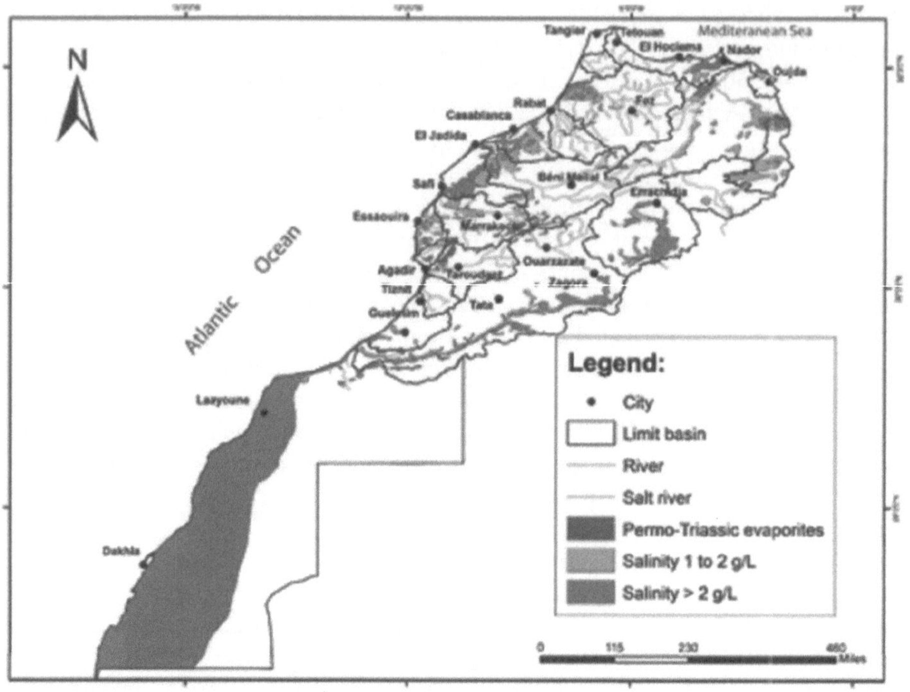

FIGURE 4.4 Spatial distribution of saline aquifers and rivers in Morocco (Hssaissoune et al., 2020).

be as high as 50%. It has been reported that more than 37,000 ha of irrigated land is already affected by salinization (DDGI, 2006). According to FAO (2005),water salinization is the main challenge facing Morocco in terms of water resources quality; out of the 4 billion m^3 of groundwater resources about 1.1 billion m^3 or 25% has a salt concentration of between 1 and 2 g L^{-1}, and about 1 billion m^3 or 27.5% has a salt concentration exceeding 2 g L^{-1} (Hssaissoune et al., 2020) (Figure 4.4).

IDRC Trials (Figure 4.4) in the Southern Oasis of Tafilalt have focused on the growth of several crops (alfalfa, date palms and okra) using different systems of irrigation (furrow and drip) with conjunctive use of fresh water (from a surface water source) and saline water (groundwater with EC_w varying from 6 to 10 dS/m) (Figure 4.5).

FIGURE 4.5 Alfalfa, date palm and okra grown in south of Morocco with saline water.

TABLE 4.3
Crop Water Productivity of Different Species Irrigated with Saline Water (7 dS/m) Using Different Irrigation Systems

	Crop Water Productivity (kg/m³)	
Crops	**Drip Irrigation**	**Surface Irrigation**
Gombe	4.31	1.36
Watermelon	6.67	3.29
Cabbage	8.33	3.81
Melon	8.18	3.5
Eggplant	7.6	3.13
Pepper	8.4	4.0
Cucumber	7.0	3.13
Potato	5.95	2.43
Alfalfa	8.09	3.1

Source: Choukr-Allah and Halim,2012.

The use of drip irrigation allows an average water saving of 3225 m^3/ha for the different crops tested, resulting in a 38% water saving compared to furrow irrigation.

In the Tafilelt valley, the use of saline groundwater with drip irrigation on local cultivars has allowed farmers to produce reasonable yields. Greatest levels of productivity per cubic meter of water and highest economic gains were achieved with okra, followed by cabbage, watermelon, melon, pepper, cucumber, eggplant and alfalfa (Tables 4.3 and 4.4).

In terms of weights of fodder achieved in a test field, alfalfa was the best crop showing a better response using drip irrigation compared with surface irrigation.

4.3.8 Saudi Arabia

About 80–85% of Saudi Arabia's water supply comes from groundwater and, where groundwater extraction exceeds groundwater recharge, is classified as a non-renewable water resource (Al-Omran et al., 2012). As a result, aquifers are being depleted and the quality of groundwater is becoming more saline. For example, the EC_w of the groundwater has increased from 1.9 dS/m in 1983 to 2.8 dS/m in 1987 in the Saqaquifer. A survey of key groundwater aquifers reveals that EC_w values range from 1.6 to 8.2 dS/m with an average of 3.8 dS/m (Falatah et al., 1999 as reported by Al-Omran et al., 2012). The most popular crops grown with saline water in Saudi Arabia are wheat, sorghum, alfalfa and barley. Saline water is also used to irrigate tomato, onions and watermelon (Al-Omran et al., 2012). Cyclic reuse strategies using saline and desalinized water have been trialed, with tomato and lettuce showing that such methods can be successful for commercial production. It was concluded that

TABLE 4.4
Gross Margin and Benefit (MAD/ha) of Several Crops Irrigated with Saline Water (EC = 8 to 12 dS/m)

	Drip Irrigation			Surface Irrigation			
	Gross Margin	Costs	Benefit	Gross Margin	Costs	Benefit	Benefit Rate
Crops			(US Dollars/ha)				%
Okra	7500	879	6621	4500	710	3790	1.75
Watermelon	6000	875	5125	4200	695	3505	1.46
Cabbage	4500	887	3613	3000	715	2285	1.58
Melon	7200	880	6320	5120	690	4430	1.43
Eggplant	3800	865	2935	2800	690	2110	1.39
Pepper	6300	860	5440	4800	690	4110	1.33
Cucumber	5250	870	4380	3750	690	3060	1.43
Potato	2640	1000	1640	2040	830	1210	1.36
Alfalfa (9 cuts)	2050	1150	900	1312	820	492	1.83

Note: Benefit rate: Benefit under drip irrigation/benefit under surface irrigation.
Source: Choukr-Allah and Halim, 2012.

the country has the opportunity to expand the use of treated waste water and saline groundwater for irrigation.

The excessive use of groundwater has created major problems such as the depletion of aquifers and a deterioration of groundwater quality resulting in the conversion the fresh groundwater into saline water. The uncontrolled use of saline water by farmers for irrigating their farms increases soil salinity. Continuous addition of saline water to the soil during the irrigation process, in the absence of good irrigation management practices, undoubtedly leads to salt accumulation in the soil surface layer.

4.3.9 Sultanate of Oman

In 2012, the Ministry of Agriculture and Fisheries of Oman entered into a partnership with the International Center for Biosaline Agriculture (ICBA, 2011) to prepare a strategic plan to combat salinity and protect water resources from pollution and salinity in collaboration with other relevant partners in the Sultanate of Oman. The scope of this strategy involved a comprehensive assessment of the current status of the agricultural system in different governorates in the Sultanate of Oman. A survey of 268 farms conducted by ICBA in 2011 found that crop yields and farm profitability decreased substantially with increased soil salinity (ICBA, 2011). Moreover, many salt-sensitive vegetable crops could no longer be grown. The assessment included the extent of the salinity problem, distribution of water resources, productivity of different agricultural systems, the impact of salinity on farmers' income, policy and legislation. Furthermore, the strategy addressed socioeconomic aspects and capacity-building needs at all levels. The strategy identified alternative scenarios for sustainable water resources and production systems to bring about a more efficient and sustainable use of natural resources.

The strategy showed that the salinity of water used for irrigation in 40–50% of the farms is more than EC_w 5 dS/m (ICBA, 2011). Therefore, with the exception of a few salt-tolerant crops such as date palm and Rhodes grass, many crops cannot be successfully grown and the productivities of most other crops are far below their yield potential. Crop varieties that tolerate soil salinity provide acceptable yields in salt affected soils, when crop management practices to reduce soil salinity are employed.

The ICBA analysis also took into account farm size, the expectation being that for similar sized farms, those with higher salinity groundwater would have lower profitability. The weighted average gross margin for the farms benefitting from the best quality water (salinity less than 1500 ppm) was 2830 USD/acre per year, regardless of the size of the farm. When the salinity was low (between 1501 and 3000 ppm) the gross margin fell to 2087 USD/acre per year, 74% of the gross margin for fresh water. For water of medium salinity (between 3001 and 5000 ppm) the gross margin was 1216.8 USD/acre per year, being 43% of the Class 1 gross margin. Finally, the gross margin for water of high salinity was 1120 USD/acre per year, being 40% of the profits achievable with fresh water.

4.3.10 United Arab Emirates (UAE)

Soil and water salinity are a significant problem in many parts of the UAE due to intensive desalination, including in agriculture, and seawater intrusion into aquifers. This has resulted in some farmers abandoning their salt-degraded lands as traditional crops fail. The problem poses challenges to national efforts to enhance food security and self-sufficiency through local production. It is estimated that in the United Arab Emirates 34% of the area is salinized (EAD, 2009).

In part to address these multiple challenges, the International Center for Biosaline Agriculture (ICBA) has worked with local partners to design studies and projects that look for ways to rehabilitate salt-affected areas and make use of saline soil and water resources for food and forage production.

The Government of UAE is supporting ICBA to undertake research on the use of saline water in the agricultural sector in the UAE and at the international level. ICBA conducts research and development programs that aim to improve agricultural productivity and sustainability in marginal environments. ICBA's multi-pronged approach to strengthening the agricultural sector through expanding food production through improved and better access to technology, improved germplasm and policies, is critical to achieve greater water, environment, income and food security. During its second strategic phase (2007–2011), ICBA moved from its initial focus on salinity management to an increased emphasis on water management issues related to agriculture, including marginal water quality.

A three-year study by a team of scientists at ICBA suggested that the growth of halophytic grasses could be a good option for forage production and the rehabilitation of salt-affected lands in the UAE. What is more, they produce higher yields than some traditional grasses like Rhodes grass (*Chloris gayana*). ICBA studied three abandoned salt-degraded farms in Mezaira'a, Madinat Zayed and Ghayathi in the Emirate of Abu Dhabi. When the grasses were established in the first year of the study, the water salinity level on the farms ranged from EC_w 14.1 to 17.4 dS/m. The team tested four halophytic perennial forage grass species namely *Distichlis spicata*, *Paspalum vaginatum*, *Sporobolus virginicus* and *S. arabicus* for yield and water productivity. Harvested three times a year, the grasses produced average dry biomass yields of 32.6–40.7 t/ha. Average yields in terms of water productivity of 1.7–2.4 kg of dry matter per cubic meter of water were observed, which is better than that previously reported for Rhodes grass under less saline conditions.

On-farm trials in the Western Region, Abu Dhabi, UAE showed that *Paspalum vaginatum* produced the highest biomass yields (53.3 t/ha), followed by *D. spicata*, *S. virginicus* and *S. arabicus*. In terms of water saving – the new grasses can produce the same amount of forage with 44% less water compared to Rhodes grass (Rao et al., 2017).

Sesbania a short-lived perennial legume and moderate salt tolerant species (threshold EC_w 5 dS/m or 3500 ppm), yielded up to 175 t/ha/year (3 cuts) when irrigated with water of EC_w 3 dS/m.

Salicornia is a halophytic species, extremely tolerant to salinity and can be irrigated with sea water. It has several uses, including the consumption of young stems - eaten pickled or as a garnish in fresh salads. Seeds contain 30% oil of high quality hence the species has great potential for the production of edible oil and as for bioenergy feedstock production with seawater irrigation.

4.3.11 Yemen

Irrigated agriculture accounts for about 90% of the water use in Yemen. Salinity varies across the country and surface waters are generally of much higher quality than groundwater sources (Al-Sabri and Halim, 2012). For example, the salinity in many dams varies between EC_w 0.8 and 1.2 dS/m, except those downstream of large cities where the EC_w can range between 2.0 and 2.9 dS/m. Groundwater quality, on the other hand, is more complex in nature. In many of the highland and lowland basins the EC_w can range from 2.0 to 5.0 dS/m, particularly near the Wadis. But the salinity in groundwater near coastal areas can be as high as 8.0–14.0 dS/m due to seawater intrusion from excessive pumping. Vegetables and fruits are the primary crops that are irrigated in the country. However, irrigation with saline water is mainly used on salt-tolerant crops in the coastal plains, but effluents from wastewater treatment plants are also of poor quality and used for irrigation. Generally, the main crops irrigated with saline water are forages, grains (millet and sorghum), cotton, tobacco, sesame, dates and tomatoes.

Saline waters are available as surface and groundwater and are mainly used by the rock cutting industry in the highlands, as well as for irrigating some tolerant crops mainly in coastal plains. However, the extensive withdrawal of groundwater causes salinity to increase in several parts particularly in the coastal areas. The availability of saline water has not been quantified over the entire country. However, the use of saline water for agriculture in Yemen is about 300 million m^3/year, mostly in the Tehama region.

4.4 CONCLUSION

The NENA region has limited freshwater resources and there is a great need to use saline water to meet food and feed demands. In most countries of the NENA region, future projections suggest that it will be necessary to use salt-affected lands in order to meet the food and fiber needs for an expanding population. The severity of salinization differs from one country to the other. However, the use of saline water requires appropriate management to minimize the negative impact of salinity on soil, plant and the environment. About 11.2% of NENA soils are affected by various levels of salinity and sodicity. Human-induced soil salinization in the region is rapidly increasing, both in irrigated and non-irrigated lands. Salinization drastically reduces crop yields, forcing communities to abandon their agricultural lands. Its negative impacts extend to environmental health and local economies.

Different types of saline water reuse exists (agriculture drainage water, groundwater and treated wastewater) and are widely used. However, one should bear in mind the negative impacts of such waters, such as the increase in soil salinity, yield reductions and cost. Saline water could be used for irrigation directly or desalinated or mixed with treated wastewater. It is highly recommended that good agricultural practices (GAPs) be compiled based on research results using saline water in pilot sites in the NENA region, with the aim of developing guidelines for the safe use of this water.

Guidelines are very important to assist stakeholders and farmers in the use of saline water for irrigation while safeguarding the environment, conserving natural resources, increasing crop productivity/quality and enhancing farm income.

Collaborative research is needed to adapt national programs and policies to turn low-quality water into resources, and to develop the capacity of member countries in the use of saline water. Participatotory research approaches are highly desirable so that extension services can provide local farmers with the know-how on best management practices for irrigation. This information can also be relevant and useful for meeting environmental requirements.

Given the scarcity of water resources, a new paradigm is required that will consider saline water resources as an asset to be managed as part of each country's integrated water resources management framework. This asset would not only increase the availability of water for specific purposes that is hygienically safe, ecologically sustainable and beneficial for the society as a whole, but would also contribute to adaptation to climate change and mitigation of its impacts through the reduction in greenhouse gases.

More effort should be directed towards the establishment of new management and agricultural strategies for the use of saline water that sustain crop production and safeguard the environment. Therefore, in assessing the suitability of saline water for irrigation, it is important to consider the following:

- Crop tolerance to salinity must be known.
- Management practices to prevent or minimize salt accumulation in the soil profile should be put in place.
- Advanced irrigation and drainage technology that are suitable for the use of saline water need to be adopted.
- Saline drainage water can also be used for growing fish, shrimp and algae.

REFERENCES

Abdul Halim RK, Halim MK (2012) Status of new development on the use of brackishwater for agricultural production in the Near East: Iraq country report. United Nations Food and Agricultural Organization (FAO), Regional office for the near east (RNE).

Abou-Hadid AF (2003) The use of saline water in agriculture in the Near East and North Africa Region: present and future. *Journal of Crop Production* **7**, 299–323.

Abo Soliman MS, Halim MK (2012) Status of new development on the use of brackish water for agricultural production in the Near East: Egypt country report. United Nations Food and Agricultural Organization (FAO), Regional office for the near east (RNE). doi: 10.2172/1169680.

Abu-Zeid M (2013) Coping with water scarcity in Near East and North Africa: shifting gear. Keynote Speech. FAO Land & Water Days, 15–18 December, 2013, Amman, Jordan. https://www.slideshare.net/FAOoftheUN/plenary1-keynote-speech-16dec2013az

Abu-Zeid M, Hamdy A (2008) Coping with water scarcity in the Arab world. 3rd International Conference on Water Resources and Arid Environments and 1st Arab Water Forum, 16–19 November, 2008. Saudi Arabia.

Achour H, Halim MK (2012) Status and new development on the use of brackish water for agricultural production in the Near East. Tunisia Country Report. United Nations Food and Agricultural Organization (FAO), Regional office for the near east (RNE).

Al-Omran AM, Aly AA, Halim MK (2012) Status and new development on the use of brackish water for agricultural production in the Near East. Saudi Arabia Country Report. United Nations Food and Agricultural Organization (FAO), Regional office for the near east (RNE).

Al-Rifaee MK (2013) Salinity: Jordan Valley Basin. Power point presentation during Salinity management workshop for the CGIAR Research Program on Water, Land, and Ecosystem (WLE) Jordan, 7–8 December. https://www.researchgate.net/publication/277803170_Salinity_The_Jordan_Valley_basin.

Al-Sabri A, Halim MK (2012) Status and new development on the use of brackish water for agricultural production in the Near Eas*t*. Yemen Country Report.

Ammari TG, Tahhan R, Abubaker S, Al-Zu'bi Y, Tahboub A, Ta'any R, Abu-Romman S, Al-Manaseer N, Stietiya MH (2013) Soil salinity changes in the Jordan Valley potentially threaten sustainable irrigated agriculture. *Pedosphere* **23**, 376–384.

CAMRE/UNEP/ACSAD (1996) State of Desertification in the Arab Region and the Ways and Means to deal with it. Council of Arab Ministers Responsible for the Environment (CAMRE), United Nations Environment Programme (UNEP), Arab Center for Studies of Arid Zones and Drylands (ACSAD), Syria, Damascus. 444 pp.

Cheragi SAM, Halim MK (2012) Status and new development on the use of brackish water for agricultural production in the Near East. Iran Country Report. United Nations Food and Agricultural Organization (FAO), Regional office for the near east (RNE).

Choukr-Allah R, Halim MK (2012) Status and new development on the use of brackish water for agricultural production in the Near East. Morocco Country Report. United Nations Food and Agricultural Organization (FAO), Regional office for the near east (RNE).

Daddi Bouhoun M, Saker ML, Hacini M, Boutoutaou D, Ould El Hadj MD (2013) The soil degradation in the Ouargla basin: a step towards the desertification of the palm plantations (north East Sahara Algeria). *International Journal of the Environment and Water* **2**, 93–98.

DDGI (2006). La mise en en oeuvre d'un système d'observation optimisé pour le suivi et le contrôle de l'impact de l'irrigation sur les ressources en sols et eaux dans les périmètres irrigués: Loukkos, Moulouya, Haouz, Souss Massa et Moyen Sebou dites des périmètres de grande irrigation, Direction du Développement et de la Gestion del'Irrigation.

EAD (2009) Soil survey of Abu Dhabi Emirate. Vol. 5. United Arab Emirates, Environment Agency Abu Dhabi.

Falatah AM, Al-Omran AM, Nadeem MS, Mursi MM (1999) Chemical composition of irrigation groundwater used in some agriculture region of Saudi Arabia. *Emirates Journal of Agricultural Science* **11**, 1–23.

FAO, AWC (2018) Guidelines for Brackish Water Use for Agricultural Production in The Nena Region. Regional Initiative on Water Scarcity for the Near East and North Africa.

FAO (2018) Handbook for saline soil management. www.fao.org/publications.

FAO,(2015) Status of the World's Soil Resources Report – Main Report. http://www.fao.org/documents/card/en/c/c6814873-efc3-41db-b7d3-2081a10ede50/.

FAO (2008) Water Reports 34. Irrigation in the Middle East region in figures AQUASTAT Survey.

FAO. 2017. Near East and North Africa Regional Overview of Food Insecurity 2016. Cairo, pp. 35. https://www.climamed.eu/wp-content/uploads/files/Near-East-and-North-Africa-Regional-Overview-of-Food-Insecurity.pdf.

FAO (2005) L'irrigation en Afrique en chiffres – Enquête AQUASTAT 2005. Food and Agriculture Organization of the United Nations: p 18.

German TechincalCoorporation (GTZ) (2003) Guidelines for brackish water irrigation in the Jordan Valley, brackish Water Project, Jordan Valley Authority (JVA), (GTZ), November 2003.

Hachicha Mohamed (2007) Les sols salés et leur mise en valeur en Tunisie. Sécheresse vol. 18, n° 1, janvier, février, mars 2007. file:///C:/Users/pc/Downloads/Secheresse2007_Hachicha.pdf

Hazbavi H, Silabi S (2021). Innovations of the 21st century in the management of Iranian salt-affected lands. In: Future of Sustainable Agriculture in Saline Environments. Edited by Negacz, Vellinga, Barrett-Lennard, Choukr-Allah, Elzenga. CRC Press |Taylor & Francis Group

Heydari N (2019) Water Productivity Improvement Under Salinity Conditions: Case Study of the Saline Areas of Lower Karkheh River Basin, Iran In book: Multifunctionality and Impacts of Organic Agriculture https://www.researchgate.net/publication/335549949_Water_Productivity_Improvement_Under_Salinity_Conditions_Case_Study_of_the_Saline_Areas_of_Lower_Karkheh_River_Basin_Iranhttp://dx.

Hssaisoune M, Bouchaou L, Sifeddine A, Bouimetarhan I, Chehbouni A (2020) Moroccan groundwater resources and evolution with global climate changes. *Geosciences*, **10**, 81.

ICBA (2011) Oman Salinity Strategy.Dubai, UAE. http://biosaline.org/projects/oman-salinity-strategy.

Kumar P, Sarangi A, Singh DK, Parihar SS, Sahoo RN (2015) Simulation of salt dynamics in the root zone and yield of wheat crop under irrigated saline regimes using SWAP model. *Agricultural Water Management* **148**, 72–83.

Lahouati R, Halim MK (2012) Status of new developments on the use of brackish water for agricultural production in the Near East: Algeria country report. United Nations Food and Agricultural Organization (FAO). Regional office for the Near East (RNE).

Lei Li C, Zhao J, Xu Y, Li XS (2015) Effect of saline water irrigation on soil development and plant growth in the Taklimakan Desert Highway shelterbelt. *Soil and Tillage Research* **146**, 99–107.

Naeimi M, Zehtabian G (2011) The review of saline water in desert management *International Journal of Environmental Science and Development* **2**, **6**, 474–478.

Qadir M, Qureshi AS, Cheraghi SAM (2008) Extent and characterisation of salt affected soils in Iran and strategies for their amelioration and management. *Land Degradation and Development* **19**, 214–227.

Partow H (2001) The Mesopotamian Marshlands: Demise of an Ecosystem Early Warning and Assessment Technical Report. UNEP/DEWA/TR.01-3 Rev. 1, Division of Early Warning and Assessment, United Nations Environment Program, Nairobi, Kenya.

Rahi KA, Halihan T (2018) Salinity evolution of the Tigris River. *Regional Environmental Change* **18**, 2117–2127. https://link.springer.com/article/10.1007%2Fs10113-018-1344-4

Rao NK, McCann I, Shahid SA, Ur Rahman KB, Al Araj B, Ismail S (2017) Sustainable use of salt-degraded and abandoned farms for forage production using halophytic grasses. *Crop and Pasture Science* **68**, 483–492.

Rhoades JD (1972) Quality of water for irrigation. *Soil Science* **113**, 277–284.

Rhoades JD, Kandiah A, Mashali AM (1992) The use of saline waters for crop production. FAO Irrigation and Drainage Paper 48. Food and Agricultural Organization of the United Nations, Rome 133 pp.

Saad El-Dein AA, Galal ME (2017) Prediction of reclamation processes in some saline soils of Egypt. *Egypt Journal of Soil Science* **57**, 293–301.

5 Salinization Threats to Agriculture across the North Sea Region

Iain Gould, Jeroen De Waegemaeker, Domna Tzemi, Isobel Wright, Simon Pearson, Eric Ruto, Leena Karrasch, Laurids Siig Christensen, Henrik Aronsson, Susanne Eich-Greatorex, Gary Bosworth, and Pier Vellinga

CONTENTS

5.1 INTRODUCTION

Agricultural production faces unprecedented challenges in the 21st century (Foley et al. 2011). By the end of the century, the global population may reach upwards of 11 billion (UN 2017), with evolving dietary requirements adding further pressures on land resources (Bodirsky et al. 2015). All of this will evolve against a backdrop of a changing climate that could have severe implications for yields and production

DOI: 10.1201/9781003112327-5

(Agovino et al. 2019). The sustainable management of agricultural land, and soils, will be key to addressing this challenge. Soils produce around 95% of our food, and if managed well, even have some capacity to mitigate the harmful effects of flooding and drought, whilst also sequestering carbon (FAO 2015). Nevertheless, global soil health is under threat, with around a third of the world's soils already suffering from degradation (FAO 2015); in many parts of the world, a major driver of this degradation is salinization (Rengasamy 2006; Qadir et al. 2014; FAO 2015).

High levels of salts in soil have direct effects on crop yields by impacting osmotic potential thus reducing plant water uptake (Abrol et al. 1988), but also have severe consequences for longer-term soil function and agricultural production (Pitman and Läuchli 2002). Sodium ions (Na^+) bind to the exchange sites on clay particles, increasing the chance of clay dispersal (Abrol et al. 1988). Once dispersed, soils are susceptible to structural degradation, resulting in surface slaking and reduced infiltration rates (Paes et al. 2014). Soil dispersal can also expose previously-occluded soil organic matter to decomposition, altering the microbial structure and carbon cycling of a soil (Rath and Rousk 2015). Salts can accumulate at the soil surface and root zone following evaporation of soil water, whereas an increased flushing with fresh water can aid salt removal.

It is estimated that salt affected soils cover 932.2 Mha globally, with Europe contributing about 30.7 Mha or 3.3% of total global saline and sodic soils (Rengasamy 2006). Well documented regions of concern range from central Asia, North and South America, Australia, the Middle East and parts of Africa and Southern Europe. These constitute mainly arid and semi-arid regions where salinization is intensified by high temperatures and rapid evaporation, surface water resources are scarce, and irrigation utilizing water sources of high ionic strength is widely practiced (Endo et al. 2011; Cui et al. 2019).

Salinity is not solely confined to arid and semi-arid regions and can still manifest as a threat to soils in areas of higher rainfall with greater flushing rates, most notably in coastal zones (Tóth et al. 2008; Jones et al. 2012; Daliakopoulos et al. 2016). Under future climate predictions, one area at particular risk of salinization is the North Sea region (Figure 5.1), with its combination of low-lying land, productive farming in coastal regions and reliance on a regulated water supply to maintain crop requirements. Throughout the North Sea region, localized pockets of land have been subjected to salinization processes, such as from seawater inundation (Gerritsen 2005; Wadey et al. 2015), however, the occurrence of salinization in the region is sporadic. The risk to agricultural soils inherently differs as a result of varying climatic and geological conditions and management factors such as flood defence, land drainage and extent of irrigation (Daliakopoulos et al. 2016). Existing frameworks summarize the salinization process on a global and wider European scale (Tóth et al. 2008; Daliakopoulos et al. 2016). These are comprehensive but not specific to the northern coastal regions surrounding the North Sea, a maritime region where the multiple salinization threats are a function of marine and coastal dynamics. Furthermore, in coastal regions, multiple processes often occur within the same locality. As a result, novel frameworks need to be employed to summarize the salinization risks solely in the North Sea region.

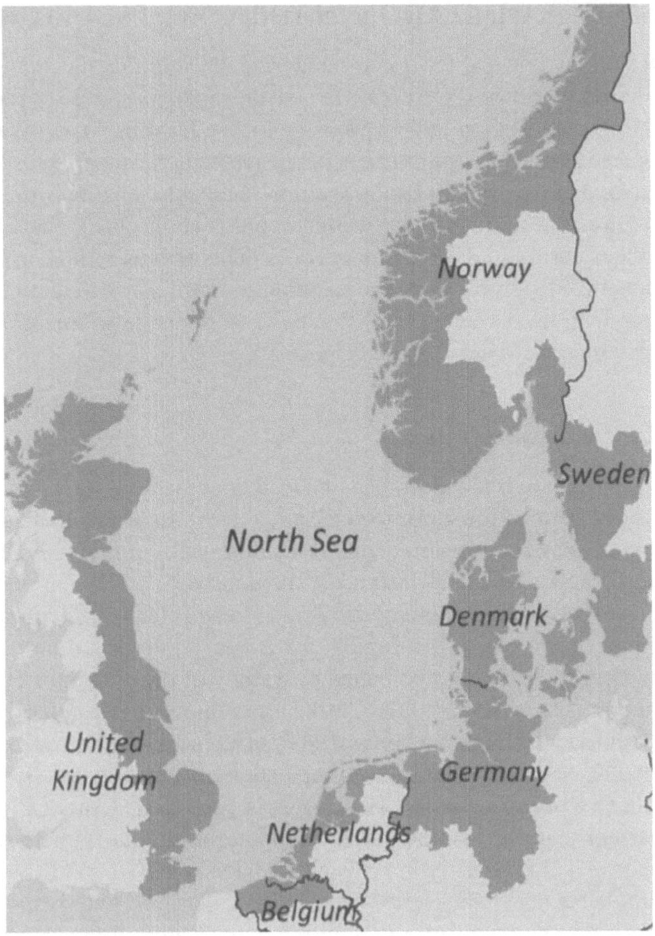

FIGURE 5.1 The North Sea region.

Given future climatic predictions, we anticipate greater threat of salinity to agricultural systems around the North Sea. Salinization will impact on agricultural systems in the region through a range of mechanisms from flooding, saline intrusion of groundwater, irrigation with brackish water or airborne salinity (De Waegemaeker 2019). Agricultural land managers will need to explore ways to adapt, but at present, very little research exists that investigates the issue of salinity in North Sea coastal region agriculture.

In this chapter, we provide an overview of salinity occurrence in the North Sea countries with the aim of identifying knowledge gaps for developing a strategy for agricultural salinity adaptation in the region. On account of limited existence of quantitative, comparable, salinity data across the region, this review presents an exploration of salinization in the region, rather than a comprehensive review of every country.

5.2 CAUSES OF SALINIZATION IN THE NORTH SEA REGION

We first propose a suitable framework to categorize the common mechanisms of land salinization shared by all of the North Sea coastal regions. In a 2019 report for the Interreg VB North Sea Region SalFar project ("Saline Farming"), De Waegemaeker (2019) developed a framework of four key pathways, which would cause salinization of agricultural land across the North Sea region. This framework simplifies previous work by Manca et al. (2015) and Daliakopoulos et al. (2016) that elaborate on salinization issues across multiple climate zones. This review builds on the original De Waegemaeker (2019) framework and in doing so, we discuss the four salinization pathways: (1) irrigation, (2) aerosol, (3) flood and (4) seepage salinization; and how they may evolve in a changing climate (Figure 5.2.).

5.2.1 Irrigation Salinization

Salinization by irrigation is a widespread and well documented problem across many arid and semi-arid parts of the globe (Rengasamy 2006). In such climates, the strong evaporative forces leave salt residue from irrigation water at the soil surface, leading to soil salinization. In many instances, the source of irrigation water is from groundwater reserves, which have a higher ionic strength than surface waters. In the maritime climate of the North Sea region, we expect more of a dominance of salt flushing from natural rainfall as confirmed by a study of the dynamics of soil salinization in Denmark (Christensen 2021). Unlike arid regions, irrigation salinization in the North Sea region will likely only manifest in instances where an originally freshwater resource has become salinized to some extent, or where growers utilize more brackish sources in instances where freshwater is becoming scarce. Climate models predict warmer and drier summers, and milder and rainier winters in Northern Europe (Palmer and Räisänen 2002; Rowell and Jones 2006). With growing variability in rainfall events (Pendergrass et al. 2017), farmers may resort to further reliance on irrigation systems. As the requirements of agriculture, households and industry compete for an ever more limited water resource, the utilization of brackish water reserves, if deemed viable, may be an option for growers.

5.2.2 Aerosol Salinization

Coastal farmland is also at risk of airborne salinization (Rozema et al. 1983; McCune 1991). Key factors affecting the process are meteorological – namely the speed and direction of the wind (Franzén 1990). Wind speed needs to exceed 4 metres per second in order for the wind to take up droplets of seawater (O'Dowd and de Leeuw 2007). As such, aerosol salinization in the North Sea region may be more common in areas exposed to higher winds, for example prevailing winds, and less so in sheltered areas (Franzén 1990). The North Sea region is generally subject to westerly winds, and thus we may anticipate more aerosol salinization at the eastern end of the Sea. The majority of airborne salts are deposited within 1 km of the coast (Gustafsson and Franzén 1996). Topographic obstacles to airborne transmission, such as a dune belt, could offer protection of the hinterlands from aerosol salinization. Nevertheless,

FIGURE 5.2 Four salinization processes of the North Sea region. (De Waegemaeker 2019).

aerosol deposition of salts has been known to travel much further inland (Balance and Duncan 1985). In a changing climate, the degree of aerosol salinity will be determined by how wind patterns may manifest.

5.2.3 Flood Salinization

Projections from the IPCC point to a worst-case scenario of sea level rise of between 61 and 110 cm by 2100 (IPCC 2019). Such a rise, alongside a predicted increase in storm surge frequency, will put coastal areas at risk of widespread coastal flooding and the inundation of agricultural soils with saltwater (Nicholls and Cazenave 2010; Brecht et al. 2012; Salehin et al. 2018). The North Sea region will be no exception, and in fact, may be particularly exposed to this rise in sea level (Vousdoukas et al. 2017); the incidence of coastal flooding is therefore forecast to rise in the region (European Environment Agency 2019). Furthermore, changing weather patterns could lead to more unpredictability in storm surge events (Woth et al. 2006), of the scale that has severely damaged North Sea agriculture in the past (Steers et al. 1979).

The severity of flood salinization will not only depend on environmental factors, such as the salinity of the water source, the duration and the extent of the flood, alongside the land's capacity to recover, but also on post-flood management factors such as the availability of machinery, cultivation and crop choice. Around the coasts of the Netherlands, Belgium and the UK, the North Sea has a typical salt content of 3.5% (Raats 2015), whilst closer to the Baltic, and in estuarine areas, the salinity levels are reduced by dilution with freshwaters. In terms of flood duration, some coastal floods can last only a few hours, whilst others caused by sea defence failures in very low-lying areas can last weeks or months, even resulting in land abandonment (Fagherazzi et al. 2019). Even a short-duration flood event can have devastating impacts on the land (Durant et al. 2018). Finally, the soil type and land management will also determine how persistent salts may remain in the soil, and also the speed of recovery e.g. salt flushing. Clay soils not only have slower infiltration, and thus reduced salt flushing rates, but also can exhibit more structural damage from sodification (Abrol et al. 1988). Taking into account all three of these factors, some areas of the North Sea region may recover quickly, with salinity only being a problem in the short duration following flooding, whilst in other situations, salinity could impact on crop yields for a long time, or even lead to permanent changes in land management (Gould et al. 2020).

5.2.4 Seepage Salinization

Salinization by seepage involves the subsurface movement of saline or brackish water to bring it into contact with surface waters, shallow groundwater reserves or the soil root zone via saline intrusion. Saline intrusion can impact on agricultural lands by permeating into freshwater sources, such as surface or groundwater resources utilized for crop production. Additionally, in certain instances, the saline interface could rise up to within crop root zone itself (Stofberg et al. 2017). The rate of intrusion depends on two key factors: (i) the upward encroachment of saline water and (ii) the downward pressure of freshwater (de Louw et al. 2011). If more upward pressure is exerted from beneath, as a result of rising sea levels, storm surge events or sea level rise (Masterson and Garabedian 2007; Werner et al. 2013), the saline interface is forced upwards. Consequently, if less freshwater pressure is being placed from above, for example from increased freshwater abstraction or reductions in rainfall recharge (European Environment Agency 2019), the saline interface will also rise towards the surface. Michael et al. (2017) recently termed the pressure on coastal aquifers "coastal groundwater squeeze", which threatens coastal groundwater resources by overuse and contamination. In a changing climate, these drivers – from above and from below – will bring the issue of seepage salinization to increasing significance in coastal agricultural areas of the North Sea region.

5.3 COUNTRY CASE STUDIES

5.3.1 United Kingdom

Many coastal areas of Eastern England have a long history of reclamation, drainage and conversion to productive arable land (Hazelden and Boorman 2001). As a result, much of the UK's most productive agricultural land is found in these low-lying areas of fertile soils, areas also occupying flood risk zones, with water levels managed by Internal Drainage Boards. North Sea storm surges flooded east coast farmland in events in 1953, 1978 and 2013 (Steers et al. 1979; Baxter 2005; Spencer et al. 2015). Farms most at risk to these saline inundations can be areas of particularly high value: growing potentially less salt tolerant crops, such as salads and potatoes, exposing vulnerabilities to regional agricultural economies if coastal flood incidence rises (Gould et al. 2020). In many areas along the North Sea coast, there is continual debate over the cost/benefit of maintaining flood defences and of nature-based adaptation (Liski et al. 2019), keeping the risk of flood inundation ever present in the attention of UK coastal farmers.

In addition to flooding, there is concern regarding seawater intrusion into some of the UK's coastal aquifers, such as the chalk aquifers, a valuable freshwater resource (MacAllister et al. 2018). These deep aquifers are not considered a direct pathway to agricultural soil salinity (Cooper et al. 2010), but agricultural abstraction licenses are in place for other groundwater reserves and surface waters in coastal regions to supply irrigation systems (Weatherhead et al. 2014), areas in which seepage salinization could compromise public and private abstraction.

The extent of salt affected soils in the UK is unknown, although considered not to be insignificant (Loveland et al. 1986). Given the aforementioned mechanisms, the potential for salinization is geographically extensive, and not localized to any one agricultural region along the North Sea coastline. As recently as 2013, coastal flooding inundated farmland in areas around the Humber, south Lincolnshire, Norfolk, Suffolk, Essex and Kent (NFU 2013; Wadey et al. 2015) (Figure 5.3a). Eastern coasts of Scotland have been subject to periodic flood events, and coastal defences are in place to protect property and farmland (Hickey 1997). The Norfolk Broads (which have RAMSAR and SAC designations) have been subjected to increased salinity, both from changes to local water management and also as an impact of storm surges (Roberts et al. 2019). Former salt marsh lands of Essex, Kent and other parts of Southern and Eastern England, areas which have undergone past conversion to farmland, can exhibit a degree of salinity, and faced with greater coastal flooding, their longevity as productive farmland may be in question (Hodgkinson and Thorburn 1995; Hazelden and Boorman 2001); the financial impacts of a flood on prime agricultural land could cost up to £5,000 per hectare (Gould et al. 2020). Furthermore, farmland in South Lincolnshire and East Anglia receives less rainfall compared to the rest of the UK (Mayes and Wheeler 2013) and thus has a demand for irrigation (Rey et al. 2016) with less salt flushing potential from natural rainfall. By 2050,

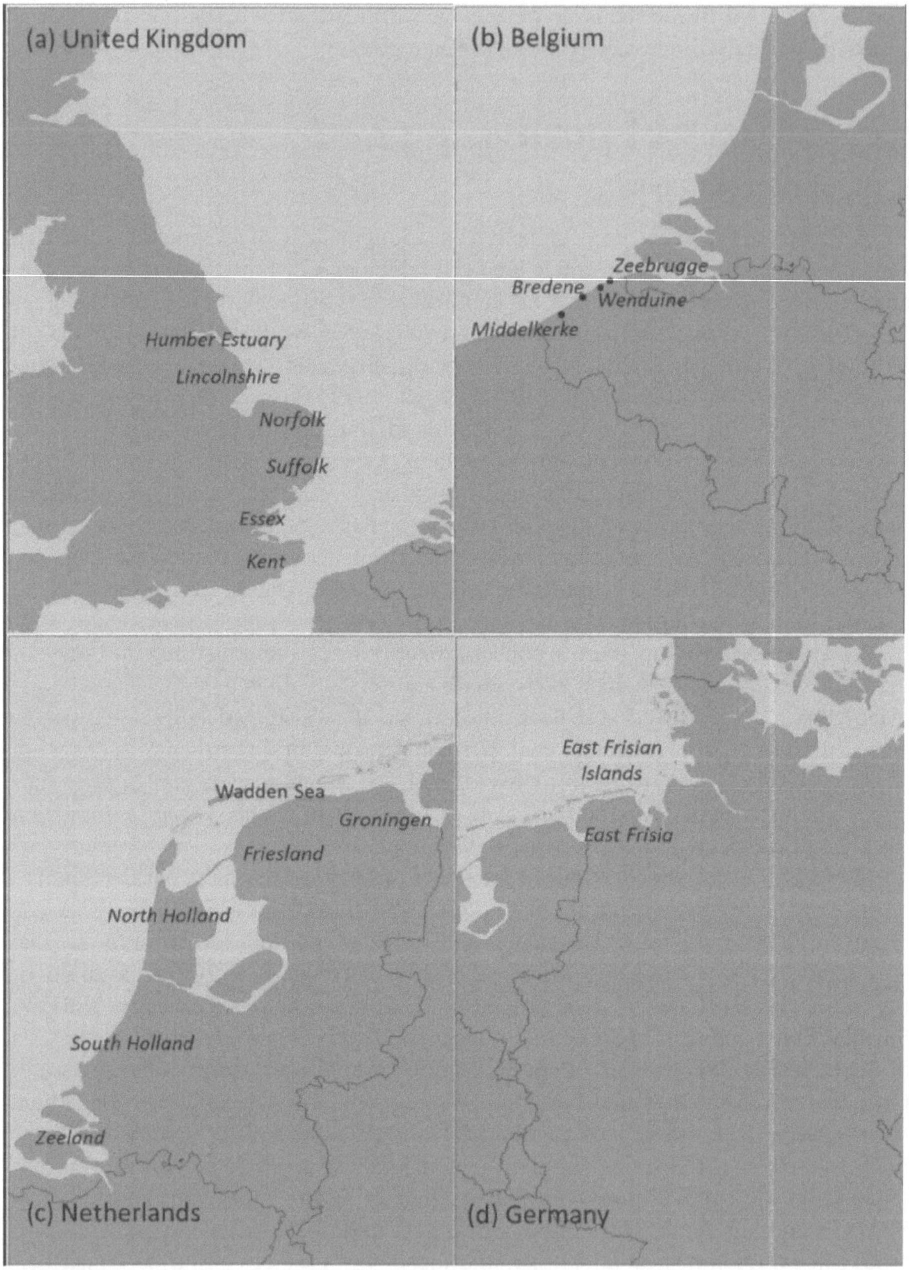

FIGURE 5.3 Locations discussed in the United Kingdom (a); Belgium (b); the Netherlands (c); Germany (d).

the water requirements of agriculture in these regions will face increased pressure (Environment Agency 2020). Some growers in these areas, faced with limitations in irrigation supply and confronted with brackish surface and groundwater, are starting to explore trials with brackish water irrigation.

5.3.2 Belgium

Most of the Belgian coastal plain was reclaimed in the 11th and 12th centuries (Tavernier et al. 1970), and surface water today is maintained by a system of drainage and pumping of surface waters to manage winter and summer levels. The phreatic aquifer at the Belgian coast constitutes a freshwater lens situated on top of saline groundwater (Vandenbohede et al. 2015). The cartography of the freshwater lens showcases the local geology; the lens is thickest at the sandy ridges and thinnest in the lowest-lying areas (Vandenbohede et al. 2010; VMM 2014; Delsmans et al. 2019). In some exceptional cases, mainly grassland, the saline groundwater interface is located at less than 2 metres depth. A comparison of the current situation to research from the 1970s indicates that the freshwater lens at the Belgian coast is relatively stable at present (Vandenbohede et al. 2010). Nonetheless, water managers need to be vigilant as climate change exacerbates seepage salinization.

The closer to the Belgian coast, the more sea level rise puts pressure from below on this freshwater lens (Lebbe et al. 2008). At two sections of the coast, namely between Middelkerke and Bredene, and between Wenduine and Zeebrugge (Figure 5.3b), the dune belt is narrow (50–150 metres wide). Here, the small dune area offers flood protection but it is not sufficient to prevent the seepage of seawater to the adjacent polder area, as the fresh groundwater reserve under this narrow dune belt is shallow (Oude Essink 2001; Lebbe et al. 2008). In addition to this upward encroachment near the North Sea, there is pressure from below on the freshwater lens along the Boudewijn canal, which connects the city of Bruges to the port (VMM 2014). In this case, saline water seeps from the canal to the adjacent polders.

The water availability in Belgium is likely to change with future climate predictions (Tabari et al. 2015) leading to periodic reductions in downward pressure on the freshwater lens. Recently the province of West-Flanders was confronted with prolonged droughts, for example in the summer of 2017, 2018 and 2019 (CIW 2018, 2019, 2020). These extreme weather events foreshadow the impact of climate change on local water management and agriculture. In the aforementioned droughts, all available water was directed to the rivers and canals in order to ensure navigability and, as a result, there was no water left to increase the level of surface waters in the polders. In addition, farmers were prohibited from pumping groundwater for irrigation in order to stall seepage salinization.

5.3.3 The Netherlands

Agriculture in the Netherlands has a long and rich history of land reclamation, drainage and water management (Hoeksema 2007), resulting in the characteristic polder landscape we know today. With around a quarter of its land surface below sea level (Huisman et al. 1998), the Netherlands has seen its share of historic coastal floods leading to soil salinization (Raats 2015). However, after every flood dikes were raised above the levels of the last flood. In the aftermath of devastating floods in the early to mid 20th century, significant investment was allocated to the construction of a series of dams and flood barriers in the following decades (Raats 2015). As such, much of the coastline is protected by defences based on withstanding a 1 in 4,000 to 1 in 10,000-year event, constituting a much more robust coastal flood defence system than other North Sea countries. In fact, it is estimated that without this extensive defence network, 65% of the country's land surface would suffer regular flooding (Huisman et al. 1998). As a consequence of this investment, the likelihood of saline inundation of Dutch farmland is much less than in other countries, but it can never be ruled out (Bouwer and Vellinga 2007; Vousdoukas et al. 2016).

The more pressing issue concerning exposure of agricultural systems to salinity in the Netherlands is through the impacts of saline groundwater seepage. The major part of the low-lying coastal area has a history of being flooded with every tide. As a geological relict, the groundwater at around 4 metres or below exhibits a degree of salinity, either brackish or saline. Compared to other North Sea countries, the Netherlands benefits from more extensive research activity exploring salinity and groundwater dynamics (de Louw et al. 2010; Oude Essink et al. 2010; de Louw 2013). Beauchampet (2019) provides a comprehensive review of agriculture and salinization from groundwater in the Netherlands, which we briefly summarize in this case study. In the coastal areas of the Netherlands, model simulations predict up to twofold rise in salt levels as a result of groundwater seepage from sea level rise pressures by 2100 (Oude Essink et al. 2010). The predicted increase in saline seepage will not only come from rising saline interfaces, but also by lowering ground levels due to subsidence (Oude Essink et al. 2010). As a result, the most vulnerable areas to salinization are located in the reclaimed lands of the coastal areas. Here, groundwater lies very close to the surface (Velstra et al. 2009), and productive agricultural systems may be increasingly reliant on freshwater from surface waters for production (Nillesen and van Ierland 2006). In future, such lands would require sufficient flushing with freshwater in order to keep the saline interface at bay. Faced with ever increasing sea level-induced pressures from below, and the potential for less rainfall or more freshwater abstraction from above, keeping the saline interface at sufficient depth will prove to be a challenge for Dutch agriculture in the 21st century (Velstra et al. 2009).

5.3.4 Germany

Following global sea level rise in the Holocene, salinization of the German North Sea coast stretched up to 20 km inland (Martens and Wichmann 2011). Centuries of storm surges, land reclamation and extensive dike building created the current German North Sea coastline we see today (Vollmer et al. 2001). It stretches about 1,300 km (including islands), with a closed dike line offering protection for the land behind. Without these dikes, in some places up to 9 metres in height, and coastal protection system, the low-lying hinterland would be flooded with the tides. Large areas of the region lie below sea level and the coastline is vulnerable in a changing climate (Sterr 2008).

The most vulnerable regions to groundwater salinization are the barrier islands, such as the East Frisian Islands (Figure 5.3d) (Röper et al. 2012; Seibert et al. 2018), and the low-lying marsh lands such as East Frisia. In these regions, German agriculture is threatened by flood inundation during storm surges, which can lead to rapid salinization of groundwater lenses (islands) or groundwater aquifers (mainland); whilst increasing seawater pressure and decreasing freshwater resources lead to comparatively slower seepage salinization (Werner et al. 2013). Recovery of infiltrated freshwater bodies in such areas takes years to decades (Anderson 2002; Holding and Allen 2015; Post and Houben 2017).

Groundwater in parts of North Germany's coastal areas is also subject to the "coastal groundwater squeeze" (Daliakopoulos et al. 2016; Michael et al. 2017) and as such, salinization of groundwater resources in Northern Germany becomes a growing issue for public water supply. Although currently much of the groundwater monitoring focuses on other anthropogenic inputs such as nitrates and pesticides, in the long-term groundwater salinization is expected to become a critical threat to the utilization of groundwater sources (Grube, 2000). The situation is exacerbated by increasing population, tourism and general water demand (Michael et al. 2017) and consequences of sea level rise, flooding and droughts (Jurasinski et al. 2018).

5.3.5 Denmark

The Danish coastline stretches 7,400 km from the North Sea in the west to the Baltic in the east, comprising the Jutland peninsula alongside hundreds of islands, exposed to a gradient of salinity from west to east (Hansen et al. 2011). With a few exceptions of land reclamation by dike construction in fjords, such as Lammefjorden in 1873, the landscape of Denmark is generally above mean sea level. The southern part of the North Sea coast of Jutland constitutes the northern extent of the Wadden Sea and marshes protected from flooding by dikes established in the past 100 years. North of Esbjerg the coast is protected by an extensive belt of sand dunes. This belt of sand dunes has been subject to significant erosion during the past 50 years and in major sections is only maintained by protective initiatives such as beach enrichment and repetitive reshaping of the remaining sand dune barrier. Flooding in Denmark falls into two categories: flooding in the marsh of the North Sea coast is a rare event with only limited effects during the past 100 years, while flooding of coasts in the inner sea (Baltic Sea and Kattegat) including flooding of arable land is reported with increasing frequency.

Groundwater salinity in Denmark is currently monitored in sources of drinking water. An increase in salinity has been seen in coastal regions, most significantly in coasts of the inner sea including Zealand and small islands such as Læsø, Endelave, Sejerø (Figure 5.4a) (summarized by Kristiansen et al. 2011). These areas also correspond with reported areas of saline intrusion noted by Fenger et al. (2008). There is also evidence in Danish coastal regions of legacy impacts of sea spray on forest soils (Pedersen and Bille-Hansen 1995), suggesting airborne salinity as another potential pathway of salinization in coastal Denmark. Root zone soil salinity has only recently been subject to study in Denmark (Christensen 2021), which revealed that there is no accumulation of marine salt in soil in Denmark, most likely due to flushing from

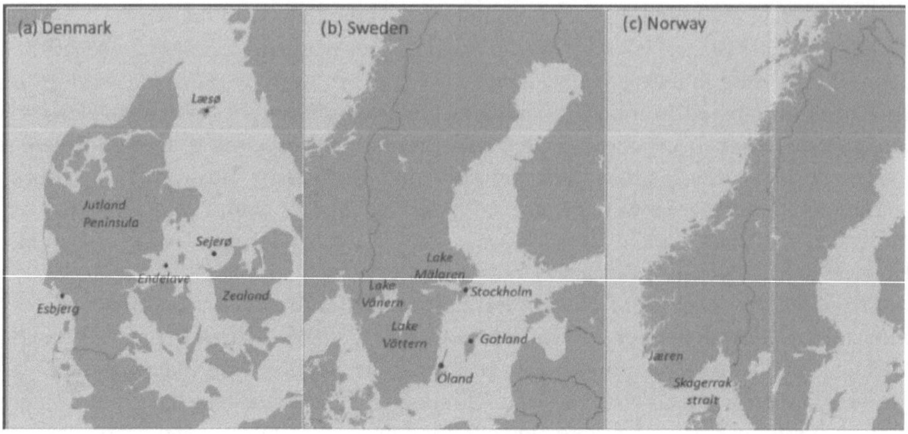

FIGURE 5.4 Locations discussed in Denmark (a); Sweden (b); Norway (c).

sufficient rainfall. However, temporary soil salinity can be observed as a result of flood or seepage from the coast as well as aerosol from the sea.

The Danish case also offers an historic insight into how local communities can adapt to, and even exploit, salinization for economic benefit. Saline groundwater on the island of Læsø (Jørgensen 2002) allowed large-scale production of salt on the island between the 12th and 16th centuries, bringing wealth to the island, salt being an essential commodity for food preservation in Scandinavia and Northern Europe.

5.3.6 Sweden

Groundwater salinization in Sweden has been explored in several studies, most of which date back to the late 1980s and early 1990s. The most comprehensive works (Knutsson and Fagerland 1977; Lindewald 1981; Olofsson 1996) investigating the occurrence of groundwater salinization in Sweden utilize data sourced from the Geological Survey of Sweden (SGO). According to Olofsson (1996), groundwater salinization in Sweden is attributed to several factors including the intrusion of seawater in coastal areas, fossil seawater, chemical interaction between the groundwater and the aquifer, as well as anthropogenic sources such as runoff from waste deposits and the use of de-icing salts on roads.

Lindewald (1981) presented evidence of groundwater salinity in Sweden using data from the SGO. The results showed that 780 wells were identified as salinized, with Chloride concentrations of 300–7000 mg per litre. In a later study by Olofsson (1996), as many as 13,000 wells were said to have a "salty taste", with Chloride levels of 300 mg/L or higher. The majority of wells exhibiting salinity could be found along the whole Swedish coastline (including Öland and Gotland), and into central Sweden within a 200 km wide zone from the Swedish west coast via Lake Vänern, and from Lake Vättern to Lake Mälaren near Stockholm (Figure 5.4b) (Olofsson 1996). Saline groundwater occurred in different aquifers in Sweden and could be found in sedimentary as well as in crystalline bedrock as well as in sand and gravel deposits (Lindewald 1981). Lindewald (1981) linked the cause of salinization on the west coast with the transgression from the Atlantic Ocean.

Approximately 1,800 km of the Swedish coast is at risk of climate-driven erosion, land degradation, loss of natural habitats, and infrastructure depreciation (SOU 2007). The coast of Southern Sweden is particularly exposed to erosion and coastal flooding; which are expected to increase in the coming 100 years (SOU 2007). Like much of the North Sea region, sea level rise in combination with increasing demand on groundwater may intensify the salinization of groundwater along with the coastal areas of Sweden bordering the North Sea.

5.3.7 Norway

Unlike many other North Sea countries, groundwater only plays a relatively small, although increasing, role in Norway's drinking water supply on account of the large reserves of surface water available (NGU 2014). Groundwater is thus not monitored in as much detail as in some other countries, resulting in limited data for groundwater salinity measures. As a result also, seawater intrusion of groundwater has seen little attention in monitoring and research. However, despite a generally low awareness concerning potential agricultural problems related to seawater intrusion, the Geological Survey of Norway lists reducing groundwater removal to avoid saltwater intrusion in coastal areas among possible measures to maintain good groundwater quality.

For Norway, climate-driven sea level rise will dominate over land subsidence in the coming century, although predicted sea level changes are expected to be below the global mean (Simpson et al. 2015). Coastal areas in South-Western and Western Norway will most likely experience larger relative sea level increases than in more northerly areas (with the exceptions of some islands) or in the Skagerrak strait (Simpson et al. 2015). In terms of potential salinization of agricultural land, though the risk may be low in general, some areas along the North Sea coast may be prone to saline inundation, as well as saline intrusion into groundwater, as reflected by high electrical conductivities and salt concentrations measured in a coastal location within the national groundwater surveillance grid (Orresanden; Seither et al. 2016). One of these areas is Jæren in South West Norway (Figure 5.4c), the country's largest lowland plain with significant agricultural production. While agricultural land in this region mainly remains several metres above sea level, extreme storm surges may still flood farmland (Eich-Greatorex et al. 2020).

Future salinization threats in Norway may thus manifest from inundation due to storm events, and also aerosol and seepage salinization may be relevant in some areas. Additionally, interest in irrigation with brackish water is rising. However, Norway will likely be one of the least impacted nations in the North Sea region.

5.4 DISCUSSION

It is clear from the exploration of our country case studies that even in a region of relatively similar climatic conditions, the issue of salinization is diverse and complex. Nevertheless, the country studies do highlight some common themes – namely that sea level rise, and increased demand for freshwater, will induce pressures on groundwater resources across the coastal region. As such, seepage salinization will manifest and could either lead to increased salinity in the root zone, for example in the Netherlands, Denmark or Belgium where evidence shows saline groundwater is relatively shallow in the soil profile; or by upward movement of saline water mixing with surface waters or irrigation sources. Despite all countries facing some degree of seepage salinization, it is the relatively slow nature of seepage salinization, in comparison to flooding, which provides one of the major policy challenges facing the region. Subsequent impacts take time, and may not become evident for many years. Therefore, the problem may be classified as a "creeping catastrophe" (Schneider et al. 2013) posing challenges to a better understanding of its dimensions and societal responses.

Unlike seepage, current flood salinization threats differ vastly between counties in the region as a result of geographical and management factors. Countries that have invested more in coastal defences, such as the Netherlands, or have mainly higher and steeper topography, such as Norway, are unlikely to see much flood salinization at present, although a future risk to these countries should not be ruled out. The Netherlands further offers us an example of how strategic investment can help reduce flood salinization risks; although this comes at substantial financial cost. Conversely, the mechanisms of salinization to the UK, Germany and Denmark are manifold: these countries have experienced flooding in recent history; are likely confronting the "coastal groundwater squeeze" (Michael et al. 2017), and have been subject to the airborne deposition of salts. Taking this all into account portrays a very complex picture throughout the North Sea region, highlighting the interdisciplinary nature required to address the future salinization challenge.

Of the four mechanisms outlined in the framework (Figure 5.2), irrigation salinization may be a likely pathway of future salinization, but not something of present concern given the lack of existing data on the subject. However, whilst conducting this review, and the wider SalFar project, we noted some curiosity and willingness from growers to explore new opportunities with brackish water irrigation across the region, and several anecdotal reports of such practices. For growers, brackish water irrigation may lead to reductions in water costs, or saving viable crops in times of drought. Much of the data on the salt tolerance of conventional crops has emerged from trials in more arid regions, and on older varieties (Ayers and Westcot 1985; Tanji and Kielen 2002). More research is needed to identify whether such crops may be able to withstand more salinity than first thought, if grown in a climate of more precipitation such as the North Sea region. Recent work on the island of Texel in the Netherlands has explored salt tolerance of conventional crops commonplace in European agriculture (de Vos et al. 2016). Their findings, albeit conducted in a sandy soil environment with greater flushing rates, suggest that several typical crops may withstand more salinity in their irrigation water than first thought. However, such practices are not without risk. The potential for long-term soil damage from brackish irrigation in maritime climates remains unknown. Further investigation with long-term salinity field trials is required in order to explore both crop impact and the soil physical, chemical and biological response under a range of different management regimes.

What is evident from our exploration of the North Sea region is not only a general lack of sufficient data on salinity, but also an inconsistency of measurement between countries. In the Danish islands, we obtained data on percentage (%) salinity. In Belgium, saline groundwater data has been reported in g/L total dissolved solids. In other countries, monitoring of Chloride levels is more routine for water companies. Many farmers and growers use portable and commercial probes to measure salinity in units of ppm. Across all countries, data on water salinity was more available than soil data. The lack of data and consistency between nations is not unsurprising, given that salinization has been of little historic concern in the region compared to other parts of the world. However, over the coming century the issue of salinity will become ever more prescient, and systematic monitoring, mapping and data collection, consistent across all countries of the region, will be essential if the region is to adapt to such threats.

Despite the complexity around the region, and the lack of available data, we predict increased occurrence of salinization to the region in the coming century. Faced with this, coastal land management and agriculture will need to adapt. Key to this will be the complex role of water management bodies. A combination of coastal protection, groundwater abstraction regulation, drainage networks and attenuation and storage areas already constitute much of the North Sea region's multifaceted approach to coastal land management. This will become ever more critical in the coming century, where sea level rise will put greater demands on coastal defences, freshwater abstraction rates will see greater demand in times of water scarcity, and approaches to drainage and pumping will need to address seasonal water shortages. An example of one possible solution might be the designation of water retention areas, both above and below surface, in order to store water instead of pumping it into the North Sea, e.g. large attenuation areas or irrigation reservoirs. In doing this, excess freshwater over winter can be retained to later supply irrigation in the summer months (Karrasch et al. 2017).

Without investment in water management and protection, farmers may need to explore adaptations. Whether it be extending rotation and introducing more grazing, or growing more salt tolerant crops, potentially even halophytes (Rozema and Schat 2013), coastal growers will need to make evidence-based decisions. As such, the potential for "Saline Agriculture" (Ladeiro 2012) – food production that accepts, and adapts to, a degree of salinity in the system – requires further investigation from the research, policy and agricultural communities. In each context, however, we must remember that farmers' decision-making is also complex. Scientists and economists may develop models that suggest optimal choices based on salinization-risk and market analyses, but local traditions, social networks and family imperatives all shape behaviour in different ways. Some of these personal factors will influence a farmer's willingness to innovate, impacting on their "perceived room for manoeuvre" (Methorst et al. 2017). Other factors will see strategic changes emerging from a range of other sources, including both local networks and extensive supply chain connections, thus the communication of science and policy must engage a diversity of networks in order to bring about behavioural change.

5.5 CONCLUSION

Unlike the arid and semi-arid regions, there is limited data available on salinization, or its potential, in the North Sea region. However, under future climate projections, we anticipate the risk of salinization to agriculture in the region to dramatically increase. The risks and mechanisms of salinization across all North Sea nations are extremely diverse and can vary greatly from one country to another. If the region is to develop resilience to salinization in our agricultural systems, it requires more comprehensive knowledge about salinization. This could be realized through more extensive mapping and monitoring, and further research into how farms can respond and adapt, potentially opening greater scope for "saline agriculture".

ACKNOWLEDGMENTS

This review was funded as part of the Interreg VB North Sea Region programme as part of the SalFar project, and the authors would like to thank all SalFar partners, in addition to the multiple farmers, stakeholders and other practitioners we have worked with around the North Sea region. We would also like to thank Carole Ampe (Flemish Land Agency, VLM) for reviewing an earlier version of the manuscript. The authors also gained valuable knowledge from the SALTSA project (supported by the DFG, grant number GZ: SI 728/5-1). We also thank the anonymous reviewers for comments on an earlier draft.

REFERENCES

Abrol, I. P., Yadav, J. S. P., and Massoud, F. I. 1988. Salt-affected soils and their management. In: Soils Bulletin. FAO Rome, Italy.

Agovino, M., Casaccia, M., Ciommi, M., Ferrara, M., and Marchesano, K. 2019. Agriculture, climate change and sustainability: The case of EU-28, *Ecological Indicators* 105: 525–543. https://doi.org/10.1016/j.ecolind.2018.04.064

Anderson, W. P. 2002. Aquifer salinisation from storm overwash. *Journal of Coastal Research* 18(3): 413–420.

Ayers, R. S., and Westcot, D. W. 1985. Water quality for agriculture. FAO Irrigation and drainage paper (29) Rev. 1

Balance, J. A., and Duncan, J. R. 1985. Wind-borne transport and deposition of sea-salt in New Zealand. *New Zealand Journal of Technology* 1: 239–244.

Baxter, P. J. 2005. The east coast big flood, 31 January–February 1, 1953: A summary of the human disaster. *Philosophical Transactions of the Royal Society A: Mathematical, Physical and Engineering Sciences* 363(1831): 1293–1312. https://doi.org/10.1098/rsta.2005.1569

Beauchampet, I. C. 2019. Salinization and agriculture in the Netherlands: benchmarking stakeholder perspectives. MSc Environment and Resource Management, Vrije Univeriteit Amsterdam, September 2019. https://www.waddenacademie.nl/fileadmin/inhoud/pdf/06-wadweten/Scripties/Beauchampet_Salinization_NL.pdf

Brecht, H., Dasgupta, S., Laplante, B., Murray, S., and Wheeler, D. 2012. Sea- level rise and storm surges. *The Journal of Environment & Development* 21(1): 120–138. https://doi.org/10.1177/1070496511433601

Bodirsky, B. L., Rolinski, S., Biewald, A., Weindl, I., Popp, A., and Lotze-Campen, H. 2015. Global food demand scenarios for the 21st century. *PLoS ONE* 10(11): e0139201. https://doi.org/10.1371/journal.pone.0139201

Bouwer, L., and Vellinga, P. 2007. On the flood risk in the Netherlands. In: Begum, S., Stive, M. J. F., Hall, J.W. (eds) Flood Risk Management in Europe. Advances in Natural and Technological Hazards Research, vol 25. Springer, Dordrecht. https://doi.org/10.1007/978-1-4020-4200-3_24

Christensen, L. S. (2021) Dynamics of soil salinity in Denmark. This present book.

CIW. 2018. Evaluatierapport waterschaarste en droogte 2017, report by Coördinatiecommissie Integraal Waterbeheer (CIW), available at www.integraalwaterbeleid.be

CIW. 2019. Evaluatierapport waterschaarste en droogte 2018, report by Coördinatiecommissie Integraal Waterbeheer (CIW), available at www.integraalwaterbeleid.be

CIW. 2020. Evaluatierapport waterschaarste en droogte 2019, report by Coördinatiecommissie Integraal Waterbeheer (CIW), available at www.integraalwaterbeleid.be

Cooper, D., Foster, C., Gooday, R., Hallett, P., Hobbs, P., Irvince, B., Kirkby, M., Morrow, K., Ragab, R., Rawlins, B., Richards, M., Smith, P., Spurgeon, D., and Tye, A. 2010. Modelling the impact of climate change on soils using UK climate projections. Scientific Report. DEFRA Project SPo571

Cui, G., Lu, Y., Zheng, C., Liu, Z., and Sai, J. 2019. Relationship between soil salinization and groundwater hydration in Yaoba Oasis, Northwest China. *Water* 11:175. doi:10.3390/w11010175

Daliakopoulos, I., Tsanis, I., Koutroulis, A., Kourgialas, N., Varouchakis, A., Karatzas, G., and Ritsema, C. J. 2016. The threat of soil salinity: A European scale review. *Science of the Total Environment* 573: 727–739.

de Louw, P. G. B., Essink, G. O., Stuyfzand, P. J., and Van der Zee, S. E. A. T. M. 2010. Upward groundwater flow in boils as the dominant mechanism of salinization in deep polders, The Netherlands. *Journal of Hydrology* 394(3–4): 494–506.

de Louw, P. G. B., Eeman, S., Siemon, B., Voortman, B. R., Gunnink, J., van Baaren, E. S., and Oude Essink, G. H. P. 2011. Shallow rainwater lenses in deltaic areas with saline seepage. *Hydrology and Earth System Sciences* 15: 3659–3678. doi:10.5194/hess-15-3659-2011

de Louw, P. G. B. 2013. Saline seepage in deltaic areas: Preferential groundwater discharge through boils and interactions between thin rainwater lenses and upward saline seepage (Doctoral dissertation). Vrije Universiteit, Amsterdam, the Netherlands.

de Vos, A., Bruning, B., van Straten, G., Oosterbaan, R., Rozema, J., and van Bodegom, P. 2016. Crop salt tolerance under controlled field conditions in The Netherlands based on trials conducted at Salt Farm Texel. December 2016, Salt Farm Texel, Den Berg.

De Waegemaeker, J. 2019. SalFar framework on salinization processes. A comparison of salinization processes across the North Sea Region, a report by ILVO for the Interreg VB North Sea Region project Saline Farming (SalFar).

Delsmans, J., van Baaren, E., Vermaas, T., Karaoulis, M., Bootsma, H., de Louw, P., Pauw, P., Oude Essink, G., Dabekaussen, W., Van Camp, M., Walraevens, K., Vandenbohede, A., Teilmann, R., and Thofte, S. 2019. TOPSOIL Airborne EM kartering van zoet en zout grondwater in Vlaanderen (FRESHEM Vlaanderen), A report for the Flemish Environmental Agency (VMM), Aalst.

Durant, D., Kernéïs, E., Meynard, J-M., Chosis, J. P., Chataigner, C., Hillaireau, J-M., and Rossignol, C. 2018. Impact of storm Xynthia in 2010 on coastal agricultural areas: The Saint Laurent de la Prée research farm's experience. *Journal of Coastal Conservation* 22: 1177–1190. https://doi.org/10.1007/s11852-018-0627-8

Eich-Greatorex, S., Malesevic, T., and Almås, ÅA. R. 2020. Sea level rise and potential impact of saltwater on agriculture along the Norwegian North Sea coast – Case study Jæren. MINA/NMBU report, in prep.

Endo, T., Yamamoto, S., Larringa, J., Fujiyama, H., and Honna, T. 2011. Status and causes of soil salinization of irrigated agricultural lands in Southern Baja California, Mexico. *Applied and Environmental Soil Science*. http://dx.doi.org/10.1155/2011/873625

Environment Agency. 2020. Meeting our Future Water Needs: A National Framework for Water Resources. 16 March 2020, Version 1. Environment Agency, Bristol.

European Environment Agency. 2019. Global and European sea-level rise. Indicator Assessment. European Environment Agency. Copenhagen, Denmark.

FAO. 2015. Status of the World's Soil Resources Food and Agriculture Organisation of the United Nations Rome, Italy.

Fagherazzi, S., Anisfeld, S. C., Blum, L. K., et al. 2019. Sea level rise and the dynamics of the marsh-upland boundary. *Frontiers in Environmental Science* 7. 10.3389/fenvs.2019.00025

Fenger, J., Buch, E., Jakobsen, P., and Vestergaard, P. 2008. Danish attitudes and reactions to the threat of sea-level rise. *Journal of Coastal Research* 24(2): 394–402.

Foley, J., Ramankutty, N., Brauman, K., et al. 2011. Solutions for a cultivated planet. *Nature* 478: 337–342. https://doi.org/10.1038/nature10452

Franzén, L. 1990. Transport, deposition and distribution of marine aerosols over southern Sweden during dry westerly storms. *Ambio* 19(4): 180–188.

Gerritsen, H. 2005. What happened in 1953? The Big Flood in the Netherlands in retrospect. *Philosphical Transaction of the Royal Society A* 3631271–1291. http://doi.org/10.1098/rsta.2005.1568

Gould, I. J., Wright, I., Collison, M., Ruto, E., Bosworth, G., Pearson, S. 2020. The impact of coastal flooding on agriculture: A case- study of Lincolnshire, United Kingdom. *Land Degradation and Development* 1–15. https://doi.org/10.1002/ldr.3551

Grube, A. T. 2000. Geogene Grundwasserversalzung in den Poren-Grundwasserleitern Norddeutschlands und ihre Bedeutung für die Wasserwirtschaft, DVGW-Technologiezentrum Wasser (TZW).

Gustafsson, M., and Franzen, L. 1996. Dry deposition and concentration of marine aerosols in a coastal area, SW Sweden. *Atmospheric Environment* 30(6): 977–989.

Hansen, J. W., Windelin, A., Göke, C., et al. 2011. Notat 1.1 - Fysiske og kemiske forhold. DCE - Nationalt Center for Miljø og Energi, Aarhus Universitet. Notat udarbejdet for Naturstyrelsen

Hazelden, J., and Boorman, L. A. 2001. Soils and 'managed retreat' in south east England. *Soil Use and Management* 17: 150–154.

Hickey, K. R. 1997. *Documentary records of coastal storms in Scotland 1500-1991 A*.D. Volume 1. Unpublished Thesis. Coventry, UK:Coventry University. Accessed https://curve.coventry.ac.uk/open/file/aa6dfd04-d53f-4741-1bb7-bdf99fb153be/1/hick1comb.pdf

Hodgkinson, R. A., and Thorburn, A. A. 1995. Factors influencing the stability of salt affected soils in the UK - criteria for identifying appropriate management options. *Agricultural Water Management* 29: 327–338.

Hoeksema, R. J. 2007. Three stages in the history of land reclamation in the Netherlands. *Irrigation and Drainage* 56: S113–S126. doi:10.1002/ird.340

Holding, S., and Allen, D. M. 2015. Wave overwash impact on small islands: Generalised observations of freshwater lens response and recovery for multiple hydrogeological settings. *Journal of Hydrology*, 529: 1324–1335.

Huisman, P., Cramer, W., Van Ee, G., Hooghart, J. C., Salz, H., and Zuidema, F. C. 1998. *Water in the Netherlands*. Netherlands Hydrological Society, Rotterdam.

IPCC. 2019. The Ocean and Cryosphere in a Changing Climate. Accessed 26th October 2019 at: https://report.ipcc.ch/srocc/pdf/SROCC_SPM_Approved.pdf

Jones, A., Panagos, P., Barcelo, S., et al. 2012. The state of soil in Europe. A Contribution of the JRC to the European Environment Agency's Environment State and Outlook Report (European Commission: Luxembourg) Available at https://ec.europa.eu/jrc/en/publication/reference-reports/state-soil-europe-contribution-jrc-european-environment-agency-s-environment-state-and-outlook

Jørgensen, N. 2002. Origin of shallow saline groundwater on the Island of Læsø, Denmark. *Chemical Geology* 184: 359–370. 10.1016/S0009-2541(01)00392-8.

Jurasinski, G., Janssen, M., Voss, M., Böttcher, M. E., Brede, M., Burchard, H., et al. 2018. Understanding the coastal ecocline: Assessing sea-land-interactions at non-tidal, low-lying coasts through interdisciplinary research. *Frontiers in Marine Science* 5: 342.

Karrasch, L., Maier, M., Kleyer, M., and Klenke, T. 2017. Collaborative landscape planning: Co-design of ecosystem-based land management scenarios. *Sustainability* 9(9): 1668.

Knutsson, G., and Fagerland, T. 1977. Grundvattentillgångar i Sverige (Groundwater resources in Sweden), SGU.

Kristiansen, S. M., Christensen, F. D., and B. Hansen, B. 2011. Does road salt affect groundwater in Denmark? *Geological Survey of Denmark and Greenland Bulletin* 2: 45–48. Open Access: www.geus.dk/publications/bull

Ladeiro, B. 2012. Saline agriculture in the 21st century: Using salt contaminated resources to cope food requirements. *Journal of Botany* 1–7. doi:10.1155/2012/310705

Lebbe, L., Van Meir, N., and Viane, P. 2008. Potential implications of sea-level rise for Belgium. *Journal of Coastal Research* 24(2): 358–366

Lindewald, H. 1981 Saline groundwater in Sweden. Intruded and relict groundwater of marine origin, 7th Salt-Water Intrusion Meeting, 1981. 24-32

Liski, A. H., Ambros, P., Metzger, M. J., Nicholas, K. A., Wilson, A. M. W., and Krause, T. 2019. Governance and stakeholder perspectives of managed re-alignment: Adapting to sea level rise in the Inner Forth estuary, Scotland. *Regional Environmental Change* 19:2231–2243. https://doi.org/10.1007/s10113-019-01505-8

Loveland, P., Hazelden, J., Sturdy, R., and Hodgson, J. 1986, Salt- affected soils in England and Wales. *Soil Use and Management* 2: 150–156. doi:10.1111/j.1475-2743.1986.tb00700.x

MacAllister, D. J., Jackson, M. D., Butler, A. P., and Vinogradov, J. 2018. Remote detection of saline intrusion in a coastal aquifer using borehole measurements of self- potential. *Water Resources Research* 54: 1669– 1687. https://doi.org/10.1002/2017WR021034

Manca, F., Capelli, G., and Tuccimei, P. 2015. Sea salt aerosol groundwater salinization in the Litorale Romano Natural Reserve (Rome, Central Italy). *Environmental Earth Sciences* 73: 4179–4190

Martens, S., and Wichmann, K. 2011. Grundwasserversalzung in Deutschland. In: WARNSIGNAL KLIMA: Genug Wasser für alle? 3.Auflage- Hrsg. Lozán, JLH, Graßl, P, Hupfer, L, Karbe & Schönwiese, CD: 203-207.

Masterson, J. P., and Garabedian, S.P. 2007. Effects of sea- level rise on ground water flow in a coastal aquifer system. *Groundwater* 45: 209–217. doi:10.1111/j.1745-6584.2006.00279.x

Mayes, J., and Wheeler, D. 2013. Regional weather and climates of the British Isles – Part 1: Introduction. *Weather* 68: 3–8. doi:10.1002/wea.2041

McCune, D. C. 1991. Effects of airborne saline particles on vegetation in relation to variables of exposure and other factors. *Environmental Pollution* 74(3): 176–203.

Methorst, R. G., Roep, D., Verhees, F. J. H. M., and Verstegen, J. A. A. M. 2017. Differences in farmers' perception of opportunities for farm development. *NJAS – Wageningen Journal of Life Sciences* 81: 9–18. https://doi.org/10.1016/j.njas.2017.02.001

Michael, H. A., Post, V. E., Wilson, A. M., and Werner, A. D. 2017. Science, society, and the coastal groundwater squeeze. *Water Resources Research* 53(4): 2610–2617.

National Farmers Union (2013). Briefing. December 2013 tidal surge floods – Agricultural impact. Version 2. Retrieved from https://www.nfuonline.com/assets/21657

NGU, Geological Survey of Norway. 2014. Groundwater. Accessed 19 December 2018 from http://www.miljostatus.no/tema/ferskvann/grunnvann/Rapport. In Norwegian.

Nicholls, R. J., and Cazenave, A. 2010. Sea- level rise and its impact on coastal zones. *Science* 328(5985): 1517–1520. https://doi.org/10.1126/science.1185782

Nillesen, E. E. M., and E. C. van Ierland (eds), 2006. Climate change scientific assessment and policy analysis. Climate adaptation in the Netherlands. Report Wageningen UR 500102 003, 118.

O'Dowd, C., and de Leeuw, G. 2007. Marine aerosol production: A review of the current knowledge. *Philosophical Transactions of the Royal Society of London. A* 365:1753–1774.

Olofsson, B. 1996. Salt groundwater in Sweden – occurrence and origin. *Salt Water Intrusion Meeting (SWIM)*. 16–21.

Oude Essink, G. H. P. 2001. Improving fresh groundwater supply – problems and solutions. *Ocean and Coastal Management* 44: 429–449

Oude Essink, G. H. P., van Baaren, E. S., and de Louw, P. G. B. 2010. Effects of climate change on coastal groundwater systems: A modeling study in the Netherlands. *Water Resources Research* 46: W00F04. doi:10.1029/2009WR008719

Paes, J. L. d. A., Ruiz, H. A., Fernandes, R. B. A., Freire, M. B. G. d. S., Barros, M. d. F. C., and Rocha, G.C. 2014. Hydraulic conductivity in response to exchangeable sodium percentage and solution salt concentration. *Revista Ceres* 61: 715–722.

Palmer, T., and Räisänen, J. 2002. Quantifying the risk of extreme seasonal precipitation events in a changing climate. *Nature* 415: 512.

Pedersen, L. B., and Bille-Hansen, J. 1995. Effects of airborne sea salts on soil water acidification and leaching of aluminium in different forest ecosystems in Denmark. In: Nilsson, L. O., Hüttl, R. F., Johansson, U. T. (eds) Nutrient Uptake and Cycling in Forest Ecosystems. Developments in Plant and Soil Sciences, vol 62. Springer, Dordrecht.

Pendergrass, A. G., Knutti, R., Lehner, F., et al. 2017. Precipitation variability increases in a warmer climate *Scientific Reports* 7: 17966. https://doi.org/10.1038/s41598-017-17966-y

Pitman, M. G., and Läuchli, A. 2002. Global impact of salinity and agricultural ecosystems. In: Läuchli, A., Lüttge, U. (eds) Salinity: Environment – Plants – Molecules. Springer, Netherlands Dordrecht, 3–20.

Post, V. E., and Houben, G. J. 2017. Density-driven vertical transport of saltwater through the freshwater lens on the island of Baltrum (Germany) following the 1962 storm flood. *Journal of Hydrology* 551: 689–702.

Qadir, E. Quillérou, Nangia, V., et al. 2014. Economics of salt-induced land degradation and restoration. *Natural Resources Forum* 38: 282–295.

Raats, P. A. C. 2015 Salinity management in the coastal region of the Netherlands: A historical perspective. *Agricultural Water Management* 157: 12–30. https://doi.org/10.1016/j.agwat.2014.08.022

Rath, K. M., and Rousk, J. 2015. Salt effects on the soil microbial decomposer community and their role in organic carbon cycling: A review. *Soil Biology and Biochemistry* 81: 108–123.

Rengasamy, P. 2006. World salinization with emphasis on Australia. *Journal of Experimental Botany* 57 (5): 1017–1023. https://doi.org/10.1093/jxb/erj108

Rey, D., Holman, I. P., Daccache, A., Morris, J., Weatherhead, E. K., and Knox, J. W. 2016. Modelling and mapping the economic value of supplemental irrigation in a humid climate. *Agricultural Water Management* 173: 13–22. https://doi.org/10.1016/j.agwat.2016.04.017.

Roberts, L. R., Sayer, C. D., Hoare, D., et al. 2019. The role of monitoring, documentary and archival records for coastal shallow lake management. *Geography and Environment* e00083. https://doi.org/10.1002/geo2.83.

Röper, T., Kröger, K. F., Meyer, H., Sültenfuss, J., Greskowiak, J., and Massmann, G. 2012. Groundwater ages, recharge conditions and hydrochemical evolution of a barrier island freshwater lens (Spiekeroog, Northern Germany). *Journal of Hydrology* 454: 173–186.

Rozema, J., and Schat, H. 2013. Salt tolerance of halophytes, research questions reviewed in the perspective of saline agriculture. *Environmental and Experimental Botany* 92: 83–95. doi:10.1016/j.envexpbot.2012.08.004.

Rowell, D. P., and Jones, R. G. 2006. Causes and uncertainty of future summer drying over Europe. *Climate Dynamics* 27: 281–299.

Rozema, J., Van Manen, Y., Vugts, H. F., and Leusink, A. 1983. Airborne and soilborne salinity and the distribution of coastal and inland species of the genus Elytrigia. *Acta Botanica Neerlandica* 32: 447–456.

Salehin, M., et al. (2018) Mechanisms and drivers of soil salinity in coastal Bangladesh. In: Nicholls, R., Hutton, C., Adger, W., Hanson, S., Rahman, M., Salehin, M. (eds) Ecosystem Services for Well-Being in Deltas. Palgrave Macmillan, Cham.

Schneider, V., Leifeld, P., and Malang, T. 2013. Coping with creeping catastrophes: National political systems and the challenge of slow-moving policy problems. In: Siebenhner, B., Arnold, M., Eisenack, K., and Jacob, K. H. (eds) Long-Term Governance of Social-Ecological Change. Series: Environmental politics (21). Routledge, New York, pp. 221–238. ISBN 9780415633529

Seibert, S. L., Holt, T., Reckhardt, A., et al. 2018. Hydrochemical evolution of a freshwater lens below a barrier island (Spiekeroog, Germany): The role of carbonate mineral reactions, cation exchange and redox processes. *Applied Geochemistry* 92: 186–208.

Seither, A., Gundersen, P., Jæger, Ø., and Sæther, O. M. 2017. Nationwide Soil Water and Groundwater Observation Network (LGN) – Past and future after 39 years of operation. NGU-Report no. 2016.039, ISSN: 2387-3515 (online). In Norwegian.

Simpson, M., Nilsen, J. E., Ravndal, O.,et al. 2015. Sea Level Change for Norway: Past and Present Observations and Projections to 2100. 10.13140/RG.2.1.2224.9440.

SOU 2007. Swedish Official Report on Climate and Vulnerability. Regeringskansliet, Stockholm. 60.

Spencer, T., Brooks, S. M., Evans, B. R., Tempest, J. A., and Möller, I. 2015. Southern North Sea storm surge event of December 5, 2013: Water levels, waves and coastal impacts. *Earth-Science Reviews* 146: 120–145. https://doi.org/10.1016/j.earscirev.2015.04.002

Steers, J. A., Stoddart, D. R., Bayliss-Smith, T. P., Spencer, T., and Durbidge, P. M. 1979. The storm surge of January 11, 1978 on the East Coast of England. *The Geographical Journal* 145(2): 192– 205. https://doi.org/10.2307/634386

Sterr, H. 2008. Assessment of vulnerability and adaptation to sea-level rise for the coastal zone of Germany. *Journal of Coastal Research*, 380–393.

Stofberg, S. F., Essink, G. H. P. O., Pauw, P. S., et al. 2017. Fresh Water Lens Persistence and Root Zone Salinization Hazard under Temperate Climate. *Water Resource Management* 31: 689–702. https://doi.org/10.1007/s11269-016-1315-9

Tabari, H., Taye, M. T., and Willems, P. 2015. Actualisatie en verfijning klimaatscenario 's tot 2100 voor Vlaanderen - Appendix 2: Nieuwe modelproject voor Ukkel op basis van globale klimaatmodellen (CMIP5) en actualisatie klimaatscenario's, p.107, a report by KU Leuven for Vlaamse Milieumaatschappij, Aalst

Tanji, K., and Kielen, N. C. 2002. Agricultural Drainage Water Management in Arid and Semi-arid Areas. FAO, Italy.

Tavernier, R., Ameryckx, J., Snacken, F., and Farasyn, D. 1970. Atlas van België blad 17 – Kust, Duinen, Polders, p. 32, a report by Nationaal Comité voor Geografie (Commissie voor de Nationale Atlas), Brussels.

Tóth, G., Montanarella, L., and Rusco, E. 2008. Threats to Soil Quality in Europe. EUR 23438 EN. Institute for Environment and Sustainability, Land Management and Natural Hazards Unit, Office for the Official Publications of the European Communities, Luxembourg.

United Nations Department of Economic and Social Affairs, Population Division. 2017. World Population Prospects: The 2017 Revision, DVD Edition.

Vandenbohede, A., Courtens, C., Lebbe, L., and de Breuck, W. 2010. Fresh-salt water distribution in the central Belgian coastal plain: An update. *Geologica Belgica* 13(3): 163–172.

Vandenbohede, A., Walraevens, K., and De Breuck, W. 2015. What does the interface on the fresh – saltwater distribution map of the Belgian coastal plain represent? *Geologica Belgica* 18(1): 31–36.

Velstra J., Hoogmoed M., and Groen, K. 2009. Inventarisatie maatregelen omtrent interne verzilting. In: Leven met zout water (203). Acacia Water.

VMM (2014) Verziltingskaart Vlaanderen, dataset by Vlaamse Milieumaatschappij, Aalst, available online at dov.vlaanderen.be

Vollmer, M., Guldberg, M., Maluck, M., Marrewijk, D., and Schlicksbier, G. 2001. Landscape and cultural heritage in the Wadden Sea Region – Project report. Wadden Sea Ecosystem No. 12. Common Wadden Sea Secretariat, Wilhelmshaven.

Vousdoukas, M. I., Voukouvalas, E., Annunziato, A., Giardino, A., and Feyen, L. 2016. Projections of extreme storm surge levels along Europe. *Climate Dynamics* 47: 3171–3190.

Vousdoukas, M., Mentaschi, L., Voukouvalas, E., Verlaan, M., and Feyen, L. 2017. Extreme sea levels on the rise along Europe's coasts. *Earth's Future* 5(3): 304–323.

Wadey, M. P., Haigh, I. D., Nicholls, R. J., et al. 2015. A comparison of the 31 January–1 February 1953 and 5–6 December 2013 coastal flood events around the UK. *Frontiers in Marine Science* 2: 84. doi: 10.3389/fmars.2015.00084

Weatherhead, E. K., Knox, J. W., Daccache, A., et al. 2014. Water for Agriculture: Collaborative Approaches and On-farm Storage. FG1112 Final Report to Defra, Cranfield University.

Werner, A. D., Bakker, M., Post, V. E. A., et al. 2013. Seawater intrusion processes, investigation and management: Recent advances and future challenges. *Advances in Water Resources* 51: 3–26. https://doi.org/10.1016/j.advwatres.2012.03.004

Woth, K., Weisse, R., and Van Storch, H. 2006. Climate change and North Sea storm surge extremes: An ensemble study of storm surge extremes expected in a changed climate projected by four different Regional Climate Models. *Ocean Dynamics* 56: 3–15. 10.1007/s10236-005-0024-3.

6 Economic Impact of Soil Salinization and the Potential for Saline Agriculture

Eric Ruto, Domna Tzemi, Iain Gould, and Gary Bosworth

CONTENTS

6.1 INTRODUCTION

Soil salinization, defined as the accumulation of water-soluble salts in the soil to a level that impacts on agricultural production, environmental health, and economic welfare (FAO 2011), is a global problem and one of the major causes of land degradation. A major United Nations Environment Programme (UNEP) study GLASOD (Global Assessment of Soil Degradation), which was a first

DOI: 10.1201/9781003112327-6

attempt to produce a world map on the status of human-induced soil degradation, identified soil salinization as one of the major types of soil degradation (Oldeman et al. 1991).

The drivers or types of soil salinization have generally been characterized as either primary or secondary (Daliakopoulos et al. 2016). Primary salinization is the accumulation of salts in the soil profile through natural processes. Secondary (or human-induced) salinization, on the other hand, is driven by human interventions, mainly irrigation with saline water often coupled with poor drainage systems, over-exploitation of groundwater and seawater ingress into coastal land that may be exacerbated by climate change and sea-level rise.

Soil salinization is a significant constraint to agricultural production globally. For example, FAO and ITPS (2015) estimates that increasing soil salinity problems are taking up to 1.5 million ha of farmland out of agricultural production each year and compromising the yield potential of a further 20 to 46 million ha. Furthermore, projected changes associated with climate change are likely to exacerbate the risks associated with salinization (Koutroulis et al. 2013). Climate change is also expected to lead to a reduction in potential yields of major crops (such as wheat) around the world which has implications for global food security. Food security is an important policy issue as espoused by UN Sustainable Development Goals (SDGs): "End hunger, achieve food security and improved nutrition and promote sustainable agriculture" (SDG 2) (UN 2015). Lipper et al. (2014) posit that the expansion of the level of agricultural output will require greater use of inputs at an increasing cost and innovations in "climate-smart" agricultural practices such as saline farming.

Despite the significance of soil salinization, there is sparse information on its impact on agriculture (and economies) in Europe and globally. This is partly because of the unavailability of reliable data on the extent and severity of salinization, which limits the biophysical modelling of impacts of salinization and concomitant economic impacts. For example, industry and policymakers need information on the economic costs of salinization to guide investment decisions and strategies for the amelioration of salinization related impacts and to set priorities for innovative adaptation strategies such as the development of saline agriculture.

The aim of the chapter is to provide a framework for economic risk assessment in regions where salinity poses a significant threat to agricultural production and the local/national economy. The analysis of the costs of salinization should provide a "baseline" for economic impacts of salinization (on agriculture and the wider economy) which helps to inform the assessment of adaptation measures including the potential for saline agriculture. This topic remains largely unaddressed by the literature and this chapter helps to fill in a significant gap.

The rest of the chapter is structured as follows. Section 6.2 reviews key literature on economic impacts of soil salinity. Section 6.3 presents our conceptual and methodological framework for assessing the economic costs of salinization. This leads to Section 6.4 in which we present empirical results of farm-level, regional (case study) and wider economy impacts of salinity, structured around a typology of salinization processes (irrigation, seepage and flood salinization). Section 6.5 concludes the chapter.

6.2 ECONOMIC IMPACTS OF SOIL SALINITY

The biophysical effects (e.g. yield losses) of soil salinization are relatively well documented. Although there is a wide variation between and within crop types, farm-level studies show crop yield losses on salt-affected lands of 40–63% in India, 36–69% in Pakistan and 71–86% in Kazakhstan (Qadir et al. 2014).

One of the first studies on global costs of salinity was conducted by Ghassemi et al. (1995), who assessed that the global income losses due to salinity at about USD 11.4 billion per year in irrigated areas and USD 1.2 billion per year in non-irrigated areas. Building on Ghassemi et al. (1995), a comprehensive meta-analysis conducted by Qadir et al. (2014) estimated the annual (inflation adjusted) income losses from salt-affected irrigated areas as USD 27.3 billion, based mainly on crop yield losses. The authors based their calculations on an Food and Agriculture Organization of the United Nations (FAO) estimated globally irrigated area of 310 million hectares (Mha) (FAO 2011) with an estimated 20% of this area being salt affected (62 Mha). Based on these estimates, the annual cost of salinity related land degradation was approximated as USD 441 per ha in 2013. It is noted, however, that these estimates on the global cost of salinized land degradation are mainly based on crop yield losses. These costs are expected to be even higher when other cost components are taken into consideration, such as the environmental costs associated with salt-affected lands and the potential social cost on farm businesses. On the other hand, adaptation measures such as the use of salt-tolerant crops may be expected to ameliorate some of the impacts of salinization.

Economic studies on the impact of soil salinization in Europe are limited. One of the early studies in Europe was conducted by Zekri and Albisu (1993) who studied the economic effect of salinity at the farm level in Berdenas, an area of 56,760 ha of irrigated land situated north of Zaragoza and south of Navarra in Spain. The objectives of the research were to assess soil salinity levels, to simulate the future situation without the effects of salinity and to estimate soil reclamation costs and benefits. They employed an interactive multi-objective mathematical programming methodology, optimizing three different objectives: (a) maximizing total farm gross margin, (b) maximizing labor used and (c) minimizing labor seasonality in order to avoid periods of unemployment during the year and minimizing risk. The study showed considerable benefits from soil reclamation at a level equivalent to 69 million €, with 799 jobs generated. More recently, a study conducted by Montanarella (2007) in three European countries (Spain, Hungary and Bulgaria) estimated annual costs of soil salinization in the range of €158–321 million, mainly as a result of agricultural yield losses.

A review of the literature shows that most studies focus on the cost of salinity in irrigation systems. A majority of these studies estimate the cost of salinization from biophysical output losses (mainly crop yield losses) for a range of salt-affected irrigation lands (Qadir et al. 2014). However, some economic studies take account of additional costs (e.g. remediation of salt degraded land) or additional inputs (and costs) used to mitigate some of the impacts of salt related land degradation, which would otherwise not be used for non-degraded land. The consensus in the literature is that preventing salinization would result in considerable savings, mainly from reduced yield losses and opportunity costs.

TABLE 6.1
Economic Costs of Salt-Induced Land Degradation in Different Parts of the World

Study Authors	Country	Methodology	Equivalent in Million USD per Year
Marshall and Jones (1997)	Australia	Opportunity costs based on dose response method and mitigation costs	0.83
Janmaat (2004)	India	Opportunity costs (forgone agricultural income)	46
Marshall (2004)	Australia	Transaction costs	20.03
John et al. (2005)	Australia	Opportunity costs	0.09
Aslam and Prathapar (2006)	Pakistan	Opportunity costs	267
McCann and Hafdahl (2007)	Australia	Transaction costs	102
Winpenny et al. (2010)	Spain	Mitigation costs	810

Source: Negacz (2018).

Table 6.1 summarizes estimates of economic costs (yield loss and additional costs) of salinity in different parts of the world. As may be expected, most studies on the economic impact of salt-induced land degradation have been conducted in countries where salinity is a major problem, notably Australia, India, the United States, Iraq, Pakistan, Kazakhstan, Uzbekistan and Spain. Salinity-related economic analyzes particularly have a long history in Australia, where salinity is a prominent problem.

Studies on economic costs of salinization attributable to climate change are limited. One exception is PESETA (**P**rojection of **E**conomic impacts of climate change in **S**ectors of the **E**uropean Union based on bo**T**tom-up **A**nalysis) a major EU-funded project on the impacts of climate change in Europe covering 25 countries (Richards and Nicholls 2009; Bosello et al. 2012). This study examined the direct biophysical impacts of climate change and sea-level rise on: (i) increased erosion, (ii) increased flood risk and inundation, (iii) coastal wetland loss and change and (iv) (surface) salinization costs. The higher order costs of these impacts were then assessed using a computable general equilibrium (CGE) modelling framework with country-level detail to assess the wider economic implications. Focusing on the salinization part of the study, the results show that salinity intrusion costs range from €577 to 610 million per year and are projected to significantly increase with sea-level rise and over time across all scenarios investigated in the study. The study further notes that adaptation is crucial to keep the negative impacts of sea-level rise at an "acceptable" level.

6.3 METHODOLOGICAL FRAMEWORK TO ASSESSING ECONOMIC IMPACT OF SALINIZATION

6.3.1 Overall Approach and Conceptual Framework

The impacts of salinization on agriculture depend on a wide range of related factors. This includes the *type* of salinization (the process that causes salinization), the

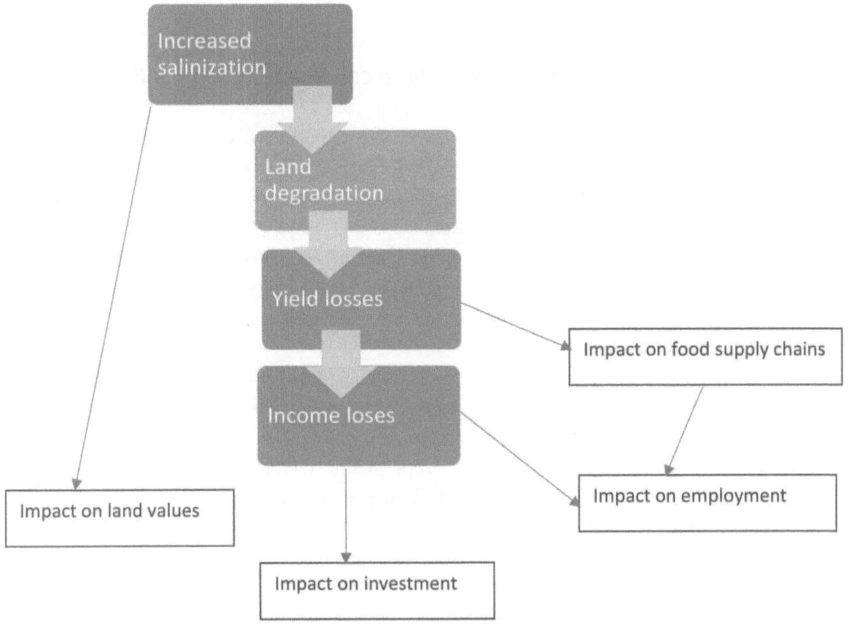

FIGURE 6.1 Stylized framework for assessing farm scale and wider impact of salinization.

degree of salinization (the present state of salinization), the types of crops grown in the affected region, the value of those crops, shocks (such as climate change induced sea-level rise) and any farm-level decisions to ameliorate the impacts of salinization (which may include the planting of salt-tolerant crops).

This section develops a modelling framework that attempts to incorporate these variables to allow farm-level and wider level evaluations of the economic risks of soil salinization. The chain of causes and effects that must be appraised is represented diagrammatically in Figure 6.1. As depicted in the figure, the economic analysis focuses on the scale of impact along each bold arrow. The wider economic impacts can also be estimated at the regional levels by using appropriate multipliers and other local evidence.

To operationalize the framework, we employed multistage empirical modelling and scenario analysis to represent the chain of causes and effects of salinization on crop yields and "downstream" economic impacts at the farm and regional or wider scales. As alluded to earlier, these impacts critically depend on the *type* and *degree* of salinization, among other factors. In our approach, the type of salinization follows a typology identified by De Waegemaeker (2019) i.e. irrigation, seepage, flood salinization and aerosol (or airborne) salinization. However, we do not include aerosol salinization in our analyzes partly because of the unavailability of data on the impact or significance of this type of salinization in our study area. The degree of salinization was developed from detailed scenario analysis informed by a critical review of the literature and analysis of data from a survey of partners in Interreg VB North Sea Region SalFar project, a project co-funded by Interreg VB North Sea Region Programme.

TABLE 6.2
A Summary of Methodology and Data Sources Used to Assess Economic Impacts of Salinization

Farm Scale Impacts

Step 1: Develop a typology of salinization based on salinization processes identified in De Waegemaeker (2019) i.e. irrigation, seepage and flood salinization.

Step 2: For each type of salinization process, develop a range of salinity scenarios informed by a critical review of the literature (e.g. van Straten et al. 2019) and data from a survey of SalFar project partners.

Step 3: Collate a representative list of crops grown in the North Sea Region using information from the survey of SalFar project partners.

Step 4: Conduct a yield gap analysis to estimate production penalties (relative yield) of specific crops under each type of salinization process and salinity scenarios, using crop salt tolerance parameters provided by Salt Farm Texel (de Vos et al. 2016).

Step 5: Estimate the yield loss (tons/ha) of specific crops under each type of salinization process and salinity scenarios, using EUROSTAT (https://ec.europa.eu/eurostat) country-level data (2019) on average yield per ha for each crop.

Step 6: Based on the estimated yield gaps per ha, calculate the gross value of production attributable to the estimated yield gaps, under each type of salinization process and salinity scenarios, using EUROSTAT (https://ec.europa.eu/eurostat) data (2019) on average prices of specific crops.

Regional/Economy-wide Scale Impacts

Step 7: Estimate the area affected or at risk of each type of salinization process, using GIS mapping of areas at risk (where available) or expert opinion, combined with typical crop composition using satellite remote sensing data, where available.

Step 8: Extrapolate crop yield loss to areas at risk of salinity under each type of salinization process, using EUROSTAT (https://ec.europa.eu/eurostat) data (2019) on average yield per ha for the regionally representative crop composition.

Step 9: Estimate expected financial losses, extrapolated to areas affected or at risk of salinization, under each type of salinization process, using EUROSTAT (https://ec.europa.eu/eurostat) data (2019) on average prices of the regionally representative crop composition.

Step 10: Scale up output losses to calculate impacts to the wider economy, using appropriate multipliers where data are available.

Our approach encompasses a series of logical steps, bringing together data from several sources as summarized in Table 6.2.

6.3.2 Salinity Processes and Scenarios

Salinity measurement is based on the electrical conductivity of the soil saturation extract (EC_e) in deciSiemens per meter (dS/m) and chloride concentrations (de Vos et al. 2016). The soil is considered saline when the ECe is 4 dS/m or higher (Table 6.3). Depending on the level, salinity may have a profound influence on plant productivity, as shown in the table below and described in detail in de Vos et al. (2016).

To facilitate comparability and compatibility, we employed a typology of salinization developed by De Waegemaeker (2019) as a basis of our economic analysis: irrigation salinization, flood salinization and seepage salinization and aerosol salinization.

TABLE 6.3
Soil Salinity Classes and Effect on Crop Growth

Soil Salinity Class	Salinity (EC_e in dS/m)	Effect on Plants
Non saline	0–2	Salinity effects negligible
Slightly saline	2–4	Yields of sensitive crops may be restricted
Moderately saline	4–8	Yields of many crops are restricted
Strongly saline	8–16	Only tolerant crops yield Satisfactorily
Very strongly saline	>16	Only a few very tolerant crops yield satisfactorily

Source: Adapted from Van Orshoven et al. (2012).

It may be noted that this typology categorizes the processes that create saline soil conditions and not the resulting saline soil conditions. Due to the unavailability of data on the actual degree of salinity, we use scenario analysis to estimate the potential economic impact of salinization. To calibrate the analysis of economic impacts, we developed a range of salinity scenarios, from slightly saline to strongly saline. This was informed by a critical review of the literature (e.g. van Straten et al. 2019) and data from a survey of SalFar project partners. Table 6.4 summarizes salinity scenarios used in the analysis.

For irrigation salinization, we used four different salinity levels of irrigation water. The salinity levels of irrigation water were chosen based on the study by Van Straten et al. (2019).

For seepage salinization we used two groundwater salinity scenarios. The calibration of the levels of groundwater salinity scenarios was based on data on actual salinity of groundwater obtained from the province of Groningen (measured as chloride (Cl) concentrations). Looking at the Cl groundwater concentrations across the province of Groningen we chose the concentrations corresponding to four percentiles

TABLE 6.4
Salinity Scenarios Employed in Economic Analysis

Salinization Process	Description	Salinity Scenario Levels (EC_w in dS/m)
Irrigation salinization (IS)	Salinization that results from irrigation of non-saline agricultural soils with salt or brackish water.	4, 8, 12, 16
Seepage salinization (SS)	Salinization that results from the rise of salt rich groundwater. The salt rich groundwater may be hydrologically linked to nearby seawater.	0.02, 0.09, 0.2, 0.7
Flood salinization (FS)	Salinization that occurs as soils are flooded by brackish or salt-rich water. Flood risk may be exacerbated by climate change	7.1, 6.08, 5.06, 3.03, 4.04

0%, 25%, 50% and 75% (corresponding to 6, 26, 64 and 215 mg/l respectively) of the Cl distribution (or ECw values of 0.02, 0.09, 0.2, 0.7 dS/m equivalent). However, in the empirical analysis, we focused only on the salinity scenario levels that had a significant impact on yields (i.e. 215 mg/L). The result of the other salinity levels had only a marginal or no impact on crop yield.

Finally, for flood salinization, we considered that seawater flooding impacts on yield can occur over many years. Therefore, to assess total yield loss (current and future years) as the soil recovers, we firstly calculated the response of different crop types (relative yields) to salt-affected land. We did this by predicting salt soil levels in recovery years. However, for farm-scale assessments, this method could be adapted by basing on known, or historic salt levels. We assumed the complete loss of the standing crop during the flood (zero yield in flood year) followed by a "sliding" recovery approach during the following years, where the rate of recovery was a function of the salt tolerance per crop type based on predicted salt soil levels. Thus, the model considered that highly tolerant crops recover yield on inundated fields at a faster rate than sensitive crops. Salt recovery time depends on soil type; for example, a well-drained sandy soil may recover back to post-flood production in 2 years, whereas a heavier, poorly drained soil may take up to 7 years. As such, without knowledge of site specific drainage regimes, we modelled six recovery scenarios on a scale of 2–7 year soil recovery.

To evaluate the impact of soil salinity and facilitate comparisons, where appropriate, we converted irrigation water salinity (i.e. electrical conductivity of irrigation water, EC_w) into corresponding soil salinity (EC_e) using procedures developed in Ayers and Westcot (1985) and Grattan (2002). Where soil salinity was measured in chloride, we converted soil salinity in chloride concentrations (mg/l) into equivalent EC (in ds/m) measurements, using established correlations in the literature (e.g. de Vos et al. 2016).

6.3.3 Economic Model: Impact of Salinity on Crop Yield and Output

Crop salt tolerance can be measured on the basis of two parameters: (a) the threshold salinity that is expected to cause the initial significant reduction in the maximum expected yield and (b) the percentage of yield expected to be reduced for each unit of added salinity above the threshold value (i.e. slope) (Shannon and Grieve 1998). Using these parameters, the first step in economic analysis was to estimate the crop relative yields based on the following model (Maas and Hoffman 1977; Tanji and Kielen 2002):

$$Yr = 100 - b(ECe - a) \tag{6.1}$$

where *Yr* is the relative crop yield relative to the potential (under no salinity); *a* is the crop salinity threshold in dS/m; *b* is the slope expressed in percent per dS/m; and *ECe* is the predicted (or measured) salinity level (dS/m) of the soil. Values for *a* and *b* for each crop are traditionally based on FAO salt tolerance data which cover a comprehensive list of crops, albeit rather dated and were based on experiments mainly conducted in non-temperate environments (Maas and Hoffman 1977; Tanji and Kielen 2002). However, in our analysis we used an updated set of parameters provided in de Vos et al. (2016) which were derived from experiments in Europe (Salt Farm Texel), albeit covering a limited range of crops. Finally in our analysis, values for *ECe* were based on soil salinity scenarios discussed in the previous section (Table 6.4).

To assess impacts to yields and crop tonnage, reference data for yield per hectare were obtained from EUROSTAT for the year 2019 (https://ec.europa.eu/eurostat). The total tonnage lost of each crop in each year was calculated using the following formula:

$$LY_x = (\mathrm{h} \times Y_{FM}) \times \left(\frac{100 - Yr_x}{100} \right) \tag{6.2}$$

where LY_x is the loss in yield (tons); h is the hectare coverage of each crop; Y_{FM} is the yield per hectare values for each crop; and Yr_x is the relative yield, based on the salinity and crop tolerance derived in equation 6.1. These were converted to financial losses using data for prices per ton of each crop obtained from EUROSTAT (https://ec.europa.eu/eurostat). Crops were chosen based on a review of economic importance of various crops in Europe, information on the most commonly grown crops in the North Sea Region of Europe and information from the survey of SalFar project partners. A refined list of crops for analysis included potato[1], barley, sugar beet, wheat, maize, ryegrass, carrot, onion, lettuce and cabbage.

Finally, the farm-level impacts (yield and financial losses) were scaled up to a wider (regional) level, where data were available. This depended on the availability of reliable data on the extent and severity of salinization (or areas at risk of salinization) as well as detailed data on crop composition and distribution.

6.4 RESULTS AND DISCUSSION: ECONOMIC IMPACTS OF SALINIZATION

Economic impacts of salinity can be assessed at different scales or levels: farm, regional and economy-wide scales. We begin with farm-level impacts by estimating relative (and absolute) yield and financial losses of specific crops under different salinization processes and salinity scenarios.

The analysis will show the potential economic impact of different salinization processes on crop yields. This can inform an assessment of crops that would be more affected by soil salinity and the countries that would undergo larger financial losses depending on the economic importance of the crops grown. We then extrapolate the impacts to the regional level (i.e. beyond the farm level), illustrated with case studies across the North Sea Region.

6.4.1 Farm-Level Economic Impacts of Salinization

6.4.1.1 Impacts of Irrigation Salinization

To assess the impact of irrigation salinization, we estimated the relative yields of key crops under a range of salinity levels and crop salt tolerance parameters

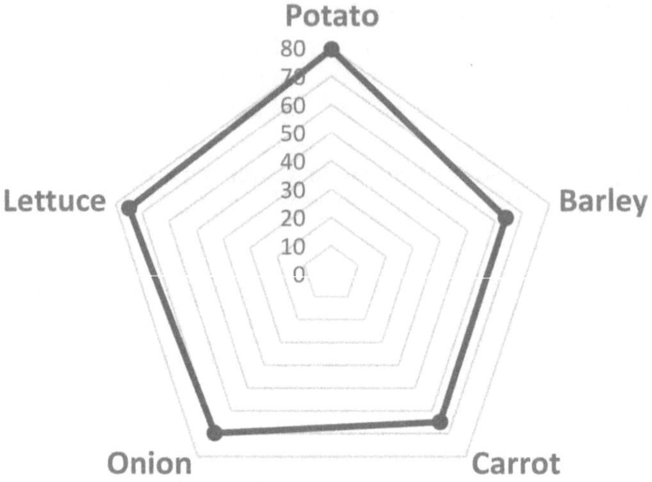

FIGURE 6.2 Relative yield of key crops under irrigation salinity.

given in de Vos et al. (2016). To represent the range of salinity impacts, we present the results of saline irrigation water of EC_w 4 dS/m. Relative yields range from 64% (barley) to 80% (potatoes), indicating potatoes are comparatively more salt-tolerant and barley is the least salt-tolerant (Figure 6.2).

In relation to yield and financial losses, we used the salinity effects on potato and barley as an example and compared yield and financial penalties across the North Sea Region countries (Figure 6.3). For instance, if potato was irrigated with EC_w 4 dS/m,

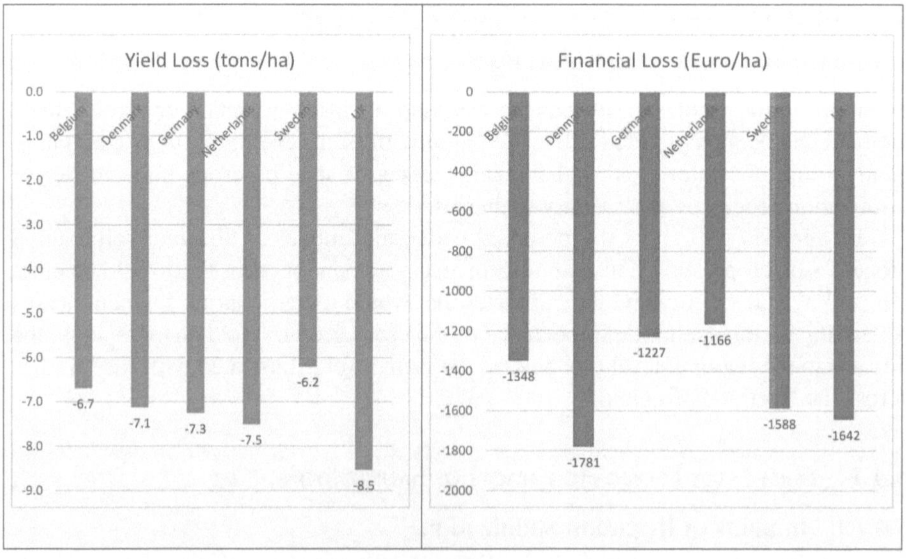

FIGURE 6.3 Irrigation salinization: Yield and financial losses of potato.

TABLE 6.5
Irrigation Salinization: Yield and Financial Losses of Barley across Countries

Country	Yield Loss (tons/ha)	Financial Loss (€/ha)
Belgium	2.8	391.67
Denmark	1.6	230.72
Germany	2.1	416.00
Netherlands	2.5	483.60
Sweden	1.1	141.00
United Kingdom	2.0	240.55
Norway	1.4	-

the yield losses ranged from 6.2 tons/ha (Sweden) to 8.3 tons/ha (the United Kingdom). We then converted these yield losses into financial penalties using crop price data from EUROSTAT (https://ec.europa.eu/eurostat).

The results ranged from €1478 to €2259; Denmark incurred the highest financial loss followed by the UK, while the Netherlands would be the least financially affected but would incur the second largest yield loss per ha after the UK. For Norway, prices were not available in EUROSTAT, hence we could only say that it would incur the least potential yield losses per ha. Similarly, comparing the financial losses under the other three irrigation levels across the countries, Denmark followed by the UK were the most affected by potato yield losses.

Estimating the impact of irrigation salinity (EC 4 dS/m) on barley showed yield penalties ranging from 1.1 tons/ha (Sweden) to 2.8 tons/ha (Belgium) with financial losses ranging from €141/ha (Sweden) to €483.60/ha (Netherlands). The results are summarized in Table 6.5. Belgium followed by the Netherlands, would undergo the highest yield losses among the countries, while the largest financial losses would occur in the Netherlands.

6.4.1.2 Impacts of Seepage Salinization

To assess the impact of seepage salinization, we used salinity (chloride concentration of 215 mg/L) scenarios of groundwater, assuming that groundwater reaches the root zone of the crops. However, the results show that all salinity scenarios would have no significant impact on the yield of all the crops investigated as shown in Table 6.6[2]. Further investigation using FAO salinity tolerance data shows that the only crops that would be affected are carrot and onion. For this type of salinization, we were not able to estimate potential yield and financial losses for each country for carrot and onion because EUROSTAT does not provide data for the prices and yields of vegetables.

6.4.1.3 Impacts of Flood Salinization

In the case of flood salinization, we estimated the relative yields and potential yield losses assuming a flooding event. Following Gould et al. (2020), we assumed the complete loss of the standing crop during the flood followed by a sliding recovery approach during the following years, where the rate of recovery was a function of the salt tolerance per crop type based on predicted salt soil levels.

TABLE 6.6
Seepage Salinization: Relative Yields for All Crops

Crops	Relative Yield (%) (Based on Texel Salt Farm Salinity Tolerance Parameters (de Vos et al. 2016))	Relative Yield (%) (Based on FAO Salinity Tolerance Parameters (Tanji and Kielen 2002))
Potato	100	100
Barley	100	100
Sugar beet	*	100
Wheat	*	100
Maize	*	100
Ryegrass	*	100
Carrot	100	95.83
Onion	100	98.44
Lettuce	100	100
Cabbage	100	100

* Texel Salt Farm salinity tolerance parameters (de Vos et al. 2016) were unavailable for these crops.

Hence, in the first year after the flood, we assumed zero yields while in the second recovery year we assumed soil salinity with an EC_e of 7.1 dS/m, a typical post-flood salinity level recorded in previous saline flooding research in the UK North Sea coastal systems (Hazelden and Boorman 2001; Gould et al. 2020). Taking as an example potato yields grown in the second recovery year after a potential flood, we compared the results across the North Sea Region countries. As shown in Figure 6.4, yield losses for potato ranged from 7.86 tons/ha (Sweden) to 10.81 tons/ha (UK) while financial losses ranged from €1,478/ha (Netherlands) to €2,259/ha (Denmark). Similar to the case of irrigation salinization, results showed that Denmark would incur the largest financial losses if potato was grown in a field 2 years after a flood event and the UK would incur the highest yield losses per ha.

Results for barley (Table 6.7), showed that Belgium would incur the highest yield losses per ha, losing 460.05 €/ton and the Netherlands would lose 608.4 €/t. Comparing potato and barley financial losses per ton, it is apparent that countries or/ and regions where potato is the principal crop would undergo more severe financial losses in a case of flooding than areas which primarily grow barley.

6.4.2 Regional Economic Impact of Salinization

In this section, we scale up salinity impacts to the wider (regional) level, focusing on selected case study areas in the North Sea Region of Europe, where data were available. We present three case studies on regional economic impact of the main types of salinization: (a) irrigation salinization–Netherlands (Groningen) (b) seepage salinization–Belgium (Oudlandpolder) and (c) flood salinization–UK (Lincolnshire).

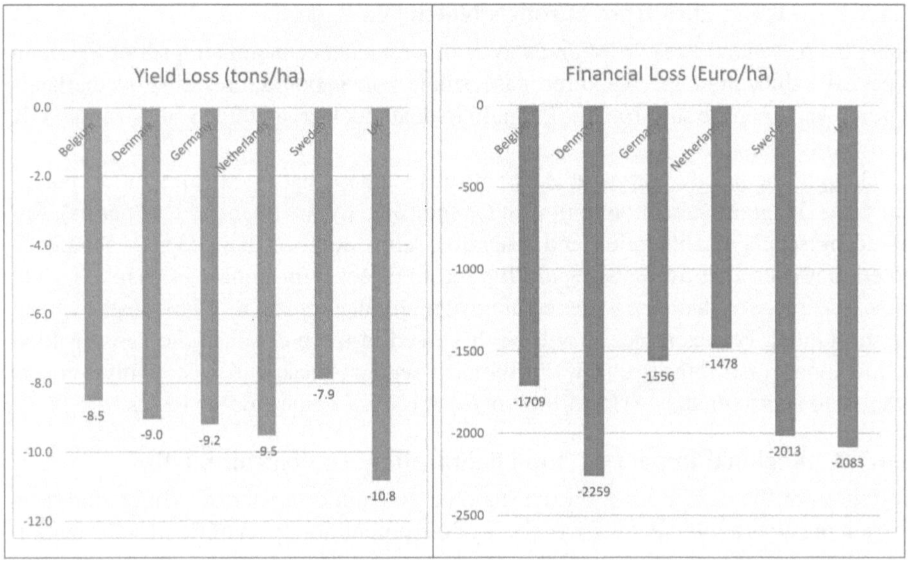

FIGURE 6.4 Yield and financial losses for potato under flood salinization across all countries.

Although the potential for salinization is geographically extensive, and not localized to any one region along the North Sea coastline, we focused on the three case studies in our empirical analysis partly because of limited availability of data on salinity risks in the region and because of anecdotal evidence of significant risks of salinization in these case study areas. For example, coastal flooding risks (and associated salinity risks) are significant within Greater Lincolnshire (UK), a low-lying, highly productive agricultural land with a history of flooding, including as recently as the year 2013. It is in this vein that the case of regional economic impact of flood salinization is based on a recent study conducted by Gould et al. (2020) on the impact of coastal flooding on agriculture in Lincolnshire, UK.

TABLE 6.7
Yield and Financial Losses for Barley under Flood Salinization across All Countries

Country	Yield Loss (tons/ha)	Financial Loss (€/ton)
Belgium	3.48	460.06
Denmark	1.96	289.80
Germany	2.61	522.00
The Netherlands	3.12	608.40
Sweden	1.37	177.27
United Kingdom	2.55	302.17
Norway	1.77	-

6.4.2.1 Case Studies from Europe's North Sea Region

Table 6.8 presents the results of an analysis of a regional economic impact of the main types of salinization for the three case studies: irrigation salinization–Netherlands (Groningen), seepage salinization–Belgium (Oudlandpolder) and flood salinization–UK (Lincolnshire).

Though not strictly comparable, the results suggest that flood salinization potentially has the greatest economic impact (as indicated by the financial loss per ha), followed by seepage salinization and irrigation salinization in that order. It should be noted, however, that these losses are limited to direct farm impact in terms of yield losses, i.e. these exclude the wider economy "multiplier" or supply chain costs that can be substantial. For example, as will be discussed in more detail in the case of flood salinization in Lincolnshire (UK) in the next section, these wider economy impacts amount to approximately €115 million in GVA (Gross Value Added) losses.

6.4.2.2 Regional Impact of Flood Salinization: Lincolnshire, UK

To represent the case studies on the regional economic impact of salinization, this section presents a more detailed analysis of the potential impact of flood salinization

TABLE 6.8
Regional Economic Impact of Salinization: North Sea Region Case Studies

Salinization Process	Case Study	Area at Risk (ha)	Methods	Estimated Loss in Yield (tons)	Estimated Financial Loss (Euro)	Financial Loss per ha (Euro/ha)
Irrigation salinization	Holland (Groningen)	17,526	GIS mapping of affected areas and analysis of regionally representative cropping composition and distribution. Groundwater salinity data (Cl) provided by the Province of Groningen	147,992	34,947,861	1,994.06
Seepage salinization	Belgium (Oudlandpolder)	11,938	Mapping of affected areas and analysis of regionally representative cropping composition and distribution. Groundwater salinity data (EC) provided by Belgium	147,663	27,381,670	2,293.66
Flood salinization	UK (Lincolnshire)	108,238	Climate (flood modelling) and salinization impact mapping based on GIS and satellite data analysis of cropping composition.	2,022,385	279,548,899	2,582.72

in Lincolnshire, UK, a region where coastal flooding presents a significant risk to agriculture. Again, the results presented here are based on recent research on the impact of coastal flooding on agriculture in Lincolnshire, UK conducted by Gould et al. (2020). In the study, economic and yield losses were estimated based on a combination of predicted flood models, typical crop composition using satellite remote sensing data and soil type/drainage potential of a flood event for a given coastal region. In particular, the study defined three flood scenarios reflecting: (i) current breach risk, (ii) future breach risk and (iii) a "big" flood event (see Gould et al. 2020 for details).

The primary focus of this chapter, however, is on the current breach risk. For all breaches, we assumed that the post-breach regime was to repair the breach and continue the existing defence strategy. To assess current areas exposed to sea bank breach risk, we used breach scenarios obtained from the UK Environment Agency. These flood scenarios are used to inform the UK flood defence strategy. They modelled the ingress of flood water for a 1 in 200 years breach (72 hours duration) of sea defences under 2006 climate conditions. These are the most recent breach scenarios data released by the Environment Agency, and as such we describe these as "current." We used breach scenarios from 67 individual locations spanning a 105 km stretch of the Lincolnshire coastline (Figure 6.5). To account for localized differences in tidal behaviour, we grouped these 67 model scenarios into four Coastal Zones (CZs) as shown in Figure 6.5. Using the Land Cover Plus data, average crop composition per breach area

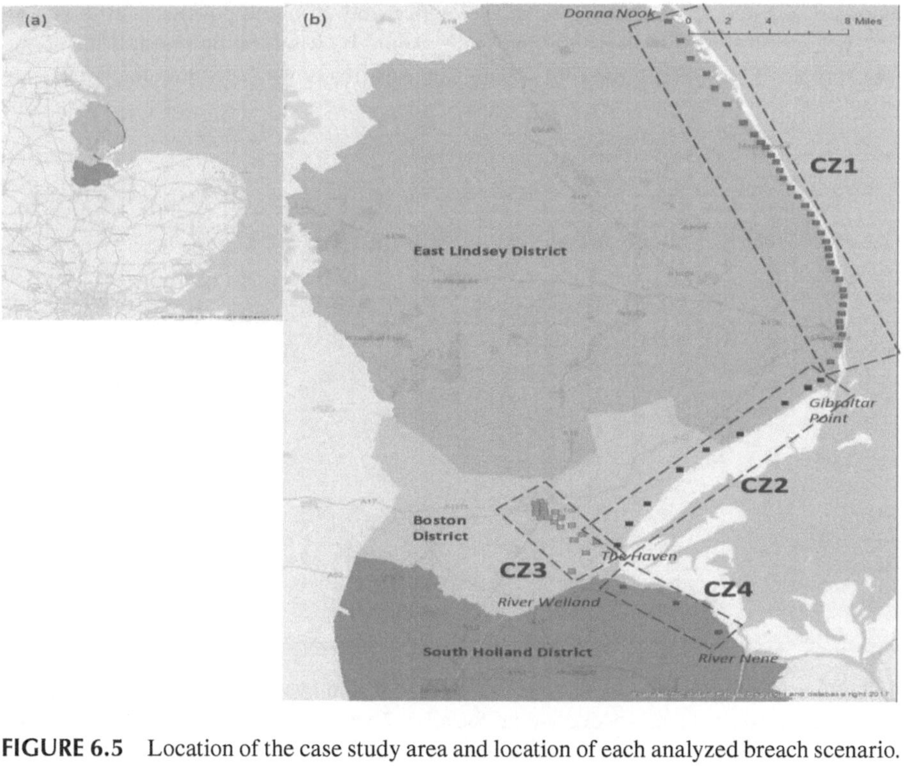

FIGURE 6.5 Location of the case study area and location of each analyzed breach scenario.

was calculated for each of the four CZs, giving a typical breach crop composition for each stretch of coastline.

To assess total yield loss (current and future years) as the soil recovers, we firstly calculated the response of different crop types (relative yields) to salt-affected land. In this chapter, we did this by predicting salt soil levels in recovery years. However, for farm-scale assessments, this method could be adapted based on known or historic salt levels. We assumed the complete loss of the standing crop during the flood (zero yield in flood year) followed by a sliding recovery approach during the following years, where the rate of recovery was defined as a function of the salt tolerance per crop type based on predicted salt soil levels. Thus, the modelling approach captured the fact that highly tolerant crops would recover yield on inundated fields at a faster rate than sensitive crops.

To assess impact, reference data for yield per hectare were obtained from the John Nix Farm Management Pocketbook (Redman 2016), an information source for financial assessments of UK farmland. These were readily converted into output losses in monetary terms using crop price data obtained from EUROSTAT.

Figure 6.6 diagrammatically shows the yield and financial losses, aggregated across all the coastal zones over the full soil recovery time for all 1–7 years salt recovery time scenarios (1–7 years). Total yield losses over the recovery period were estimated to be up to 418,866 tons while the output losses per ha averaged £5,636 over the recovery period.

To investigate heterogeneity in yield and output losses across CZs, we turn to disaggregated analysis of impacts. Figure 6.7 displays the total yield losses (tons) across CZs. The results reveal a spatial heterogeneity in yield recoveries and hence yield losses across regions (CZs) due to differences in salt tolerance and crop composition across zones. CZs, where salt sensitive crops are dominant, would be worst hit by flood salinization. For example, it was found that CZs, where salt sensitive crops are dominant, suffered a

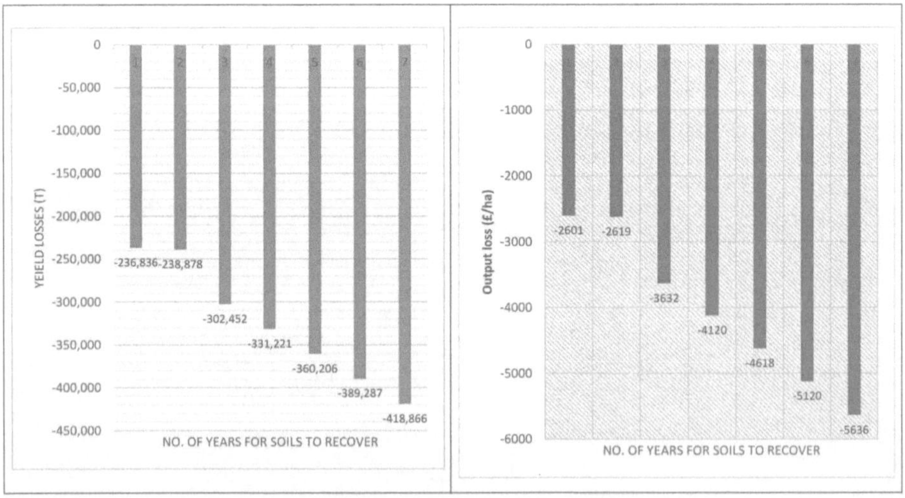

FIGURE 6.6 Regional economic impact of flood salinization: total yield and output losses per ha over full soil recovery time for all 1–7 years salt recovery time scenarios in Lincolnshire, UK.

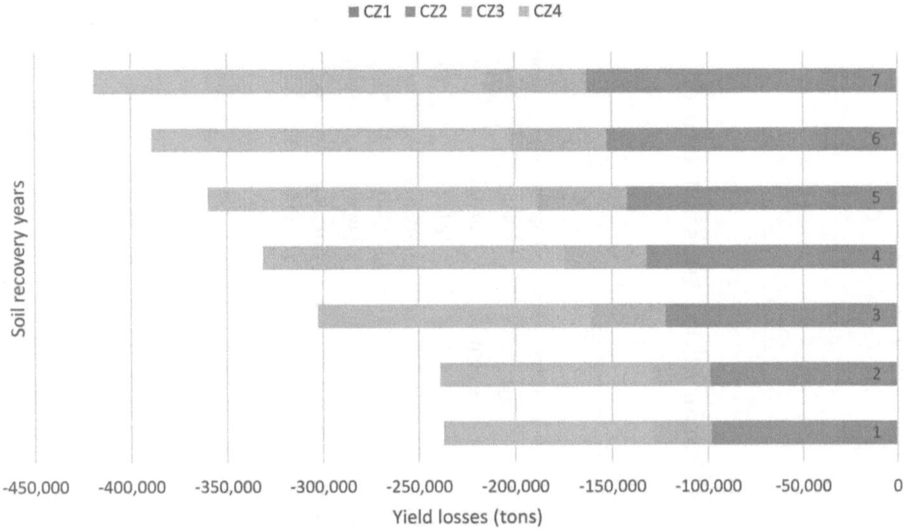

FIGURE 6.7 Regional economic impact of flood salinization; total yield losses across coastal zones in Lincolnshire, UK.

88% yield loss compared to a 27% yield loss in more "tolerant" CZs. This implies greater potential for salt-tolerant crops in these areas, particularly in early recovery phases as a remediation or adaptation option for salt degraded land.

Table 6.9 reports the total yield losses, output losses and output losses per ha over the full soil recovery time for all 1–7 years salt recovery time scenarios (1–7 years) for each coastal zone. This is based on the average breach crop composition in each coastal zone (CZ1–CZ4). The results show that in the first (flood) year alone, a single breach could deprive farms of a total yield of 31,778 tons in CZ1, 66,051 tons in CZ2, 30,671 tons in CZ3 and 108,336 tons in CZ4. When yield losses were converted into potential losses in monetary terms, this translated to £2,684,625 per breach in CZ1, £9,608,181 in CZ2, £4,183,383 in CZ3 and £15,264,116 in CZ4.

The results in Table 6.9 further show a non-linear yield recovery (i.e. differences in yield and output losses between years are not uniform) which may be related to the salt tolerance of the typical crop composition. Within 2–3 years, beet, wheat, grass and barley will return to 100% yields, whilst yield losses will remain in potatoes and vegetables for longer. As such, in the earlier recovery years (e.g. years 2–3) of CZs dominated with more salt-tolerant crops, gains in yield recovery may appear to be more rapid than in later years. This is true for CZ1, where the greatest yield losses were for more salt-tolerant crops, whereas in the other three zones, the greatest losses were for more salt sensitive crops.

The more salt sensitive crops typical of our study region tend to have higher commercial value. Such crops suffer more damage and have greater financial loss, exacerbating the financial flood impact. When total output losses were converted to pounds sterling per hectare of agricultural land flooded (over the entire recovery duration), the highest values were found in CZ2 (£3,257 to £7,510 per ha), followed by CZ4 (£2,912 to £6,533 per ha), then CZ3 (£2,867 to £6,380 per ha), with CZ1 having the lowest (£1,368 to £2,119).

TABLE 6.9
Total Yield Losses, Output Losses and Output Losses per ha over Full Soil Recovery Time for All 1–7 Years Salt Recovery Time Scenarios (1–7 years)

			No. of Years for Soils to Recover					
		Flood Year	2	3	4	5	6	7
Yield Losses (t)	CZ1	31,778	31,825	36,095	36,959	37,863	38,985	40,225
	CZ2	66,051	66,659	85,991	95,112	104,271	113,442	122,702
	CZ3	30,671	30,879	38,691	42,109	45,550	49,041	52,589
	CZ4	108,336	109,515	141,675	157,041	172,522	187,819	203,350
Output Losses(£)	CZ1	2,684,625	2,689,932	3,236,549	3,458,680	3,690,058	3,916,679	4,158,767
	CZ2	9,608,181	9,680,736	13,823,189	15,853,098	17,921,588	20,013,414	22,150,743
	CZ3	4,183,383	4,209,929	5,919,917	6,746,154	7,589,012	8,439,952	9,311,382
	CZ4	15,264,116	15,409,027	21,675,166	24,737,134	27,865,800	31,013,861	34,246,798
Output Losses per ha (£/ha) over full recovery duration	CZ1	1,368	1,371	1,650	1,763	1,881	1,996	2,120
	CZ2	3,257	3,282	4,687	5,375	6,076	6,785	7,510
	CZ3	2,867	2,885	4,057	4,623	5,200	5,783	6,380
	CZ4	2,912	2,940	4,135	4,719	5,316	5,917	6,533

This suggests CZ1, where grazing is more commonplace and there is less vegetable and potato production, is a more resilient coastal zone to the long term impacts of flooding.

Finally, we turn to the impacts of coastal flood salinization on the wider agri-food economy, drawing from the economic modelling results in Gould et al. (2020). It is acknowledged that biophysical impact of flood salinization is not limited to farmland (crop yields) but will have cascading negative consequences both backward (e.g. fertilizer, machinery suppliers) and forward (e.g. processing, distribution) along the supply chain. Based on the outputs of the model, Table 6.10 reports the results of a broader assessment of the impacts of a coastal flood salinization to the wider agri-food economy based on the flood year data alone.

The results suggest significant economic losses; total job losses and GVA across CZs is, respectively, approximately 944 and £69 million. Figure 6.8 summarizes the disaggregated impacts by sector, displaying total impacts across CZs. This figure shows that the greatest comparative losses are borne by food processing (£42 Million) followed by direct farm impacts in terms of loss in total Gross Margins (GM). These sectors similarly suffer higher losses in jobs; food processing jobs and direct farm losses amount to 348 and 407 respectively.

These costs are expected to be even higher when other cost components are added, e.g. environmental costs associated with salt-affected lands and the potential social cost of impaired farm businesses. Saline agriculture, as an adaptation strategy, has the potential to ameliorate these impacts. Future studies could assess the magnitude of the benefits afforded by saline agriculture adaptation. For example, increasing drought combined

TABLE 6.10
Wider Economy Impacts of Flood Salinization in Lincolnshire, UK: Jobs and Costs to Gross Margins (GM) or Gross Value Added (GVA) throughout the Food Value Chain

	At Risk	CZ1	CZ2	CZ3	CZ4	Total
Direct Farm	Jobs	45	111	49	202	*407*
Impacts	GM	£ 1,341,985	£ 3,339,480	£ 1,482,514	£ 6,058,340	*£12,222,319*
Impact on	Jobs	5	23	10	34	*72*
Suppliers	GVA	£ 287,134	£ 1,340,610	£ 577,601	£ 1,968,726	*£4,174,071*
Food	Jobs	38	95	42	173	*348*
Processing	GVA	£ 4,615,120	£11,484,552	£ 5,098,403	£ 20,834,779	*£42,032,854*
Food	Jobs	10	24	11	44	*89*
Marketing	GVA	£ 859,198	£ 2,138,082	£ 949,171	£ 3,878,815	*£7,825,266*
Food	Jobs	3	7	3	13	*26*
Logistics	GVA	£ 256,614	£ 638,574	£ 283,486	£ 1,158,473	*£2,337,147.00*
Total	Jobs	101	261	116	466	*944*
	Jobs per ha	0.07	0.09	0.08	0.09	*0.08*
	Direct Losses	£ 7,360,050	£ 18,941,297	£ 8,391,175	£ 33,899,133	*£68,591,655*
Multipliers	*Jobs*	*145*	*376*	*167*	*671*	*1359*
	GVA	*£10,598,472*	*£ 27,275,468*	*£ 12,083,292*	*£ 48,814,752*	*£98,771,984*

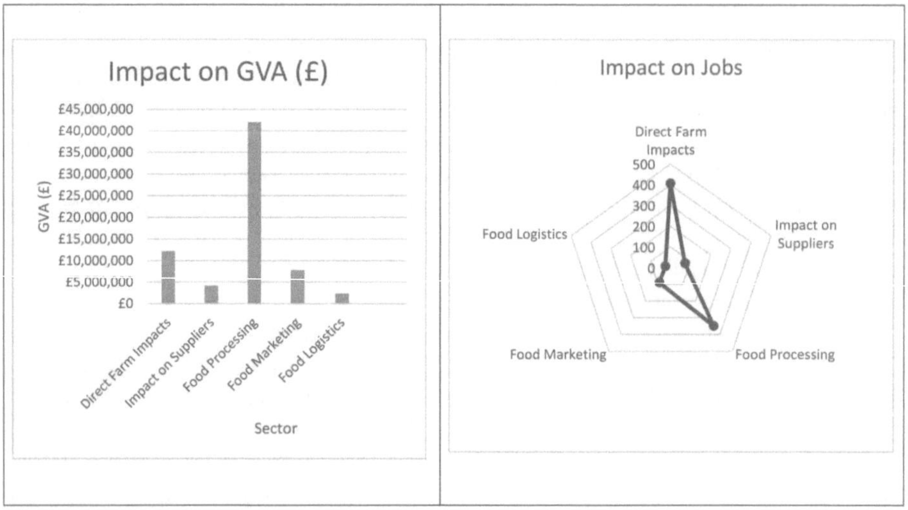

FIGURE 6.8 Wider economy impacts of flood salinization in Lincolnshire, UK.

with projected sea-level rises will lead to more sustained threats from salinization and create a sustained, long-term opportunity for salt-tolerant crop varieties.

6.5 CONCLUSION

This chapter provides an economic framework for risk assessment in regions where salinity poses a significant threat to agricultural production and the local/national economy. The chapter first reviewed the key literature on economic impacts of salinization and presented a conceptual methodological framework that could be applied to assessing such impacts, focusing on three typologies of salinization: irrigation salinization, seepage salinization and flood salinization. We conceptualized impact at different scales; farm-level, regional and wider economy scales. We then applied the framework, first to estimate crop yield and financial losses due to each salinity process. Subsequently, we scaled up the impact to regional or wider levels using data on affected areas and information on crop composition and distribution, where available. The analysis shows that there is significant economic impact of salinization.

Further, we find that the magnitude of the impact of salinization critically depend on a range of factors which include; the type of salinization process, the degree/severity of salinity, the types (and value) of crops grown, farm-level decisions/choices such as the use of salt-tolerant crops and other adaptation mechanisms as well as external shocks such as sea-level rise due to climate change. These factors may also be linked to spatial differences. For example, in a flood salinization case study in Lincolnshire, we found marked differences in flood resilience and the concomitant economic impact of salinity across CZs. The case study empirical results should provide is a "baseline" for economic costs of salinization that may inform future assessment of the potential of adaptation measures such as saline farming.

Although it is widely acknowledged that salinization poses a significant problem to agriculture and the European economy now and in the future under climate change, there are limited data available on the extent and severity of salinization. This hinders accurate assessments of the biophysical and economic impacts of salinity and the potential for saline farming. Information on the economic risks and costs of salinization would be important inputs into priority setting and the formulation of policies aimed at building resilience to salinization in agricultural systems, including development of saline agriculture. There is an urgent need, therefore, to strengthen systems and mechanisms for monitoring soil salinity and associated risks.

ENDNOTES

1. There is a nuanced distinction between seed potato and potato for consumption. In this paper, "potato" refers to potato for consumption.
2. These results are limited to the impact of seepage salinity scenarios investigated in the study and do not suggest that seepage salinity is not a problem in the case study areas. For example, there is anecdotal evidence of significant crop losses due to seepage salinity in some regions of the Netherlands, particularly during the dry summers.

REFERENCES

Aslam, M., and Prathapar, S. A. 2006. Strategies to Mitigate Secondary Salinization in the Indus Basin of Pakistan: A Selective Review. Research Report 97. IWMI, Colombo, Sri Lanka.

Ayers, R. S., and Westcot, D. W. 1985. Water quality for agriculture. FAO Irrigation and Drainage Paper No. 29, Rome.

Bosello, F., Nicholls, R. J., Richards, J., Roson, R., and Tol, R. S. 2012. Economic impacts of climate change in Europe: Sea-level rise. *Climatic Change* 112: 63–81.

Daliakopoulos, I., Tsanis, I., Koutroulis, A., Kourgialas, N., Varouchakis, A., Karatzas, G., and Ritsema, C. 2016. The threat of soil salinity: A European scale review. *Science of the Total Environment* 573: 727–739.

de Vos, A., Bruning, B., Van Straten, G., Oosterbaan, R., Rozema, J., and Van Bodegom, P. 2016. Crop salt tolerance under controlled field conditions in The Netherlands, based on trials conducted at Salt Farm Texel. Salt Farm Texel.

De Waegemaeker, J. 2019. SalFar framework on salinization processes. A comparison of salinization processes across the North Sea Region. A report by ILVO for the Interreg VB North Sea Region project Saline Farming (SalFar).

FAO. 2011. *The State of the World's Land and Water Resources for Food and Agriculture (SOLAW) – Managing Systems at Risk.* Food and Agriculture Organization of the United Nations, Rome and Earthscan, London.

FAO and ITPS. 2015. *Status of the World's Soil Resources (SWSR) – Main Report.* Food and Agriculture Organization of the United Nations and Intergovernmental Technical Panel on Soils, Rome, Italy.

Ghassemi, F., Jakeman, A. J., and Nix, H. A. 1995. *Salinisation of Land and Water Resources: Human Causes, Extent, Management and Case Studies.* CAB International.

Gould, I. J., Wright, I., Collison, M., Ruto, E., Bosworth, G., and Pearson, S. 2020. The impact of coastal flooding on agriculture: A case study of Lincolnshire, United Kingdom. *Land Degradation & Development* 31(2): 1545–1559.

Grattan, S. 2002. Irrigation Water Salinity and Crop Production, Publications by University of California, Division of Agriculture and Natural Resources.

Hazelden, J., and Boorman, L. A. 2001. Soils and 'managed retreat' in south east England. *Soil Use and Management* 17: 150–154.

Janmaat, J. 2004. Calculating the cost of irrigation induced soil salinisation in the Tungabhadra project. *Agricultural Economics* 31(1): 81–96.

John, M., Pannell, D., and Kingwell, R. 2005. Climate change and the economics of farm management in the face of land degradation: Dryland salinity in Western Australia. *Canadian Journal of Agricultural Economics* 53(4): 443–459.

Koutroulis, A. G., Tsanis, I. K., Daliakopoulos, I. N., and Jacob, D. 2013. Impact of climate change on water resources status: A case study for Crete Island, Greece. *Journal of Hydrology* 479: 146–158.

Lipper, L., Thornton, P., Campbell, B. M., Baedeker, T., Braimoh, A., Bwalya, M., and Torquebiau, E. F. 2014. Climate - smart agriculture for food security. *Nature Climate Change* 4: 1068–1072.

Maas, E. V., and Hoffman, G. J. 1977. Crop salt tolerance - current assessment. *Journal of Irrigation and Drainage Divison* 103: 115–134.

Marshall, G.R. 2004. From words to deeds: enforcing farmers' conservation cost–sharing commitments. *Journal of Rural Studies* 20(2): 157–167.

Marshall, G. R., and Jones, R. E. 1997. Significance of supply response for estimating agricultural costs of soil salinity. *Agricultural Systems* 53(2–3): 231–252.

McCann, L. M. J., and Hafdahl, A.R. 2007. Agency perceptions of alternative salinity policies: The role of fairness. *Land Economics* 83(3): 331–352.

Montanarella, L. 2007. Trends in land degradation in Europe. In Climate and Land Degradation, ed. Sivakumar, V. K., and Ndiang'ui, N., 83–104. Springer, Berlin.

Negacz, K. 2018. Potential of Salt-tolerant Crop Production on Salinized, Degraded Lands. Masters dissertation, Vrije Universiteit Amsterdam, Netherlands.

Oldeman, L. R., Hakkeling, R. T. A., and Sombroek, W. G. 1991. World Map on Status of Human - Induced Soil Degradation. UNEP/International Soil Reference and Information Centre (ISRIC), Nairobi, Kenya.

Qadir, M., Quillérou, E., Nangia, V., et al. 2014. Economics of salt-induced land degradation and restoration. *Natural Resources Forum* 38(4): 282–295.

Redman, G. 2016. John Nix Farm Management Pocketbook (47th edition), Agro Business Consultants, Melton Mowbray.

Richards, J.A., and R. J. Nicholls. 2009. Impacts of climate change in coastal systems in Europe. PESETA-Coastal Systems study. EUR - Scientific and Technical Research Reports, Publications Office of the European Union.

Shannon, M., and Grieve, C. 1998. Tolerance of vegetable crops to salinity. *Scientia Horticulturae* 78: 5–38.

Tanji, K., and Kielen, N. 2002. Agricultural drainage water management in arid and semi-arid areas. FAO Irrigation and Drainage Paper 61. FAO, Italy.

UN. 2015. Transforming our world: the 2030 Agenda for Sustainable Development: https://www.refworld.org/docid/57b6e3e44.html [accessed 10 September, 2020].

van Straten, G., de Vos, A., Rozema, J., Bruning, B., and van Bodegom, P. 2019. An improved methodology to evaluate crop salt tolerance from field trials. *Agricultural Water Management* 213: 375–387.

Van Orshoven, J., Terres, J., and Tóth, T. 2012. Updated Common Bio-physical Criteria to Define Natural Constraints for Agriculture in Europe. EUR - Scientific and Technical Research series. Publications Office of the European Union, Luxembourg.

Winpenny, J., Heinz, I., and Koo-Oshima, S. 2010. *The wealth of waste: The economics of wastewater use in agriculture*. Food and Agriculture Organization of the United Nations, Rome, Italy.

Zekri, S., and Albisu, L. 1993. Economic impact of soil salinity in agriculture. A case study of Bardenas area, Spain. *Agricultural Systems* 41(3): 369–386.

7 Cost or Benefit? *Estimating the Global Economic Potential of Saline Agriculture*

Katarzyna Negacz and Pier Vellinga

CONTENTS

7.1 INTRODUCTION

Although there is no agreement on a single definition of degraded land, researchers consider it to be land that has lost some degree of its natural productivity due to human-caused processes (WRI 2018). Globally, salinity is regarded as one of the most widespread causes of soil degradation. One of the promising adaptation measures to this pressure is saline agriculture, understood to be "profitable and improved agricultural practices using saline land and saline irrigation water with the purpose to achieve better production through the sustainable and integrated use of genetic resources (plants, animals, fish, insects and microorganisms) avoiding expensive soil

DOI: 10.1201/9781003112327-7

recovery measures" (Aslam et al. 2009; Ladeiro 2012). Our previous study (Negacz et al. 2019) has shown that there are 420 Mha of saline soils in the world, of which 16 Mha have at least 500 mm of water available annually (including rainfall and irrigation) that could be potentially used for agriculture. However, there is a lack of a comprehensive overview of the economic potential of saline degraded lands. To address this challenge, this study provides one of the first estimates of the economic potential of saline degraded lands for food production.

One of the first assessments of global costs of salinity was proposed by Ghassemi et al. (1995), who calculated that the global income loss due to salinity was about 11.4 billion USD per year in irrigated agriculture and 1.2 billion USD per year in non-irrigated areas. Further research focusing on the cost side of agricultural production was conducted by Qadir et al. (2014) who presented a comprehensive overview of costs of salinization. Other publications have often adopted a case study approach resulting in economic values for specific regions or sites (Kabir et al. 2018a,b,2017).

The main research question of this study is: what is the potential of saline degraded lands for food production, given the growing knowledge on salt-tolerant crops? We address it by estimating the economic values of saline degraded lands through the ecosystem services valuation methods (being food provision in our case). Ecosystem valuation helps in improved natural resource allocation by showing the full social costs and benefits of goods and services provided by ecosystems (Van Beukering et al. 2015). Providing an answer to our research question enables further exploration of the total value of saline agriculture on the global level and assessing its impact on food production.

7.2 RESEARCH METHODS

Food provision is an ecosystem service which allows applying direct market valuation methods. For this project, three valuation methods were applied to obtain the most comparable results.

7.2.1 META-ANALYSIS METHOD

Meta-analysis is a research method employing analysis of the data from independent primary studies referring to a chosen topic (Koetse et al. 2015). As a first step of the meta-analysis, we conducted an in-depth literature review to identify the data sources for further investigation. This revealed that the literature addressing salinization potential could be categorized into three groups: (1) evaluating geophysical properties and processes including climate, (2) studies looking at salinity effects on plants and (3) salinity management techniques based on case studies. Articles including valuation appeared rarely. Secondly, we identified studies which report the added value of saline agriculture, operationalized as additional income or profit, for the most salinized areas on each continent. Studies on aquaculture, saline pastures, biofuels, halophytes and greenhouse experiments were excluded as they address different markets. As a result, we selected six studies for further comparison. Third, the values reported in each study were converted to USD per hectare and adjusted for inflation with 2019 as the reference year. Finally, the average value was calculated based on the chosen studies. The average added value per hectare was multiplied by 16 Mha, a total surface area of saline degraded lands with an $EC_e \geq 4$ dS/m and with water availability over 500 mm annually (Negacz et al. 2019). Soil salinity is usually expressed as the electrical conductivity of the saturation extract of the soil (the EC_e) expressed in dS/m (UoC 2020).

7.2.2 Cost-Based Method

The cost-based valuation was derived from values obtained in the meta-analysis done by Qadir et al. (2014). This study involved 14 articles reporting costs of salinization. The reported costs of saline land degradation were mostly crop yield losses, in terms of biophysical output (e.g. t/ha) and/or in monetary terms (e.g. USD per hectare). The costs included in that analysis were opportunity costs, production losses, replacement costs, transaction costs, market prices for required services and mitigation costs (Qadir et al. 2014). Building on these results, we adjusted the average cost for inflation with 2019 as the reference year. Similarly, as for added value, the average cost per hectare was multiplied by 16 Mha, the total surface area of saline degraded lands with an EC_e above 4 dS/m and with water availability over 500 mm annually (Negacz et al. 2019).

7.2.3 Market Prices Method

The market price method uses the prices of goods on markets to determine the value of ecosystem services, which in this case is food provision through saline agriculture. This method focuses on the quantity (Q) and quality of goods, represented as a field composition in our case (w). We present the value as revenue (R) which equals the quantity of the production multiplied by the price (P): $R = Q \times P$. Since the costs of saline agriculture are rarely reported, the costs could not be calculated. The process was: first, to determine the quantity we selected six conventional vegetable crops (potatoes, carrot, onion, lettuce, cabbage and barley) of which varieties exist that show good yield potential at the salinity levels relevant to this study (de Vos et al. 2016). The yields of these crops were obtained from FAOSTAT (2018). Second, we established a composition on our hypothetical global field (share of the crops on 1 ha). Based on FAOSTAT, we assigned a global area of cultivation per crop. Further, we allocated the corresponding weights to the crops (w1 to w6). Third, we derived average producer prices per ton for each crop based on available data in FAOSTAT. Fourth, we calculated the total revenue (TR) according to the equation: $TR = (q_1 \times w_1 \times p_1) + \ldots + (q_6 \times w_6 \times p_6)$. The calculations are presented in Table 7.1. The total revenue was multiplied by 16 Mha, the total surface area of saline degraded lands above ECe 4 dS/m with water availability over 500 mm annually (Negacz et al. 2019). Based on the total revenue, we calculated the revenue with 90% and 50% yield resulting from the impact of salinity.

TABLE 7.1
Calculations for the Global Revenue Based on the Market Prices (in USD)

Crop	Yields (t/ha) (q)	Share in Global Area of Cultivation for Six Analyzed Crops (w)	Yield (t/ha) (q) Adjusted per Weight (w)	Prices (p) (USD/t)	Revenue per Crop (q × p)	Revenue Scaled (q × w × p)
Potatoes	21	0.23	4.89	423	8859	2066
Carrot	35	0.02	0.53	667	23573	354
Onion	19	0.07	1.28	492	9438	631
Lettuce	21	0.02	0.36	1399	30024	506
Cabbage	29	0.03	0.92	561	16128	516
Barley	3	0.64	1.88	283	834	531
		Total average yield per hectare	9.86		Total revenue per hectare	4604

7.3 RESULTS

7.3.1 Meta-Analysis

Our meta-analysis revealed that economic analysis of saline agriculture is scarce. While studies often report changes in yields, they rarely refer to economic categories, such as cost, revenues and profits. Table 7.2 presents six studies from four continents which allow the calculation of an average added value (income or profit) per hectare. They range from 209 to 654 USD, with the Australian case being an outlier; in this example, a calculation was made of the benefits of saline agriculture to a hypothetical farm that also grew other crops and pastures. Across all cases, the average added value per hectare was 383 USD.

TABLE 7.2
Meta-analysis of the Added Value of Saline Agriculture

Author	Year	Country	Area Studied (hectare)	Calculated Added Value per Hectare in 2019 (USD)
Lefkoff and Gorelick (1990)	1990	USA	820	396
Wang et al. (2013)	2013	China	367	570
De Vos et al. (2021)	2021	Bangladesh	n.a.	209
Vyshpolsky et al. (2008)	2008	Kazakhstan	3	654
Ali et al. (2001)	2001	Egypt	400	227
Khan et al. (2003)	2003	Australia	306	239
Average				383

The average value added per hectare multiplied by the global area of saline soils above ECe 4 dS/m with water availability of 500 mm per year (16 Mha) gave a total of 6 billion USD of potential value-added per year.

7.3.2 Cost-Based Valuation

A recent comprehensive meta-analysis on the costs of salinization has been conducted by Qadir et al. (2014). These authors estimated an inflation-adjusted cost of salt-induced land degradation in 2013 as 441 per hectare. This estimate adjusted for inflation in 2019 equals 497 USD per hectare. For a total area of 16 Mha saline soils, this equals 8 billion USD of costs for salt-induced land degradation. The analysis of costs shows that preventing salinization would result in considerable savings, coming from avoidance of yield loss, mitigation and opportunity costs understood as the value of the trade-off when a decision is made. It may be also considered as an extra income that would remain if the soils were not saline.

7.3.3 Market Prices

Based on crop yields from FAO sources and current market prices we estimated revenue for 16 Mha. The total revenue from this land assuming 100% yield would be 74 billion USD per year. Decreasing yields by 50% or 90% would decrease revenues by 37 or 66 billion USD per year, respectively. These results suggest a large revenue potential, which needs to be contrasted with costs of implementing saline agriculture on items such as seeds and fertilizers. However, no information on these was identified.

In summary, the global economic potential of saline agriculture can be presented as three values shown in Table 7.3.

TABLE 7.3
Comparison of Economic Valuation of Saline Agriculture

Method	Value for 16 Mha in billion USD
Meta-analysis value (added value)	6
Cost-based value (total costs)	8
Market price (revenue)	37–66

These aggregate numbers obtained through different methods show that saline agriculture can be a source of benefits, not only expenses, which we discuss in Section 7.4.

7.4 DISCUSSION

This discussion focuses on three general themes, the threshold of added value per hectare, costs included in the cost-based valuation and several challenges in the market price analysis.

7.4.1 Inconsistent Agreement

A lack of articles including economic valuation of the potential of saline degraded lands turned out to be a major issue for our meta-analysis. The available publications are usually inconsistent in terms of the metrics used and they apply diverse approaches to estimating costs and benefits. This made it especially difficult to bring the reported values to a common denominator and resulted in some more general statements (e.g. measuring added value instead of profit). However, among all the diverse values and methods, there was a certain agreement when it came to the added value of saline agriculture per hectare. The values varied between ~200 and 600 USD depending on the country. We excluded from the analysis a study with 76,000 USD per hectare from United Arab Emirates (Robertson et al. 2019), as it focused on halophytes which have a different market and considerably higher prices than other crops. Lastly, we also decided to exclude studies which provided values for biofuels, pastures, forage and aquaculture, as all these involved different markets.

7.4.2 Costs of not Applying Saline Agriculture

The cost-avoided valuation method in case of saline agriculture can bring highly diverse results based on the inputs, and whether irrigated or non-irrigated land is considered. Amongst others, the costs included by Qadir et al. (2014) were opportunity costs (forgone agricultural income or alternative activities), production losses (crop response function to salt level), replacement costs (preventing and repairing land degradation), transaction costs (costs of developing and implementing four land and water management plans), market prices (groundwater recharge credits trading, land rents) and mitigation costs (desalination plant). All these cost categories vary in order of magnitude and the solutions that apply. Further studies could produce estimates based on different cost categories. In terms of irrigated and non-irrigated areas, according to the UNCCD Global Land Outlook report on Food Security and Agriculture, at least 20% of irrigated lands are salt-affected and this number may increase to 50% by 2050, (UNCCD 2017). The average cost suggested by FAO is 245 USD per hectare (inflation-adjusted 255 USD per hectare), which would result in 4 billion USD of costs globally when extended over 16 Mha. For both costs and benefits, it would be beneficial to calculate them for different salinity classes as each level of salinity has its own economics. However, a current level of data availability and granularity does not allow for more in-depth analysis.

7.4.3 Market Prices

Our analysis produced an average revenue per hectare, taking into account crop yields, their response to salinity and the market prices of six selected crops. The crop selection and field composition would vary considerably per country, therefore we scaled it to the global area used for each crop cultivation. Similarly, the producer (off-farm) prices are country-specific. Therefore, the international prices and produce quantity resulting from field composition indicate more a direction than a specific number. However, revenues represent only one side of the equation. The costs of implementing saline agriculture on a scale beyond experiments or pilots are presently unknown. The costs of seeds and fertilizers and initial investments in equipment and training led by a multidisciplinary team would vary greatly between locations and amongst crops.

7.4.4 Non-Monetary Values

Finally, very few studies refer to the non-monetary values of improved agriculture in degraded environments for local communities and the environment. These values, often included in ecosystem service valuations, could add to the full picture and potential of saline agriculture. Examples of the studies which pioneer in this regard are the work of Kabir (2016) and de Vos et al. (2021) presented in this book. Including indirect values such as social and environmental costs and benefits in the analysis could be an interesting direction for future research.

7.8 CONCLUSION

This study aimed to investigate the economic value of saline degraded lands based on three valuation methods. Our analysis has confirmed that the values vary considerably between countries, but certain trends can be seen. The potential added value from applying saline agriculture and current costs borne due to salinization are relatively similar, ranging between 200 and 650 USD per hectare. Most of the past studies have focused on the high costs induced by salinity, including loss of yield, additional soil treatment costs and forgone opportunity costs. However, the evidence from the research included in the meta-analysis suggests that profits can be obtained thanks to the cultivation of salt-tolerant crops. Finally, potential revenues can be calculated based on average yields and prices.

Further research should focus on the costs of saline agriculture application beyond trials and pilots. To get more insight into projects' feasibility and profitability, we recommend collecting more precise financial data on fixed and variable costs, and benefits, including non-direct values. These could be obtained through a survey among the stakeholders to assess social and environmental costs and benefits, as well as the willingness to pay for the necessary investments, saline agriculture's products or accept the changing environment.

Since the potential of salt-tolerant crops cultivation seems evident, it is important to enumerate reasons why it is not yet widespread. These include a lack of knowledge and understanding of soil and water management methods, varying costs and conditions among the countries, lack of focus on salt-tolerant varieties among the breeders, difficulties with knowledge transfer to farmers living in remote areas, and the lack of financing for initial saline agriculture investments. The means to overcome these obstacles should be explored in future studies.

The results of this research project offer one of the first estimates of economic potential that could be useful for various groups of stakeholders. For policymakers, it shows that the cultivation of salt-tolerant crops can be part of the answer to food security issues at a local, regional and global scale. The use of saline degraded soils allows for the cultivation of previously empty lands and may help to prevent land-use change in other areas, which may have higher biodiversity or provide other ecosystem services. Saline agriculture may be also a way to diversify the farmers' portfolio, innovate and obtain additional financial resources.

ACKNOWLEDGMENTS

This research was conducted with support of the Wadden Academy, as a contribution to the European Union Interreg VB North Sea Region Salfar project, co-funded by the North Sea Region Programme 2014–2020.

REFERENCES

Ali, Amal Mohamed, HM Van Leeuwen, and RK Koopmans. 2001. "Benefits of Draining Agricultural Land in Egypt: Results of Five Years' Monitoring of Drainage Effects and Impacts." *International Journal of Water Resources Development* 17 (4). Taylor & Francis: 633–46.

Aslam, Zahoor, AR Awan, M Rizwan, Asia Gulnaz, and Kauser A. Malik. 2009. "*Saline Agriculture Farmer Participatory Development Project in Pakistan: Punjab Component (Technical Report, 2002–2008)*." Published by Nuclear Institute for Agriculture and Biology (NIAB), Faisalabad, Pakistan 46.

John, Michele, David Pannell, and Ross Kingwell. 2005. "Climate Change and the Economics of Farm Management in the Face of Land Degradation: Dryland Salinity in Western Australia." *Canadian Journal of Agricultural Economics/Revue Canadienne d'agroeconomie* 53 (4): 443–59. doi:10.1111/j.1744-7976.2005.00029.x.

Food and Agriculture Organization of the United Nations. (2018). *FAOSTAT statistical database.* Rome, FAO.

Ghassemi, F., Jakeman, A. J., & Nix, H. A. (1995). *Salinisation of land and water resources: human causes, extent, management and case studies.* CAB international.

Kabir, Jahangir, Rob Cramb, Donald S. Gaydon, and Christian H. Roth. 2017. "Bio-Economic Evaluation of Cropping Systems for Saline Coastal Bangladesh: II. Economic Viability in Historical and Future Environments." *Agricultural Systems* 155 (July): 103–15. doi:10.1016/j.agsy.2017.05.002.

Kabir, Md Jahangir. 2016. "The Sustainability of Rice-Based Cropping Systems in Coastal Bangladesh: Bio-Economic Analysis of Current and Future Climate Scenarios."

Kabir, Md. Jahangir, Rob Cramb, Donald S. Gaydon, and Christian H. Roth. 2018a. "Bio-Economic Evaluation of Cropping Systems for Saline Coastal Bangladesh: III Benefits of Adaptation in Current and Future Environments." *Agricultural Systems* 161 (March): 28–41. doi:10.1016/j.agsy.2017.12.006.

Kabir, Md. Jahangir, Donald S. Gaydon, Rob Cramb, and Christian H. Roth. 2018b. "Bio-Economic Evaluation of Cropping Systems for Saline Coastal Bangladesh: I. Biophysical Simulation in Historical and Future Environments." *Agricultural Systems* 162 (May): 107–22. doi:10.1016/j.agsy.2018.01.027.

Khan, Shahbaz, Emmanuel Xevi, and W. S. Meyer. 2003. "Salt, Water, and Groundwater Management Models to Determine Sustainable Cropping Patterns in Shallow Saline Groundwater Regions of Australia." *Journal of Crop Production* 7 (1–2): 325–40.

Koetse, Mark J, Roy Brouwer, and Pieter J.H. Van Beukering. 2015. "Economic Valuation Methods for Ecosystem Services." *Ecosystem Services: From Concept to Practice*, 108–31.

Ladeiro, Bruno. 2012. "Saline Agriculture in the 21st Century: Using Salt Contaminated Resources to Cope Food Requirements." *Journal of Botany* 2012: 1–7. doi:10.1155/2012/310705.

Lefkoff, L Jeff, and Steven M. Gorelick. 1990. "Benefits of an Irrigation Water Rental Market in a Saline Stream- aquifer System." *Water Resources Research* 26 (7). Wiley Online Library: 1371–81.

Negacz, Katarzyna, Pier Vellinga, and Arjen de Vos. 2019. "Saline Degraded Lands Worldwide: Identifying Most Promising Areas for Saline Agriculture." *Paper Presented at the Saline Futures Conference.*

Qadir, Manzoor, Emmanuelle. Quillérou, Vinay Nangia, Ghulam Murtaza, Murari Singh, Richard Thomas, Pay Drechsel, and Andrew Noble. 2014. "Economics of Salt-Induced Land Degradation and Restoration." *Natural Resources Forum* 38 (4): 282–95. doi:10.1111/1477-8947.12054.

Robertson, S. M., Dionysia Angeliki Lyra, J. Mateo-Sagasta, S. Ismail, and M. J. U. Akhtar. 2019. "Financial Analysis of Halophyte Cultivation in a Desert Environment Using Different Saline Water Resources for Irrigation." In *Ecophysiology, Abiotic Stress Responses and Utilization of Halophytes*, edited by Mirza Hasanuzzaman, Kamrun Nahar, and Münir Öztürk, 347–64. Springer, Singapore. doi:10.1007/978-981-13-3762-8_17.

UNCCD. 2017. "Global Outlook Report." Bonn: United Nations Convention to Combat Desertification.

UoC, University of California Agriculture and Natural. 2020. "Salinity Measurement and Unit Conversion." https://ucanr.edu/sites/Salinity/Salinity_Management/Salinity_Basics/Salinity_measurement_and_unit_conversions.

Van Beukering, Pieter JH, Roy Brouwer, and Mark J. Koetse. 2015. "Economic Values of Ecosystem Services." *Ecosystem Services. From Concept to Practice. Cambridge University Press, Cambridge*, 89–107.

Vos, Arjen de, Andres Parra González, and Bas Bruning. 2021. "Case Study: Putting Saline Agriculture into Practice – A Case Study from Bangladesh." In *Future of Sustainable Agriculture in Saline Environments*. Routledge, Taylor & Francis Group, London.

Vos, Arjen de, Bas Bruning, Gerrit van Straten, Roland Oosterbaan, Jelte Rozema, and Peter van Bodegom. 2016. "Crop Salt Tolerance under Controlled Field Conditions in The Netherlands, Based on Trials Conducted at Salt Farm Texel." Salt Farm Texel.

Vyshpolsky, Frants, Qadir, Manzoor, Karimov, Akmal, Mukhamedjanov, Khamit, Bekbaev, Ussen, Paroda, Raj, Aden Aw-Hassan & Karajeh, Fawzi (2008). "Enhancing the Productivity of High-magnesium Soil and Water Resources in Central Asia through the Application of Phosphogypsum.". *Land Degradation & Development*, *19*(1), 45–56.

Wang, Jiali, Xianjin Huang, Taiyang Zhong, and Zhigang Chen. 2013. "Climate Change Impacts and Adaptation for Saline Agriculture in North Jiangsu Province, China." *Environmental Science & Policy* 25: 83–93.

WRI. 2018. "What Is Degraded Land? | World Resources Institute." https://www.wri.org/faq/what-degraded-land.

8 Challenges and Opportunities for Saline Agriculture in Coastal Bangladesh

Atiq Rahman and Md. Nasir Uddin

CONTENTS

DOI: 10.1201/9781003112327-8

8.1 INTRODUCTION

Populations living in deltas (low-lying coastal floodplains of the world's major rivers) are increasingly vulnerable to risks from tropical cyclones, coastal floods, storm surges, sea-level rise (SLR), salinity, shoreline erosion and accretion, and pollution (IPCC 2014; Barbier 2015). Agriculture in delta landscapes is particularly vulnerable to climate change. The Intergovernmental Panel on Climate Change (IPCC) has indicated that in South Asia, agricultural crop yield could be reduced by up to 30% by 2050 because of a changed climate (IPCC 2014).

Bangladesh is located in the tropical zone and is highly prone to natural disasters like riverine and tidal floods, tropical cyclones, storm surges, heat stress, hailstorms, lightning strikes, drought, SLR, and salinity intrusion on land and in water (Rasid and Paul 2013). The coastal community of Bangladesh is highly dependent on agricultural production from crops, fish, and livestock. Local climatic factors are favorable for a wide range of crop cultivation and production in coastal areas (Uddin et al. 2019). However, recent studies have revealed that extreme climate events (e.g., coastal floods, cyclones, storm surges, and SLR) are increasing every year, devastating lives and livelihoods, and decreasing agricultural production in coastal and island Bangladesh (Uddin et al. 2019). This decrease in food production and availability represents a challenge and threat to the capacity of coastal communities to exercise their right to food.

SLR-induced salinity in soil and water threatens both crop yields and economic development in coastal Bangladesh. River water and groundwater are both influenced by rainfall, river flow, upstream withdrawal of water, salinity, cyclonic storm surges, and tidal flood. Over the last few decades, the salinity of rivers in southern districts has increased by about 45% and over 20 million people are affected by salinity in the water along the coastal region of Bangladesh (Dasgupta et al. 2018). The increased frequency of cyclonic events and intensity of associated storm surges intensifies the risk of salinity intrusion in coastal areas. There is a strong association between the presence of salinity and the storm surges that cause flooding along the Bangladesh coast (Paul and Rashid 2017).

A number of adaptation measures have been undertaken to increase agricultural production in affected communities. These include the cultivation of vegetables on floating beds, the planting of faster maturing crop species and varieties with institutional support, the promotion of alternative livelihoods, and the planting of saline tolerant trees on embankments (Aryal et al. 2020). A few recent studies have shown that community-based adaptation (CBA) systems reduce the risk and increase the benefits to smallholder farmers (Schipper et al. 2014).

Few papers have described the SLR-induced salinity and its impacts on coastal agriculture practices. Based on primary and secondary data, this chapter, therefore, explores salinity risk areas using long-term climate risk analysis and coping measures that communities in the coastal areas of Bangladesh use to minimize their vulnerability to coastal floods, cyclone hazards, and salinity intrusion. We integrate and analyze expert judgments along with community perceptions on SLR-induced salinity and its direct effects on local agricultural systems. We show how climate-induced salinity impacts agriculture, and identify opportunities for future coastal saline agricultural management.

8.2 METHODOLOGICAL APPROACH

8.2.1 Key Feature of Case Study Area

Bangladesh has 19 coastal districts (and 147 subdistricts), which extend over 47,150 sq km (around 32% of the total land area). About 35 million people (6.9 million households), or 28% of the country's total population live in these areas (BBS 2011). The coastal zone can be classified based on three characteristics: the level of tidal fluctuation, the salinity of surface and groundwater, and the risks from cyclones, storm surges, and tidal influences (Brammer 2014). The 19 coastal districts have been further divided into interior (7 districts, 48 subdistricts) and exposed (12 districts, 99 subdistricts) zones, with regard to distance from the coast or estuaries, under the Integrated Coastal Zone Management Project (ICZMP) of the Water Resources Planning Organization (WARPO). The coastal zone of Bangladesh is divided into three regions, the western zone (Ganges tidal plain), the central zone (Meghna deltaic plain), and eastern zone (Chittagong coastal plain). The coastal zone is characterized by a vast network of rivers and channels, an enormous discharge of water with huge amount of suspended sediment, many islands, a strong tidal influence, and tropical cyclones and storm surges (CCC 2016). About 70% of total farmers in coastal areas are sharecroppers while more than 53% of the total coastal population are functionally landless, live below the poverty line, and have no cultivable land (CCC 2016).

8.2.2 Identifying the Drivers of Coastal Salinity Intrusion

This study reviewed the secondary literature and interviewed 10 key experts to quantify the drivers of salinity intrusion in the coastal belt of Bangladesh. Using a semi-structured questionnaire, experts were asked about: (a) causes of salinity, (b) salinity intrusion in coastal agriculture land, (c) the historical trend of coastal salinity intrusion, (d) the major impacts of soil salinity on coastal agriculture, and (e) saline agriculture management for sustainable development. In addition, we conducted a detailed spatial analysis of SLR, sea-level temperatures, and cyclones in coastal areas.

8.2.2.1 Preliminary Analysis of Sea-Level Rise (SLR) and Ocean Warming

Techniques were used to analyze and map SLR and track ocean warming in the Bengal basin. Daily tidal gauge SLR records for the period of 1980–2010 were collected from the hydrography department of Bangladesh Inland Water Transport Authority (BIWTA). To derive the annual SLR trends of ground observation stations, we used a simple linear regression model ($Y = \beta X + \alpha$) in the R platform. To explore the sea surface temperature (SST) in the Bay of Bengal, we used the monthly

HadISST V1.1 (1° × 1°) NetCDF dataset of 39 years (1980–2019), retrieved from the NCAR global climate guide data hub (NCAR 2020). Using these data, we developed a simple linear regression to identify the trend in ocean warming (SST) for the Basin area. Additionally, based on the ASTER NASA LPDAAC (30m) GDEMV3, we identified the SLR (0cmSLR, 40cmSLR, 80cmSLR, 1mSLR) affected crops land of coastal subdistricts using ArcGIS 10.5 platform.

8.2.2.2 Tropical Cyclonic Storm Surge Analysis (1901–2020)

Data on historical tropical cyclone tracks (1901–2020) were collected from the NOAA National Centers for Environmental Information IBTrACS data sets. We extracted the wind speed and pressure of individual tracks along with their coordinates. The dataset contained comprehensive information on each tropical cyclone and storm, a synoptic history, meteorological statistics, casualties, and damages. To estimate the damage area of individual cyclones, we created a buffer area based on wind speed and pressure around the tracks in ArcGIS 10.7.1 platform. Then the storm risk map was prepared using frequency and kennel density functions. Damage to agriculture due to storms was collected from national disaster census (BBS 2015) and national published daily newspapers. Considering the secondary damage data and generated risk data, the subdistricts were ranked based on the loss and damage scenarios and the prepared cyclone and storm spatial risk map for the coastal regions.

8.2.3 Spatial Analysis of Coastal Soil Salinity

Subdistrict-level soil salinity data were collected from the subdistrict supplements published by Soil Resource Development Institute (SRDI) of Bangladesh. On average 41 samples were collected from each subdistrict (SRDI 2020). The data were classified into four categories: very slightly saline (< 2 ppt), slightly saline (3–6 ppt), strongly saline (7–12 ppt), very strongly saline (> 12ppt), and the soil salinity stress areas were mapped using the ArcGIS platform.

8.2.4 Identifying the Challenge and Opportunity of Saline Agriculture

Community-based information and the grey literature were reviewed to collect key relevant information including the challenges and opportunities for coastal saline agriculture, coastal SLR and cyclonic storm surge induced salinity, and impacts on crop production. Local-level information was collected from 34 Focus Group Discussions (FGDs) held at the most vulnerable subdistricts of 15 most vulnerable coastal districts. The clusters were selected through Geographic Information System- GIS mapping based on a gradient of salinity. All these FGDs were conducted based on the checklist developed to gain in-depth information on coastal climate change-induced extremes, impacts on agriculture, and types of community led and exogenous adaptation measures practiced. A total of 15 stakeholder consultation workshops were conducted in the study districts. The synthesis research findings from subdistrict-level group discussions were presented in district-level consultation workshops. The findings from the local-level discussions and interviews were validated through these workshops. Based on the secondary and field observations, we prepared an opportunities matrix for future saline agriculture management in coastal Bangladesh.

8.3 RESULTS AND DISCUSSION

The study identified four climate change-induced factors that have direct impacts on coastal agricultural land and production in the study areas: SLR, cyclonic storm surges, tidal surge/waves, and coastal flooding. Salinity is one of the main constraints for regular crop production in the southern region. This study also discussed the opportunities for future coastal saline agriculture management.

8.3.1 Drivers of Coastal Salinity in Bangladesh

From expert observations and the literature, we found that the main causes of increased soil salinity along the coast are: the withdrawal of fresh river water from upstream, irregular rainfall, the introduction of brackish water for the culture of shrimp, the faulty management of sluice gates and polders, the regular intrusion of tidal saline water during high tide in the unprotected lands, and the capillary rise of soluble salts from shallow groundwater towards the soil surface. About 1 Mha in the southern area is affected by soil salinity. Some of the new lands in Satkhira, Patuakhali, Barguna, Barisal, Jhalokati, and Pirojpur districts have been affected significantly by different degrees of soil salinity during the last few decades. Several recent studies (Dasgupta et al. 2014; Salehin et al. 2018) have identified multiple reasons for salinity intrusion in the coastal area of Bangladesh. Broadly, these include changes to natural, socio-economic and political systems (Mahmuduzzaman et al. 2014). The literature review helped to quantify a few key points which play a role in increasing salinity intrusion; these are firstly the biophysical context of coastal Bangladesh, secondly climate change-induced extremes (e.g., SLR, increased numbers of tropical cyclones, storm surges, and coastal floods – Dasgupta et al. 2015; Haque 2006), thirdly changes to the socio-economic system such as increased shrimp cultivation in agricultural land (Clarke et al. 2015), and fourthly the establishment of the Farakka Barrage in the upstream Ganges River which caused increased seawater intrusion into the basin area of coastal Bangladesh (Rahman and Rahaman 2018).

8.3.1.1 Mapping of Sea-Level Rise and Sea Surface Temperature (SST)

Analysis of data from tidal gauges revealed a significant rising trend in sea level in south coastal regions of Bangladesh during the period 1980–2010 ($P < 0.000$), with a rate of rise of 10.4 mm per year. Monthly analysis showed that the lowest trend (3.1 mm/year) was in May, and the highest trend (10.9 mm/year) was in November. These results are consistent with other previous studies (Brammer 2014), but they are higher than the reported average for north India (1.3 mm/year – Unnikrishnan and Shankar 2007) and the global average of 3.6 mm/year (Rignot et al. 2009). Different coastal regions of Bangladesh have differing rising trends (Table 8.1). The highest increasing trend (21.0 mm per year) was in the western region (Khepupara), whereas the rates of SLR on the central and eastern coast were 7.2 and 8.4 mm/year respectively.

TABLE 8.1
Summary of Sea-level Rise (SLR) Trend Analysis in Four Coastal Stations of Bangladesh

Tide Gauge Stations	Latitude (°)	Longitude (°)	Regions	Trends (mm/yrs.)	*P*-value
Hiron Point	21.80	89.47	Western	Y = 4.91x – 516.40	0.001
Khepupara	21.99	90.22	Western	Y = 21.0x – 2903.34	0.000
Char Changa, Hatiya	22.23	91.01	Central	Y = 7.22x – 190.382	0.049
Conx's Bazar	21.43	92.00	Eastern	Y = 8.39x – 4027.29	0.000

Our study showed that the south coastal ocean is rapidly warning. Near the south western coast (around Hiron Point), the SST has a highly significant ($P < 0.001$) increasing decadal trend of over 1.3°C in the monsoon (April, May) and about 0.7°C in the post-monsoon period (October, November). Near the south eastern coast, SST has significant rising trends (0.8°C per 10 years in the monsoon and about 0.5°C per 10 years in the post-monsoon period).

The increasing trends of SLR have devastating effects on coastal ecosystems including increased erosion, the flooding of wetlands, and the contamination of agricultural soils with salt. Figure 8.1B shows the districts of Bangladesh and the extent of inundation that would occur with a 40, 80, or 100 cm rise in sea level. It can be seen that a 1 m rise in sea level would inundate the exposed coast and coastal islands and remove about 30% of the monsoon cropland of Bangladesh accounting for cropland in 103 tidal subdistricts in 15 districts (i.e., Satkhira, Khulna, Bagerhat, Barisal, Barguna, Patuakhali, Bhola, Noakhali, Shariatpur, Chandpur, Lakshmipur, and Feni). An 80 cm rise in sea level would inundate over 21% of the summer rice cropland in 88 subdistricts of the interior coast of Bangladesh (Figure 8.1B).

8.3.1.2 Long-Term Tropical Cyclonic Storm Surge Mapping (1901–2020)

An analysis of 245 cyclones over 119 years was undertaken in this study. We analyzed the density, synoptic behavior, wide speed, and pressure of 245 individual tropical cyclones from 1901 to 2020 to identify the risk-prone areas at the subdistrict scale (Figure 8.2A). We found 180 tracks for category 1 storms (119–153 km/h), 13 tracks in category 2 (154–177 km/h), 11 tracks for category 3 (178–208 km/h), 15 tracks for category 4 (209–251 km/h) and six tracks for category 5 storms (252 km/h or higher) (Figure 8.2A). In addition, we classified 483 tropical storms (pressure, wind speed and surge height) in which ~63% storm surges occurred during September–November and ~38% occurred in April–July (1901–2020). The subdistrict scale damage mapping showed that the south central and south eastern part of the coast had the highest tropical cyclone risk. Figure 8.2B shows that the 41 subdistricts of high cyclone risk are in the districts of Barguna, Bhola, Jhalokati, Patuakhali, Feni, Noyakhali, and Chittagong. Our analysis shows that ~79% of cyclone tracks hit the south western coast (Barguna and Khulna regions) in the month of April– July. The subdistricts of Chittagong and Patuakhali districts experienced the highest (96) and

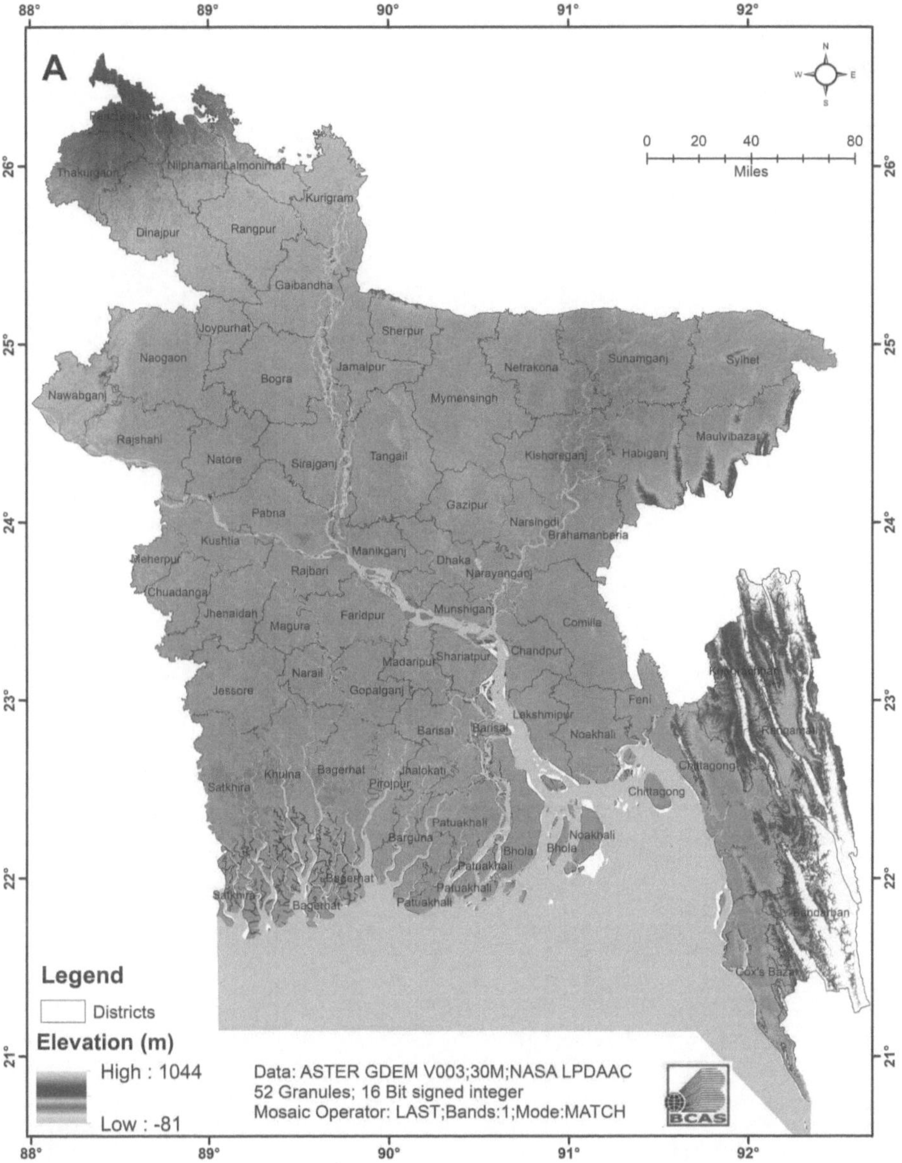

FIGURE 8.1 Mapping of ASTER NASA LPDAAC (30m) elevation along with districts boundary (A), sea-level rise affected subdistricts (using ocmSLR, 40cmSLR, 80cmSLR, and 1mSLR line graph) along with tidal water level (WL) observatories and tidal limit (B).

(Continued)

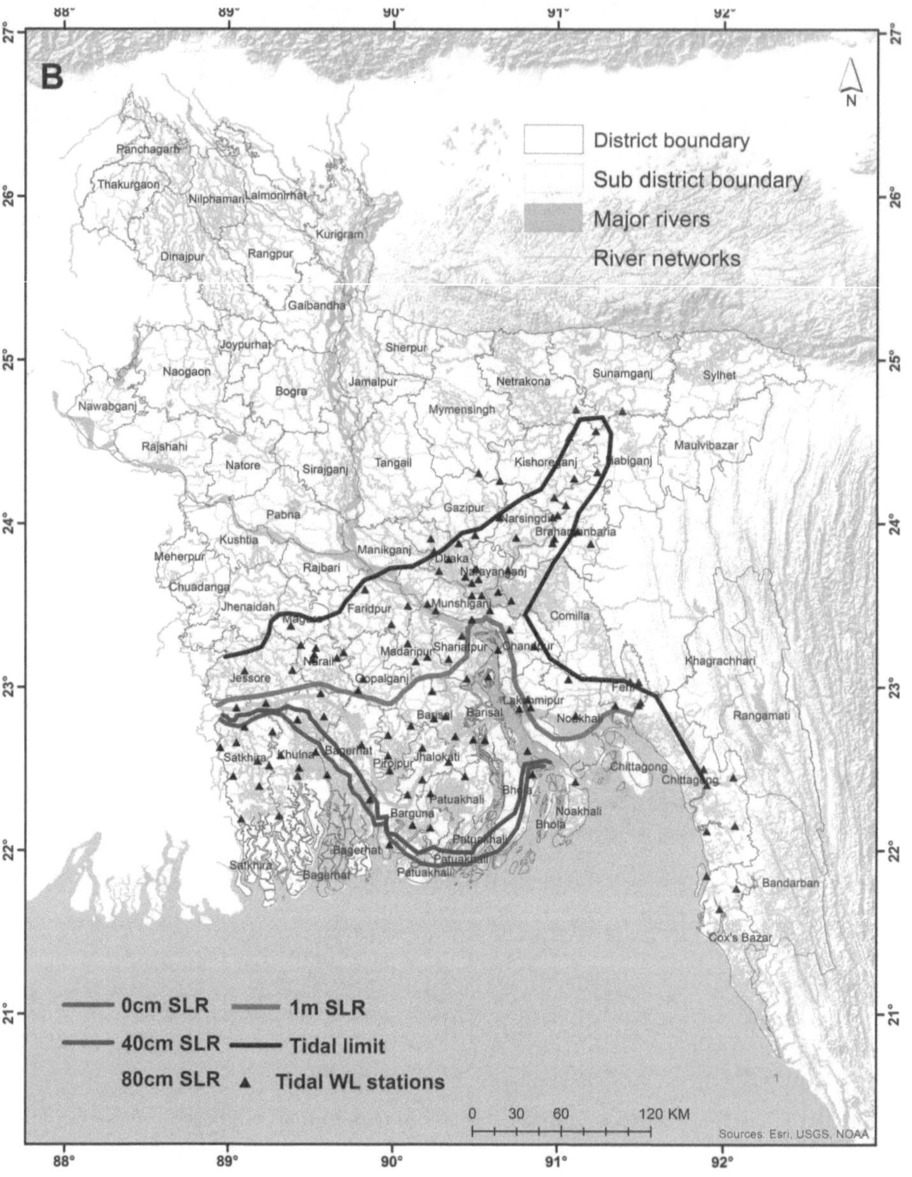

FIGURE 8.1 (Continued)

lowest (27) number of cyclones respectively (Figure 8.2B). From the analysis of long-term cyclone tracks, we found that over 89% cropland was affected and damaged about every three-year due to cyclonic storm surges in coastal areas. The last few major cyclones along with surge height (from Indian Meteorological Department, IMD), the affected districts, subdistricts, affected croplands, and the total loss and damage are summarized in Table 8.2.

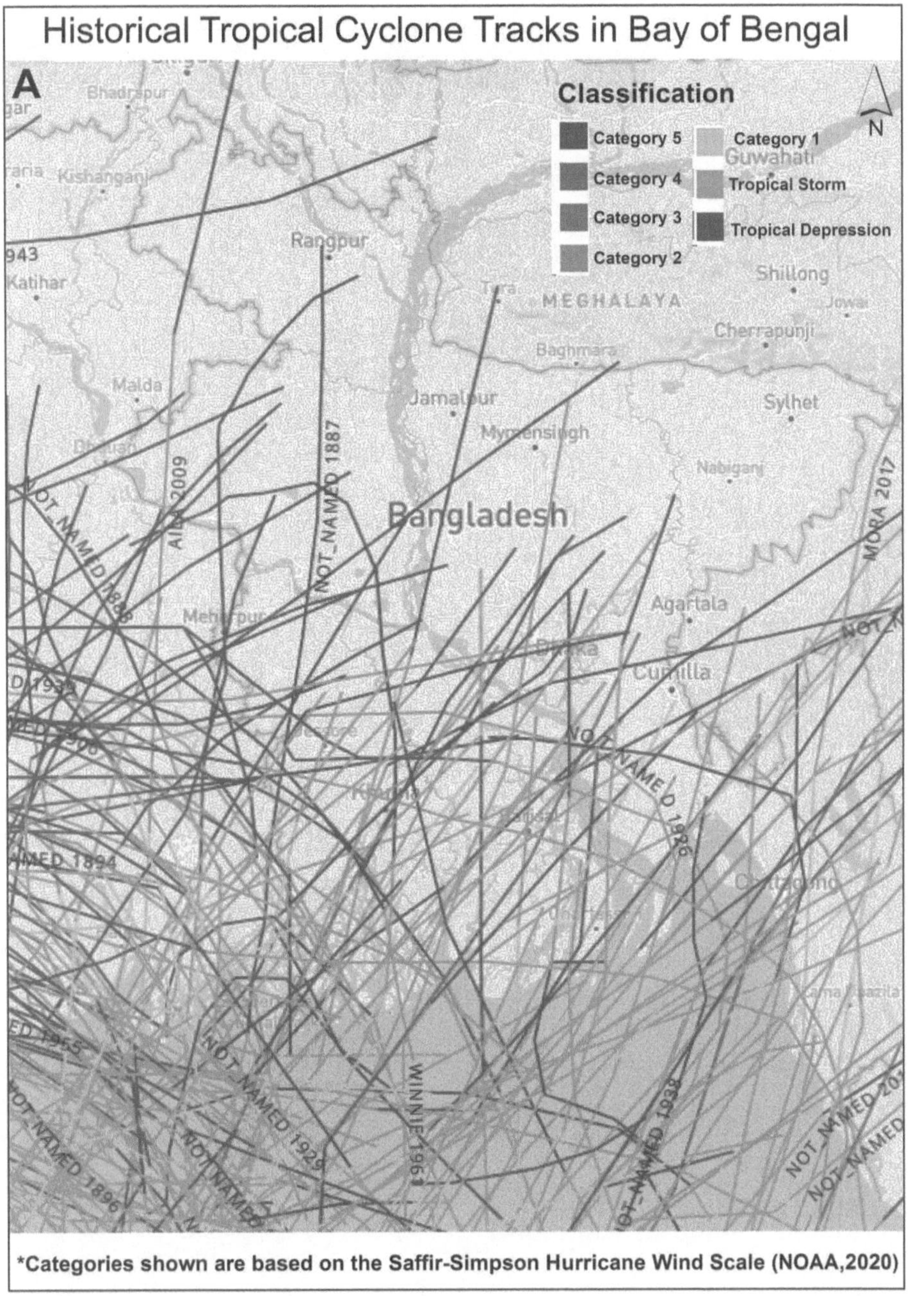

FIGURE 8.2 Historical tropical cyclone tracks (1901–2020) along with Saffir–Simpon classification (A), cyclone risk index map indicating the degree of risk at subdistrict level (B).

(Continued)

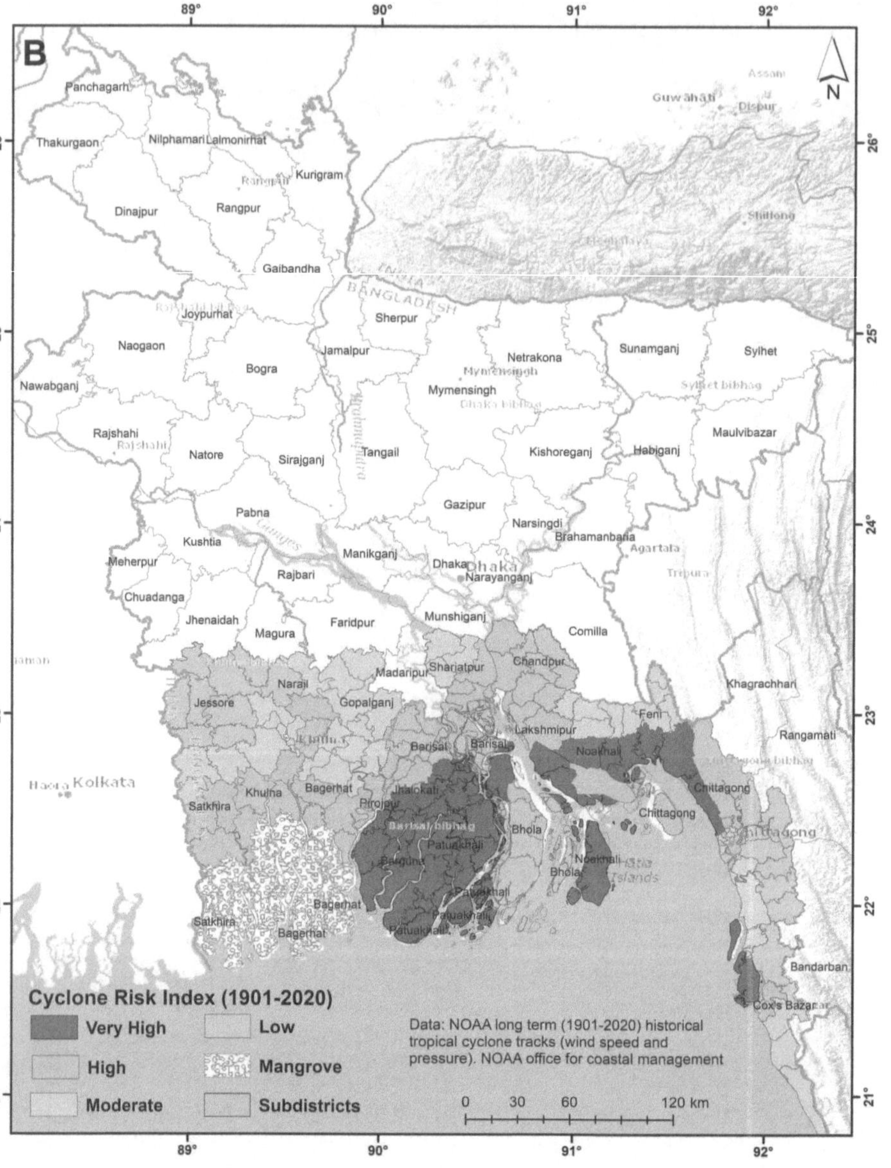

FIGURE 8.2 (Continued)

8.3.2 Spatial Analysis and Mapping of Soil Salinity

The salinity in the soil largely determined the crop productivity and potential land use in the coastal areas. Spatial analyses of variation in soil salinity in the topsoil and the subsoil shows that all coastal subdistricts are affected by salinity. Based on measures of soil EC_e, these coastal subdistricts are classified into four major

TABLE 8.2
Eight Major Severe Cyclones along with Storm Surge Height, Affected Districts, Farmland, and Total Loss and Damage

Cyclone Name	Surge Height (m) (IMD)*	Affected Areas	Affected Farmland (ha) (DAE)**	Loss and Damage (US $M)
Amphan (May 2020)	3–5	Over 100 villages of nine districts in the coastal divisions of Khulna and Barisal	176,000	130
Fani (April 2019)	1.5	35 districts	63,000	63.6
Bulbul (November 2019)	1.2–1.5	Khulna, Bagerhat districts	289,006	33
Mora (May 2017)	1.2–1.5 m	Chittagong, Cox's Bazar, and Rangamati	NA	9 deaths, 52,000 houses Damaged
Roanu (May 2016)	2	Sandwip, Hatia, Kutubdia, Sitakundu and Feni, Chittagong, and Cox's Bazar	NA	31.8
Mahasen (May 2013)	2	Chittagong, Patuakhali, Noakhali district	NA	49,000 houses destroyed and 45,000 houses partially destroyed
Aila (May 2009)	3	15 districts of south western part	60,000	~1,000
Sidr (November 2007)	3–5	Sharankhola, Patuakhali, Barguna, and Jhalokati	~1,000,000	2,310

* Storm surge height data taken from India Meteorological Department (IMD) dataset,
** affected croplands and damage estimated by Department of Agricultural Extension (DAE) of Bangladesh
Source: Daily newspapers.

divisions (a) very strongly saline (>15 dS/m), (b) strongly saline (8–15 dS/m), (c) slightly saline (3–7 dS/m), and (d) very slightly saline (< 2 dS/m). The result shows most of the exposed coastal subdistricts of Satkhira, Barguna, and Chittagong districts and south western coastal areas are very strongly saline (Figure 8.3). Group discussions revealed that salinity increased in the dry season because less water flows from major rivers like the Ganges and Meghna, and there are frequent storm surges along with man-made salinity through intrusion of saline water for shrimp culture in Khulna and other coastal regions. In addition, salinity has increased from eastern coastal belts to the western coastal belt mainly due to the very low flow of upstream water from the Ganges and its tributaries during November–May because of water withdrawal at the Farraka Barrage on the Ganges in West Bengal of India

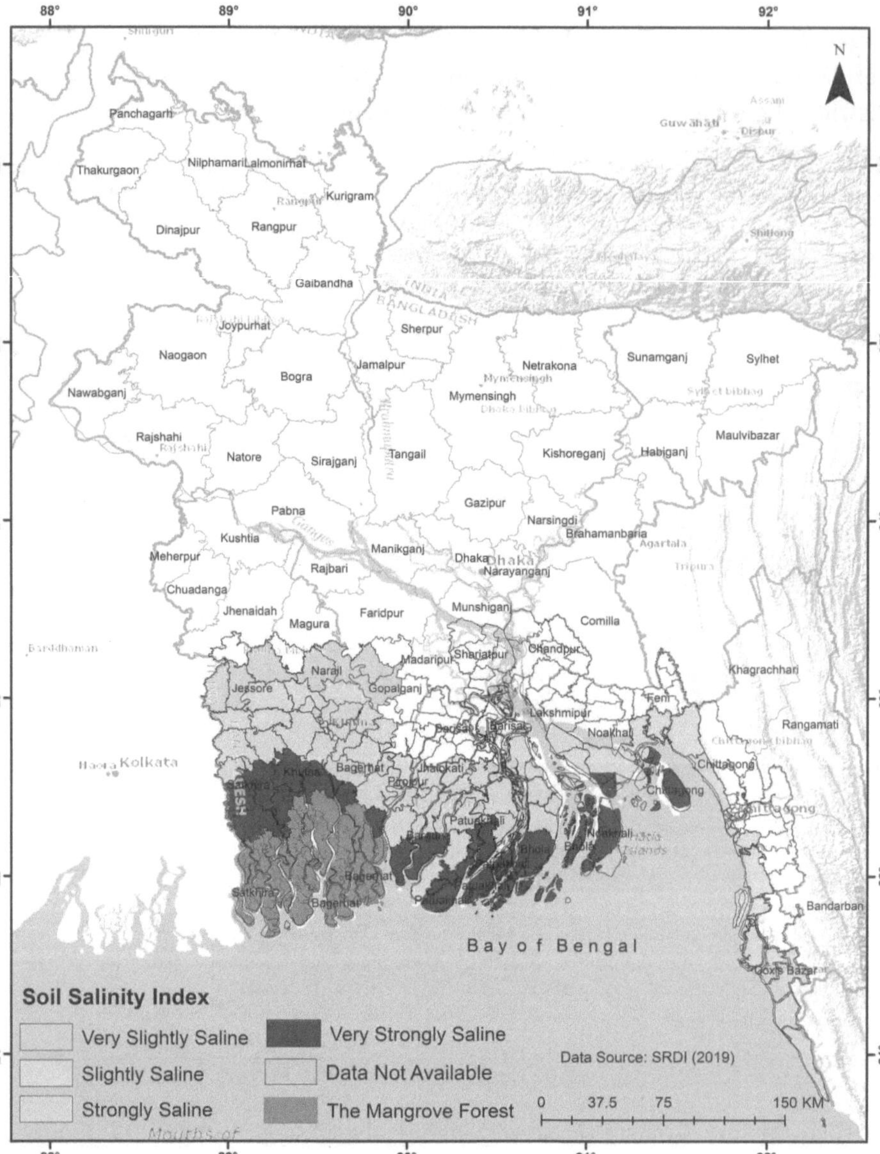

FIGURE 8.3 Soil salinity map of coastal Bangladesh.

(commissioned in 1974). The coastal area of Bangladesh has relatively less cropping intensity due to the rise of salinity during dry season and other constraints such as low soil fertility, river erosion, floods, late draining condition, heavy clay soils, the scarcity of irrigation water, exposure to cyclone storm surges, difficulties in communication, and remoteness from urban markets.

8.3.3 Constraints of Coastal Agricultural Development

Our study suggests that the agricultural systems and cropping intensities in over 79 subdistricts of Khulna, Jessore, Satkhira, Bagerhat, Barguna, Narail, Gopalganj, Patuakhali, Barisal, Jhalokati, and Pirojpur districts and of coastal islands are seriously affected by saline stress, and as a consequence agricultural activities have been changing in coastal regions. Over the last 20 years, most of the rice land in coastal regions has been replaced by rice–shrimp farming (rice in the rainy season and shrimp in the dry season), indicating a sharp increase in shrimp cultivation over the coastal areas. Crop production in the study areas has declined following shrimp cultivation. In addition, the yield of wheat, jute, and sugarcane have been affected seriously, and it is now not possible to grow these crops because of soil salinization. Other constraints affecting coastal agricultural development include the limited availability of good-quality groundwater, a severe scarcity of quality irrigation water during the dry season, prolonged artificial waterlogging with saline water for agriculture, the presence of toxic potential acid sulfate soils in some areas, and the relatively high flooding depth during monsoon season for high yield variety rice crops.

8.3.3.1 Salinity Impacts on Agriculture Systems

SST has a strong influence on coastal climate and salinity (Ji et al. 2019). The gradual increasing scenarios of salinity intrusion into the coastal areas (soil, river, and groundwater) of Bangladesh is very threatening to the primary production system, coastal biodiversity, and human health (Amores et al. 2013; Islam et al. 2020; Uddin et al. 2019). According to the Soil Resources Development Institute (SRDI), the total amount of salt-affected land in Bangladesh was ~83.3 Mha in 1973, which increased to ~102 Mha in 2000, to ~105.6 Mha in 2009, and is still continuing to increase (Chen and Mueller 2018). Salinity intrusion directly affects the livelihoods of farmers including rice cultivators and fisherfolk. Vegetation, soil quality, and infrastructure in these areas are also affected by salinity. The net cropped area in coastal Bangladesh has been decreasing over the last few years due to several factors and studies have identified salinity as the main cause for yield reduction in coastal agriculture (Baten et al. 2015). The extent and intensity of salinity are projected to increase due to climate change-induced saltwater intrusion. Other statistics that point to the trauma of salinity include: (a) salinity has decreased the production of wheat by 4.42 Mt per year in coastal Bangladesh (Habiba et al. 2014), (b) 19 of 40 local rice varieties are already extinct and about four to five varieties have become rare in coastal areas, and (c) between 1975 and 2006 the number of cultivated winter vegetables has declined in coastal areas (Rahman et al. 2004).

8.3.3.2 Salinity Impacts on Fisheries Resources

Increased coastal salinity has affected fish yields, leading to substantial reductions in the inland open water fishery. Shrimp farming occurs on about 138,600 ha of coastal land in Satkhira (42,550 ha), Khulna (36,500 ha), and Bagerhat (49,550 ha) districts. Apart from these, the production of native freshwater fish species (e.g., rui, katla, carp, boal, tengra, golsha tengra, koi, shing, taki, khalisha, potka, kani magur, salbaim – all local names) is gradually declining due to increased salinity (Alam et al. 2017; Habiba et al. 2014).

8.3.3.3 Salinity Impacts on Livestock Systems

The 2014 IPCC assessment report showed that the projected impacts of climate change and extremes will affect livestock and livestock production systems. Increased salinity has significant negative impacts on livestock products in low-lying coastal floodplain. Our study has shown that in the last few decades, the area of grazing land and the intensity of grazing has decreased due to increased salinity and the loss of agricultural land in coastal areas. Qualitative assessments suggest that the aggregated impacts of coastal climatic factors including SLR-induced salinity, coastal floods, cyclonic storm surges, and waterlogging will substantially decrease domestic cattle production in coastal areas. Increased temperatures will also have adverse effects. In general, livestock perform best at temperatures between 10 and 30°C; at temperatures above 30°C, cattle, sheep, goats, pig, and chickens reduce their feed intake by 3–5% for each 1°C increase (Thornton et al. 2015). Our group discussions show that salinity will seriously affect the productivity and species composition and dynamics (quality and abundance of feed sources), resulting in overall livestock and poultry productivity decreases with increasing salinity. Increases in salinity in the study area will be negatively correlated with the changes in animal diets and the reduced nutrient availability for domestic cattle.

8.3.3.4 Salinity Impacts on Human Mobility

A number of studies have found that soil salinity has significant impacts on seasonal and internal migration in coastal Bangladesh (e.g., Chen and Mueller 2018). Coastal indigenous communities with traditional ways of life face unprecedented impacts from climate change-induced disasters. Households experienced both short- and long-term displacement as response to extreme climatic change events. Communities have their own indicators for predicting and adapting to these changes. Individuals, households, and community in coastal areas are already implementing traditional ways of adaptation to secure their livelihoods and development by trying to obtain better agricultural crops, and safe drinking water. Climate-induced migration, has negative impacts on community-based approaches used in coping with these changes. Human mobility from coast to urban cities has reduced the primary productivity and negatively affected agricultural development in coastal regions. As a result, climate-induced migration, is being recognized as a major potential threat to indigenous knowledge systems and strategies. Often men leave first, putting an extra domestic burden on the women left behind. This can lead to the loss of local innovation and locally led sectoral adaptation practices.

8.3.4 Opportunities for Future Saline Agriculture

In view of these climate change-induced extremes, farmers bear significant losses every year. Our study shows that a range of adaptations are required to ensure their survival and safeguard their livelihoods. Most of the adaptation plans were identified from government extension agents, non-government (NGO) project-based activities, and the findings of action research by research organizations and NGOs. The present study has developed a database on salinity tolerance of agricultural practices for community development, policy suggestions, and coastal crop and climate risk management. Based on an analysis of field observation and secondary literature, we broadly classified the coastal adaptation into following subsections.

8.3.4.1 New Technology Development

Different research institutes including the Bangladesh Rice Research Institute (BRRI), Bangladesh Institute of Nuclear Agriculture (BINA), and the Bangladesh Agriculture University (BAU) have developed and promoted salt-tolerant rice varieties (see Table 8.3 for details) and tidal and monsoon flood-tolerant varieties (BRRI rice 51 and 52; BINA rice 11 and 12). A number of studies conducted by BARC, BRRI, and DAE have focused on crop suitability in the coastal zone for the dry (rabi season). These have extended multi-crop farming systems with salt-tolerant vegetables, pulses, and oilseeds. The available salt-tolerant rabi field crops are sweet potato, green gram, linseed, groundnut, millet, sunflower, soybean, triticale, wheat, cowpea, mungbean, mustard. Vegetable and fruits include batisak, chilli, spinach, kangkong, garlic, china sak, Indian spinach, okra, water melon, red amaranth, and sunflower in the salt-affected coastal zone (Paul et al. 2020). Research has focused

TABLE 8.3
Summary of Improved Salt-Tolerant Crop Varieties along with Intervention Areas

Crops	Improved Agricultural Crop Varieties	Interventions Districts (Sub-Districts)
Rice	BRRI dhan 47, 53, 54, 55, and 61; BINA dhan 8 and 10	Khulna (Dumuria, Bhatiaghata, Dacope); Satkhira (Shaymanagar, Debhatta, Assasuni); Bagerhat (Fakirhat, Kachua, Chitalmari); Jhalokhati (Kathalia, Rajabari, Amtali); Barguna (Pathargata,), Patuakhali (Galachipa, Kolapara); Pirojpur (Nazipur, Bhandaria); Barisal (Wazipur, Agaailjhara); Bhola (Char fession, Tajumuddin, Daulatkhan); Laksmhipur (Sadar, Ramgati); Noakhali (Companyganj, Subornochar); Feni (Sonagazi, Chokoria); Cox's Bazar (Moheskhali, Bashkhali); Chittagong (Sandwip)
Vegetable	High yield variety water gourd, bitter gourd, cowpea, cucumber, red amaranth, jhinga, Indian spinach	
Sweet Potato	Cardinal, Diamont, BARI Sweet Potato-8,9 and local varieties	
Pulse	BARI mung-1, 2,3,4,5,6 and BARI Khesari-6, BINA Mung-3, Local Khesari	
Wheat	Shorab, Gourab, Prodip	
OilSeed (mustard)	BARI Sharisa-15	
Barlie	Bari Barli-6	
Kawon	Bari Kawon-2, Bari Kawon-3	
Tomato	Bari Tomato-4, 5, 10, Bari Hybrid tomato-3, 4, 8	
Gonjon Til	Shova	
Soybean	Bari Soyabin-6	
Coconut	Bari coconut-1, Bari coconut- 2	

TABLE 8.4
Locally Led Adaptations Being Practiced in Coastal Areas

No	Community-Based Adaptation Techniques	Location(s)
1	Integrated farming approach (rice, fish, poultry, vegetable cultivation)	Mongla, Bagerhat, Chila Union
2	Floating vegetable gardening (Floating bed used for crop cultivation)	Satkhira, Khulna, Paickgacha Pirojpur, Goplaganj, Faridpur, Jalakathi, Barisal
3	Homestead garden and vegetable on raised plinth	Satkhira, Khulna, Paickgacha
4	Dyke cropping techniques	Borguna, Patuakhali, Satkhira, Bagerhat, Khulna, Paickgacha, Koyra
5	Pitcher irrigation	Satkhira, Khulna, Dacope, Koyra
6	Fish cage farming	Satkhira
7	Crab fattening techniques	Mongla, Bagerhat

on the development of early planting varieties to avoid high temperature months, and the development of high-yielding, submergence-tolerant, and short-duration (110–120 days) rice varieties including rice varieties that are resistant to pests and diseases, salinity, inundation, drought, and temperature stress.

8.3.4.2 Community-Based Adaptation (CBA) in Coastal Bangladesh

Salinity has been engulfing new areas in the coastal region of Bangladesh. The capacity for CBA (Ayers and Forsyth 2009; Schipper et al. 2014) has increased through learning by doing and action research. In collaboration with government organizations and NGOs, the Bangladesh Centre for Advanced Studies (BCAS) and International Institute for Environment and Development (IIED) has identified a number of CBAs in Asia, Africa, Europe, and Australia that can be used as models for actions in Bangladesh (BCAS 2012). Through community consultation workshop and group discussion, we have tabulated a number of coastal communities led agricultural adaptation techniques (Table 8.4).

8.3.4.3 The Sundarbans Mangrove Forest and Agriculture

Coastal ecosystem services are becoming increasingly vulnerable to natural disasters like cyclones. The Sundarbans mangrove forest of Bangladesh is the largest mangrove forest in the world. It works as a buffer, protecting coastlines and attenuating storm surges and wind speed (Akber et al. 2018). It, therefore, plays a vital role in reducing the vulnerability of coastal communities to tropical cyclones (Barua et al. 2010). Our investigation has established that about 21% of cyclonic storm surges hit the Sundarbans with low consequent damage to communities. From group discussion and consultation, we suggest that there is a need for better mangrove management and coastal afforestation/reforestation which can reduce cyclonic damage and improve agricultural systems in coastal areas.

8.3.4.4 Coastal Embankment and Agricultural Development

Bangladesh has 5,017 km of embankments to protect the polders in coastal areas from regular natural disasters and boost agricultural production (Brammer 2014;

Islam et al. 2013; Mallick et al. 2011). However, due to the poor maintenance of embankments, in many places there have been increases in salinity intrusion into fields (Mahmuduzzaman et al. 2014; Nowreen et al. 2014). Poor maintenance of embankments has also increased the frequency of flash floods in coastal areas (Choudhury et al. 2004). In addition, shrimp farmers cut embankments to allow saline water into their shrimp fields which makes embankments vulnerable and easily further damaged due to tidal pressure. This pressure is highest particularly during high tides associated with the full moon (Auerbach et al. 2015; Hossain et al. 2008; Saari and Rahman 2003). Our study indicates that government organizations and NGOs should take care of embankments, raising these where required, restoring river banks, providing safe water supplies to communities, and installing tube wells and rainwater harvesting facilities in vulnerable villages.

8.4 CONCLUSIONS – A WAY FORWARD

Our study has found that coastal areas are being adversely affected by coastal flood, cyclones, storm surges, salinity ingression, and extreme events associated with climate change. The lives and livelihoods of the common people are severely affected by climate change stresses in all the study areas. The poor, women and marginal communities are particularly exposed and vulnerable to the impacts of these hazards. Agricultural land in coastal areas has been damaged due to the soil salinization, flooding, the rising seawater level, and scarcity of freshwater. Despite efforts to adapt, communities are on a sliding downward spiral due to increasing salinity.

Our study has shown that during the dry (rabi) season irrigation with more saline canal water has significant adverse effects on the yield of maize, watermelon, and pumpkin (Murad et al. 2018). Our study has suggested that a disaster early warning system, better coastal afforestation and reforestation, embankment improvement, the growth of flood and salt-tolerant crops, and resilient crop management practices could help reduce the loss of agricultural production. Authorities should ensure the availability of quality agriculture inputs and equipment including seeds, fertilizers and pesticides, power tillers, pumps, and spray machines. A key intervention is the implementation of community led adaptation techniques.

It is evident from the analysis of this chapter that there is an urgent need for organized, structured, and rigorous scientific research in Bangladesh that is conducted in partnership with affected communities. However, these issues are not confined to Bangladesh: they are worldwide. Overall, the global research base for saline agriculture is still weak and at an early stage. The challenges to agriculture from salinity are increasing with the increase in SLR, and with climate change, cyclones are increasing in frequency, and intensity across the world. Saline agriculture affects several different types of regions and ecosystems. These include the coastal areas, deltas interacting with marine water, small islands of many countries, and the Small Island Developing States (SIDS). All of these are particularly vulnerable to the ingress of increasing salinity. Salinity affects surface land, water, and groundwater systems. No country can undertake all the different components, enhanced resilience, growth and productivity in crops, vegetables, fisheries, livestock, and poultry, which are most needed for the survival and sustainability of coastal communities.

There is a strong need to develop global saline agriculture research systems with regional hubs and country-based programs to advance and exchange research, undertake field trials, and advance existing, new, and more appropriate crops to ensure the survival, sustainability, and consumption of agriculture products from the coastal saline agriculture systems.

For these, significant organizational planning, technical capacities, infrastructure, and financing would be required. These aspects are being dealt with in the first paper of this book (Vellinga et al. 2019). It is most urgent that we rapidly improve saline agriculture systems. Otherwise, huge populations and many communities will be forced to migrate or be forcibly displaced. In many countries, such as Bangladesh, such migration would create huge additional problems for both guest and host communities. National agricultural research systems-NARS, appropriate national coastal agencies, international organizations such as FAO, WFP, IFAD, IRRI, World Bank group, and other international financial institution, such as Asian, African, Latin American Development Banks will need to be involved in a national, regional, and global initiatives on saline agriculture research, development, implementation, marketing, and financing. The building of capacity and resilience in affected communities will be the key objectives of these initiatives.

ACKNOWLEDGMENTS

This manuscript was prepared based on the output of multiyear and multiple projects of the BCAS. The authors are grateful to the Department of Environment (DoE) for providing the necessary support and related project reports, the Bangladesh Water Development Board (BWDB) for providing the salinity data, the Water Resources Planning Organization (WARPO) for coastal land data, the Department of Agricultural Extension (DAE) for information about coastal agriculture, the Bangladesh Agricultural Research Institute (BARI) for information on crop salt-tolerant varieties, the Bangladesh Agricultural Research Council (BARC) for crop and soil data, and the Soil Resource Development Institute (SRDI) for soil salinity data. We are grateful to local-level NGOs particularly to communities and their participants in assisting our research.

REFERENCES

Akber, M. A., M. M. Patwary, M. A. Islam, M. R. Rahman. 2018. Storm protection service of the Sundarbans mangrove forest, Bangladesh. *Natural Hazards* 94, 405–418.

Alam, M. Z., L. Carpenter-Boggs, S. Mitra, et al. 2017. Effect of salinity intrusion on food crops, livestock, and fish species at Kalapara Coastal Belt in Bangladesh. *Journal of Food Quality* 2017, ID 2045157, 23–46

Amores, M. J., F. Verones, C. Raptis, R. Juraske, S. Pfister, F. Stoessel. 2013. Biodiversity impacts from salinity increase in a coastal wetland. *Environmental Science and Technology* 47, 6384–6392.

Aryal, J. P, T. B. Sapkota, D. B. Rahut, T. J. Krupnik, S. Shahrin, M. Jat. 2020. Major climate risks and adaptation strategies of smallholder farmers in coastal Bangladesh. *Environmental Management* 66, 105.

Auerbach, L., J. S. Goodbred, D. Mondal C. et al. 2015. Flood risk of natural and embanked landscapes on the Ganges–Brahmaputra tidal delta plain. *Nature Climate Change* 5(2), 153–157.

Ayers, J., and T. Forsyth. 2009. Community-based adaptation to climate change. *Environment: Science and Policy for Sustainable Development* 51, 22–31.

Barbier, E. B. 2015. *Climate change impacts on rural poverty in low-elevation coastal zones*, United States: The World Bank.

Barua, P., S. Chowdhury, S. Sarker. 2010. Climate change and its risk reduction by mangrove ecosystem of Bangladesh. *Bangladesh Research Publication Journal* 4, 208–225.

Baten, M. A., L. Seal, K. S. Lisa. 2015. Salinity intrusion in interior coast of Bangladesh: Challenges to agriculture in south-central coastal zone. *American Journal of Climate Change* 4, 248.

BCAS.2012. Bangladesh Centre for Advanced Studies, Community Based Adaptation (CBA): Early learning from CBA conferences, approaches, practices, challenges and way forward (CBA1 to CAB 6), 29-50, Dhaka, Bangladesh.

BBS. 2011. Bangladesh Bureau of Statistics (BBS), population & housing census 2011, Statistics Division, Ministry of Planning, Government of the People's Republic of Bangladesh.

BBS. 2015. Bangladesh Bureau of Statistics (BBS), Bangladesh disaster related statistics 2015: climate change and natural disaster perspectives, Statistics and Informatics Division, Ministry of Planning, Government of the People's Republic of Bangladesh.

Brammer, H. 2014. Bangladesh's dynamic coastal regions and sea-level rise. *Climate Risk Management* 1, 51–62.

CCC. 2016. Assessment of Sea Level Rise on Bangladesh Coast through Trend Analysis, Climate Change Cell (CCC), Department of Environment, Ministry of Environment and Forests, Bangladesh.

Chen, J., and V. Mueller. 2018. Coastal climate change, soil salinity and human migration in Bangladesh. *Nature Climate Change* 8, 981–985.

Choudhury, N.Y, A. Paul, B. K. Paul. 2004. Impact of costal embankment on the flash flood in Bangladesh: a case study. *Applied Geography* 24, 241–258.

Clarke, D., S. Williams, M. Jahiruddin, K. Parks, M. Salehin. 2015. Projections of on-farm salinity in coastal Bangladesh. *Environmental Science: Processes & Impacts* 17, 1127–1136.

Dasgupta, S., M. M. Hossain, M. Huq, D. Wheeler. 2014. *Climate change, soil salinity, and the economics of high-yield rice production in coastal Bangladesh*, United States: The World Bank.

Dasgupta, S., M. M. Hossain, M. Huq, D. Wheeler. 2015. Climate change and soil salinity: The case of coastal Bangladesh. *Ambio* 44, 815–826.

Dasgupta, S., M. M. Hossain, M. Huq, D. Wheeler. 2018. Climate change, salinization and high-yield rice production in coastal Bangladesh. *Agricultural and Resource Economics Review* 47, 66–89.

Habiba, U., M. A. Abedin, R. Shaw, A. W. R. Hassan. 2014. Salinity-induced livelihood stress in coastal region of Bangladesh. *Water insecurity: A social dilemma*, United Kingdom: Emerald Group Publishing Limited.

Haque, S. A. 2006. Salinity problems and crop production in coastal regions of Bangladesh. *Pakistan Journal of Botany* 38, 1359–1365.

Hossain, M., M. Islam, T. Sakai, M. Ishida. 2008. Impact of tropical cyclones on rural infrastructures in Bangladesh. *Agricultural Engineering International: CIGR Journal* 2, 1–13.

IPCC. 2014. Intergovernmental Panel on Climate Change (IPCC) Working Group II. Climate Change 2014: Impacts, Adaptation, and Vulnerability. Philadelphia: Saunders. www.ipcc.ch/report/ar5/wg2/.

Islam, M, Arifuzzaman, H. M. Shahin, S. Nasrin. 2013. Effectiveness of vetiver root in embankment slope protection: Bangladesh perspective. *International Journal of Geotechnical Engineering* 7, 136–148.

Islam, M. A., N. Warwick, R. Koech, M. N. Amin, L. L. D. Bruyn. 2020. The importance of farmers' perceptions of salinity and adaptation strategies for ensuring food security: Evidence from the coastal rice growing areas of Bangladesh. *Science of the Total Environment* 727, 138674.

Ji, C., Y. Zhang, Q. Cheng, Y. Li, T. Jiang, X. San Liang. 2019. Analyzing the variation of the precipitation of coastal areas of eastern China and its association with sea surface temperature (SST) of other seas. *Atmospheric Research* 219, 114–122.

Mahmuduzzaman, M., Z. U. Ahmed, A. Nuruzzaman, F. R. S. Ahmed. 2014. Causes of salinity intrusion in coastal belt of Bangladesh. *International Journal of Plant Research* 4, 8–13.

Mallick, B., K. R. Rahaman, J. Vogt. 2011. Coastal livelihood and physical infrastructure in Bangladesh after cyclone Aila. *Mitigation and Adaptation Strategies for Global Change* 16(6), 629–648.

Murad, K. F. I., A. Hossain, O. A. Fakir, S. K. Biswas, K. K. Sarker, R. P. Rannu, J. Timsina. 2018. Conjunctive use of saline and fresh water increases the productivity of maize in saline coastal region of Bangladesh. *Agricultural Water Management* 204, 262–270.

NCAR. 2020. National Center for Atmospheric Research Staff (Eds). Last modified 27 Feb 2020. "The Climate Data Guide: SST data: HadiSST v1.1." Retrieved from https://climatedataguide.ucar.edu/climate-data/sst-data-hadisst-v11.

Nowreen, S., M. R. Jalal, A. K. M. Shah. 2014. Historical analysis of rationalizing South West coastal polders of Bangladesh. *Water Policy* 16, 264–279.

Paul, B. K., and H. Rashid. 2017. Climatic hazards in coastal Bangladesh. *Non-structural and structural solutions*, United Kingdom: Elsevier.

Paul, P. L. C., R. W. Bell, E. G. Barrett-Lennard, E. Kabir. 2020. Variation in the yield of sunflower (Helianthus annuus L.) due to differing tillage systems is associated with variation in solute potential of the soil solution in a salt-affected coastal region of the Ganges Delta. *Soil and Tillage Research* 197, [104489]. https://doi.org/10.1016/j.still.2019.104489.

Rahman, M., M. S. Uddin, M. Uddin, S. A. Bagum, N. Halder, M. Hossain. 2004. Effect of different mulches on potato at the saline soil of southeastern Bangladesh. *Journal of Biological Sciences* 4, 1–4.

Rahman, M. M. and M. M. Rahaman. 2018. Impacts of Farakka barrage on hydrological flow of Ganges River and environment in Bangladesh. *Sustainable Water Resources Management* 4, 767–780.

Rasid, H., and B. Paul. 2013. *Climate change in Bangladesh: confronting impending disasters*, United States: Lexington Books.

Rignot, E., I. Allison, N. L. Bindoff, et al. 2009. The Copenhagen diagnosis. *Updating the world on the latest climate science*, The University of New South Wales Climate Change Research Centre, Sydney.

Salehin, M., M. M. A. Chowdhury, D. Clarke, S. Mondal, S. Nowreen, M. Jahiruddin. 2018. Mechanisms and drivers of soil salinity in coastal Bangladesh. *Ecosystem services for well-being in deltas*. Cham: Palgrave Macmillan, 333–347.

Saari, M., and S. Rahman. 2003. Development of the coastal embankment system in Bangladesh. *Soft shore protection*, Springer, 115–125.

Schipper, E. L. F., J. Ayers, H. Reid, S. Huq, A. Rahman. 2014. *Community-based adaptation to climate change: Scaling it up*, Routledge.

SRDI. 2020. Soil Resource Development Institute, subdistricts supplements 2020, Soil Resource Development Institute, Ministry of Agriculture, 'Mrittika Bhaban', Government of the People's Republic of Bangladesh.

Thornton, P. K., R. B. Boone, J. Ramírez Villegas. 2015. Climate change impacts on livestock. Working Paper No. 120. CGIAR Research Program on Climate Change, Agriculture and Food Security (CCAFS). Copenhagen, Denmark

Uddin, M. N., A. S. Islam, S. K. Bala, G. T. Islam, S. Adhikary, D. Saha. 2019. Mapping of climate vulnerability of the coastal region of Bangladesh using principal component analysis. *Applied Geography* 102, 47–57.

Unnikrishnan, A., and D. Shankar. 2007. Are sea-level-rise trends along the coasts of the north Indian Ocean consistent with global estimates? *Global and Planetary Change* 57, 301–307.

Vellinga, P., A. Rahman, B. Wolthuis, K. Negacz, R. Choukrallah, E. G. Barrett-Lennard, T. Elzinga. 2019. Saline agriculture, building communities pushing innovation, local and global. Future saline conference, 10–13 September 2019, Leeuwarden, the Netherlands.

9 Innovations of the 21st Century in the Management of Iranian Salt-Affected Lands

Zeinab Hazbavi and Mostafa Zabihi Silabi

CONTENTS

9.1 INTRODUCTION

All life on Earth is anchored by soil and sustained by water. The European Commission (EC 2006) has highlighted seven functions of soil including biomass production (viz., agriculture and forestry), the storing, filtering and transforming of nutrients, substances and water, as a pool for biodiversity (such as in habitats, species and genes), a physical and cultural environment for humans and human activities, a source of raw material, acting as a carbon pool and as an archive of geological and archaeological heritage. Water is the most widespread substance that plays a crucial role both for the environment and in human life (Stec 2020). However, a range of natural and human-based factors are affecting soil and water functions (Sadeghi et al. 2019; Hazbavi et al. 2020), making them saline in many parts of the world. About 1125 Mha of the Earth's surface in more than 100 countries is affected by salinity, and these numbers are continually growing (Ivushkin et al. 2019). The salinity

DOI: 10.1201/9781003112327-9

problem has been threatening sustainable land development and is one of the significant causes of land degradation from ancient times until now. Salinity is a universal concern for the 21st century because of its accelerating impact on the quality of life, particularly food security. The Middle East, including Iran (i.e., the Islamic Republic of Iran), has significant areas of salt-affected soils (Ivushkin et al. 2019). After India and Pakistan, Iran is one of the most vulnerable countries to salinity (Moameni 2011). Numerous studies have been done to assess, monitor, evaluate, map and combat the salinity of soil and water resources in Iran (e.g., Tavousi et al. 2018; Hazbavi et al. 2019; Salehi and Dehghani 2019; Bagheri 2019; Fathizad et al. 2020; Mahmoodi-Eshkaftaki and Rafiee 2020). The present review summarizes Iranian activities in the salinity area over the first two decades of the 21st century. This research presents a useful road map for future policy making at regional, national and global levels.

9.2 SALT-AFFECTED REGIONS IN IRAN

Iran is located in West Asia; it has an area of 164.8×10^6 ha (163.6×10^6 ha land and 1.2×10^6 ha water bodies) (Qadir et al. 2008). Iran is home to several salt diapirs, lakes and marshes, including Lake Urmia, one of the largest hypersaline lakes in the world. The main soil types of Iran are Xerosols, Arenosols, Regosols, Solonchalks and Lithosols (Siadat 1998). The mass of saline waters in the country is also significant. Shiati (1998) reported that of the country's 100 billion m^3 of water resources, about 11 billion m^3 have a salinity of over 1500 mg L^{-1}. Several rivers of the south, southwest and central parts of the country are saline (Ranjbar and Pirasteh-Anosheh 2015).

Figure 9.1 shows the spatial variation of soil salinity in Iran (Banie 2001). Soil salinity hotspots occur on the Khuzestan Plain, central plateau, southern coastal plain and in the inter-mountain valleys. In addition, about 50% of irrigated lands fall into different categories of salt-affected soils. The salinization of land and water resources in Iran has been the result of natural and anthropogenic conditions. The arable soils of Iran show some evidence of a long history of anthropological interference. Around 11% of Iran is cultivated land, which also includes 8.1 Mha of irrigated agriculture (ICID 2002; Qadir et al. 2008). The latest assessments have shown a 6% increase in the total cultivated area since the late 1990s (Qadir et al. 2008). Average crop yields are higher with irrigation than under dry-farming conditions, and this has led to a 3.8% expansion of irrigated farming systems between the 1980s and early 1990s. During the second national development plan (1993–98), there was an increase of 0.5×10^6 ha in irrigated land which had consequences for salinization. The nonexistence of appropriate water management tools resulted in the low effectiveness of water use; the overall efficiency was ~30%. The level of land affected by salinity also varies in the different provinces of the country. In Khuzestan and Fars provinces, salinity accounts for 32.9 and 16.4% of the land area, respectively; but in East Azarbaijan Province, only 1.36% is affected. Figure 9.2 shows a general view of the soil salinity status in irrigated lands of Iran. Soils with EC_e values of 0–4, 4–8, 8–16, 16–32 and more than 32 dS m^{-1} are, respectively, designated as soils with no salinity, slight, moderate, strong and very strong salinity (World Bank 2005).

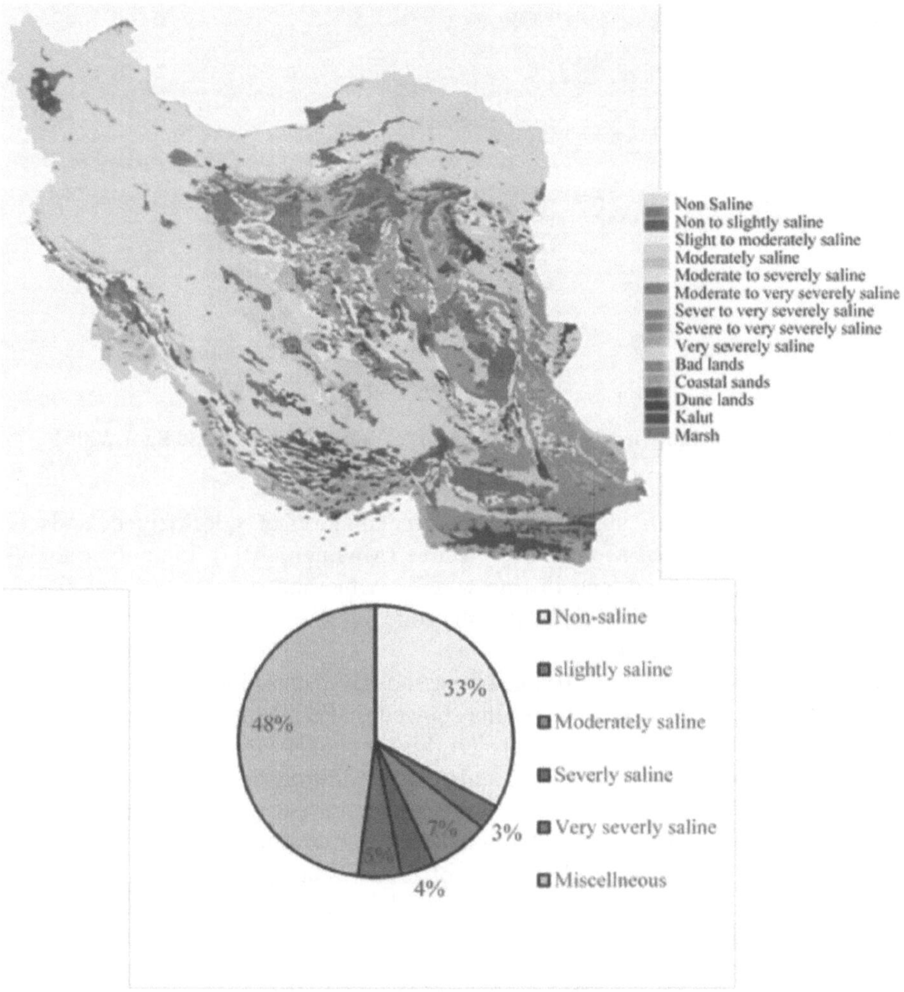

FIGURE 9.1 Spatial variation of salinity in Iran. (Source: Banie 2001.)

The oldest report on salinity in Iran comes from Dewan and Famouri (1964) who indicated that saline and alkaline soils accounted for about 12.5% of the country's total area. Other reports have referred to salinity accounting for 25–27 Mha (15–17% of total land area) (Sayyari and Mahmoodi 2002) and 34 Mha (20.6% of the total land area), including 25.5 Mha of low to medium salinity and 8.5 Mha of high salinity (Moameni et al. 1999). Of the total of 6.8 Mha of salt-affected agricultural land, about 4.3 Mha have only salinity limitations whereas the other 2.5 Mha have other additional limitations including a susceptibility to erosion and a shallow depth to water table. In most salt-affected soils in agricultural areas, the depth to the water table is deeper than the root zone. (Areas in which the depth to water-table falls within the root zone are only ~8.4% of the total salt-affected

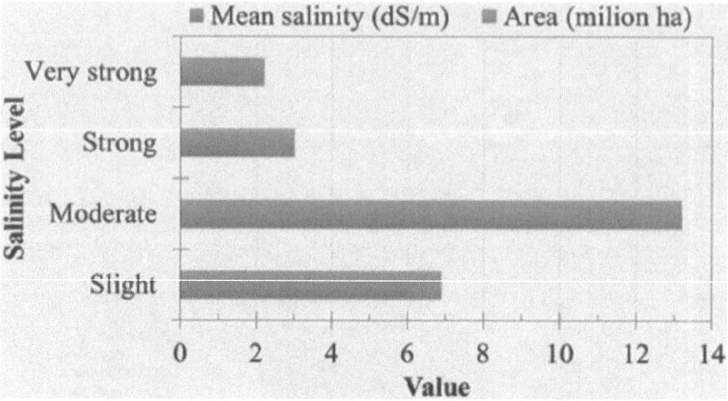

FIGURE 9.2 Soil salinity levels of irrigated lands of Iran. (Source: World Bank 2005.)

agricultural lands). This suggests that the reclamation of salt-affected lands is technically practical and economically viable (Moameni 2011). In another study, the total area of irrigated land in Iran was 7.3 Mha, and the total area of arable land with varying degrees of salinity of the soil, water, or both was estimated to be 3.3 Mha (Banaei et al. 2004).

Most recently, Moameni (2011) has assessed the geographical distribution and salinity levels of Iran's soils. The results showed that 6.8 Mha of agricultural land have soils with different salinity levels. Of this, about 3.4 Mha have no restrictions other than salinity, and about 2.5 Mha also have additional restrictions caused by variation in texture, depth, drainage coefficient, poor permeability and high risk of erosion; about 0.5 Mha (8.4%) have shallow groundwater problems in the root zone (Moameni 2011). The salinization of land resources in Iran has been the consequence of both natural and human activities (Siadat et al. 1997; FAO 2000; Qadir et al. 2008).

9.3 MANAGEMENT STRATEGIES FOR SALT-AFFECTED AREA

Different research institutes and universities have conducted research on the management of soil and water salinity in Iran for more than half a century. There are many effective ways for improving salt-affected lands, such as water leaching, drainage water management, soil management, chemical remediation, phytoremediation, growth and genetic improvement of salt-tolerant crops and use of irrigation scheduling practices. The first field works to evaluate different techniques for improving salt-affected soils were started in the 1970s by the Soil and Water Research Institute (SWRI; www.swri.ir). The Agriculture Biotechnology Research Institute of Iran (ABRII) was established in 1999 to develop and use modern agricultural technologies to solve agricultural problems, improve food security and community nutritional health and protect essential resources and environmental substrates for sustainable development. This institute carried out a comprehensive project on *Salicornia* to produce forage, oil and other biological sub-products and mitigate against dust storms. In addition, numerous soil and water scientists, conservationists and ecologists were engaged in projects focused on salinity mitigation, control and management. In 2000, the National Salinity Research Center (NSRC; www.nsrc.areeo.ac.ir) was established to focus on all research activities in the field of salinity. This center now tackles diverse aspects of soil and water salinity problems in different agro-ecological regions. The main areas included in salinity management projects have been in Yazd, Golestan, Fars, Khurasan, Khuzestan, Markazi, Hormozgan (Bushehr), Moghan, Azerbaijan, Esfahan and Qom provinces (Heydari 2019).

9.3.1 Utilization of Floodwaters

The use of seasonal floods in agriculture and horticulture has a long history in Iran and elsewhere in the world. Sabzevar and Khosroshahi (2010) investigated the effects of low-quality floodwater on the desert area of the KaleShoor River, Sabzevar. They concluded that flood irrigation was able to leach gypsum from the root zone into deeper soil layers. This approach was also of benefit in restoring and improving pastures in different areas. By creating a diversion bar on the margin of the BarAbad Desert in Sabzevar, rier, seasonal floods of the KaleShoor River with a salinity of 7–14 dS m^{-1} were used to revitalize and modify pastures revegetated with *Atriplex canescens* through irrigation once a year. With this treatment, forage production increased from 10 to more than 700 kg ha^{-1}. This practice allowed for the development of dairy farming and decreased rural migration. Though droughts and environmental conditions have caused some parts to be dry in some years, the projected area was increased by more than 3000 ha over 10 years, and species such as *Nitraria schoberi*, *Seidlitzia rosmarinus* and *Haloxylon* spp. were cultivated (Filehkesh and Hashemi Nezhad 2017). Floodwater spreading and floodwater farming have also improved crop production in Africa (Asch and Woperei 2001), America (Nabhan 1979), China (Guo et al. 2008; Seydehmet et al. 2019) and Spain (Hooke and Mant 2002).

9.3.2 Leaching

The use of soil leaching can decrease the adverse effects of salt on crop establishment (Heydari 2019). This approach is especially relevant to soils with high levels of clay or hardpans in the Khuzestan, Isfahan and Tabriz plains. Khoshgoftarmanesh and Shariatmadari (2002) did a ground survey to explore the leachability of saline soils in Qom Province. Soils initially had EC_e values of 67.1 and 54.7 dS m^{-1} in the surface and sub-surface layers; leaching water came from rainfall and snowmelt. With leaching after one year, EC_e values in soil were less than EC_w of the irrigation water. Severe symptoms of delayed germination and the burning of leaf margins were observed in control plots without leaching, whilst improved crop yields were reported for plots treated with salt leaching.

Moameni and Stein (2002) and Qadir et al. (2008) concluded that incorporating leaching with an improved drainage system could sustainably ameliorate most of the soils of Iran. Azadegan (2008) noted the benefits of leaching, drainage and sulfur application in improving the quality of *pistachio* orchards growing on saline soils in Garmsar County, Semnan Province. Bazzaneh and Rezaei (2017) evaluated the leaching of saline-sodic soils using different levels of pure and acidic water in some parts of the Mahabad Plain (West Azerbaijan Province) irrigation and drainage network. In this study, to evaluate the possibility of modification of saline-sodic soils, different treatments with pure and acidified waters at three levels of application (25, 50 and 75 cm) were repeated in a randomized complete block design in three days. The results showed that the lowest and highest percentages of salinity and alkalinity improvement were related to the application of 25 cm pure water or 75 cm acidic water with 24.7 and 41.2% decreases in mean values of exchangeable sodium and 25.9 and 69.0% decreases in EC_e values compared to initial soil conditions, respectively.

The international scientific literature suggests that the leaching requirement varies with the level of salinity in the irrigation water and with crop salt tolerance (Kolahchi and Jalali 2007; Mostafazadeh-Fard et al. 2009; Heydari 2019). In other regions of the world, this technique has been studied and applied to achieve varying target outcomes. For instance, Corwin et al. (2007) described the use of steady-state and transient models to characterize leaching requirements for soil salinity control. Ning et al. (2020) evaluated the irrigation water salinity and leaching fraction on the water productivity of barley, bean, wheat and maize crops in China. The required leaching fraction depended on the salt tolerance of the crop and the salinity of the irrigation water. Despite the advantages of leaching, its application has not always been successful or accepted by decision-makers. Ning et al. (2020) noted that the risk of environmental contamination might be increased due to the leaching of pesticides, nutrients and trace elements. Furthermore, it is worth noting that drought and water crises exist in many regions because there is limited access to freshwater. Using this method for salinity abatement may, therefore, gain less attention in the future.

9.3.3 Reusing Drainage Waters

The use of drainage and saline waters for agricultural irrigation or other objectives involves several management methods (Nasrollahi et al. 2017). Choosing the appropriate method for releasing or reusing drainage waters mainly depends on the quality of the water concerned. Water with an EC_w of up to 8 dS m^{-1} can be used for the irrigation of salt-resistant crops such as barley, sugarbeet, rapeseed and cotton (Homaei 2003). Nasrollahi et al. (2017) examined the effects of drip irrigation using saline water on corn crops at the research farm-scale, at Shahid Chamran University in Khuzestan Province. Drip irrigation with proper management appeared suited to the reuse of vast quantities of drainage water. Sharifipour et al. (2017) investigated the reuse of saline drainage water using a system dynamics analysis tool (Vensim) in the western part of the Karun River, a region where the land is ~99% saline. Their results showed that, for at least the next 20 years, the quality of drainage water released from south Karkhe and west Karun River Basins will not be suitable for the cultivation of saline resistant crops except for sugarcane. However, the drainage effluent from these areas will be suitable for developing vegetation in rangelands. The most critical challenge in controlling dust production in the south of Khuzestan Province is the supply of water to these areas; vegetation development with these effluents will be possible shortly.

Rhoades and Dinar (1991) conducted a study in the United States on the reuse of agricultural drainage water to increase water supplies for irrigation. They first intercepted and isolated the drainage water from the good-quality water; this water was then reused for the irrigation of suitably salt-tolerant crops, decreasing the volume of secondary drainage water that needed to be disposed of. Hussain et al. (2019) have reviewed the use and management of non-conventional water resources such as saline water, wastewater and graywater for the rehabilitation of the arid and semi-arid MENA (the Middle East and North Africa) regions. They concluded that the recovery of marginal and saline degraded lands via appropriate planning and the sustainable use of different sources of non-conventional water could be an economic and environmental investment.

9.3.4 Dry Drainage

Dry drainage refers to the planting of irrigated crops in slightly higher areas adjacent to lower fallow land. "Drainage" can occur if the irrigation results in the development of a shallow water-table, but the water actually evaporates from the adjacent fallow area (Figure 9.3). This method, initially introduced by Gowing and Wyseure (1992) can contribute to the removal of excess water and salt from irrigated land and control soil salinity (Wu et al. 2009; Mostafazadeh-fard and Ghasemi 2016). Dry drainage is an appropriate alternative management tool where artificial drainage is not applicable (Wu et al. 2009). Dry drainage has been recently introduced into some regions of Iran such as Tehran and Isfahan as a novel method for the drainage of agricultural areas. This method has potential for arid and semi-arid areas with a shallow water-table and high potentials for evaporation

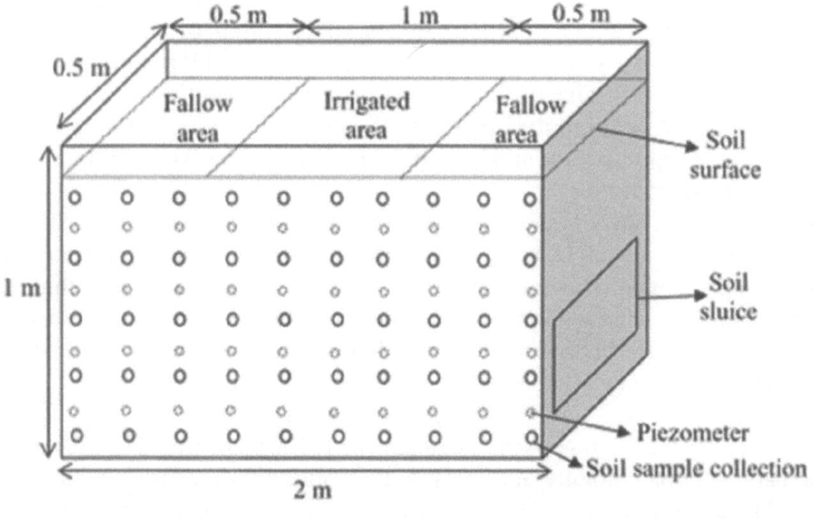

FIGURE 9.3 Diagram of the dry drainage simulation model. (Source: Ansari et al. 2017.)

and capillary rise (Soltani et al. 2017; Ansari et al. 2017, 2019). Mostafazadeh-Fard and Ghasemi (2016) tested the effects on dry drainage of depth to water-table, irrigation water salinity, groundwater salinity and soil salinity. Their results indicated a significant reduction in soil salinity in a study area without polluting surface or underground water. This technology has potential for areas that have an impermeable layer near the soil surface and a hydraulic gradient between the irrigated and non-irrigated (fallow) fields.

In China Wang et al. (2019) also investigated the efficiency of the dry drainage method for soil salinity control at a field location over five years (2007–2011). The established dry drainage system played an important role in draining excess water and salt. Konukcu et al. (2006) evaluated the application of dry drainage in the Lower Indus Basin in Pakistan. This area was characterized by shallow saline groundwater, intensive irrigation and high evaporation. The results showed a satisfying balance between water and salt when the cropped area and sink area were approximately equal, and the water-table depth was around 1.5 m.

9.3.5 Saline Aquaculture

Nakhaei et al. (2018) conducted a project to use brackish and saline water resources in Kerman Province from 2008 to 2009 in the city of Ravar-Shahrbabak-Sirjan-Zarand and Rafsanjan. They concluded that the culture of rainbow trout, sturgeon, seabream, milkfish, carp, tilapia and artemia are feasible. Typical pictures of the artemia ponds established during the project are shown in Figure 9.4. In the study region, there are many saline water sources which are not suitable for agriculture, drinking or industry. In addition, with the movement of salty water towards freshwaters, the volume of freshwater has also been reduced. The results of this research showed that the salinity of these waters varies from 5.12 to 96 dS m^{-1}. Regardless of source, the water ranged in temperature (lowest to highest) from 13.5 to 24.5°C, and the pH varied from 6.8 to 8.6. Manaffar et al. (2020) studied the reproduction and life span characterization of *Artemia urmiana* in Lake Urmia at five sampling stations of Heydarabad, Golmankhaneh, Bari, Kabodan and Eslami. The different environmental conditions at the various locations in the lake caused differences in the growth, survival and reproduction of the fish; best reproduction and life span occurred at the Bari, Golmankhaneh and Heydarabad locations.

At a global level, aquaculture provides more than 50% of food needs, and is estimated to expand to about 93.2×10^9 kg by 2030; aquaculture will therefore be a leading food supply source (Hawrot-Paw et al. 2020). To this end, several technologies are developing to increase saline aquaculture production. For instance, in Poland Hawrot-Paw et al. (2020) used a growth medium containing *Chlorella minutissima* and verified its efficiency in saline aquaculture. In India Debroy et al. (2020) cultured shrimp (*Litopenaeus vannamei*) as a potential species for sustainable production in inland saline waters. High growth rate, tolerance to high stocking density, tolerance to fluctuations in temperature and salinity, utilization of a low protein diet, a high meat yield, easy spawning and high disease resistance were noted as advantages of this shrimp for saline aquaculture.

FIGURE 9.4 *Artemia* culture using saltwater (a) and a sample of the cultured *artemia* (b). (Source: Nakhaei et al. 2018.)

9.3.6 Halophytes as Salt-Tolerant Crops

The importance of salt-tolerant plant species for the restoration of salt-affected soils has been evaluated in several parts of the country (Koocheki 1996). Halophytes are important species in the rangeland communities of salt-affected areas. While close interactions have been recognized between livestock and halophytic communities as forages, these linkages have not been sufficiently employed at scale. Utilizing halophytes for various goals (e.g., for fodder, manufacturing and domestic applications) has been well-received by indigenous people. Qadir et al. (2008) reported that 16 halophytic plant families contain 92% of the halophytic species identified in the country. *Atriplex* is a prospective and economic forage source and is an agent in combating desertification in arid lands (Koocheki 1996; 2000).

The physiological characteristics of nine halophytic forage plants were evaluated at the Research Station of Chah Afzal, Yazd in 2011–2012. There were significant physiological differences between species, which may be related to their ability to produce forage. Amongst the species studied, *Kochia indica*, *Sesbania aculeata* and

Atriplex halimus were selected as superior species due to their higher forage production and low ash, sodium and chloride concentrations (Banakar et al. 2012). Soleimani and Najafifar (2017) investigated the effects of several hand-planted halophytes on the salinity and sodium in soils in southeastern Fars Province. Their results showed that regardless of the depth and distance apart of the plant species, the EC_e of the soil in the shade of plants was significantly higher than outside the shade. The effects were more significant at 0–10 cm than at 10–40 cm depth. Various research has also been conducted on the nutritional value of halophytes, including *Gamanthus gamocarpus*, *Petrosimonia glauca*, *Salsola crassa*, *Halotis occulta*, *Halocaris sulphurea*, *Alhagi* sp., *Gundelia tournefortii* and *Kochia prostrata*. Because of their valuable nutrients, and chemical composition, in arid and semi-arid regions they can act as alternatives to wheat straw and dry hybrid straw and help maintain livestock, such as *Baluchi* sheep (Bashtani 2017). To feed the increasing nourishment needs of the world, different actions are adapted in various countries, e.g., India (Kashyap et al. 2020), Vietnam (Paik et al. 2020), China (Nadeem et al. 2020), Italy (Incrocci et al. 2020), Tunisia (Bani et al. 2020) and Oman (Al-Farsi et al. 2020) to find novel solutions to manage, adapt and improve tolerance of various crops against salinity conditions.

9.3.7 Saline Agriculture

Saline soils and water are being used in Iran for saline agriculture (Khorsandi 2016). This concept and related techniques such as biosaline agriculture, seawater agriculture, haloculture, haloengineering and halophyte farming, are also being increasingly adopted in Iran. These, along with other methods to combat salinity, offer a range of benefits such as producing sustainable and economic products, erosion control, land reclamation, environmental quality improvement and improving the socio-economic status of local communities; they are in line with the objectives of the Water-Energy-Food Nexus concept, and take into consideration basic human needs (Khorsandi and Siadati 2017).

The feasibility of halophytic plants as new crops was examined for the first time in Iran through the international project (i.e., INT/5/144 "Sustainable Utilization of Saline Groundwater and Wastelands for Plant Production") (Khorsandi and Siadati 2017). The foundations of haloculture were initiated with the first "biosaline agriculture" research project in Iran (Pedraza 2009). The general principles of haloculture according to the capacities, conditions and needs of Iran are summarized in Figure 9.5.

Because of its range of agro-climatic environments, Iran has a rich genetic diversity of flora, fauna and microorganisms for economic haloculture under different saline situations in a diversity of geographical settings. One example is a successful program of halophytic tree cultivation in the highly saline lands of the Chah Afzal Area, Yazd Province (Khorsandi 2016; shown in Figure 9.6). The generation of energy and management of water are two of the critical activities that occur with haloengineering. Reducing dependency on fossil fuels and on the limited freshwater resources of the country are two goals that could be achieved (Khorsandi 2016; Khorsandi and Siadati 2017).

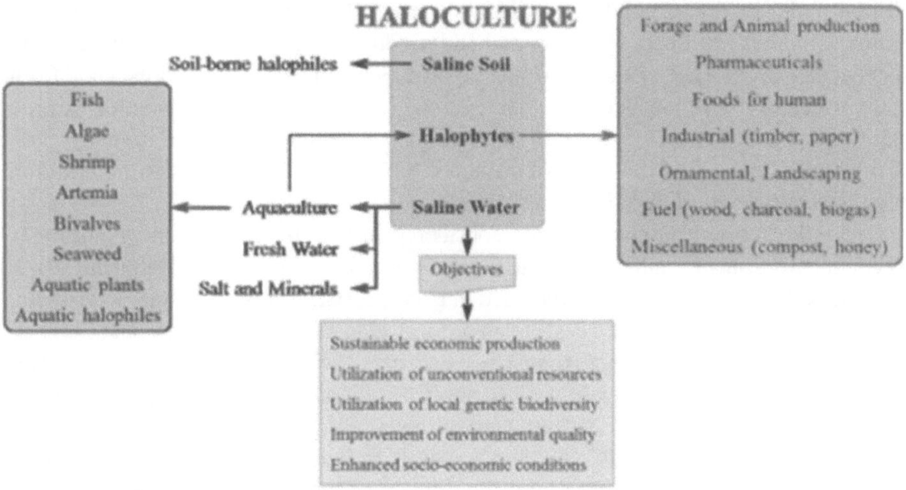

FIGURE 9.5 General view of haloculture principles and objectives. (Source: Khorsandi 2016.)

A review of literature (Ranjbar and Pirasteh-Anosheh 2015) has shown that halophytes studies began in the last half-century but have substantially increased since 2001. The research focus has been mainly on *Kochia scoparia*, *Portulaca oleracea*, *Chenopodium album*, *Chenopodium quinoa* and *Salicornia* spp.

Quinoa (*Chenopodium quinoa*) from the Amaranthaceae family is a discretionary halophyte and is classified as a pseudo-cereal. The cultivation and growth of

FIGURE 9.6 Haloculture services: (a) bare salty land, and (b) restored salty land using halophytes. (Source: Yazd Province, Iran; Khorsandi 2016.)

quinoa under different conditions has occurred in European countries, the United States, Canada, Morocco, Pakistan and India. Quinoa has different mechanisms to survive salt stress. One of the exciting adaptations of this plant is the presence of salt bladder cells on leaf surfaces and panicles, which can absorb excess salt from the plant cells (Salehi and Dehghani 2019). Quinoa is currently cultivated in many parts of the world because of the value of quinoa seed and its high production potential in harsh environments (Khalili et al. 2019).

The cultivation of quinoa has been considered in Iran in recent years (Figure 9.7), where it is promising and resistant to saline conditions (Salehi and Dehghani 2019; Tavousi et al. 2018). This plant is cultivated in Iran in Sistan and Baluchestan, Khuzestan, South Kerman and Karaj provinces, which showed adaptation in November, late October, early October and the Middle of August, respectively (Tavousi et al. 2018). The main factors of this sound adaptation could be attributed to the appropriateness of temperature and precedent soil moisture. Seed yields of quinoa of 3 t ha^{-1} have been obtained under irrigation with saline water (EC_w 14 dS m^{-1}) in Iran. Important selection issues for quinoa are photoperiod sensitivity, and salinity and drought tolerance, but the selection of better genotypes for different climates may be possible depending on the availability of sufficient genetic diversity. Quinoa genotypes have produced 4 t ha^{-1} under rainfed conditions (200 mm rainfall) (Salehi et al. 2018).

The initial establishment of quinoa is affected by irrigation water salinity, initial soil salinity, air temperature, soil oxygen and soil moisture. It is highly tolerant of salinity during germination and can germinate at EC_e values up to 40 dS m^{-1}. However, its susceptibility or tolerance to salinity depends strongly on cultivar.

FIGURE 9.7 Quinoa at the pollination period. (Source: Tavousi et al. 2018.)

Researchers have suggested suitable quinoa cultivars and appropriate areas for quinoa cultivation in the country (Tavousi et al. 2018; Salehi and Dehghan 2019; Bagheri 2019).

In order to examine the effect of electrical conductivity of water (EC_w) and phosphorus levels on some of the characteristics of quinoa plant in greenhouse conditions, a factorial experiment was conducted by Khalili et al. (2019) in a completely randomized design with three replications. The results showed that increasing EC_w to 15 dS m^{-1} (equivalent to EC_e 30.4 dS m^{-1}), decreased plant height by 18.65% and decreased panicle length by 52.4%. When EC_e values reached 183 dS m^{-1}, the sodium concentration in the plants increased by 18.5% compared with the control. Increasing phosphorus application to the soil to 100 kg ha^{-1} increased plant height by 12.3%, increased panicle length by 8.8%, and increased phosphorus concentration in shoots by 12.5%, compared to the control. A comparison of the average interactions between salinity of irrigation water and phosphorus on plant height showed that a salinity of 3 dS m^{-1} increased plant height by 15.1%, compared with the control. Adding phosphorus (100 kg per hectare triple-super phosphate) reduced the effect of salinity stress on the plant (Khalili et al. 2019).

Talebnejad and Sepaskhah (2018) conducted experiments over three years under greenhouse conditions with shallow saline groundwater in Fars Province, to determine the phenology, and responses of quinoa ("Titicaca, no. 5206") to salinity and water stress. Their findings suggested that quinoa could complete its growth cycle and generate a seed yield under Iranian conditions. With full irrigation (800 mm) there was a seed yield of 2.21 Mg ha^{-1}; with a 70% decline in irrigation there was a 40% decrease in seed yield (to 1.16 kg m^{-3}).

Akhani et al. (2003) described the *Salicornia persica* AKHANI sp. novo (Chenopodiaceae) from the central inland salt marshes of Iran in Esfahan, Fars and Yazd provinces. Zare and Keshavarzi (2007) conducted other research to identify the morphological characteristics of the Salicornieae (Chenopodiaceae) tribe native to Iran.

The Nanobiotechnology Research Institute has conducted a comprehensive plan on the ability of agricultural halophytes to produce forage, oil and other biological sub-products and prevent dust storms. Some important characteristics mentioned for the Salicornieae are including the possibility of irrigation with unconventional saline water and seawater (with a concentration of 27.734–33.281 dS m^{-1}) producing at least 15 t ha^{-1} dry fodder, a potential yield of 1.5 t ha^{-1} of high-quality edible oil, and a potential yield of 700–1800 L ha^{-1} biofuel, application as a meal in poultry nutrition, productive employment creation, high salt storage capacity in aerial parts, saline soils emendation and potential application in bioremediation of heavy metals and petroleum contaminants from soil. Figure 9.8 shows some views of the Salicornieae produced in this project (http://www.abrii.ac.ir/en/).

9.3.8 Executive Operations around Lake Urmia

Lake Urmia (LU) is the one of the largest salt lakes in the world, and is located in northwestern Iran. The Lake Urmia Basin (LUB) has an area of 520 Mha, or 3.15% of the country, and supports a population of more than three million people.

FIGURE 9.8 Implantation of superior Salicornia genotypes at the Nanobiotechnology Research Institute. (Source: http://www.abrii.ac.ir/en/.)

Fourteen permanent rivers provide the water (3.1 billion m^3 annually) needed to maintain LU's water balance. Over the past two decades, unsustainable human activities have put increasing pressure on the natural resources of the LUB and the lake has decreased in size. Intensification of agricultural activities, the development of water resources and the destruction of natural resources, have increased salinity in the water to more than 300 g L^{-1}; recent droughts have worsened the situation (Fanni and Maroofi 2017). The Urmia Lake Restoration National Committee (ULRNC; www.ulrp.ir) was founded in July 2013. It implemented measures during the years 2014 and 2015 to reclaim saline land and control the saline dust blowing off the lake. These measures have included the allocation of 55 billion Rials for

FIGURE 9.9 Production of Atriplex, Nitraria and Tamarix seedlings for planting in the areas around Lake Urmia. (Source: www.ulrp.ir.)

FIGURE 9.10 Surface runoff and sewage control operations in saline lands around Lake Urmia. (Source: www.ulrp.ir.)

FIGURE 9.11 Construction of a quantification station of wind erosion and salt dust events. (Source: www.ulrp.ir.)

LUB restoration objectives, identifying critical areas and sites for geology and soil profiling operations on 7600 ha, combating desertification on 3200 ha using native species (*Atriplex*, *Nitraria*, *Tamarix*; Figure 9.9), improving surface runoff and sewage control operations (Figure 9.10). Land restoration processes and processes to increase plant cover have included the use of curved pits (150 ha), contour furrows (1450 ha) and pitting (1000 ha), seeding (2600 ha) and drainage channels (7500 m), the construction of windbreaks using Tamarix planting (528 ha), fence building (3200 ha), grazing management (200,000 ha) and the conservation of forests (300 ha). In addition, a new station to quantity wind erosion and salt dust events has recently been created. This station is equipped with sediment traps (set of three centralized and 64 decentralized traps) and can collect wind deposits at different altitudes (Figure 9.11).

9.4 CONCLUSION

We have discussed some of the crucial research conducted on the management of salt-affected lands in Iran. Table 9.1 summarizes the most important studies. The widely used management activities during the two last decades in Iran were leaching, drainage and irrigation water management, soil amendments application, haloculture, dry drainage, crop rotation systems and saline aquaculture. Use of best management practices (BMPs) in the management of salt-affected areas will be essential in ensuring sustainability and food security without compromising soil and water quality. The key to salinity management centers on assessing, managing and monitoring, whereby the efficiency of management options is monitored and assessed, and modifications are made consequently. Rational salinity management practices must allow for both environmentally and economically sustainable yields and for the restoration of soil and water health. The poor management of salt-affected resources in low-income countries is intrinsically linked to financial and managerial limitations and poor practices in situations of a changing climate, poor infrastructure and poor marketing arrangements.

TABLE 9.1
Summary of Research Done on the Management of Saline Soil and Water Resources of Iran during the Period 2001–2020

Year	Researcher/s	Approach	Case Study	Land Use	Texture	EC (dS m^{-1})	Experiment Type
2002	Khoshgoftarmanesh and Shariatmadari	*Leaching*	Qom Province	Agriculture	-	67.1	Field
2008	Azadegan	Drainage and sulfur application	Garmsar region, Semnan Province	Orchards	Loam-Clay Loam	8.9	Field
2010	Feizi et al.	Leaching	Isfahan Province	Agriculture	Clay	7.7	Field
2011	Yazdanpanah and Mahmoodabadi	Amendments application and leaching	Kerman Province	Agriculture	Sandy Silty	19/81	Laboratory-Soil column
2012	Yazdanpanah et al.	*Soil amendments*	Kerman Province	Agriculture	Loam	19.81	Laboratory-Soil column
2015	Feizi and Saadat	Irrigation management with salt water	Isfahan Province	Agriculture	Clay	6.8	Field
2015	Saghafi et al.	Rhizobacteria and saline water application	Karaj, Alborz Province	Agriculture	Loam	1.28	Pot
2016	Boostani et al.	Rhizobacteria (PGPR) and Arbuscular mycorrhizae fungi (AMF) application	North of Khuzestan Province	Agriculture	Silty Clay Loam	2	Greenhouse
2016	Khorsandi	Haloculture	Chah Afzal, Yazd Province	Agriculture	-	-	Field
2016	Mostafazadeh-fard and Ghasemi	Dry drainage	Isfahan Province	Agriculture	Sandy Loam Clay	3 and 7	Laboratory
2017	Ansari et al.	Dry drainage	Isfahan Province	Agriculture	Loam	3	Laboratory
2017	Bazzaneh and Rezaie	Leaching	Mahabad, West Azerbaijan Province	Agriculture	Silty Clay	8.4 38.9	Field
2017	Asadi Kapourchal and Homaee	Crop rotation and leaching	Ramhormoz, Khuzestan Province	Agriculture	Silty Clay – Clay	44–45	Field

(*Continued*)

TABLE 9.1 (*Continued*)
Summary of Research Done on the Management of Saline Soil and Water Resources of Iran during the Period 2001–2020

Year	Researcher/s	Approach	Case Study	Land Use	Texture	EC (dS m^{-1})	Experiment Type
2017	Nasrollahi et al.	Drip irrigation with saline water	Ahwaz, Khuzestan Province	Agriculture	Silty Loam	2.5	Experimental site
2017	Soltani et al.	Dry drainage	Aburaihan Faculty, University of Tehran	Agriculture	Loam	5.1–7.4	Farm
2018	Nakhaei et al.	Saline aquaculture	Kerman Province	Water resources close to agriculture and orchards	Different mainly with high sand	7.3–147.2	Field
2018	Hassantabar Shobi et al.	Dissolved organic carbon application	Mazandaran Province	Agriculture	Clay	13.85	Incubation
2018	Talebnejad and Sepaskhah	Quinoa cultivation	Fars Province	Agriculture	-	20	Greenhouse
2018	Ansari et al.	Dry drainage	Isfahan Province	Agriculture	Loam	3	Laboratory
2019	Rousta et al.	Humic Acid compounds application	Yazd Province	Agriculture	Sandy Clay Loam	15.24	Field
2019	Moradi et al.	Organic carbon application	Urmia, West Azerbaijan Province	Agriculture	Sandy Loam	0.58	Pot
2019a	Sharifipour et al.	Leaching and water and drainage water application	South of Khuzestan Province	Agriculture	Clay	-	Field
2019b	Sharifipour et al.	*Drainage water management*	South of Khuzestan Province	Agriculture	Clay	-	Field

Challenges in salinity research in Iran, include the absence of critical managerial initiatives such as a well-cleared strategic blueprint, a sound research foundation and a lack of teamwork. There will be an increased focus on "precision agricultural" approaches in the future.

ACKNOWLEDGMENTS

We warmly thank Dr. Farhad Khorsandi (Retired Professor of Soil Science, Islamic Azad University. Dārāb, Iran and independent research scientist) for the time he devoted to improve the final manuscript and for his valuable advices. We also extend our gratitude to the editorial team and anonymous reviewers for substantial comments on earlier version of this chapter.

REFERENCES

Akhani, H. 2003. *Salicornia persica* Akhani (Chenopodiaceae), a remarkable new species from Central Iran. *Linzer Biologische Beiträge* 35:607–612.

Al-Farsi, S. M., A. Nawaz, S. K. Nadaf, A. M. Al-Sadi, K. H. Siddique, and M. Farooq. 2020. Effects, tolerance mechanisms and management of salt stress in lucerne (Medicago sativa). *Crop and Pasture Science* 71:411–428.

Ansari, S., J. Abedi-Koupai, B. Mostafazadeh-Fard, M. Shayannejad, and M. R Mosaddeghi. 2019. Investigation of effect of dry drainage on the transport and distribution of cations using a physical model. *Iranian Journal of Soil and Water Research* 49(6):1215–1225.

Ansari, S., B. Mostafazadeh-Fard, J. A. Koupai. 2017. Soil salinity control under barley cultivation using a laboratory dry drainage model. International Drainage Workshop of ICID, Ahwaz., IRAN 4-7 March 2017, 398–404.

Asadi Kapourchal, S., and M. Homaee. 2017. Using desalinization models for scheduling crop rotation of saline-sodic soils: A case study in Ramhormoz Region, Iran. *Journal of Soil and Water Resources Conservation* 6(4):91–106.

Asch, F., and M. C. S. Woperei. 2001. Responses of field-grown irrigated rice cultivars to varying levels of floodwater salinity in a semi-arid environment, *Field Crops Research* 70 (2):127–137.

Azadegan, B. 2008. Effect of drainage and sulfur on quality improvement of saline and sodic soils of pistachio orchards in Garmsar. *Irrigation and Drainage* 2(1):55–62.

Bagheri, M. 2019. *Handbook of Quinoa Cultivation. Seed and Plant Improvement Research Institute*, Alborz, 47p.

Banaei, M. H., A. Moameni, M. Baybordi, and M. J. Malakouti. 2004. *Iran Soils: New Transformations in the Identification, Management and Operation. Soil and Water Research Institute*, Tehran, DC

Banakar, M. H., G. H. Ranjbar, and V. Soltani. 2012. Physiological response of some forage halophytes under saline conditions. *Environmental Stresses Crop Sciences* 5(1):55–65.

Bani, A., I. Daghari, A. Hatira, A. Chaabane, and H. Daghari. 2020. Sustainable management of a cropping system under salt stress conditions (Korba, Cap-Bon, Tunisia). *Environmental Science and Pollution Research* 1–8.

Banie, M. H. 2001. *Soil Map of IRAN: Land Resources and Potentialities. Soil and Water Research Institute*, Tehran, DC.

Bashtani, J. 2017. The importance of halophytes plants in animal nutrition. First National Congress of Haloculture, November 22–23, 2017, Yazd, 18 p.

Bazzaneh, B., and H. Rezaie. 2017. Assessment of leaching of the salt-affected soils using different levels of pure and acidic water in selected regions of Mahabad irrigation and drainage Network. *Applied Soil Research* 5(1):104–114.

Boostani, H. R., M. Chorom, A. Moezzi, A. Karimian, N. Enayatizamir, and M. Zarei. 2016. Effect of plant growth promoting *Rhizobacteria* (PGPR) and *Arbuscular Mycorrhizae Fungi* (AMF) application on distribution of zinc chemical forms in a calcareous soil with different levels of salinity. *Soil Management and Sustainable Production* 6(1):1–24.

Corwin, D. L., J. D. Rhoades, and J. Šimůnek. 2007. Leaching requirement for soil salinity control: Steady-state versus transient models. *Agricultural Water Management* 90(3):165–180.

Debroy, S., T. Paul, and A. Biswal. 2020. Shrimp culture in inland saline waters of India: A step towards sustainable aquafarming. *Food and Scientific Reports* 1(4):84–88.

Dewan, M. L., and J. Famouri. 1964. *The Soils of Iran. FAO*, Rome, DC.

European Commission (EC): Communication from the commission to the council, the European Parliament, the European economic and social committee and the committee of the regions 2006. Thematic Strategy for Soil Protection, COM 231 Final, *Brussels*. https://eur-lex.europa.eu/legal-content/EN/TXT/PDF/?uri=CELEX:52006DC0231&from=EN

Fanni, Z., and A. Maroofi. 2017. The drought's effects of Urmiyeh Lake on natural and human disasters/vulnerability of peripheral areas. *Environment* 2(57):1–16.

Fathizad, H., M. A. H. Ardakani, H. Sodaiezadeh, R. Kerry, and R. Taghizadeh-Mehrjardi 2020. Investigation of the spatial and temporal variation of soil salinity using random forests in the central desert of Iran. *Geoderma* 365:1–13.

FAO. 2000. Global network on integrated soil management for sustainable use of salt-affected soils. *Country Specific Salinity Issues-Iran. FAO*, Rome. http://www.fao.org/ag/agl/agll/spush/degrad.asp?country¼iran (accessed July 20, 2020).

Feizi, M., M. A. Hajabbasi, and B. Mostafazadeh-Fard. 2010. Saline irrigation water management strategies for better yield of safflower (*'Carthamus tinctorius' L.*) in an arid region. *Australian journal of crop science* 4(6):408–414.

Feizi, M., and S. Saadat. 2015. The effect of irrigation management with salt water in a crop rotation period. *Water and Irrigation Management* 5(1): 11–25.

Filehkesh, I., and Y. Hashemi Nejad. 2017. Application of salty flood for restoration of saline- sodic lands, a pattern for increasing forage production in desert ranges (Case study: Barabad, Sabzevar). First National Congress of Haloculture *November 22–23*, 2017: Yazd, 10 p.

Gowing, J. W., and G. C. L. Wyseure. 1992. Dry-drainage, a sustainable and cost-effectible solution to water logging and salinization. *Proceedings of* 5th *International Drainage workshop*. Lahore Pakistan: ICID-CIID, 1992. 6.26–6.34.

Guo, H., F. Zeng, S. K. Arndt, J. Zeng, W. W. Xing, and B. Liu. 2008. Influence of floodwater irrigation on vegetation composition and vegetation regeneration in a Taklimakan desert oasis. *Chines Science Bulletin* 53:156–163.

Hassantabar Shobi, S., F. Sadegh-Zadeh, M. A. Bahmanyar, and B. Jalili. 2018. Reclamation of saline-sodic soil with clay texture using dissolved organic carbon. *Soil Management and Sustainable Production* 8(1):159–173.

Hawrot-Paw, M., A. Koniuszy, M. Gałczyńska, G. Zając, and J. Szyszlak-Bargłowicz. 2020. Production of microalgal biomass using aquaculture wastewater as growth medium. *Water* 12(1):106. https://doi.org/10.3390/w12010106

Hazbavi, Z., M. R. Hosseini, Khalili, B., Moghadam, R. Mostafazadeh, and A. Jafarpoor. 2019. Salinity issue in agriculture and solutions scoping study in Iran, September 10-13, 2019, *Leeuwarden*, the Netherlands, 1 p.

Hazbavi, Z., S. H. R. Sadeghi, M. Gholamalifard, and A. A. Davudirad. 2020. Watershed health assessment using the pressure–state– response (PSR) framework. *Land Degradation and Development* 31:3–19.

Heydari, N. 2019. Water productivity improvement under salinity conditions: Case study of the saline areas of lower Karkheh River Basin, Iran. *In: Multifunctionality and impacts of organic agriculture* IntechOpen. doi: 10.5772/intechopen.86891.

Homaei, M. 2003. *The Response of Plants to Salinity. Iranian National Committee on Irrigation and Drainage*, Iran, DC.

Hooke, J. M., and J. Mant, 2002. Floodwater use and management strategies in valleys of southeast Spain. *Land Degradation and Development* 13:165–175.

Hussain, M. I., A. Muscolo, M. Farooq, and W. Ahmad. 2019. Sustainable use and management of non-conventional water resources for rehabilitation of marginal lands in arid and semiarid environments. *Agricultural Water Management* 221:462–476.

ICID (International Commission on Irrigation and Drainage). 2002. Irrigation and food production information about ICID network countries. http://www.icid.org/index_e.html (accessed July 20, 2020).

Incrocci, L., P. Marzialetti, G. Incrocci, A. Di Vita, J. Balendonck, C. Bibbiani, S. Spagnol, and A. Pardossi. 2019. Sensor-based management of container nursery crops irrigated with fresh or saline water. *Agricultural Water Management* 213:49–61.

Ivushkin, K., H. Bartholomeus, A. K. Bregt, A. Pulatov, B. Kempen, and L. De Sousa. 2019. Global mapping of soil salinity change. *Remote Sensing of Environment* 231: 111260.

Kashyap, P. L., M. K. Solanki, P. Kushwaha, S. Kumar, and A. K. Srivastava. 2020. Biocontrol potential of salt-tolerant *Trichoderma* and *Hypocrea* isolates for the management of tomato root rot under saline environment. *Journal of Soil Science and Plant Nutrition* 20:160–176.

Khalili, S., A. Bastani, and M. Bagheri. 2019. Effect of different levels of irrigation water salinity and phosphorus on some properties of soil and quinoa plant. *Soil and Water Sciences* 33(2):155–167.

Khorsandi, F. 2016. Haloculture: strategy for sustainable utilization of saline land and water resources. *Iranian Journal of Earth Sciences* 8(2):164–172.

Khorsandi, F., and S. M. H. Siadati. 2017. Haloengineering and its significant role in sustainable development of salt affected ecosystems. *First National Congress of Haloculture*, Yazd, November 22-23, 2017, Yazd:13 p.

Khoshgoftarmanesh, A. H., and H. Shariatmadari. 2002. Reclamation of bare saline soils by leaching and barley production. *Paper presented at the 17th World Congress of Soil Science*. August 14-21 2002, Bangkok, Thailand.

Kolahchi, Z., and M. Jalali. 2007. Effect of water quality on the leaching of potassium from sandy soil. *Journal of Arid Environments* 68(4):624–639.

Konukcu, F., J. W. Gowing, and D. A. Rose. 2006. Dry drainage: A sustainable solution to waterlogging and salinity problems in irrigation areas?, *Agricultural Water Management*, 83(1–2): 1–12.1

Koocheki, A. 1996. The use of halophytes for forage production and combating desertification in Iran. *In* Choukr-Allah, R: Malcom, C. V; Hamdy (eds) *Halophytes and Biosaline Agriculture. A. Marcel Dekker. Inc.*, USA:263–274.

Koocheki, A. 2000. Potential of saltbush (*Atriplex spp.*) as a fodder shrub for the arid lands of Iran. In G. Gintzburger (ED.), *Fodder shrub development in arid and semiarid zones* (pp. 178–183).

Mahmoodi-Eshkaftaki, M., and M. R. Rafiee. 2020. Optimization of irrigation management: A multi-objective approach based on crop yield, growth, evapotranspiration, water use efficiency and soil salinity. *Journal of Cleaner Production* 252:119901.

Manaffar, R., N. Abdolahzadeh, G. MoosaviToomatari, S. Zare, P. Sorgeloos, P. Bossier, and G. Van Stappen 2020. Reproduction and life span characterization of Artemia urmiana in Lake Urmia, Iran (Branchiopoda: Anostraca). *Iranian Journal of Fisheries Sciences* 19(3):1344–1358.

Moameni, A. 2011. Geographical distribution and salinity levels of soil resources of Iran. *Soil Research* 24(3):203–215.

Moameni, A., H. Siadat, and M. J. Malakouti. 1999. The extent distribution and management of salt affected soils of Iran. FAO global network on integrated soil management for sustainable use of salt affected soils. Izmir, Turkey, 5.

Moameni A., and A. Stein. 2002. Modeling spatio-temporal changes in soil salinity and waterlogging in the Marvdasht Plain, Iran. *In Paper presented at the 17th World Congress of Soil Science* August 14–21 2002, Bangkok.

Moradi, S., M. H. Rasouli-Sadaghiani, E. Sepehr, H. Khodaverdiloo, and M. Barin. 2019. The role of organic carbon in the mineralization of nitrogen, carbon and some of nutrient concentrations in soil salinity conditions. *Soil Management and Sustainable Production* 9(3):153–169.

Mostafazadeh-Fard, B., H. Mansouri, S. F. Mousavi, and M. Feizi. 2009. Effects of different levels of irrigation water salinity and leaching on yield and yield components of wheat in an arid region. *Journal of Irrigation and Drainage Engineering* 135(1):32–38.

Mostafazadeh-Fard, B., and E. Ghasemi. 2016. Soil salinity control using dry drainage concept. *International Journal of Advances in Science Engineering and Technology* 3:68–71.

Nabhan, G. P. 1979. The ecology of floodwater farming in arid southwestern North America, *Agro-Ecosystems* 5(3):245–255.

Nadeem, M., Li, J. Yahya, M. Wang, M. Ali, A. Cheng, A. X. Wang, and C. Ma. 2020. Grain legumes and fear of salt stress: Focus on mechanisms and management strategies. *International Journal of Molecular Sciences* 20:799. https://doi.org/10.3390/ijms20040799

Nakhaei, N., M. Ramin, L. Yazdanpanah, H. Naghavi, M. Masomi, A. Zeioddini, and M. Alizade. 2018. Identification the useless source of brackish waters for aquaculture in Kerman Province. Ministry of Jihad-e-Agriculture, Agricultural Research, Education & Extension Organization, *Iranian Fisheries Science Research Institute – Kerman Agricultural and Natural Resources Research and Education Center Project*, Project No.: 4-54-12-89043: 49 p.

Nasrollahi, A. H., S. Boroomand Nasab, and A. R. Hooshmand. 2017. Management practices using of agricultural drainage water with drip irrigation for crop production and lands sustainability in arid and semi-arid areas. *International Drainage Workshop of ICID*, Ahwaz., IRAN 4-7 March 2017: (pp.607-616)

Ning, S. R, B. B. Zhou, Q. J. Wang, and W. H. Tao. 2020. Evaluation of irrigation water salinity and leaching fraction on the water productivity for crops. *International Journal of Agricultural and Biological Engineering* 13(1):170–177.

Paik, S., D. T. P. Le, L. T. Nhu, and B. F. Mills. 2020. Salt-tolerant rice variety adoption in the Mekong River Delta: Farmer adaptation to sea-level rise. *PLoS ONE* 15(3):e0229464. https://doi.org/10.1371/journal.pone.0229464

Qadir, M., A. S. Qureshi, and S. A. M. Cheraghi. 2008. Extent and characterisation of salt-affected soils in Iran and strategies for their amelioration and management. *Land Degradation & Development* 19(2):214–227.

Ranjbar, Gh., and H. Pirasteh-Anosheh, 2015. A glance to the salinity research in Iran with emphasis on improvement of field crops production. *Iranian Journal of Crop Sciences.* 17(2):165–178.

Rhoades J. D., and A. Dinar. 1991. Reuse of agricultural drainage water to maximize the beneficial use of multiple water supplies for irrigation. In: Dinar A., Zilberman D. (eds) *The Economics and Management of Water and Drainage in Agriculture. Springer*, Boston, MA.

Rousta, M. J., K. Enayati, M. Soltani, F. Ghane, N. Besharat, and E. Neshat 2019. The effects of humic acid application on yield and yield components of wheat and some chemical properties of a saline-sodic soil. *Soil Management and Sustainable Production* 8(4):95–110.

Sabzevar, A. D., and M. Khosroshahi. 2010. The effects of the use of low quality flood on desert area. *Iranian Journal of Range and Desert Research* 17(1):127–148.
Sadeghi, S. H. R., Z. Hazbavi, and M. Gholamalifard. 2019. Interactive impacts of climatic, hydrologic and anthropogenic activities on watershed health. *Science of the Total Environment* 648:880–893.
Saghafi, D., H. A. Alikhani, and B. Motesharezadeh. 2015. Effect of plant growth promoting rhizobia on ameliorating the effects of salt stress in canola (*Brassica napus L.*). *Soil Management and Sustainable Production* 5(1):23–41.
Salehi, M., and F. Dehghani. 2019. *Guide to Planting, Growing, and Harvesting Quinoa in Saline Conditions. Journal of Agricultural Education*, Tehran, DC.
Salehi, M., V. Soltani, and F. Dehghani. 2018. Research finding and development program of quinoa production under saline and low-yielding dryland of Iran. *First National Congress of Haloculture*, 22-23 November 2017, Yazd, 7p.
Sayyari, M., and S. Mahmoodi. 2002. An investigation of reason of soil salinity and alkalinity on some part of Khorasan Province (Dizbad-e Pain Region). *17th WCSS, (pp.14-21)*, August 2002. Paper No. 1981, Thailand.
Seydehmet, J., G. H. Lv, and A. Abliz. 2019. Landscape design as a tool to reduce soil salinization: The study case of Keriya Oasis (NW China). *Sustainability* 11:2578. doi:10.3390/su11092578.
Sharifipour, M., A. R. Hooshmand, A. Naseri, A. Hassanoghli, and H. Moazed. 2019a. Effect of leaching method and water and drainage water application order on desalinization and desodification of heavy soils. *Iranian Journal of Soil and Water Research.* doi: 10.22059/ijswr.2019.286228.668273.
Sharifipour, M., A., Liaghat, A. Naseri, H. Nozari, M. Hajishah, M. Zarshenas, H. Hoveizeh, and M. Nasri. 2019b. Drainage water management of irrigation and drainage networks of south west Khuzestan. *Iranian Journal of Soil and Water Research.* doi: 10.22059/ijswr.2019.274004.668107.
Sharifipour, M., H. Nozari, A. Liaghat, A. Naseri. 2017. Plan of reusing of effluent waters of irrigation and drainage Networks at West region of Karun River. *First National Congress of Haloculture*, November 22–23 2017, Yazd, 7 p.
Shiati, K. 1998. Brackish water as a source of irrigation: behavior and management of salt-affected reservoirs (Iran). In*: 10th Afro-Asian Conf.* Bali, Indonesia.
Siadat, H. 1998. Iranian agriculture and salinity. In Proc. Conf. *New Technologies to Combat Desertification*, October (pp. 12–15).
Siadat, H., M. Bybordi, and M. J. Malakouti. 1997. Salt affected soils of Iran: A country report. *In International symposium on "Sustainable management of salt affected soils in the arid ecosystem*, September, Cairo, Egypt.
Soleimani, R., and A. Najafifar. 2017. The effect of several transplanted halophyte species on soil salinity and sodium adsorption ratio. *First National Congress of Haloculture*, November 22-23 2017, Yazd, 8 p.
Soltani, M., A. Rahimikhoob, A. Sotoodeh Nia, and M. Akram. 2017. Evaluation of HYDRUS_2D software in simulating dry drainage. *Water Research in Agriculture* 31(4):595–607.
Stec, A. 2020 Water Resources. In: *Sustainable Water Management in Buildings. Water Science and Technology Library*, Vol. 90. *Springer*, Cham.
Talebnejad, R., and A. R. Sepaskhah. 2018. Quinoa: a new crop for plant diversification under water and salinity stress conditions in Iran. *ISHS Acta Horticulturae* 1190:101–106. doi:10.17660/actahortic.2018.1190.17.
Tavousi, M., Gh. Lotfali, and A. Aine. 2018. *Quinoa cultivation and related research results. Agricultural Education Publication*, Karaj, DC.
Wang, C., J. Wu, W. Zeng, Y. Zhu, and J. Huang. 2019. Five-year experimental study on effectiveness and sustainability of a dry drainage system for controlling soil salinity. *Water*, 11(1):111. https://doi.org/10.3390/w11010111

World Bank. 2005. Islamic Republic of Iran; Cost assessment of environmental degradation. Report No. 32043-IR. http://documents.worldbank.org/curated/en/401941468284096627/text/320430IR.txt (accessed July 20, 2020).

Wu, J., L. Zhao, J. Huang, J. Yang, B. Vincent, S. Bouarfa, and A. Vidal. 2009. On the effectiveness of dry drainage in soil salinity control. *Science in China Series E: Technological Sciences* 52:3328–3334. https://doi.org/10.1007/s11431-009-0341-8

Yazdanpanah, N., and M. Mahmoodabadi. 2011. Time monitoring of leachate quality during reclamation process of saline-sodic soil using soil column. *Soil Management and Sustainable Production* 1(1):1–22.

Yazdanpanah, N., S. Pazira, A. Neshat, and M. Mahmoodabadi. 2012. Effect of different amendments on some physical and chemical properties of a saline-sodic soil. *Arid Biome Scientific and Research* 2(1):83–97.

Zare, G., and M. Keshavarzi, 2007. Morphological study of Salicornieae (Chenopodiaceae) native to Iran. *Pakistan Journal of Biological Sciences* 10(6):852–860.

10 An Approach to Monitoring of Salt-Affected Croplands Using Remote Sensing Data

The Case Study in the Nukus District (Uzbekistan)

Maria Konyushkova, Alexander Krenke, Gulchekhra Khasankhanova, Nizamatdin Mamutov, Victor Statov, Anna Kontoboytseva and Yevgenia Pankova

CONTENTS

10.1 INTRODUCTION

Soil salinity is one of the main threats to agriculture in the arid zones of Central Asia, especially in Uzbekistan and Turkmenistan, where salt-affected soils comprise the substantial part of these countries (Figure 10.1). In Uzbekistan, the area prone to salinity in the upper meter of soil profile is 35% of the area of the country and reaches 50% of the area of irrigated cropland (Bucknall et al., 2003; Vargas et al., 2018).

DOI: 10.1201/9781003112327-10

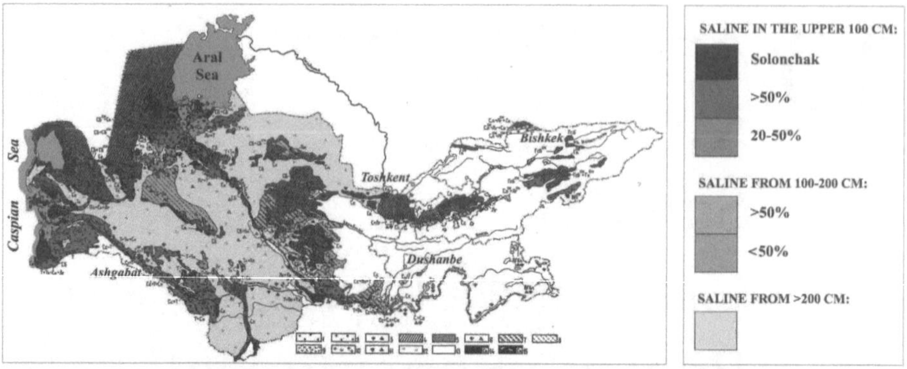

Country	Area of salt-affected soils (% of the country's area)*	Area of salt-affected soils at the irrigated croplands (% of the country's irrigated cropland)**
Uzbekistan	15.6 Mha (35%)	2.14 Mha (50%)
Turkmenistan	17.3 Mha (35%)	1.67 Mha (96%)
Kyrgyzstan	1.2 Mha (6%)	0.12 Mha (11%)
Tajikistan	0.6 Mha (4%)	0.11 Mha (16%)

Map source: Pankova et al. 1996. * Vargas et al. (Eds.) 2018. ** Bucknall et al. 2003.

FIGURE 10.1 Distribution of salt-affected soils in the countries of Central Asia.

The monitoring of soil salinity in Uzbekistan lies mainly under the responsibility of the Ministry of Water Resources and its regional hydrogeological expeditions.[1] The monitoring is organized within the irrigation zones relative to the network of water distribution (Water Users Associations, or WUA). As a result, each WUA in theory represents a small catchment, which manages the irrigation system in a proper way. More detailed information about WUAs is given in Yalcin and Mollinga (2007). Each WUA is sampled for soil salinity once per 3 years with the norm of sampling 100–170 sampling points (SP) per square kilometer for 1:2000 scale, 40–80 SP for 1:5000, and 20–35 SP for 1:10000 scale (Soil Survey for Amelioration Construction, 1985). The sampling depths are 0–25, 25–50, 50–75, 75–100, 100–150, and 150–200 cm. Such a sampling standard, if recalculated into the area justified by ground truth data, would give 80 × 80 m (for the maximum number of SP) to 225 × 225 m (for the minimum number of SP). Judging from the detailed spaceborne imagery (Figure 10.2), it is obvious that how the pattern of soil salinity within such an area can be complex. At present, even the minimum number of SP is usually not taken for a variety of reasons.

It is essential that remote sensing imagery be introduced into the system for monitoring soil salinity of the irrigated lands in Uzbekistan as it can give invaluable and prompt information about the pattern of soil salinity. The approaches to monitoring of soil salinity based on remote sensing data are abundant and well represented in the scientific literature (e.g., Ivushkin et al., 2019; Pankova et al., 2018; Rukhovich et al., 2016). Still, in Uzbekistan, there is a gap between science and practice which should be overcome in the near future. At present, Uzbekistan is passing through a

FIGURE 10.2 Soil heterogeneity within the area justified by one sampling point. (From minimum to maximum required amount of samplings.)

set of reforms in agriculture aimed at modernizing the whole agricultural system. So, now is the most appropriate time to introduce the digital technologies of soil monitoring based on remote sensing data and proximal soil sensing.

The goal of this study was to develop and justify the procedure of processing remote sensing data such from Landsat to integrate these or similar into the national system of soil monitoring including soil salinity monitoring. Such an approach can help achieve the sustainable management of irrigated lands in Uzbekistan.

10.2 MATERIALS AND METHODS

The test area was the Nukus district of the Republic of Karakalpakstan (Uzbekistan) which is located on the right bank downstream of the Amu-Darya river (Figure 10.3). This area is heavily prone to salinity and has 82% of salt-affected soils within the irrigated cropland. There is no rainfed cropland in this region. The predominant crops are monocultures of cotton and some paddy rice and winter wheat.

A set of 18 no-cloud Landsat 8 OLI scenes for the growing season of 2017–2019 were used, namely those collected on 10/05/17, 18/06/17, 04/07/17, 05/08/17, 06/09/17, 01/10/17, 04/05/18, 05/06/18, 07/07/18, 24/08/18, 02/09/18, 11/10/18, 07/05/19, 08/06/19, 03/07/19, 04/08/19, 28/09/19, and 14/10/19. The band numbers 2–7 of Landsat-8 OLI were taken for analysis.

Along with the reflectance in spectral bands of Landsat-8 imagery, the NDVI vegetation index *(formula 1)* and S4 salinity index *(formula 2)* were used for analysis. The spectral S4 index was calculated according to Abbas et al. (2013) and was used

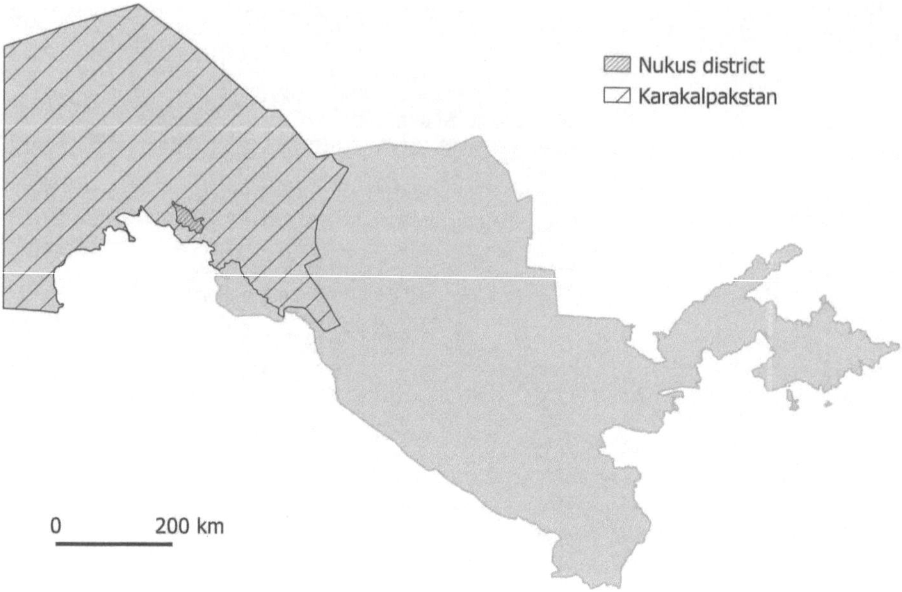

FIGURE 10.3 Location of the study area. (Nukus district of Karakalpakstan.)

in our research as it proved to be well correlated with soil salinity both in the study of Abbas et al. (2013) in Pakistan and in our studies in Uzbekistan.

$$NDVI = \frac{NIR - R}{NIR + R} \tag{10.1}$$

$$S4 = \sqrt{B \times R} \tag{10.2}$$

where *B*, *R*, *NIR* are reflectance in blue, red, and near-infrared bands, respectively.

Most calculations were performed over the whole set of pixels within the Nukus district using STATISTICA software.

The map of croplands was downloaded from the LPDAAC GFSAD collection (https://lpdaac.usgs.gov/products/gfsad30eucearumecev001/) (Phalke et al., 2017).

The resulting map was compared with the data on the areas of salt-affected soils in WUAs of the Nukus district of Karakalpakstan as of 01/11/2017. These estimates were based on the chemical analysis of 1894 SP across the irrigated cropland of the Nukus district (22,930 ha), or 7–13 SP per km^2 (depending on WUA), or 1 SP per 8–15 ha.

10.3 METHODOLOGY

The whole procedure of processing the remote sensing data was aimed at the gradual dimensionality reduction[2] until the invariant feature space was found according to the approach described in Puzachenko (2009). A 3-year time span was taken to smooth the variations caused by different weather conditions. With 18 dates and 6 bands for each date, 108 variables in total were processed in this analysis.

The first stage was the principal component analysis (PCA) of 6 bands of one date. 1st and 2nd factors (components) described more than 90% of variability and were taken for further analysis. For 18 dates, 36 (i.e., 2 × 18) new variables were calculated as a result of this first stage.

At the second stage, one more PCA over these new 36 variables was performed. The 1st and 2nd factors (components) derived at this second stage described the invariant (intrinsic) state of the image pattern.

At the third stage, the *k*-means classification was performed using two variables (two first components) obtained at stage 2. Different metrics (Euclidean, Mahalanobis and Manhattan distances, scalar product, and others) were tested and that with maximum entropy was chosen as it reflected the most unbiased classification. Given the iterative nature of the selected classification method, the stop point for separation of further classes was chosen when the increment of entropy started to decrease. This method was dichotomic and produced 2-power number of classes (2, 4, 8, 16, etc.); the visual filtration of noisy classes was therefore performed and discriminant analysis was used to fill in the resulting gaps.

At the fourth stage, the discriminant analysis with the final classes obtained at stage 3 and all initial variables (in our case, 6 bands × 18 dates, or 108 in total) was performed. The resulting confusion matrix showed the performance of classification by discriminant functions. At this stage, the probability of correct classification of each pixel was visualized on the map. The pixels with high probability were correspondent with the cores of the classes whereas the pixels with low probability corresponded to the peripheries of the classes.

We supposed that the stable (invariant) classes derived in the end of the fourth stage of this procedure corresponded to agroecological groups of soils. As the main threat to soil productivity in this area is salinity, we expected that these classes would correspond well with the salinity status of soils.

At the last stage of our analysis, we performed the reverse analysis collecting the values of NDVI and S4 indices within each class to see what is happening with vegetation (NDVI) and surface behavior (S4).

10.4 RESULTS AND DISCUSSION

The resulting map of agroecological groups of soils at the irrigated cropland of the Nukus district of Karakalpakstan is shown in Figure 10.4. As a result, nine stable classes were identified. Classes 1–4 were mainly found on cropland whereas classes 6–9 were located outside the cropland; they were therefore excluded from further analysis. Class 5 was found in both cropland and non-cropland.

The overall confusion matrix of class identification showed a very good performance of classification, 67.2%, especially for classes 1, 2, and 6–9 (Table 10.1).

In order to interpret the invariant classes (groups) of spaceborne imagery, the analysis of NDVI and S4 within classes 1-5 on the cropland was performed (Figure 10.5). It can be seen that despite the dynamic character of NDVI and S4 in time, the classes were organized in an ordinal manner, that is, the classes with higher NDVI and lower salinity index had higher NDVI and lower salinity irrespective of

TABLE 10.1
Confusion Matrix of Classification by Discriminant Functions. The Grey Color Marks the % of Correct Identifications, and the Light Grey Color Marks the % of Identification as a Neighboring Class

			Classifications: Rows (Observed) Columns (Predicted)								
Class	N	Percent	1	2	3	4	5	6	7	8	9
1	19275	64.4	64.4	33.8	0.9	0.2	0.4	0.12	0.05	0.00	0.0
2	138583	67.8	4.2	67.8	25.5	1.8	0.6	0.08	0.02	0.00	0.0
3	206363	56.9	0.5	18.5	56.9	20.9	3.0	0.29	0.04	0.00	0.0
4	201577	56.3	0.1	1.9	20.3	56.3	20.1	1.30	0.10	0.00	0.0
5	137290	56.6	0.0	0.2	1.9	23.1	56.6	17.32	0.69	0.05	0.0
6	91569	66.1	0.0	0.0	0.1	0.9	17.9	66.09	14.62	0.34	0.0
7	81478	73.0	0.0	0.0	0.0	0.1	0.9	17.49	72.98	8.49	0.1
8	75377	68.4	0.0	0.0	0.0	0.0	0.1	0.44	19.21	68.38	11.9
9	207680	93.7	0.0	0.0	0.0	0.0	0.0	0.01	0.33	5.92	93.7
Total	1159192	67.2									

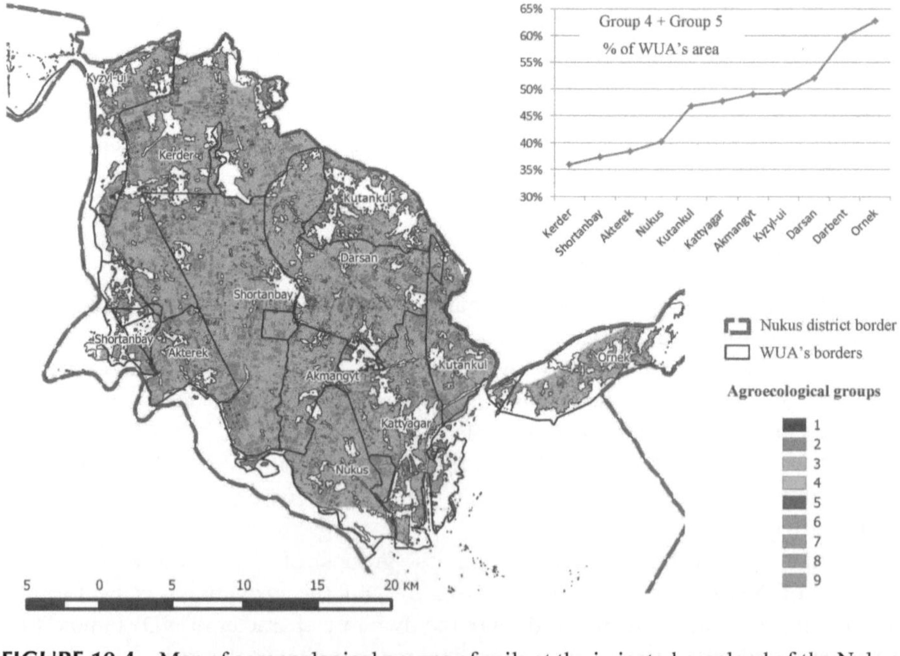

FIGURE 10.4 Map of agroecological groups of soils at the irrigated cropland of the Nukus district of Karakalpakstan derived from the Landsat data. The comments are given in the text.

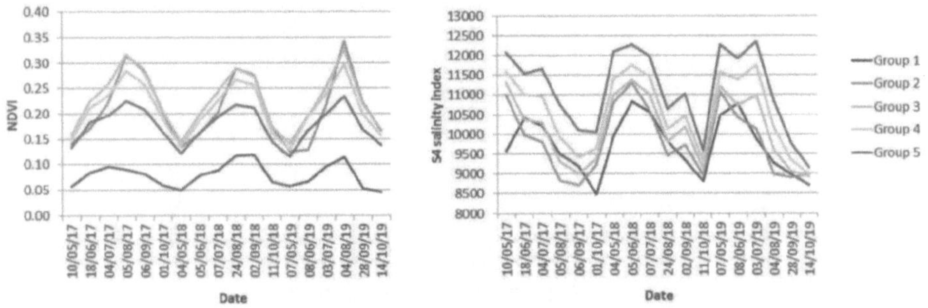

FIGURE 10.5 Temporal dynamics of NDVI and S4 salinity index by agroecological groups of soils at the croplands of the Nukus district. Each date has a total of 2,711,070 pixels.

the season. Judging from Figure 10.5, group 1 was constant water. Groups 2 and 3 were very close to each other in terms of NDVI and salinity and they had highest biological productivity and lowest salinity. There was a slight difference in salinity between these two groups which allows us to conclude that group 2 was of nonsaline soils whereas group 3 had slightly saline soils. The slight degree of salinity does not influence the productivity of cotton, which is salt tolerant and can grow well on slightly saline soils. Group 4 is slightly different from groups 2 and 3 in terms of NDVI (less productive) but is very different in terms of S4 salinity index. We conclude that group 4 contains mainly medium saline soils which have some impact on cotton productivity. Group 5 has a substantially decreased biological productivity and highest values of salinity index comparing to all other groups.

As a result, the classes found on cropland were interpreted as:

Group 1 – constant shallow water;
Group 2 – most productive nonsaline soils;
Group 3 – productive slightly saline soils;
Group 4 – medium productive and medium saline soils;
Group 5 – poorly productive strongly saline soils.

According to Landsat data, the worst situation with soil salinity and crop productivity is observed in the Kutankul WUA (northern part), and the Ornek, Darsan, and Kyzyl-ui WUAs. The best situation within Nukus district was observed in the Kerder, Shortanbay, Akterek, and Nukus WUAs.

The data on the proportions of different categories of soils according to the estimates of regional hydrogeological expedition and of our estimates based on Landsat data are given in Table 10.2. No statistically valid correlations were found between these estimates. However, the WUAs which were estimated as being in the worst state by remote sensing data (Kutankul, Ornek, and Kyzyl-ui, with the exception of Darsan) all had increased areas of medium and strongly saline soils. The WUAs with good situations according to the remote sensing data (Kerder, Akterek, and Nukus, with the exception of Shortanbay) all had increased areas of nonsaline soils.

TABLE 10.2
The Percentage of Different Categories of Saline Soils (according to Soil Salinity Survey Data as of 01.11.2017) and Agroecological Groups of Soils within WUAs of the Nukus District

	Data from the Regional Hydrogeological Expedition as of 01.11.2017						Data from Landsat-8 OLI Multi-temporal Imagery (2017–2019)				
WUA	Surveyed Area, ha	Nonsaline	Slightly Saline	Medium Saline	Strongly Saline	Extremely Saline	Surveyed Area, ha*	Group 2	Group 3	Group 4	Group 5
Akmangyt	1027	48.10%	48.30%	3.60%	0.00%	0.00%	2128	12.78%	35.80%	35.52%	13.62%
Akterek	764	63.74%	30.24%	5.24%	0.79%	0.00%	2229	20.93%	34.52%	26.68%	11.75%
Darbent	56	0.00%	0.00%	100.00%	0.00%	0.00%	95	4.00%	28.48%	36.19%	23.62%
Darsan	2468	62.28%	31.93%	5.79%	0.00%	0.00%	6903	10.91%	32.80%	38.03%	14.10%
Kattyagar	1790	47.21%	44.30%	8.49%	0.00%	0.00%	4135	14.63%	28.66%	33.09%	14.67%
Kerder	4859	55.19%	30.77%	12.93%	1.11%	0.00%	9168	21.39%	39.52%	26.12%	9.84%
Kutankul	1153	18.56%	41.11%	29.49%	9.63%	1.21%	4350	15.05%	30.85%	30.56%	16.36%
Kyzyl-ui	1049	49.48%	28.03%	20.97%	1.53%	0.00%	1767	18.11%	24.75%	28.94%	20.24%
Nukus	1664	64.24%	30.95%	4.81%	0.00%	0.00%	3077	18.97%	38.54%	31.87%	8.33%
Ornek	710	57.89%	28.87%	13.24%	0.00%	0.00%	1916	4.49%	26.67%	43.84%	18.95%
Shortanbay	4508	50.04%	41.75%	8.21%	0.00%	0.00%	8850	23.15%	37.40%	29.28%	8.09%

*As we used the areas of croplands derived from remote sensing data (LPDAAC GFSAD collection), the total area of the cropland studied with remote sensing data is higher than that given in statistical documents (31,022 ha).

10.5 CONCLUSION

Analysis of remote sensing data gives a good opportunity for the development of a national monitoring system of soil salinity to assess the spatial pattern of different categories of saline soils allowing better assessment and decision support in the management of soil and water in irrigated croplands.

On the basis of the resulting maps, different strategies can be chosen. The first strategy which has been used traditionally in this area is aimed at the unification of the quality of soil cover by abundant watering to flush the salts into the deeper soil layers. The economic costs of improving the quality of bad (marginal) lands are high and the return is low. Another strategy can be the differentiated management of this area which is inherently prone to salinization. In case of this approach, the main efforts should be focused on good lands, where the return on activities will be greatest, whereas the poor (marginal) lands should be used for low-cost agricultural activities. Both approaches require an economic assessment based on the sound information on the areas occupied by different agroecological groups of soils, as well as on planned activities.

We propose that at the worse categories of soils which constantly show the low NDVIs and high values of salinity index (groups 4 and 5 in our map) should be considered for use in saline farming, e.g., growing halophytes for cattle. The lands with good soils which constantly show high NDVIs and low salinity index should be monitored with higher attention so that their state does not deteriorate with time.

ACKNOWLEDGMENT

The financial support for this study was provided by the subsidy of the Government of Russian Federation to the Eurasian Center for Food Security, resolution 2448-r (18/11/2016), and Global Soil Partnership (project GCP/GLO/853/RUS). The statistical modeling of agroecological conditions in the reported study was funded by the Russian Foundation for Basic Research (RFBR), project number 20-16-00211A. We are also very grateful to the Regional Hydrogeological and Amelioration Expedition of the Republic of Karakalpakstan for their assistance and consultations during the development of the procedures described in this paper. We thank Tatyana Khamzina (UZGIP) for providing the data on official statistics.

ENDNOTES

1. The land in Uzbekistan is state-owned. The land users can rent the land for 30–50 years. Also, the small household farms (dehkans) can be possessed inheritably. More details can be found in Melnikovová and Havrland (2016).
2. "Dimensionality reduction is the transformation of high-dimensional data into a meaningful representation of reduced dimensionality. Ideally, the reduced representation should have a dimensionality that corresponds to the intrinsic dimensionality of the data. The intrinsic dimensionality of data is the minimum number of parameters needed to account for the observed properties of the data" (van der Maaten et al., 2009).

REFERENCES

Abbas, A., Khan, S., Hussain, N., Hanjra, M. A., and Akbar, S. 2013. Characterizing soil salinity in irrigated agriculture using a remote sensing approach. *Physics and chemistry of the Earth, Parts A/B/C* 55–57: 43–52. https://doi.org/10.1016/j.pce.2010.12.004.

Bucknall, J., Klytchnikova, I., Lampietti, J., Lundell, M., Scatasta, M., and Thurman M. 2003. *Irrigation in Central Asia: Social, economic and environmental considerations.* Washington, DC.

Ivushkin, K., Bartholomeus, H., Bregt, A. K., Pulatov, A., Kempen, B., de Sousa, L. 2019. Global mapping of soil salinity change. *Remote sensing of environment* 231: 111260. https://doi.org/10.1016/j.rse.2019.111260

van der Maaten, L., Postma, E., and van der Herik, J. 2009. *Dimensionality reduction: A comparative review.* Ticc, Tilburg University. https://lvdmaaten.github.io/publications/papers/TR_Dimensionality_Reduction_Review_2009.pdf (accessed October 28, 2020)

Melnikovová, L., and Havrland, B. 2016. State ownership of land in Uzbekistan – an impediment to further agricultural growth? *Agricultura tropica et subtropica* 49: 5–11. https://doi.org/10.1515/ats-2016-0001.

Pankova, E. I., Aidarov, I. P., Yamnova, I. A., Novikova, A. F., and Blagovolin N. S. 1996. *Natural and human-induced salinization in the Aral Sea Basin.* Moscow: Dokuchaev Soil Science Institute. [in Russian]

Pankova, E. I., Soloviev, D. A., Rukhovich, D. I., and Savin, I. Yu. 2018. Soil salinity monitoring by the use of remote sensing (following the example of irrigated territories of Central Asia). In *Handbook for saline soil management*, ed. R. Vargas, E. Pankova, S. Balyuk, P. Krasilnikov, and G. Khasankhanova, 21–24. Rome: FAO.

Phalke, A., Ozdogan, M., Thenkabail, S. P. G., Congalton, R., Yadav, K., Massey, R., Teluguntla, P., Poehnelt, J., and Smith, C. 2017. NASA making earth system data records for use in research environments (MEaSUREs) global food security-support analysis data (GFSAD) Cropland Extent 2015 Europe, Central Asia, Russia, Middle East 30 m V001 [Data set]. NASA EOSDIS Land Processes DAAC. Accessed 2019-08-20 from https://doi.org/10.5067/MEaSUREs/GFSAD/GFSAD30EUCEARUMECE.001.

Puzachenko, M. Yu. 2009. Multi-functional landscape analysis of the southwestern part of Valdai Upland. PhD diss., Institute of Geography RAS. http://www.sevin.ru/ecosys_services/landscape/puz_autoref.pdf (accessed October 28, 2020) [in Russian].

Rukhovich, D. I., Simakova, M. S., Kulyanitsa, A. L., Bryzzhev, A. V., Koroleva, P. V., Kalinina, N. V., Chernousenko, G. I., Vil'chevskaya, E. V., Dolinina, E. A., Rukhovich, S. V. 2016. Methodology for comparing soil maps of different dates with the aim to reveal and describe changes in the soil cover (by the example of soil salinization monitoring). *Eurasian soil science* 49: 145–162. https://doi.org/10.1134/S1064229316020095.

Soil Survey for Amelioration Construction. 1985. VSN-33-2.1.02-85. Moscow: Ministry of amelioration and water of the USSR.

Vargas, R., Pankova, E., Balyuk, S., Krasilnikov, P., and Khasankhanova, G. (eds.) 2018. *Handbook for saline soil management.* Rome: FAO. http://www.fao.org/3/i7318en/I7318EN.pdf (accessed October 28, 2020).

Yalcin, R., and Mollinga, P. P. 2007. Water Users Associations in Uzbekistan: the introduction of a new institutional arrangement for local water management. Amu Darya case study – Uzbekistan, Deliverable WP 1.2.10 of the NeWater Project, Centre for Development Research, University of Bonn http://www.cawater-info.net/bk/iwrm/pdf/wua-uzb-2007_e.pdf (accessed October 28, 2020).

11 From Desert Farm to Fork

Value Chain Development for Innovative Salicornia-Based Food Products in the United Arab Emirates

Dionysia-Angeliki Lyra, Efstathios Lampakis, Mohamed Al Muhairi, Fatima Mohammed Bin Tarsh, Mohamed Abdel Hamyd Dawoud, Basem Al Khawaldeh, Meis Moukayed, Jacek Plewa, Luca Cobre, Ohod Saleh Al Masjedi, Khawla Mohammed Al Marzouqi, Hayatullah Ahmadzai, Mansoor Khamees Al Tamimi, and Wasel Abdelwahid Abou Dahr

CONTENTS

DOI: 10.1201/9781003112327-11

11.1 INTRODUCTION

Climate change projections show an increase in average temperature by 1.1–5.4°C, a decrease in precipitation of more than 20%, and an increased incidence of prolonged droughts on a global scale (IPCC 2013). These are bad omens for agricultural production and farmers' livelihoods in hot and dry areas. On top of that, aggravating stresses such as heat, drought, and salinity worsen the situation even more. There is an urgent need to adapt agriculture to the adverse impacts of climate change especially in areas with existing environmental constraints. Conventional farming should be reassessed in salt-affected areas and the utilization of the available saline land and water resources should be explored for the implementation of biosaline agricultural systems. In desert environments, saline water resources must be desalinated in order to sustain agricultural production. However, the reject brine, a dense saline concentrate produced throughout the desalination process when disposed of in the environment, may have detrimental impacts on its attributes (Morillo et al. 2014; Giwa et al. 2017). At the world scale, inland and coastal desalination plants produce 141.5 million m^3/day of brine; most of this quantity (70.3% or 100 million m^3/day) comes from the Middle East and North African regions (Jones et al. 2019). The safe disposal of the brine remains a key environmental issue since it has more inorganic salts than brackish water and can contaminate groundwater resources when disposed of inland (Jones et al. 2019). Interesting solutions on brine management have been proposed adding economic value to its use by growing salt-tolerant plants and marine species such as fish and algae that can tolerate saline and hypersaline conditions following the integrated production systems approach (Crespi and Lovatelli 2010; Morillo et al. 2014; Sanchez et al. 2015; Giwa et al. 2017; Jones et al. 2019). This integrated farming model constitutes an effective management strategy that ensures efficient disposal and reuse of the reject brine to produce food, forage, and other valuable products. Such multi-component farming schemes can minimize external inputs, thereby decreasing the ecological footprint and promote biodiversity through growing a variety of crops, while producing high crop yields. The resource-saving practices of such schemes, allow the by-product of one system to become the input for another. These systems minimize the adverse effects of intensive farming and maximize the use of the available water resources through recycling. These combined farming systems could enhance food, nutrition, and livelihood security especially in hot and dry regions.

11.2 LANDSCAPE ANALYSIS OF INTERNATIONAL PROJECTS ON BIOSALINE AGRICULTURE

There have been various successful biosaline farming projects launched around the world. For example, the SalFar project is implemented in the North Sea Region and focuses on bringing into production degraded lands due to salinization (https://northsearegion.eu/salfar/). The main cause for increased salinization in the area is the continuous rise in sea levels which further aggravates seawater intrusion into inland farming zones affecting agricultural productivity. Without taking the appropriate measures to counteract the increasing salinity, this will lead to significant production losses, and severe damage will be caused to the local coastal economies.

The project is applying innovative methods of biosaline agriculture across selected coastal areas in the North Sea Region. It is a multidisciplinary project encompassing various disciplines such as agronomists, climate experts, farmers, entrepreneurs, chefs, etc. Apart from demonstrating alternative methods of farming under saline conditions, the project is also focusing on creating new value chains and business opportunities for local coastal communities using local salt-tolerant vegetation.

Coastal desert areas are barren lands that can be used for unconventional farming using seawater for halophytes (salt-loving plants) irrigation. The Seawater Energy and Agriculture System (SEAS) that was developed by the Masdar Institute in Abu Dhabi in the United Arab Emirates (UAE) (https://www.ku.ac.ae/the-seawater-energy-and-agriculture-system-seas-gets-an-upgrade) has been targeting coastal zones in hot and dry areas. SEAS combined an integrated system of aquaculture, halo-agriculture, and mangrove silviculture to produce sustainable biofuels for aviation and other by-products such as seafood. Based on the system's operation, seawater is pumped to supply shrimp and fish ponds and the water from the aquaculture then flows to the halophytes section.

Regarding inland desalination and the use of the reject brine for integrated farming, there has been a national program in Brazil, the so-called "Aqua Doce" program, that implemented the Integrated Agri-Aquaculture System (IAAS) approach using the reject brine from desalination for marine species farming and the aquaculture effluents were utilized for the irrigation of halophyte forage shrubs such as *Atriplex* spp. (Sanchez et al. 2015). The program launched in 2004 and has benefitted more than 150,000 inhabitants of the semi-arid region in the north-east part of Brazil. Several small-size reverse osmosis (RO) desalination plants were built to serve the local rural communities to reinforce the supply of freshwater, however, the brine produced was a problematic environmental issue. The IAAS scheme succeeded in turning an environmental problem (brine disposal in inland areas) into a source of new economic activities for the cultivation of fish and halophytes. In addition, due to its characteristics and good performance in arid regions and saline waters, the cultivation of a microalgae species (*Spirulina* sp.) was proposed as an alternative to fish farming within this production scheme.

Since 2014 the International Center for Biosaline Agriculture (ICBA) has been implementing an IAAS using the reject brine from desalination to grow fish and the aquaculture effluents (rich in nutrients) are directed to grow halophytes in an open field and hydroponically (saline aquaponics) (Robertson et al. 2019). More than 15% of farmers in the Gulf region are currently using RO-units to produce freshwater for the farming of vegetables (https://www.biosaline.org/projects/integrated-aqua-agriculture-enhanced-food-and-water-security); it is therefore crucial to explore the potential of the brine by-product from desalination for food and feed production. In a desert environment where there is a lack of replenished freshwater resources, it is imperative to tap into the use of alternative water sources for food production. Biosaline farming schemes therefore constitute a good alternative for agriculture in desert areas. The IAAS developed at ICBA was funded by the EXPO LIVE program (https://www.expo2020dubai.com/en/programmes/expo-live) in two phases. The first phase focused on the improvement of the production components (fish and halophytes) of the inland and coastal integrated farms

(https://www.biosaline.org/projects/inland-and-coastal-modular-farms-climate-change-adaptation-desert-environments); the second phase targeted to create a range of halophyte-based products for human consumption and animal feed produced from inland IAAS (https://www.biosaline.org/projects/expo-live-project-phase-ii-desert-farm-fork-value-chain-development-innovative-halophyte). Coastal IAAS can also offer a wide portfolio of halophytic and fish products (Lyra et al. 2019). Apart from the cultivation of halophytes and fish in local farms in the UAE, the program has included activities related to different stages of the value chain such as the development of halophytic products and public awareness campaigns on halophytes and biosaline farming. The EXPO LIVE project constitutes a multidisciplinary project comprising a consortium of prominent national and international partners such as the Abu Dhabi Agriculture and Food Safety Authority (ADAFSA), the Khalifa Fund for Enterprise Development (KFED), the Environmental Agency in Abu Dhabi (EAD), the Max Planck Institute (MPI) in Germany and Global Food Industries/Healthy Farm, the food company that is developing the halophyte-based food products.

The various characteristics of the four projects mentioned above have been summarized in Table 11.1.

TABLE 11.1
Characteristics of the Biosaline Farming Systems Implemented by ICBA and the Masdar Institute in the UAE, the Aqua Doce Program in Brazil, and the SalFar Project in North Sea Region

	IAAS Developed at ICBA (UAE)	SEAS at Masdar Institute (UAE)	Aqua Doce Program (Brazil)	SalFar Project (North Sea Region)
Inland and coastal IAAS developed	Inland/Coastal	Coastal	Inland	Coastal
Marine species cultivation (fish, algae, cockles, shrimps, etc.)	+	+	+	+
Use of aquaculture effluents for halophytes irrigation	+	+	+	−
Modular farms include vegetable farming	+	−	+	−
Seawater use for halophytes farming	+	+	-	+
Use of reject brine from desalination	+	−	+	−
Halophytes cultivation for human consumption	+	−	−	+
Halophytes use as forage	+	−	+	−
Halophytes cultivation for biofuel production	−	+	−	−
Multipurpose halophytes used (i.e. Salicornia)	+	+	−	+
Training of farmers	+	−	+	+
Policy development for proper use of saline water	+	−	+	+
Value chain development for halophytic products	+	−	−	+
Halophytic cuisine	+	−	−	+

11.3 OBJECTIVES OF THE EXPO LIVE PROJECT – PHASE II

The EXPO LIVE project entitled "From Desert Farm to Fork: Value Chain Development for Innovative Halophyte-Based Food Products" had the overarching goal of developing the value chain of Salicornia-based food products and increasing consumers' knowledge on desert farming and the nutritional aspects of halophytes. Three specific objectives were addressed: (a) adding value to the reject brine from desalination growing Tilapia fish and Salicornia, (b) development of food products using fresh tips from *Salicornia bigelovii* as the main ingredient, and (c) environmental and economical assessments looking into the sustainability of the IAAS and biosaline component.

11.4 MATERIALS AND METHODS

11.4.1 Farms' Profile for the IAAS Implementation

Eight farms were selected in Abu Dhabi Emirate, as shown in Figure 11.1. The selection criteria for the farms were the following: (a) desalination units should be installed within the farms, (b) the salinity level of the reject brine from desalination should be higher than 20 dS/m, (c) the selected farms should be relatively close to one another, and (d) the farmers should be collaborative. The characteristics of the farms are presented in Table 11.2. In half of the farms only the Salicornia component

FIGURE 11.1 The locations of the eight farms in Al Khatim and Al Khazna villages in Abu Dhabi Emirate where the IAAS approach was implemented.

TABLE 11.2

The Characteristics of the Eight Farms in Abu Dhabi Emirate. EC_{GW} Is the Electrical Conductivity of the Groundwater; $EC_{RO\text{-}BRINE}$ Is the Electrical Conductivity of the Reject Brine from RO-desalination

Farms with Salicornia and Tilapia Component			
Farm 453	Farm 17	Farm 79	Farm 168

EC_{GW} = 22.3 dS/m $EC_{RO\text{-}BRINE}$ = 40.2 dS/m	EC_{GW} = 28.7 dS/m $EC_{RO\text{-}BRINE}$ = 30.0 dS/m	EC_{GW} = 20.2 dS/m $EC_{RO\text{-}BRINE}$ = 24.9 dS/m	EC_{GW} = 12.9 dS/m $EC_{RO\text{-}BRINE}$ = 28.6 dS/m
Farms with Salicornia Component Only			
Farm 211	Farm 658	Farm 136	Farm 364

EC_{GW} =26.8 dS/m $EC_{RO\text{-}BRINE}$ = 36.8 dS/m	EC_{GW} = 12.7 dS/m $EC_{RO\text{-}BRINE}$ = 20.8 dS/m	EC_{GW} = 21.5 dS/m $EC_{RO\text{-}BRINE}$ = 31.0 dS/m	EC_{GW} = 20.0 dS/m $EC_{RO\text{-}BRINE}$ = 21.7 dS/m

was added (farms 211, 658, 136, and 364); Salicornia was directly irrigated with the reject brine from desalination. In the other four farms both Salicornia and fish were incorporated (farms 453, 17, 79, and 168); in this case, Salicornia was irrigated with the effluents from aquaculture. The lowest and highest groundwater salinity (EC_{GW}) values ranged from 12.7 dS/m (farm 658) to 28.7 dS/m (farm 17), and the lowest and highest salinity values of the reject brine ($EC_{RO\text{-}BRINE}$) were between 20.8 dS/m (farm 658) and 40.2 dS/m (farm 453). The salinity levels of the groundwater and reject brine were prohibitive for the growth of conventional vegetables and crops, so halophytes were considered for the production of food and forage.

11.4.2 Salicornia and Fish Components

Salicornia bigelovii was sown at a rate of 0.5 g/m² in all eight farms between 16 December 2019 and 8 January 2020, as shown in Table 11.3. The actual area of cultivated land ranged from 410 m² to 820 m². Bubblers were used for irrigation, trying to simulate the tidal and flooding effects observed in Salicornia's natural habitats (tidal marshlands, mangrove swamps, etc.). Leaching fractions were also considered for the water calculations.

An affordable and simple to operate Recirculating Aquaculture System (RAS) was installed and operated 24/7 throughout the experimental period (from January till June) at the four farms (453, 17, 79, and 168). The fish species that was cultivated was Tilapia (*Oreochromis niloticus*). The fish were cultured in two circular polypropylene tanks with a total water volume of about 7.2 m³ each. Three more tanks were used: one sedimentation tank for the removal of solid particles and wastes;

TABLE 11.3
Various Data Collected from the Eight Farms Related to Sowing Date of Salicornia, Surface Area Cultivated, Water Consumption for the Whole Salicornia Growth Cycle (from Sowing till Forage Harvest), Yield of Fresh Tips and Forage Biomass Yield

	Farm	Actual Land Cultivated (m²)	Salicornia Sowing (Date)	Water Consumption for Whole Growth Cycle (from Sowing till Forage Harvest) (m³)	Fresh Tips Yield (kg per m²)	Dry Forage Biomass (kg per m²)
Salicornia and Tilapia component	Farm 453	410	16-12-2019	3296	0.19	5.2
	Farm 17	820	26-12-2019	8311	0.06	2.6
	Farm 79	605	26-12-2019	4866	0.05	3.5
	Farm 168	512	08-01-2020	3743	0.03	0.4
Salicornia component only	Farm 211	710	17-12-2019	5440	0.06	3.6
	Farm 658	780	31-12-2019	4175	0.04	0.7
	Farm 136	512	17-12-2019	4377	0.15	4.0
	Farm 364	605	17-12-2019	3467	0.36	1.0

one biofilter tank full of cable hoses as bio-media for bio-filtration; and one water tank to pump the filtered water back to the fish tanks and for Salicornia irrigation. An air blower as a source of oxygen and two water pumps were also used to support the aquaculture system. All Tilapia systems operated using the reject brine from the RO-desalination plant and the aquaculture effluents were then directed to the Salicornia plots. To ensure that the system was functioning properly, the farm staff involved in the IAAS system were trained for the various aquaculture activities such as fish feeding; measurements of nitrite, nitrate, and ammonia; and the cleaning of the two fish and sedimentation tanks.

11.4.3 Soil and Water Monitoring

The electrical conductivity of the soil saturation extract (EC_e) was measured as an indicator of salinity in soil samples collected from Salicornia plots (0–30 cm depth) at the beginning (October 2019) and the end of the experiment (July 2020) using the reject brine from desalination and aquaculture effluents for Salicornia irrigation. Other soil parameters measured were soil texture, pH, Na, K, Ca, Mg, P, N, organic matter, and carbon content; these analyses are currently being assessed.

Detailed analyses were also conducted for all four water resources available at the farm level (groundwater, desalinated water, reject brine and aquaculture effluents) at two time intervals: (a) before starting the growing season and before using the reject brine for fish farming and Salicornia irrigation, and (b) at the end of the cultivating season. The results of these additional analyses are under assessment. Various group parameters were also analyzed such as: anions (fluoride, nitrate, nitrite, sulfate, etc.), metals (aluminium, arsenic, cadmium, etc.), BTEX compounds (benzene, toluene, ethylbenzene, and xylene), inorganic parameters (electrical conductivity, pH, etc.), pesticides, phenols and the microbiological load (total bacterial count, *Escherichia coli*, total coliform).

11.5 RESULTS AND DISCUSSION

11.5.1 Salicornia Cultivation

The water consumption values for the whole growth cycle from sowing until the harvest of the Salicornia biomass to be used as forage (early July 2020) were between 3296 m^3 and 8311 m^3 (Table 11.3). It is apparent that significant quantities of water were consumed for Salicornia cultivation using bubblers for irrigation.

The harvest of fresh tips took place in early April 2020; this coincided with the strictest lockdowns in the UAE during the COVID-19 pandemic. The yield of fresh tips (the upper 10–15 cm of the Salicornia shoots) was not as high as expected and there were several challenges faced which are explained in Section 11.6. The yield ranged from 0.03 (farm 168) to 0.36 kg/m^2 (farm 364). Whole Salicornia plants were harvested at a later growth stage to be used as forage in early July 2020. The dry biomass yields were between 0.4 (farm 168) and 5.2 kg/m^2 (farm 453).

The Salicornia fresh tips were purchased by a food company (Global Food Industries/Healthy Farm) at 4.1 USD $/kg to develop Salicornia-based recipes, dishes, and products. The selling price of Salicornia forage was 354 USD $/ton at the farm gate. Salicornia forage was distributed to camel, sheep, and goat farms and was used at a rate of 30% due to its high salt content (De La Llata Coronado 1991; Glenn et al. 1992; Swingle et al. 1996; Al-Owaimer 2000).

11.5.2 Soil and Water Analyses

Regarding the starting soil salinity (samples collected in October 2019), farm 168 had the highest soil EC_e (33.8 dS/m) and farm 136 had the lowest EC_e (2.3 dS/m) (Table 11.4). By July 2020, there had been substantial (118–465%) increases in the soil salinity at four farms (farms 364, 453, 658, and 136), but lower increases (15 and 54%) at farms 168 and 211 respectively. By contrast, on farms 17 and 79 the soil salinity decreased by 34% and 2.6% respectively. Factors such as the soil properties, the leaching fractions used and the time of the soil sampling (before or after irrigation) were taken into consideration to evaluate the obtained results of the electrical conductivity. Sanchez et al. (2015) also observed that a progressive salinization of the land irrigated with the reject brine could not be prevented, even though there was a slight salt removal capacity observed by the halophytic forage cultivated. After running a 5-year study of continuous irrigation and drainage of the fields they noted that salinity progressively increased from 0.60 in 2000 to 8.2 dS/m in 2006. Proper management of the irrigation, appropriate cultivation techniques, and the use of liquid manure were imperative to improve the performance of the yield grown with the saline water. In addition, periodical flushing of the salts with freshwater was necessary to decrease the soil salinity and maintain an acceptable salt balance.

Results for water analysis are still under assessment. Overall, a microbiological load (*E. coli* – Total Coliform) was not detected in any water samples. Most of the results derived from the analyses of the inorganic parameters were characterized by lower values than the allowed detection limits.

TABLE 11.4
Electrical Conductivity of the Soil Saturation Extract (EC_e) Collected from the Top 30 cm in All Eight Farms

		EC_e (dS/m)		Increase or Decrease of EC_e
	Farm	October 2019	July 2020	between the two Samplings (%)
Salicornia and Tilapia component	Farm 453	16.8	47.4	182
	Farm 17	18.3	12.0	−34
	Farm 79	15.2	14.8	−2.6
	Farm 168	33.8	38.9	15
Salicornia component only	Farm 211	14.7	22.7	54
	Farm 658	14.2	43.6	207
	Farm 136	2.3	13.0	465
	Farm 364	5.6	12.2	118

TABLE 11.5
Tilapia Biomass (kg) Progress per Farm

	Tilapia Biomass Progress per Farm (kg)					
	Months					
Farms	Initial Stocking	February 2020	March 2020	April 2020	May 2020	June 2020 (Total)
Farm 453	78	Fish died	78 (Restocking)	88	124	167
Farm 17	85	116	155	189	221	288
Farm 79	82	92	130	114	116	152
Farm 168	78	112	147	185	220	227

11.5.3 Fish Farming

The fish was stocked at each farm in late January 2020 and the initial biomass produced ranged from 78 to 85 kg (Table 11.5). A second restocking was conducted on Farm 453 because the fish died due to a lack of oxygen caused by an electrical power outage. The fish biomass was measured at the end of every month (from February till the end of June) based on the average body weight of fish sampled. Although the initial biomass was similar in all farms, the final produced biomass varied significantly between farms and ranged from 152 (farm 79) to 288 kg (farm 17). This variation was attributed to the fact that in one farm (farm 453) restocking of new fingerlings (fish of smaller size and less weight) was done due to fish death and the lack of correct fish feeding in some others. The total fish biomass from all four farms was 834 kg at the end of the experiment. The maximum fish density was achieved in farm 17 and was 20.5kg/m^3.

Feed conversion ratio (FCR) is the efficiency in terms of how much feed is required to produce 1 kg of fish. FCR is calculated as the ratio of feed given/fish weight gain. A good FCR is close to 1. Based on the results presented in Table 11.6, the best FCR was achieved on Farm 17 with FCR values ranging from 1.21 to 1.94 throughout the farming cycle. Farm 453 also had a good FCR but in order to be compared to the other farms, two more months of rearing would have been needed. Farm 79 had

TABLE 11.6
Feed Conversion Ratio (FCR) of Tilapia Fish per Farm

	Feed Conversion Ratio of Tilapia Fish				
	Months				
Farms	February 2020	March 2020	April 2020	May 2020	June 2020
Farm 453	Fish died	0.70	3.36	1.10	1.65
Farm 17	1.21	1.72	1.94	1.51	1.24
Farm 79	1.59	1.75	-2.61	17.67	0.47
Farm 168	1.03	1.80	1.43	1.92	14.76

the worst FCR (big and negative values) due to fish spawning. As a result, the fish population increased by 25% (newborn fish) which significantly affected the consumption of feed.

11.6 MAIN CHALLENGES ADDRESSED DURING THE SALICORNIA VALUE CHAIN DEVELOPMENT

Several challenges impeded some activities during the project (Figure 11.2). Some of these were aggravated by the lockdowns imposed by the COVID-19 pandemic.

- *Salicornia fresh tips harvest:* The harvest of fresh tips was conducted manually, and a high number of workers were utilized for this purpose (Figure 11.2 – photo A). Unfortunately, there are no harvesting machines in the market customized for Salicornia tips, which means that more working hours are needed for the farm staff to harvest sufficient quantity of Salicornia fresh tips to sell. This impacted on overall productivity and profitability. In addition, Salicornia fresh shoots can range in length from 5 to 25 cm, based on the requirements and specifications each company has for its food products, thus, the final yield of fresh tips might change. Proper specifications are therefore needed to be defined in advance to characterize the proper length of Salicornia tips based on the food use either as fresh or processed. Moreover, the lack of uniform height of Salicornia plants in the field meant that the cuttings of fresh tips were of uneven length (Figure 11.2 – photo B). This lack of uniformity was attributed to the cross-pollinating nature of Salicornia and to the fact that there is no actual variety developed with more stable characteristics.
- *Quality control and sanitization process for Salicornia fresh tips:* In order for Salicornia to be used as ingredient in food products, specific protocols based on HACCP (Hazard Analysis Critical Control Points) principles (FAO/WHO Codex Alimentarius Commission 2003) and procedures should be followed. As a result, an audit was conducted on the abilities of the farms to comply with the quality protocols. Analyses were carried out to determine the presence of heavy metals and pesticide residues in Salicornia grown in the different farming environments. In addition, before bringing in Salicornia fresh tips to the processing facilities of the food company, they needed to have been washed to minimize cross-contamination issues. Salicornia was therefore cleaned and sanitized at a hygienically designed facility near to harvesting sites using clean potable water and chlorine tablets (Figure 11.2 – photos C and D). After sanitization, the Salicornia shoots were placed in clean disinfected perforated plastic crates (15 kg in a crate) and were immediately transferred into a chiller truck (4°C) (Figure 11.2 – photo E).
- *Weeds presence in Salicornia plantation:* A few farms had persistent weeds such as *Tribulus terrestris*, *Sesuvium portulacastrum*, and *Portulaca oleracea* (Figure 11.2 – photo F). As there are no herbicides registered for Salicornia cultivation, these weeds had to be removed by hand. This increased the overall expenses of production.

(a) (b) (c) (d)

FIGURE 11.2 Manual harvest of Salicornia fresh tips on one of the eight farms in Abu Dhabi Emirate in April 2020 (a); uneven Salicornia fresh tips (b); sanitizing Salicornia fresh tips with chlorine tablets (c); washing Salicornia fresh tips (d); Salicornia shoots placed in clean and disinfected perforated plastic crates (e); and weeds in Salicornia plots (f).

(Continued)

(e) (f)

FIGURE 11.2 (Continued)

- Other challenges addressed had to do with the lack of workers at the farms especially during the COVID-19 outbreak, where social distancing and limited movement measures were imposed. Other challenges faced were the instability of the electrical supply, which resulted in fish death in one farm, and a lack of water supply at other farms which led to irregular Salicornia growth.
- *Lack of public's knowledge on halophytes, desert, and biosaline farming:* In order to increase the demand for Salicornia in the UAE, consumers should be aware of the opportunities that lie within the agricultural context in a desert environment, the nutritional benefits that halophytes have and how biosaline farming can contribute in fostering the food, nutrition, and security of livelihoods on a local level. Various halophytic crops that can be used for food and feed have great potential within such a marginal farming context.

11.7 INITIATIVES TO OVERCOME THE CHALLENGES

- *Salicornia breeding:* Improved high-yielding Salicornia varieties are needed with good, stable, and uniform characteristics that will guarantee a sustainable production of Salicornia for different uses in desert areas. As a result of the increasing number of national and international requests for technical back-stopping on Salicornia and after a 7 year of field selection of Salicornia germplasm (Lyra et al. 2016, 2020), ICBA is moving forward with a breeding program on this halophytic species which involves plant selection and the use of recombination breeding methodologies.

- *Mechanization of the fresh tips and seeds harvest*: ICBA is currently collaborating with manufacturing companies to develop customized equipment for the harvest of Salicornia fresh tips, the cleaning and collection of seeds since these are currently long, tedious processes.
- *Developing quality control and sanitization protocols for halophytic produce:* Designated areas need to be prepared for the proper auditing of the Salicornia produce when it is directed for the food industry so that cross-contamination is avoided. Sanitization protocols should be also developed and suitable facilities and equipment should be prepared for such use. In addition, heavy metals, pesticide residues, and the microbiological load should be analyzed to assess the safe use of the harvested material. Farm staff also need to be trained on good post-harvest handling practices of the fresh Salicornia produce.
- *Weeds management in halophytic cultivations:* When the weed load in Salicornia plantation is high, it is really challenging to control it manually. Because Salicornia consumes a lot of water, soil moisture is abundant, thus, weeds can emerge and spread quickly, especially when the Salicornia is at a young vegetative stage. Trials on the use of pre-emergent and post-emergent herbicides for broadleaf and grass weed species need to be conducted. Efficient weed control methods for Salicornia will contribute to the cleanliness of cultivation and its effective management.
- *Training programs for farmers and farm workers on IAAS and halophytes:* In order to avoid any hindrances on the operation of Salicornia cultivation and IAAS, farm staff should be trained properly to act proactively when an unexpected issue comes up such as a power outage, intermittent water supply, etc. ICBA is developing training modules translated into local languages (currently Arabic and Urdu) for farmers and farm workers to close the knowledge gaps that currently exist on Salicornia cultivation.
- *Initiatives to increase public knowledge on halophytes, desert and biosaline farming:* Public awareness campaigns have been launched to showcase the benefits of halophytic plants, their farming potential in desert environments, and their vital role in contributing to the national food security strategy plan in the UAE. Under the EXPO LIVE project, there is an initiative called "The Halophytic Kitchen Lab" (https://www.emiratessoilmuseum.org/education-programs/university-corporate-programs/halophytic-kitchen-lab) which is tailor-made for both students and adults. This initiative includes live interactive cooking sessions using halophytes (Salicornia, quinoa, etc.) as ingredients for recipes prepared by a chef with a nutritionist providing the necessary information on the nutritional value of halophytic crops. Creating dishes, recipes, and food products based on halophytes is effective in increasing the awareness of consumers and informing the public about halophytes' nutritional value, cooking, and farming potential, especially in areas dealing with soil and water salinity issues. Halophytes, with the contribution of chefs, can be included in the cuisines and dietary patterns of the local communities strengthening the food security component at the community, region, and country level.

11.8 INNOVATIVE SALICORNIA-BASED FOOD PRODUCTS

The Salicornia fresh produce collected from the eight farms was transformed through a collaboration with a food company (Global Food Industries/Healthy Farm) into innovative halophytic products marketed locally at an initial stage. Preliminary versions of the Salicornia-based products are presented in Figures 11.3 and 11.4. The

FIGURE 11.3 Sorbet with mango, banana and Salicornia (a); Camel Laban with Salicornia (b); Lasagna with Salicornia (c); Charcoal bread with vegan Salicornia burger (d). (All the food products and recipes shown in the photos were prepared by Healthy Farm team ©.)

ratio of Salicornia as an ingredient in the recipes ranged between 20% and 40% for all the products and recipes developed. Salicornia showed great versatility in cooking options and processing possibilities for both salty and sweet dishes, liquid and solid food products. The nutritional profile of Salicornia for food use has been investigated with promising results (Patel 2016). Salicornia is characterized by a high content of minerals and high Vitamin C, especially at a later growth stage, which in combination with Zn, Mg, and Mn make it a good candidate to boost the immunity system (unpublished data). Salicornia has also a good antioxidant profile, anti-aging properties, and fertility-boosting effects (Zhang et al. 2015). Overall, the vision is to

FIGURE 11.4 (a) Falafel with Salicornia, quinoa, chickpea and kale; (b) Vegan Salicornia burger; (c) Steamed Salicornia bread; (d) Camel cheesecake with Salicornia.

(Continued)

(e) (f)

FIGURE 11.4 (e) Charcoal pizza with Salicornia; (f) Vegan Salicornia, quinoa, peas balls. (All the food products and recipes shown in the photos were prepared by Healthy Farm team ©.)

create a halophyte-based food industry with local produce in a desert environment that could be replicated in similar climatic contexts and salt-affected areas.

11.9 CONCLUSIONS

Biosaline farming is a feasible solution in marginal, coastal, and salt-affected areas utilizing unconventional water resources such as seawater, brackish groundwater, and the reject brine from desalination. Various projects on biosaline farming have been implemented on a global scale applying a diverse range and combination of production components. One of these production schemes is the IAAS. The IAAS constitute climate-resilient systems that encompass different production modules (marine species and halophytic species for various uses) capable of enhancing the food and nutrition security of the local rural communities. Agricultural biodiversity is a direction that should be primarily adopted in vulnerable regions to make farm ecosystems more resilient to adverse climate change impact. IAAS can provide additional income-generation opportunities since both crops and fish can be produced for human consumption, as well as forage for animals. This combination provides an additional and diversified income to farm businesses. The EXPO LIVE project targeted to scale-up IAAS to eight farms in the UAE, looking into how value could be added to the reject brine from desalination by growing Tilapia and *Salicornia bigelovii*. Different challenges were faced during project implementation and initiatives to overcome these have been already launched. In order for the farming of halophytes to expand, there should be a gradual increase in their demand by consumers.

However, the public is not familiar with halophytes, their nutritional value and this knowledge need to be enhanced. Awareness campaigns need to be deployed to increase halophytic products' visibility and demand, so that halophytic products become more popular to consumers. Moreover, periodic soil and water monitoring should be conducted, so that the sustainability of such biosaline ventures could be guaranteed. As the requests on biosaline farming are increasing (ICBA is currently introducing IAAS and Salicornia at the Red Sea Governorate in Egypt and Morocco), it is imperative to adopt methodologies and directions that are primarily community-driven looking into all the value chain components and the local socioeconomic and climatic context. An impressive impact could be achieved for small-holder farmers in marginal, coastal, and salt-affected regions by implementing innovative, cost-effective, and low-consumption biosaline farming models that generate multisource food and income using brackish groundwater, seawater, and the reject brine from desalination. For such a purpose, research, governmental, academia, private sector should unite forces to make biosaline farming ventures feasible and sustainable. It is apparent that the vexing issue of the rising salinity levels should be confronted in a multidisciplinary way for more tangible and effective results.

ACKNOWLEDGMENTS

The current work is part of the EXPO LIVE project activities entitled "From Desert Farm to Fork: Value chain development for innovative halophyte-based food products". The project was funded by EXPO LIVE program of EXPO2020 Dubai for the period 2019–2021. The total investment for the installation of the infrastructure for Tilapia and Salicornia cultivation at all eight farms was conducted through the financial contribution of ADAFSA and KFED. The authors would like to thank: Mr. Turki Abdallah Turki Obeid and Mr. Rami Mohammed Araft Amsd Alqerem from ADAFSA for their contribution and support in the establishment and operation of the irrigation system for Salicornia; Mr. Ahmed Yahya Mohamed Elhasan from ADAFSA for the proper follow up with the Salicornia plot trials and data collection; Ms. Arzoo Malhotra from ICBA for the development of the map presented in Figure 11.1; Mr. Balagurusamy Santhanakrishnan from ICBA for his assistance in soil sampling, data collection and field activities; and Mr. Kaleem Ul Hassan from ICBA for conducting the soil analyses. The project team would also like to thank the management of all the partners and entities involved for their great support and encouragement.

REFERENCES

Al-Owaimer, A. N. 2000. Effect of dietary halophyte *Salicornia bigelovii* Torr on carcass characteristics, minerals, fatty acids and amino acids profile of camel meat. *Journal of Applied Animal Research* 18(2):185–192.

Crespi, V. and A. Lovatelli. 2010. Global desert aquaculture at a glance. Aquaculture in desert and arid lands: development constraints and opportunities, FAO technical workshop. 6–9 July 2010, Hermosillo, Mexico, FAO Fisheries and Aquaculture Proceedings No. 20. FAO, Rome, p.25.

De La Llata Coronado, M. M. 1991. Nutritive value of *Atriplex deserticola* and *Salicornia* forage for ruminants. MSc thesis. University of Arizona.

FAO/WHO Codex Alimentarius Commission, Joint FAO/WHO Food Standards Programme and World Health Organization. 2003. *Codex Alimentarius: Food hygiene, basic texts*. Food & Agriculture Org.

Giwa, A., V. Dufour, F. Al Marzooqi, M. Al Kaabi, and S. W. Hasan. 2017. Brine management methods: Recent innovations and current status. *Desalination* 407:1–23.

Glenn, E. P., W. E. Coates, J. J. Riley, R. O. Kuehl, and R. S. Swingle. 1992. *Salicornia bigelovii* Torr.: A seawater-irrigated forage for goats. *Animal Feed Science and Technology* 40(1):21–30.

IPCC 2013: Climate change 2013: The physical science basis. In *Contribution of Working Group I to the Fifth Assessment Report of the Intergovernmental Panel on Climate Change*, edited by Stocker, T. F., Qin, D., Plattner, G. K., Tignor, M., Allen, S. K., Boschung, J., Nauels, A., Xia, Y., Bex, V., and Midgley, P.M., 1535. Cambridge University Press: Cambridge, United Kingdom and New York, NY, USA pp. doi:10.1017/CBO9781107415324.

Jones, E., M. Qadir, M. T. van Vliet, V. Smakhtin, and S. M. Kang. 2019. The state of desalination and brine production: A global outlook. *Science of the Total Environment* 657:1343–1356.

Lyra, D., S. Ismail, K. Rahman Butt, and J. J. Brown. 2016. Evaluating the growth performance of eleven *Salicornia bigelovii* populations under full strength seawater irrigation using multivariate analyses. *Australian Journal of Crop Science* 10:1429–1441.

Lyra, D, S. Ismail, and J. J. Brown. 2020. Crop potential of six *Salicornia bigelovii* populations under two salinity water treatments cultivated in a desert environment: a field study. In *Emerging Research in Alternative Crops under Marginal Environment*, edited by Hirich Abdelaziz, Choukr-Allah Redouane, 313–333. Springer: Cham.

Lyra, D. A., R. M. S. Al-Shihi, R. Nuqui, S. M. Robertson, A. Christiansen, S. Ramachandran, S. Ismail, and A. M. Al-Zaabi. 2019. Multidisciplinary studies on a pilot coastal desert modular farm growing *Salicornia bigelovii* in United Arab Emirates. In *Ecophysiology, Abiotic Stress Responses and Utilization of Halophytes*, edited by Mirza Hasanuzzaman, Kamrun Nahar, Münir Öztürk, 327–345. Springer: Singapore.

Morillo, J., J. Usero, D. Rosado, H. El Bakouri, A. Riaza, and F. J. Bernaola. 2014. Comparative study of brine management technologies for desalination plants. *Desalination* 336:32–49.

Patel, S. 2016. *Salicornia*: evaluating the halophytic extremophile as a food and a pharmaceutical candidate. *Biotech* 6(1):104.

Robertson, S. M., D. A. Lyra, J. Mateo-Sagasta, S. Ismail, and M. J. U. Akhtar. 2019. Financial analysis of halophyte cultivation in a desert environment using different saline water resources for irrigation. In *Ecophysiology, Abiotic Stress Responses and Utilization of Halophytes*, edited by Mirza Hasanuzzaman, Kamrun Nahar, Münir Öztürk, 347–364. Springer: Singapore.

Sánchez, A.S., I. B. R. Nogueira, and R. A. Kalid. 2015. Uses of the reject brine from inland desalination for fish farming, *Spirulina* cultivation, and irrigation of forage shrub and crops. *Desalination* 364: 96–107.

Swingle, R.S., E. P. Glenn, and V. Squires. 1996. Growth performance of lambs fed mixed diets containing halophyte ingredients. *Animal Feed Science and Technology* 63(1–4): 137–148.

Zhang, S., M. Wei, C. Cao, Y. Ju, Y. Deng, T. Ye, Z. Xia, and M. Chen. 2015. Effect and mechanism of *Salicornia bigelovii* Torr. plant salt on blood pressure in SD rats. *Food & Function* 6(3): 920–926.

Section II

Biosaline Agriculture in Delta and Coastal Environments

12 Saline Agriculture as a Way to Adapt to Sea Level Rise

Pier Vellinga and Edward G. Barrett-Lennard

CONTENTS

Historically the experimental production of food under saline soil conditions has been focused on dryland systems where water shortages and salinization go hand in hand. Recently water-rich delta areas have started facing similar salinization challenges. Millions of hectares of fertile coastal lands and related livelihoods maybe lost as a result of climate change, accelerated sea level rise and an increase in extreme sea level events. The chapters in the following section describe the state of the art on salinization and saline agriculture experiments in delta areas.

Our focus has been primarily on the European North Sea countries and on Bangladesh with articles describing the typical delta characteristics and ongoing experiments including stakeholder's perceptions. However, the issues raised in these case study areas can be expected to be experienced in all low-lying coastal areas. The section also includes a case study on saline food production for tourists on a small low-lying island. Most of the articles focus on well-known crops such as potatoes, rice, grains, and vegetable crops, while major opportunities may also be found in combinations of crops and fish or in fiber, fodder, or products derived from seaweeds or mangroves. Regarding saline agriculture in delta areas, research and testing is still in the early stages. However, the challenge is clear given the global projections of sea-level rise.

12.1 SEA-LEVEL RISE

Projections of global average sea-level rise published by the Intergovernmental Panel on Climate Change (IPCC 2019) present a range from 0.4 m to 1.0 m for 2100. However, there are scenarios developed by individual researchers that predict far higher levels of sea-level rise (up to 2 m or more) by the year 2100 (DeConto & Pollard 2016).

DOI: 10.1201/9781003112327-12

Over the last few decades sea-level rise has accelerated from 1.4 mm per year over the period 1901–1990, to ~3.6 mm per year over the years 2006–2015 (IPCC 2019). At a regional level there are significant differences in rates of sea level rise because of natural variability in oceanic circulation and the rebalancing of the ocean surface in response to changes in geo-gravitational forces (caused by melting ice caps) (IPCC 2019).

12.2 LOSS OF LAND

Sea-level rise and changes in extreme sea-levels will have major implications for land in low-lying coastal areas and river deltas. Salt intrusion in river delta's is very likely, not only as a result of sea-level rise, but also as a consequence of changes in river regimes, upstream dam building and subsequent sediment interception in rivers feeding the deltas as described by Sepehr Eslami et al., for the Mekong Delta (Eslami et al. 2019).

Nicholls et al. (2011) made an estimate of the area of land loss due to regular flooding. Under conditions of no-adaptation to climate change, they have estimated land losses by 2100 of 877,000 and 1,789,000 km^2 for a 0.5 and 2.0 m rise in sea-level, respectively (Figure 12.1). This amounts to ~0.6 and 1.2% of global land area. The net population displaced by this rise is estimated to be 72 and 187 million people respectively over the century (roughly 0.9–2.4% of the global population) (Nicholls et al. 2011).

12.3 ADAPTATION

Communities and countries are likely to protect their land and settlements from flooding. In fact, this has been done for hundreds of years in the North Sea countries. However, diking and draining can be expected to trigger an almost irreversible

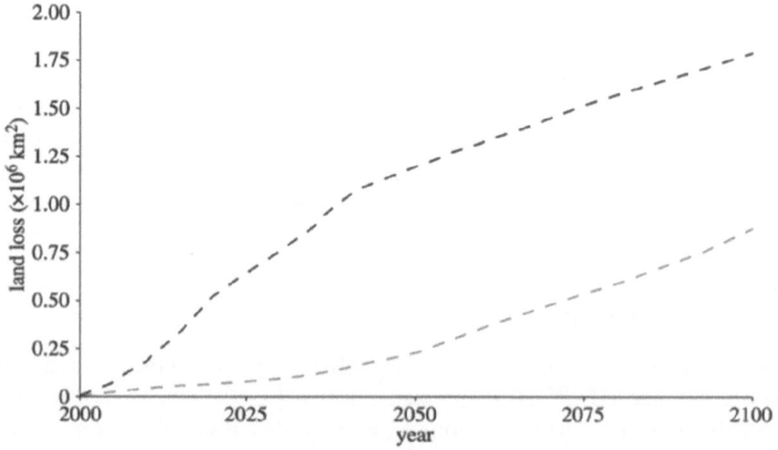

FIGURE 12.1 Global dryland losses assuming no adaptation for a 0.5 m (grey line) or 2 m (black line) rise in sea-level by 2100 as estimated by Nichols et al. (2011).

process of land subsidence and therefore the need for further dike heightening. This sequence is the reason why major parts of the agricultural coastal lands around the North Sea are now situated below mean sea-level (Chapter 13 of this book). Although higher dikes provide protection from flooding, for many geological situations they will not prevent an increase in underground saline water seepage into the low-lying agricultural areas. Once field levels are below mean sea-level it will become increasingly difficult to keep the saline water away from the agricultural soils. For this reason, coastal famers, waterboards and governments in the North Sea countries are beginning to explore a range of adaptation options including the creation of fresh water buffers and the introduction of crops and agricultural and water management practices that will allow production under saline soil and water conditions (Chapter 14 of this book).

The North Sea countries protected by dikes are now starting to explore these issues and opportunities. At the same time deltas with little or no dike protection and some experience in agriculture under brackish conditions are now also exploring their options. One way or another, all coastal areas, river deltas and small islands will need to adapt to rising sea-levels and changes in river regimes, storm regimes and sediment supply. This section presents the start of this process.

REFERENCES

Beauchampet, I. C. 2021. "Stakeholder Perspectives on the Issue of Salinization in Agriculture in The Netherlands." In *Future of Sustainable Agriculture in Saline Environments*. London: Routledge.

DeConto, R. M., & Pollard, D. 2016. Contribution of Antarctica to past and future sea-level rise. *Nature*, *531*(7596), 591–597.

Eslami, S., Hoekstra, P., Trung, N. N., Kantoush, S. A., Van Binh, D., Quang, T. T., & van der Vegt, M. 2019. Tidal amplification and salt intrusion in the Mekong Delta driven by anthropogenic sediment starvation. *Scientific reports*, *9*(1), 1–10.

IPCC, 2019: Summary for Policymakers. In: *IPCC Special Report on the Ocean and Cryosphere in a Changing Climate*. [H.O. Pörtner, D.C. Roberts, V. MassonDelmotte, P. Zhai, M. Tignor, E. Poloczanska, K. Mintenbeck, A. Alegría, M. Nicolai, A. Okem, J. Petzold, B. Rama, N.M. Weyer (eds.)]. In Press.

Nicholls, R. J., Marinova, N., Lowe, J. A., Brown, S., Vellinga, P., De Gusmao, D., … & Tol, R. S. 2011. Sea-level rise and its possible impacts given a 'beyond 4 C world' in the twenty-first century. *Philosophical transactions of the Royal Society A: mathematical, physical and engineering sciences*, *369*(1934), 161–181.

Vries, M. de, Velstra, J., Medenblik, J., Jansen, J., Smit, L., Rispens, A., & Essink, G. O. 2021. "Mitigating and Adapting Agriculture of Coastal Areas in the Netherlands Wadden Sea Region to Increasing Salinization: From a Vision towards a Practical Implementation." In *Future of Sustainable Agriculture in Saline Environments*. London: Routledge.

13 Stakeholder Perspectives on the Issue of Salinization in Agriculture in the Netherlands

Isa Camara Beauchampet

CONTENTS

DOI: 10.1201/9781003112327-13

13.1 INTRODUCTION

13.1.1 The Global Issue of Salinization

The salinization (i.e. increase in salt concentration) of fresh surface- and groundwater resources poses significant problems to farmers, industries, drinking water companies, and water managers in coastal areas all around the world (Delsman, 2015). These problems mostly relate to reduced crop yield, damaged infrastructure, adverse effects on vulnerable ecosystems, and the forced abandonment of extraction wells (Delsman, 2015). In addition, a combination of climatic and anthropogenic stresses like sea-level rise, changes in recharge and evaporation patterns, ground subsidence, population and economic growth, increasing industrial and agricultural water demands, and contamination of surface water further intensify pressures on freshwater resources and competition between the different uses in these areas (Figure 13.1) (Oude Essink et al., 2010).

With regards to agriculture, salinization of freshwater resources and agricultural land is one of the biggest threats to food production worldwide (Qadir et al., 2014), as higher salinity levels result in lower crop yields (Maas and Hoffman, 1977; Singh, 2015); furthermore, soil salinization is a global phenomenon, occurring in at least 75 countries, on more than 1 billion hectares of total land, and on 20% of irrigated land specifically (Ghassemi et al., 1995). Although recent statistics of the global extent of soil salinization do not exist (Shadid et al., 2018), Qadir et al. (2014) have estimated a daily expansion of this area by 2000 ha, and subsequent crop damage at 27.3 billion

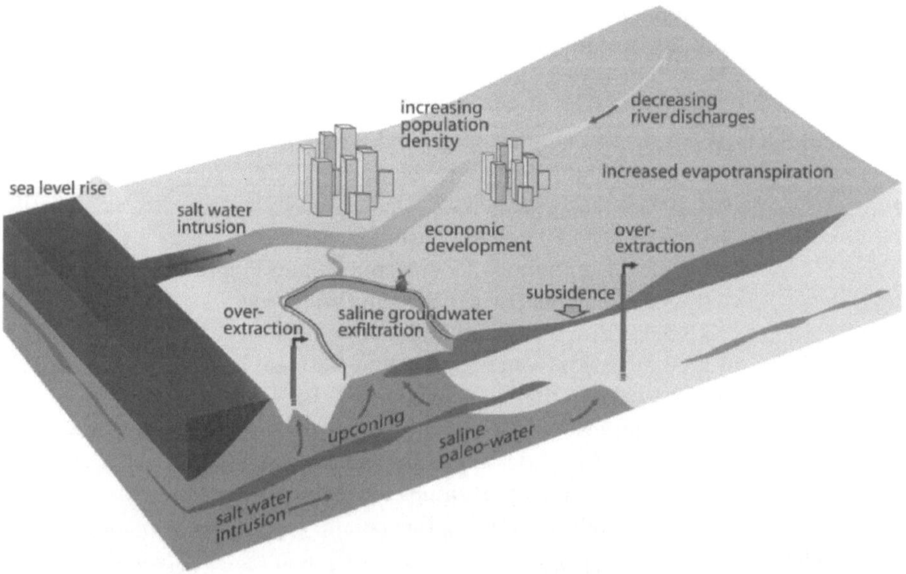

FIGURE 13.1 Overview of threats to coastal freshwater resources. (Reprinted from Deltares Select Series, 15, Delsman, J. R., Saline groundwater-Surface water interaction in coastal lowlands, 1–188, copyright (2015), with permission from IOS Press.)

US Dollars a year (Qadir et al., 2014). The issue is becoming increasingly problematic and widespread, particularly due to climate change effects like sea-level rise and more frequent and severe droughts (Singh, 2015). In the worst cases, farmers have to abandon their fields and clear new arable land, adding pressure to natural ecosystems and biodiversity (de Vos et al., 2016), thereby affecting both livelihoods and the environment. With a growing population and subsequent growing demand for agricultural products, progressing climate change, little new productive land without sacrificing valuable nature, and increasing competition for freshwater resources, salinization is a global issue that urgently requires a solution (Qadir and Oster, 2004; Singh, 2015).

13.1.2 The Issue of Salinization in the Netherlands

In the context of salinization, the Netherlands is rather unique for two reasons; firstly, about 25% of the land surface lies below mean sea level and without its dunes and dykes 65% of the country would be regularly flooded (Huisman et al., 1998), and secondly, a significant amount of the land surface (600,000 ha) consists of polders, i.e. pieces of land that have been reclaimed from a body of water (i.e. a lake, floodplain or marsh) through the creation of artificial and autonomous hydrological systems of dykes and drainage canals (Huisman et al, 1998). In areas that lie below mean sea level, saline groundwater may reach the surface by upward groundwater flow, a process which is commonly referred to as saline or brackish seepage (Oude Essink et al., 2010). This results in the salinization of surface waters and shallow fresh groundwater bodies, making the water unfit for the supply of drinking water, industrial purposes, and irrigation (de Louw et al., 2010). In addition, brackish seepage can also directly end up in the root zone and thereby cause salt stress in plants (Oude Essink et al., 2010).

A future rise in sea level is expected to increase the seepage and salt loads in surface waters and thereby reduce the availability of both fresh surface water and groundwater (Oude Essink et al., 2010). Model simulations show that with sea-level rise, salt loads from groundwater seepage will be doubled in several low-lying parts of the coastal zone of the Netherlands by 2100 (Oude Essink et al., 2010). Moreover, as the low elevation of polder systems requires perpetual drainage of water to avoid waterlogging from seepage, both direct salinization (by attracting saline water to the surface) and indirect salinization (through ground subsidence) are already common (Oude Essink et al., 2010). Therefore, most of the salinization-prone areas are located near the coast, in reclaimed lands and in previous intertidal zones, where seawater is (historically) present in the groundwater and relatively close to the soil surface (Velstra et al., 2009; Figure 13.2). Without the use of freshwater to regularly flush through the water systems and soils in these low-lying areas, the brackish groundwater would be a major limiting factor to agriculture in particular (Velstra et al., 2009). However, the combination of increasing external intrusion of seawater in groundwater aquifers and (open) waterways, decreasing river discharge, decreasing precipitation and increasing evapotranspiration in the drier seasons is limiting the availability of freshwater to do this, especially at the 'end of the pipeline' regions (Velstra et al., 2009).

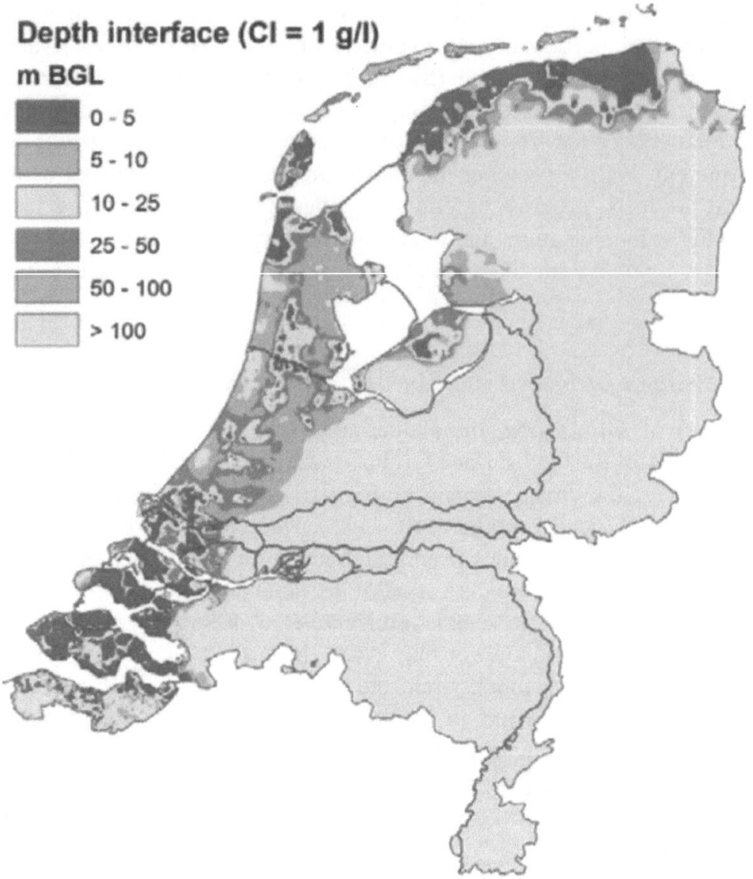

FIGURE 13.2 Depth of the fresh-salt interface in meters below the surface level, where 1 gram of chloride per litre is the concentration at which water is classified as saline in the Netherlands. (Source: de Louw, 2013.)

13.1.3 Problem Statement and Research Question

The threat of salinization in the low-lying coastal regions of the Netherlands is widespread and growing due to historically saline groundwater reservoirs, the relatively high (and increasing) sea level, ground subsidence, and changes in recharge and evapotranspiration patterns, which are not only accelerating salinization but also diminishing the supply and availability of freshwater (Velstra et al., 2009; Oude Essink et al., 2010). Agriculture in these areas is particularly vulnerable to the threat of salinization due to its exposure to brackish seepage from the sub-soils, weather events, and by being one of the last activities to receive freshwater in times of scarcity (OECD, 2014). Discovering how the Dutch agricultural sector can be made more salinization-resilient is of direct regional socio-economic importance as sensitive and intensive agriculture is especially located in salinization-prone areas (de Louw, 2013). Moreover, it can be of global importance as well since the Netherlands is one of the first deltas facing the impacts of climate change in combination with increased anthropogenic activities – due to the below sea-level position of, and high density of intense socio-economic activities in, the coastal region – thereby serving as a laboratory case for many other low-lying deltas around the world (Oude Essink et al., 2010).

However, as critical freshwater shortages have always been more of an exception to the rule – as evident from the national evaluations from water managers and users after the summer of 2018 which was the driest summer ever recorded in the Netherlands – the issue of salinization is relatively new to the Netherlands (Delta Commissioner, 2018a). Therefore, business-as-usual responses are rather unlikely to solve the problems ahead. On the contrary, new mitigation and adaptation measures are needed for 'climate-proofing' the freshwater availability in the Dutch delta, as evident from the recently established Delta Decision and Delta Plan on Freshwater Supply, which aim to secure the availability of freshwater now and in the future (Delta Commissioner, 2018b). In order to contribute to the overall knowledge gap of how the freshwater availability in the Netherlands can be made more climate-proof, this research aims to fill the gap of how the issue of salinization for agriculture in the low-lying Netherlands can be addressed. Therefore, the research question is: how can the Dutch agricultural sector be made more salinization-resilient? This will be answered through several sub-questions:

1. To what extent is salinization perceived as an issue for agriculture in the Netherlands?
2. How do current dominant land- and water-management practices relate to the issue of salinization?
3. What are the opportunities and barriers to different mitigation and adaptation measures in addressing the issue of salinization?
4. What is locking-in the status quo and what creates opportunities for a transition toward salinization-resilience[1]?

13.2 METHODOLOGY

13.2.1 Theoretical Framework and Research Method

The PRactice-Oriented Multi-level perspective on Innovation and Scaling (PROMIS) framework was applied to gain integrative perspectives on the scaling of salinization-resilient innovations for reducing the negative impacts of salinization. This framework connects the heuristic framework of the multi-level perspective on socio-technical transition (MLP) to a 'modal aspects' framework, thereby enabling the heuristic exploration of relevant, multi-faceted dimensions and dynamics involved in innovation and scaling processes (Wigboldus et al., 2016). The application of the framework aided in unraveling the different dimensions of the current agricultural and water-management system that keep it from becoming more sustainable and that affect the scaling of more sustainable technologies, practices, and policies (Wigboldus et al., 2016). Moreover, it helped to identify how a variety of dynamics in scaling interact, thereby locking current practice into its unsustainable mode or stimulating change (Wigboldus et al., 2016; Figure 13.3).

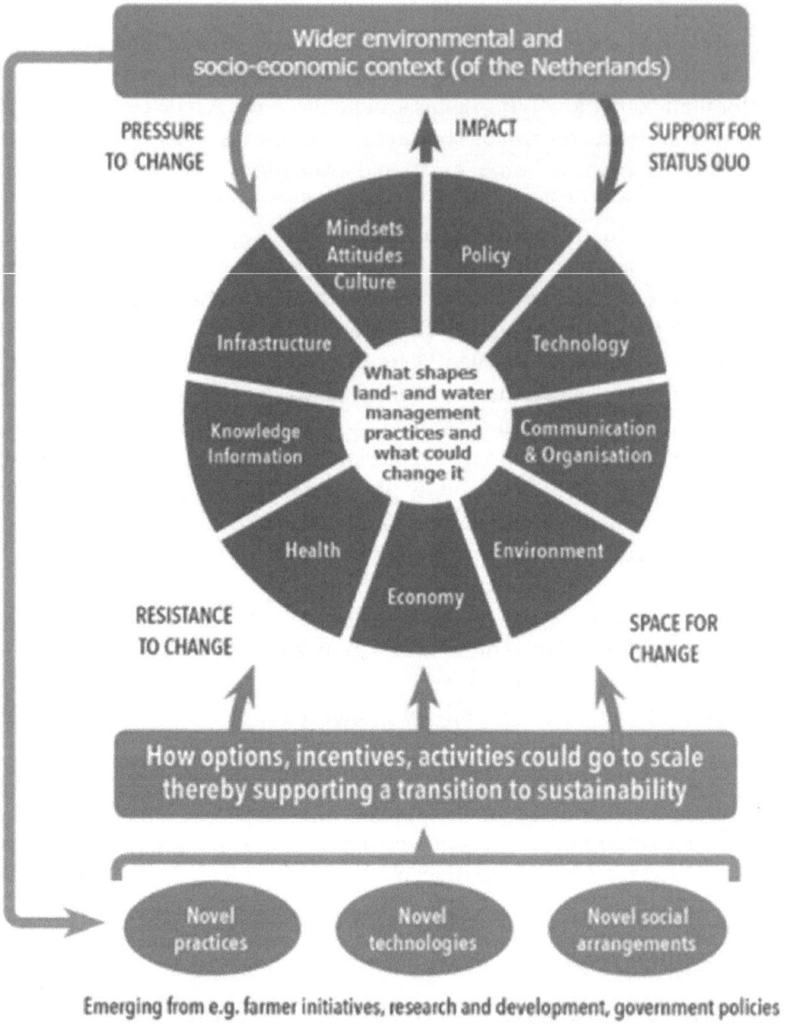

FIGURE 13.3 An integrative perspective on multi-level dynamics that have implications for opportunities to make a transition to a more salinization-resilient agricultural sector. (Adapted from Wigboldus et al., 2017.)

13.2.2 Data Collection and Analysis

Semi-structured interviews were held with stakeholders to ensure that a large range of dimensions and dynamics relevant to the scaling of salinization-resilient measures were able to emerge. This was especially important, considering that little peer-reviewed literature on the issue of salinization and (innovative) salinization-resilient measures existed to inform which specific aspects should be reviewed by the interviewees.

For the selection of stakeholders, Reed et al.'s (2009, p.1933) definition of a stakeholder as someone who is 'affected by the decisions and actions [that are taken], and who has the power to influence the outcome' was applied. Under this broad definition, agricultural businesses, waterboards, and provinces, and any other individual or organization that has the power to influence the issue of salinization could be included. To ensure a wide range of explored perspectives and interests, interviewees (50 in total) were selected from each salinization-prone province, with a balance between interviewees from both the niche level and the regime level, as well as between private and public stakeholders.

For the expert interviews, 12 Dutch experts in the field of salinization and/or salinization-resilient measures were selected. These scholars were mostly identified from the literature review and were selected on the basis of their evident expertise, i.e. either having published multiple papers on the topic and/or being referred to as an expert by the research institute they are connected to. Several experts were selected on the basis of referrals, as these were practical experts (e.g. advisors) instead of scholars. Experts were incorporated in the research for their relatively objective and/or nonpartisan evaluation of the issue of salinization and possible solutions, as compared to stakeholders.

For the stakeholder interviews, officials from six provinces and seven waterboards in salinization-prone areas were interviewed. All relevant provinces and waterboards were included to ensure that relevant regional differences like the dominant type of soils (e.g. clay or sand), agriculture (e.g. arable farming or horticulture), and water supply (e.g. presence of a river body or not) were accounted for. The experiences and perspectives of agricultural stakeholders (i.e. farmers) were represented by the farmer network organizations (12 representatives in total), as these were most aware of the current situation and issues in the sector and could thus provide an overview of the perspectives of most farmers. Furthermore, nine farmers that were experimenting with salinization-resilient measures were interviewed, as these represent the niche level and could thus identify the relevant factors and dynamics that allow or prevent the scaling of these measures. Furthermore, three plant breeders were interviewed to discover whether and how they incorporated salt-tolerance since this could tell something about the demand for and feasibility of breeding on the basis of salt-tolerance. Finally, two national governmental bodies were interviewed as well since these are also involved in the issue of salinization, mainly through coordination, legislation, and funding. Most stakeholders were identified through google searches and snowballing, i.e. selecting through the referrals of those that are interviewed already (Reed et al., 2009). The list of the interviewees can be found in Table 13.1.

The face-to-face interviews were recorded and transcribed afterward; the interviews over the phone were transcribed on the spot. The analysis of the transcriptions was done by identifying the overarching, categories, concepts, and patterns. The aspects of experienced reality from Table 13.2 were used to guide the process to ensure coherence and completeness. The subsequent emerging themes were then analyzed by use of the PROMIS framework, and categorized into the niche, regime and landscape level, including the interactions between them, and the lock-ins of the current non-salinization resilient practices; the opportunities for (more) salinization-resilience were also identified. Due to the large quantity of interviewees, the results are supported by use of a scale of how often a statement, opinion, or view emerged from the interview data, i.e. *few* (<3), *some* (4– 9), *many* (10–19), *majority* (20–34), *nearly all* (>35).[2]

TABLE 13.1
List of Interviewees

Category	Organization or Specialization
Agro-hydrological experts	Wageningen University
	Acacia Water
	Deltares
	STOWA
Agricultural experts	Wageningen University
	Salt Farm Texel (2x)
	SPNA
	Delphy (flower bulbs (3x), Zeeland, and South-Holland)
	Agrifirm
Plant breeders	Agrico Research
	HZPC
	C. Meijer B.V.
Provinces	Groningen
	Friesland
	North-Holland
	South-Holland
	Zeeland
	Flevoland
Waterboards	Noorderzijlvest
	Fryslân
	Hollands Noorderkwartier
	Rijnland
	Hollandse Delta
	Scheldestromen
	Zuiderzeeland
Central government bodies	Rijkswaterstaat
	Ministry of Infrastructure and Water Management
Agricultural network organizations	LTO North (Friesland and Groningen) 2x
	LTO North-Holland
	LTO South-Holland
	ZLTO Tholen
	ZLTO Zeeland
	LTO Flevoland
	Greenport North-Holland Noord
	Greenport Boskoop
	Royal General Union for Flower bulb culture (KAVB)
	Water Commission Northern Sand Region
Innovative farmers	Zeeland (5x): saline agriculture and freshwater measures at parcel, community, and regional level
	North-Holland (2x): freshwater measures parcel
	Groningen: freshwater measures parcel
	Friesland: freshwater measures parcel

TABLE 13.2
Aspects of Experienced Reality That Can in Various Ways Be Affected by, or Affect, Innovation and Scaling Processes

Categories of Experienced Reality	Aspects of Experienced Reality	Example of Entities That Distinguish Themselves from Other Entities Primarily along the Lines of That Aspect
Natural and physical capital	Quantitative, spatial, kinematic, physical	Numbers, location, atmosphere, climate, water, soil, natural forces, chemistry, transportation, infrastructure, buildings, equipment
	Biotic, sensitive	Plants, animals, birds, fish, organic processes, ecosystem, biodiversity, forest, desert, habitat, farm, crops, livestock, animal behavior
Human capital	Biotic, sensitive	Awareness, health, physical and mental abilities, emotion, personality, disposition, passion, observation, population dynamics, safety
	Analytical-logical	Knowledge, theory, logic, conceptual framework, science, research, education
	Formative	Construction, creativity, skill, computer software, design, power (in relationship): technology, strategy, methodology, innovation, adaptation
Social and financial capital	Lingual, social	Symbols, signs, language, communication, information, media
		Relationships, roles, social cohesion, competition, collaboration, organisation, societies, alliances, partnerships
	Economic	Resource management, conservation, stewardship, exchange of goods and services, transactions, efficiency, sustainability, economy, land use, market, value chain, firm, employment
Cultural, political and moral capital	Juridical	Rights, law, responsibility, appropriateness, policy, legal system, constitution, mandate, police, the state, democracy, ownership
	Aesthetical, ethical, certitudinal	Appeal, beauty, enjoyment, leisure, sports, art
		Attitude, care, sharing, goodwill, integrity, equity, being right, solidarity identity, belief, trust, faith, vision, commitment, aspiration, worldview, ideology, paradigm

Source: Adapted and Abbreviated from Wigboldus et al., 2016.

13.3 RESULTS

13.3.1 To What Extent Is Salinization Perceived as an Issue for Agriculture in the Netherlands?

Although *all* interviewees recognize that salinization is largely harmful to agriculture, the exact extent to which it is or will be harmful is difficult to determine because of several reasons (*majority*). Firstly, salt stress and drought stress look very similar and predominantly appear at the same time since saline water can reach the surface in dry periods (*majority*). Secondly, (the risk of) diffuse brackish seepage is largely invisible to farmers as the conditions within the parcel of land cannot be easily observed (*many*). Thirdly, the threat of salinization and salt-damage depends on many different cultivation specific factors (*majority*), like the cultivated crop (i.e. length of roots, growth stage at time of exposure, salt-tolerance), soil type, soil quality, thickness of the freshwater lens, the type of drainage, and can therefore differ per parcel and even within a parcel. Fourthly, salinization is not a linear process, it comes in peaks and fluctuates per season and year (*majority*). Fifthly, the effect of different salt concentrations on commonly cultivated crops and soils other than sand are largely unknown (*nearly all*). Sixthly, salinization risk maps cannot be directly extrapolated to the parcel level, nor estimate actual damage (*many*).

The complex nature of salinization could explain why, although there are many regions with a risk of salinization and although last year was recognized to be exceptionally dry, there are not that many signs nor reporting of widespread salinization-related issues or damage yet (*majority*). Another explanation could be that the issue has not been identified as such yet by farmers as it is not common practice to measure the salt concentrations of parcels and irrigation water (*many*). Moreover, it is rather difficult to establish maximum possible yield and to link sub-optimal yield to just one factor like salinization (*many*). *Few*, argue that the low reporting of salinization problems is the result of the issue being a taboo, e.g. as salinized lands drop significantly in financial value once discovered. Either way, there is a consensus that there is very much an awareness issue, i.e. that the threat and issue of salinization are not being recognized and/or acknowledged.

Although the issue has become more prominent in recent years and is starting to become more and more of a concern amongst farmers and (to a lesser extent) policy-makers due to climate change and especially the exceptionally dry summer of 2018 (*majority*), there is still a lack of urgency because: (1) the issue is only little or just recently starting to be experienced, recognized and/or signaled by farmers (*majority*), (2) the damage has not been significant enough yet (*majority*), and (3) it is relatively low on the priority list of current pressing issues in the agricultural sector, like droughts, flooding, nutrients, biodiversity, etc. both from the perspective of policy-makers as for farmers (*majority*).

It is generally recognized that it will take more dry periods – like the year 2018 – to make salinization into a prominent theme (*majority*). The *majority* of the interviewees, expect the latter to happen sooner or later as salinization will become more and more of a problem, mostly due to autonomous salinization, sea-level rise, decreased river runoff, and decreased precipitation and increased evapotranspiration during the summer months – although the exact extent of the issue in the future is uncertain as this is largely determined by how climate change will develop.

13.3.2 How Do Current Dominant Land- and Water-Management Practices Relate to Salinization?

In general, it is recognized that many of the current dominant agricultural land- and water-management practices are contributing to salinization. As recognized by *nearly all* interviewees, the regime of the past decades has been to dispose of freshwater as much as possible, since without it these regions would be flooded due to their position beneath sea-level (*many*), but also because floods have been, and to a large extent still are, more common and impactful (*majority*). In line with this regime, the dominant drainage method amongst farmers is conventional drainage in which water is continuously drained until the water-table is lowered to a certain depth (*nearly all*). This method, however, does not ensure that freshwater is retained for periods in which it is necessary (*majority*), as it ensures that soils continue to be drained to a certain depth, and as it drains the fresh precipitation water instead of saline groundwater, thereby drawing saline groundwater to the (sub-)surface in drier periods, possibly salinizing the root zone. The reason that this is still the dominant method of draining is because: (1) it is the customary practice, i.e. farmers do not know any better (*many*), (2) it is rather unproblematic in 'normal' years (*many*), (3) farmers are more focused on removing water instead of retaining it, for example because heavy rains are more common and impactful than drought or salinization (*majority*), (4) farmers are not that aware of a freshwater lens (*some*), (5) farmers do not know or do not acknowledge that this type of drainage can cause salinization or contribute to their drought problems (*many*), and (6) other drainage systems are more expensive (*majority*).

Another preference amongst farmers is that the water level in their parcel and their region is kept as low as possible to lower the risk of flooding during heavy rains and for them to be able to work on their fields with heavy machinery (*nearly all*). However, the water level in their parcels is determined by the water level in the ditches, and in turn determined by the compartment water level, which is a legal agreement between the farmers and the waterboards (*many*). Therefore, many farmers have installed so-called under-drainage with which they can artificially lower the water level beneath that of the regional system, whereas this also increases the risk of saline seepage (*some*). Next to the parcels being designed to drain as much water as possible, the intensification of cultivation practices over the past decades has led to the large-scale deterioration of the structure and fertility of soils, which are therefore not capable of draining large quantities of rainfall during heavy precipitation events, nor retaining freshwater when it is drier (*many*). The latter also results in an increased risk of salinization, since saline groundwater can more easily reach the root zone if the soils are dry (*many*). In regions where farmers use deep freshwater lenses to irrigate because no fresh surface water is available, there is a risk of exhausting and/or salinizing these wells, since sustainable extraction is difficult for farmers and local water managers to determine and control (*some*).

Due to the historical function of regional water systems (*many*) – and sometimes also the technological inabilities and lack of freshwater in the surrounding areas at the time which restricted a supply function (*some*) – most regional water systems have been historically designed to dispose of water instead of retaining or supplying areas with water (*many*). Next to lacking the ability to retain water, another consequence of this historical function and design is that most systems do not have separate discharge and supply channels, and/or that these are often not located in the right places. This results in the inefficient flushing of the system, i.e. large variation in chloride levels between ditches, and the general water quality being lower than it could be due to the mixing of fresh and saline water (*many*). The regional water systems that have been designed for flushing and the supply of freshwater are often the locations where more sensitive high-capital cultivations like flower bulbs have been historically located or where they start to emerge (*many*). The paradoxical situation of high-capital and high-quality water demanding crops being located in the most salinization-prone areas is a widespread phenomenon (*many*), mostly explained by the fact that the soils here are very suitable to these types of cultivations (*many*), that the distribution infrastructure is already there (*some*), and in some cases – especially flower bulbs – also because of the fact that there is a secured freshwater supply for flushing (*some*). However, the facilitation of high-quality freshwater demanding cultivations in largely salinization-prone areas results in significant challenges for waterboards during dry periods and might become too technically challenging and costly in the future (*many*).

The challenge that waterboards face of having to keep the water supply fresh mostly relates to there being 'effort obligations' of chloride norms they have to adhere to for flushing (*many*), though these norms are quite arbitrary and it is difficult to determine at what point the effort a waterboard has to take is no longer 'reasonable' (*many*). Moreover, as the waterboards are in charge of the regional water system and therefore responsible for the water quality in the ditches, farmers tend to pressure them into adhering to the norms (*many*). The latter is also the result of farmers being largely dependent on this collective service, as they have never invested in water retaining measures or their own water storage since the disposal, supply, and quality management of water have been arranged collectively in the Netherlands ever since the waterboards were established (*many*). Therefore, although several waterboards state their historical function has changed and that they are not legally obligated to supply farmers with enough freshwater at all costs, the practical reality is that they do go to great lengths to make it possible (*many*). On top of that, most farmers believe this should actually be the case because they pay waterboard taxes and as they think freshwater availability should be, or is best to be, arranged collectively (*many*). What is adding to this challenge is the fact that although the function an area gets – e.g. agriculture, residential area, nature – should be based on the water situation according to the waterboards and provinces, this is currently not yet the case in provincial spatial planning policies, implying that high-quality water demanding types of agriculture can settle in salinization-prone areas and that waterboards will have to facilitate these to the best of their ability (*many*). The latter is also the result of this spatial classification being rather broad, i.e. not differentiating between different types of agriculture (*some*).

Another dominant water-management practice at a higher management level that is perceived as unsustainable – especially by agricultural stakeholders but also by several waterboards and provinces – is the fact that the majority of freshwater that is entering the country is immediately being discharged into the North Sea (*many*). Although this is not much of an issue in 'normal' years, it can be an issue during dry summers when river discharge is low and much more water is needed to counter-pressure the salt wedge from entering the main waterways (*some*). This is where the different functions of the main waterways and inlet points start to collide, i.e. the shipping sector profits from the open connection to the sea at the Nieuwe Waterweg, whereas this increases the salt load in the main rivers to the point where the salt concentrations near the inlet points of the regional water systems can become too high for drinking water and industry, resulting in water shortages for agriculture in the area behind the inlet point as well (*some*). When this happens, the so-called climate-proof waterway which directs water from the IJsselmeer and the east of the country directly to the west via an alternative route, is expected to become the rule rather than the exception due to climate change (*some*). In addition, the IJsselmeer is seen as an important water buffer by the adjacent provinces and there will be increased competition between the different regions for using this source in the future if the supply of river water to it and/or its buffering capacity is not increased (*many*).

Farmers cope with salinization of freshwater for irrigation in various ways (*majority*); it depends on the crop they cultivate (i.e. sensitive flower bulb versus relatively tolerant sugar beet), the chloride/EC levels they are used to, their practical experience, their risk perception, the soil type, the quality of the crop or yield (the poorer, the more they are willing to take the risk), the growth stage of their crop (seedlings can take less), the time in the season (if it is at the beginning or the end of the growing cycle, do they expect rain or not), and so on. Therefore, there are large discrepancies between what different arable farmers, vegetable farmers, and flower bulb growers perceive as 'too saline' for irrigation – even between different regions – and also between what they do and what scientific experiments have determined what is possible, both positively and negatively (*many*). *Nearly all* interviewees state that it is largely uncertain what the salt tolerances of commonly cultivated crops and soils are and that farmers, therefore, maintain a conservative standard, especially considering the large financial risk they would be taking (*many*). This also implies that there are hardly any field measurements of the effect of different salt concentrations on crop yield (*some*), whereas raising these norms could potentially limit drought damage by increasing the amount of usable water in times of drought (*some*).

13.3.3 What Are the Opportunities and Barriers to Different Mitigation and Adaptation Solutions in Addressing the Issue?

13.3.3.1 The Parcel and Farm Level

Nearly all interviewees recognize that, at the parcel and farm level, salinization-resilience can be increased in the short- and medium-term by increasing the freshwater lens in the parcel. The *majority* of the interviewees see a large role for level-controlled drainage or so-called anti-salinization drainage in preventing soil and root zone salinization by conserving and even enlarging the freshwater lenses in the soils. Projects like Spaarwater in the northern regions of the Netherlands and the pilots in Zeeland are therefore perceived to be of great importance in researching what is possible but also in demonstrating to farmers what the positive effects of these alternative types of drainages are (*majority*). By buffering freshwater in the sub-soil, farmers simultaneously lower the risk of drought and salinization (*many*). Moreover, as these alternative measures are similar to what farmers are already used to, it is more attractive to make the switch (*some*). The complexity with anti-salinization drainage, however, is that it has to be carefully customized since the thickness of the freshwater lens is very parcel specific and can even differ within a parcel, and as soil structure matters as well (*many*). An obstacle to installing level-controlled or anti-salinization drains, in general, is that drains are installed only once every 20 years and that it is uncertain whether this more expensive investment is profitable since farmers cannot foresee how much they will suffer from salinization or droughts in the future (*some*). There is also the confusion/perception amongst farmers that increasing the freshwater lens through drainage can result in a wetter environment and a higher risk of flooding, which hampers the implementation of these types of drainage as well (*some*). Another obstacle to these systems is that they still rely on the availability of freshwater, as they can still run dry and/or because setting up the water in the drains with saline water is not desirable (*some*). An opportunity to further increase the freshwater lens by use of innovative drainage is by combining it with water infiltration in the sub-soil (*some*). Nevertheless, freshwater infiltration into the freshwater lens or deeper aquifer is not always possible and currently not allowed due to regulations on preventing groundwater contamination (*some*).

Another way farmers can increase the freshwater lens in their parcels quite easily is by more shallow dewatering, or by increasing the water level underneath the parcel by keeping the water up in their ditches with (simple) weirs (*some*). Nevertheless, barriers are that most farmers are more fearful of flooding or not being able to work their fields with heavy machinery (*many*) and because they think that setting up brackish water in the ditch might increase the brackish seepage in their parcel (*some*). The latter interaction is contested by what farmers do on Texel, where they set up the water early in the year already to keep the freshwater in their parcels, despite the water in the ditches being brackish or even saline (*some*). Another low-hanging-fruit for farmers with brackish ditches is to install weirs that separate and subsequently dispose of the brackish water and retain the freshwater – a rather simple and cost-effective measure (*some*). Nevertheless, this measure is not widely implemented yet, arguably since people were, or still are, unaware of the water stratification in ditches (*some*). Next to conserving and enlarging the freshwater lens, there are different low-investment measures which can improve the structure and fertility of the soil in order to retain more water and be more drought, salinization, and even flooding resilient; these include zero-tillage, supplying more organic matter, and the cultivation and use of green manure (*some*). However, it is rather unknown how farmers can do it most effectively (*some*).

Another yet uncommon practice which could lower the risk of salt stress and drought stress is the more economical use of freshwater through sub-soil drip irrigation or similar methods (*some*). However, the cost-effectiveness of such measures depends on whether, and how much, one has to irrigate as these systems are costly and time-consuming to install and only last for a (couple of) year(s) (*some*). Moreover, as drip irrigation is quite expensive, it is seen as being more suited to high-value cultivations (*some*). Nevertheless, even without investing in different types of irrigation, there is already much efficiency to gain with conventional irrigation by irrigating on the basis of data instead of experience/intuition – the latter still being the dominant basis of decision-making (*some*).

Next to the different mitigation measures at the parcel level, adaptation to brackish circumstances, i.e. (more) salt-tolerant agriculture, is also a strategy that can be considered. However, currently available research on crop salt-tolerance is rather old, is for different crops/cultivars to those that are common in the Netherlands, was conducted under a different climate, and on different soils (i.e. mostly sand), and is therefore considered to be of no or little use in the Netherlands (*many*). This also means that the exact effect of salt on crop yield is largely uncertain and that current norms used by farmers are largely based on a limited knowledge base, rough estimates, and/or intuition (*many*). Although it is recognized by the *majority* of the interviewees that it is desirable to have more research on this – especially on peak salt events and salt tolerance at different growth stages as crops are often exposed to salinization at specific moments – the main question is who should and would pay for this.

Furthermore, according to the potato breeders, it is really difficult to determine salt-tolerance, as there are large differences between the results from tests in the greenhouses (in controlled environments) and tests in the field – where the variations in results tend to be large, poorly repeatable, and rather unpredictable. Moreover, breeding takes a long time, and although research on the genetics that determines salt-tolerance can speed up this process, this is rather complicated and time-consuming as well. Another complicating factor in adapting to more brackish conditions is that farmers have to apply crop rotation and that there are large variations between the salt tolerance bandwidths of these crops, e.g. seed potato is sensitive, whereas sugar beet is rather resilient (*some*). Nevertheless, knowing the salt tolerance of different cultivars would help in making a more informed decision on whether to irrigate with more brackish water or not, and/or what crop/cultivar a farmer might be able to switch to (*many*).

However, even though a crop might be or might become more tolerant, this cannot compensate for the fact that, in general, salts affect the structure of soils, thereby resulting in lower yields as well (*many*). Nevertheless, it is recognized that it is largely uncertain at which threshold salt concentrations become too high for effective cultivation on clay, and also what one can use gypsum, green manure, and/or organic matter to compensate for the negative effect until that threshold is reached (*many*). Therefore, it is recognized that (large-scale) field experiments on clay are desirable to establish this and provide action perspectives for farmers (*many*), although so far it has been very difficult to get government funding for this, arguably because of a lack of urgency and interest (*some*). Finally, *nearly all* interviewees do not see halophyte or 'saline crops' as part of the solution, mostly due to lacking markets, the intolerance of soils, and the manual work they require. Furthermore, at least in the short- to medium-term, *some* see it is as less complicated to resist salinization from happening than to effectively adapt to it. Nevertheless, researching the salt tolerance of 'cash crops' and especially cultivars that are already marketed is seen as contributing to salinization-resilience (*many*).

13.3.3.2 Community, Compartment, and Polder Level

At the community, compartment, and polder level, freshwater can be retained through the use of weirs (*some*). Using so-called 'fresh' or 'smart weirs' like they do on Texel can also result in the freshening of the surface water of an entire compartment or even polder, as the brackish water is disposed of while the freshwater is retained (*few*). Furthermore, saline channels and ditches could be separated from the supply channels, and/or saline ditches could be closed off from the system to make flushing more efficient, but often waterboards are not aware of chloride differences at such detail (*some*). At the compartment level, the waterboard can retain water longer by (seasonally) raising and maintaining the water level, which also counter-pressures saline groundwater, thereby lowering the risk of saline seepage (*some*). However, raising the water level in a water level compartment is not simple as there is a legally bounding agreement between the farmers and the waterboard. Therefore, if part of the community wants the water level to be raised, everyone has to agree with it; if only one disagrees then it has to be adhered to again (*some*). The latter can be very difficult as there are often conflicting interests, e.g. one farmer wants the water-table to be low in order to cultivate the land, whereas another needs the water for their crops (*some*). Moreover, if it goes wrong, especially because a higher level can increase the risk of flooding, the waterboard might be held accountable by the farmers (*some*).

13.3.3.3 Regional Water System (Waterboard Level)

At the regional water system level, i.e. the waterboard, there is still some efficiency in flushing to be gained, e.g. by redesigning waterways, separating saline and fresh waterways, but also through more data-driven operations, as currently most of the operations are based on human decisions rather than measurements and forecasting (*some*). Furthermore, the chloride norms waterboards have to adhere to could become more area-specific, taking into account the variations in salt loads in different parts of the regional system and the ability to flush it through (*some*). In terms of raising the norms of inlet points, there is very little they can do, as often more functions depend on this norm (*some*).

Another (drastic) measure would be to differentiate the price farmers pay on the basis of their location in the system and/or even their extractions (*some*), as is currently the case in just one region, namely Tholen and St Philipsland in Zeeland. Nevertheless, the majority of the farmers prefer such water systems to be paid collectively, as according to them: (1) everyone profits from a 'fresh environment', (2) there are still costs even without extraction, and (3) a price incentive could result in only the farmers that have to irrigate carrying the burden of the system (*few*). This is an example of the dominant (historical) perception that arranging water supply and quality collectively is less expensive, more effective, and therefore more preferable than farmers paying a (full) users' fee or becoming (completely) self-sufficient (*many*). This is further supported by the fact that although waterboards and provinces generally promote self-sufficiency and think it is important that farmers take their responsibility in the issue, they also state that flushing is actually not that expensive whereas many self-sufficiency measures are not nearly cost-effective (*many*), exposing a tension between the distribution of societal costs and benefits.

13.3.3.4 Provincial and Sub-National Level

At the provincial level, both the interviewees from the waterboards as well as the provinces agree that the function assignment of an area should be based on the water situation, i.e. what would be the current ability of a waterboard to provide that region with freshwater of a certain quality, as well as its future ability based on salinization projections (*many*). Until now, this has not been the case, mainly because: (1) provinces are in charge of spatial planning policies whereas waterboards are in charge of the surface water management (*many*), and (2) salinization has not been a significant bottleneck so far (*some*), (3) the current water situation and projections of salinization have not been (sufficiently) mapped out (*few*), and (4) there are many other factors that a province has to take into account when assigning spatial functions (*few*). In the future, functions could be changed as an adaptation measure when it becomes too (societally) costly to facilitate certain functions in specific areas and to ensure that there is still enough freshwater of sufficient quality for regions where it is not too costly (*many*). This should also preferably include differentiation between different types of agriculture – e.g. differentiation between livestock, flower bulbs, arable crops, based on a reasonable chloride norm (*some*). However, such function differentiations are perceived to be politically difficult since it could come across as prescribing businesses what they should do, although in theory, it only determines what functions are actively being facilitated (*some*). Nevertheless, several experts, agricultural stakeholders, and governmental advisors agree that it will be impossible to facilitate freshwater agriculture everywhere in the future (*many*).

13.3.3.5 Main Water System and National Level

As previously discussed, *some* think the current distribution of the freshwater that is entering the main water system could be changed in order to supply more areas with sufficient freshwater for flushing and to increase the buffer capacity of freshwater bodies like the Volkerak-Zoommeer (which is actually still planned to become brackish for nature), Haringvliet (of which part of the sluice is left open intentionally to allow fish migration), and IJsselmeer (which is used by six provinces). Especially the external salinization at the Nieuwe Waterweg, which is in open connection to the sea for the benefits of the shipping sector, is criticized, as most of the freshwater entering the country is used on counter-pressuring the salt wedge, while still not being able to completely prevent salinization of important inlet points (*some*). Moreover, salt water enters the regional and main water system during the locking for ships, meaning that different economic considerations have to be made during times of increasing salt loads and decreasing river discharge (*some*). Although several changes to the infrastructure of the main water system are possible, they are often still not effective in the long-term and/or not cost-effective, e.g. because the shipping sector is affected by it and its economic interest tends to be larger than that of the agricultural sector (*few*).

13.3.4 What Locks-in the Status Quo and What Creates Opportunities for (More) Salinization-Resilience?

13.3.4.1 Main Lock-ins of the Status Quo

The main lock-ins of the current status quo and the opportunities for a transition toward salinization-resilience which emerged from the interviews are summarized in Figure 13.4. Currently, the majority of experts and stakeholders share the notion that a sense of urgency is lacking for addressing the issue of salinization, mostly

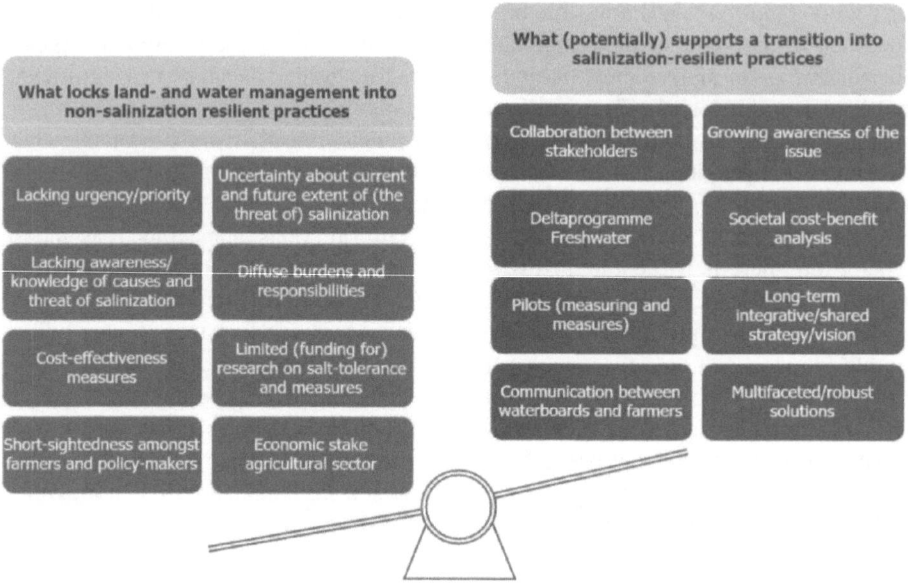

FIGURE 13.4 Key factors and actors involved in tipping the balance of factors that determine prevalent land- and water-management practices toward favoring salinization-resilient practices. (Adapted from Wigboldus et al., 2017.)

due to: (1) the lack of widespread (signaled) damage, (2) uncertainty about the extent of the issue in the (near) future, and (3) the priority of other issues. Closely related is the lacking of knowledge and awareness amongst farmers of the causes of salinization and their personal risk of salinization, which is largely preventing them from changing their current practices, as they: (1) are not aware of how their practices can induce salinization, (2) do not know if they should be taking measures, and (3) do not know what type of measures would be most effective. This also relates to the general uncertainty about the current and future extent of (the threat of) salinization as: (1) hardly any projections have been made of soil salinization, (2) salinization is a dynamic rather than a linear process, closely coupled to drought anomalies, and (3) there are large differences in expected salinization amongst the different climate change scenarios.

Another lock-in of the status quo is the limited available research on the salt-tolerance of commonly cultivated crops and soils specifically, which results in the widespread application of a precautionary principle, especially considering the financial risk a farmer would take. Moreover, the lack of funding for (practical) research and pilots, and subsequent limited knowledge on salt tolerance and measures is also locking-in current practices. This lack of funding can be tied back to the lacking sense of urgency, interest, and long-term perspective amongst policy-makers, and also the fact that it is difficult for a sector to invest collectively in a common issue when there is no or little collective research money, as this type of research is expensive, of long duration, and is in the interest of many (different) stakeholders.

Another lock-in is the economic stake of the agricultural sector in keeping certain practices and arrangements – like the flushing of the system based on the current chloride norms (or even lower) – in place, as they largely depend on such guarantees for their productivity and, therefore, profitability. This also leads to certain policies like 'function follows water' in spatial planning. Additionally, it makes certain rules and legislation, such as a prohibition for under-drainage or an obligation to install level-controlled drainage, politically sensitive/difficult. Although the economic stake of the agricultural sector could also be viewed as supportive in a transition to salinization-resilience – i.e. the future viability of an agricultural business depends on it – factors like focus on short-term gains and issues, lacking awareness/knowledge, uncertainty about the future threat of salinization, other priorities, but also the current uncertain cost-effectiveness or unprofitability of alternative measures are all reasons for not changing the status quo. The current lock-in can also be related to the diffusion of burdens, responsibilities, and even benefits, between farmers and waterboards (or society as the waterboard is financed through taxes). For instance, farmers are affected by salinization, both through brackish seepage in their parcels as well as in the ditches, and profit from the availability of freshwater, whereas the waterboards carry the responsibility of providing freshwater of a certain amount and quality. Moreover, as this creates a situation in which the (economic) incentives are not with one stakeholder, the (societal) cost-effectiveness of salinization-resilient measures for both farmers, as well as waterboards, is (even more) difficult to calculate or known to be negative.

13.3.4.2 Main Opportunities for (More) Salinization-Resilience

There are also factors that can support a transition into more salinization-resilient practices when they prevail over the lock-in factors (Figure 13.4). First of all, in recent years and especially after the summer of 2018, farmers and policy-makers have become increasingly aware of the issue of salinization, which has led to a growing interest in the topic, as evident from emerging measuring and monitoring projects and it being part of waterboard and provincial (development) programs. However, it is recognized that it takes more consecutive dry periods for salinization to become a more prominent issue. Nevertheless, projects in which farmers measure and monitor salinization at the parcel level and its effect amongst farmers can also aid in raising awareness. Furthermore, the Delta Programme Freshwater has opened up a window of opportunity for integrative, inter-stakeholder, and inter-regional addressing of freshwater issues, including salinization – although its share in the program is yet small. Another factor that is recognized by both experts and stakeholders to be crucial in stimulating this transition is more collaboration between the different stakeholders, i.e. the agricultural sector (e.g. LTO, KAVB), waterboards, provinces, and even the Ministries, Rijkswaterstaat, and knowledge institutes, as they all have a stake or (potential) role in addressing the issue, and need each other to successfully and sustainably address the issue. In the same line, increased communication between waterboards and farmers has already promoted more mutual understanding of each other's position and situation and helps in formulating an area-oriented approach. The latter is perceived to be very important by many of the interviewees, as the issue and therefore the possible solution(s) differ greatly with area. This also highlights the importance of involving multiple sectors, e.g. drinking water, industry, and nature, as they are often spatially mixed and have different interests, but could also work together on solutions. Furthermore, the majority of the interviewees call for the development of a long-term strategy, most logically on behalf of the province as this is already their formal role, but in collaboration and/or consultation with the stakeholders, in order for both waterboards and farmers to know what to expect and incorporate this in their own policies/business operations, and also to ensure that no decisions are made or pilots are initiated that might turn out to be harmful (to others) or a waste of money. Nevertheless, pilots of innovative measures like Spaarwater (i.e. anti-salinization drainage and sub-soil storage) have helped with raising awareness about the issue of salinization, and also to show the positive effects of certain measures. As action-perspectives for farmers are currently limited, multi-faceted solutions, i.e. those that address multiple issues like flooding, nutrients, and drought, at once, increase the attractiveness of such measures; again, pilots can contribute to making the positive effects visible. Finally, although difficult, some sort of analysis of the (distribution of the) costs and benefits of different measures can aid in the decision-making as to what extent the issue should be dealt with collectively or individually.

13.4 DISCUSSION

The results of this study indicate that salinization is generally perceived as a threat for agriculture in the low-lying regions of the Netherlands and that the issue is desired to be addressed sooner rather than later. This perception is in line with scientific studies that show that the problem and risk of salinization in the low-lying regions of the Netherlands is widespread and growing (Voorde and Velstra, 2009; Velstra et al, 2009; Oude Essink et al., 2010; de Louw et al., 2010; OECD, 2014; ter Maat et al., 2014). Simultaneously, however, a sense of urgency seems to be lacking amongst the majority of farmers and policy-makers, which can be mainly explained by the limited awareness and recognition of the issue, the absence of widespread damage to date, and the priority of other issues. This is also reflected by the national Delta Programme, which mentions salinization only marginally, in contrast to the issue of drought, which is highlighted as the main threat to freshwater availability (Delta Commissioner, 2018b). Because the urgency to address freshwater availability and the issue of salinization is rather limited, dominant land- and water-practices amongst farmers, waterboards, and provinces – such as the focus on water drainage rather than retention, the flushing of the water system to meet chloride norms, and not accounting for the water situation in spatial planning – that can stimulate salinization, aggravate its negative consequences, or which cannot be sustained because of salinization, continue to prevail. This is despite research dating from a decade ago that has already shown that such water-management practices are unsustainable (Voorde and Velstra, 2009) and many studies since then that have identified possible adaptation and mitigation measures, ranging from drainage systems to increased salt tolerance, and from the parcel level to the headwater system (e.g. Snellen and van Hattum, 2012; de Louw and Bogaart, 2014; Friocourt et al., 2014; Oude Essink and de Louw, 2014; Delsman, 2015; de Vos et al., 2016; Stuyt et al., 2016).

While previous studies might give the impression that there are plenty of technological solutions to address the issue of salinization, this research shows that many of these measures are in a pioneering stage and that there is a significant knowledge gap in terms of their effectiveness, efficiency, and feasibility, which could also partly explain their limited adoption. Moreover, it should also be taken into account that regimes like the prevailing land and water practices have not been deliberately shaped, but are rather the outcome of path dependencies and developed interdependencies between actors and processes that have led to a state of being locked into a status quo (Holtz et al., 2008; Fünfshilling and Truffer, 2014). In this case, the (historical) widespread occurrence and severity of flooding problems amongst farmers, for example, can explain why there is a focus on water disposal rather than retention, as well as why water-related issues are often (preferred to be) solved collectively, e.g. by the waterboard, and on the basis of a solidarity principle (by use of taxes). In the same line, interdependencies that have been formed and institutionalized in the past, such as farmers being largely dependent on waterboards for the supply of sufficient freshwater – both in volume and in quality – and waterboards being largely dependent on provinces (e.g. through spatial planning) and the central government (e.g. through the distribution of water from the main water system) for the feasibility of meeting their 'effort obligations', have led to a situation in which burdens, benefits, and responsibilities are shared, and therefore no one has a strong incentive or complete power to change the status quo. Research on socio-technological transitions (e.g. Geels, 2002, 2011; Elzen et al., 2012) however, show that the gradual stress of climate change and salinization and sudden shocks like the extraordinarily dry summer of 2018 have the ability to disturb the current regime, as is evident from the growing attention to the issue and its increasing embeddedness in programs like the Delta Programme Freshwater. A regime change to (more) salinization-resilience can happen once transition-supporting factors prevail over lock-in factors (Wigboldus et al., 2016). This research shows that a coordinated, long-term collaborative inter-stakeholder strategy on salinization has the ability to stimulate this transition. To come to such a strategy, relevant stakeholders and experts should work together to identify all relevant aspects of the issue and develop a shared vision of the future. This should be possible considering that all of these stakeholders have the opportunity to organize themselves collectively and set the political-administrative agenda.

There are multiple limitations to the findings of this study. First of all, the selected interviewees do not completely represent the targeted stakeholder groups as: (1) the interviewees from the provinces and waterboards are the (senior) advisors and not the administrators/policy-makers themselves, and (2) the interviewees from the agricultural interest groups represent the interest of their constituents and not necessarily that of all farmers. Nevertheless, as the advisors do work closely with policy-makers and provide them with information and advice, they are aware of the political situation surrounding the issue and why it is on the agenda or not, and also have the ability to shape this agenda. Moreover, although it can be argued that the agricultural interest groups might be too conservative as their role is to protect the interests of their constituents, this is also exactly what makes them representative at large. Besides, including the innovative farmers should have compensated for a too conservative formulation of the issue, although the opinions and perceptions were often in accordance as the agricultural representatives were generally well aware of the many facets of the issue. Nevertheless, this research could be complemented by a large-scale survey amongst farmers to identify the current magnitude of the issue and the bottlenecks for farmers in addressing it. Another limitation of the research is that other sectors, e.g. industry, drinking water, and nature, were beyond the scope of the research, whereas their stake in the issue does have implications for the possibilities of certain measures (and the other way around) especially considering that these functions are often spatially mixed. Furthermore, as there was no list of pertinent issues regarding the topic of salinization in the Netherlands, the analysis might be incomplete despite the large number of interviews. Nevertheless, as the aim of the study was to provide an initial wide-ranging assessment, the results can be used for a more focused analysis of selected aspects that are deemed most pertinent by this research. Finally, the results highlight a knowledge gap in the effectiveness, efficiency, and feasibility of different measures, that should be addressed by future research. Preferably, this is complemented by a societal cost-benefit analysis to inform stakeholders about the different possible pathways to salinization-resilience.

13.5 CONCLUSION

This research aimed to identify how the Dutch agricultural sector can be made more salinization-resilient, by reviewing the extent to which salinization is already perceived as an issue, how the dominant land and water practices relate to the issue of salinization, the opportunities and barriers of different mitigation and adaptation measures in addressing the issue, and what is generally preventing a transition toward salinization-resilience and the opportunities to stimulate such a transition. Based on the interviews with experts, agricultural representatives, waterboards, provinces, and innovative farmers, it can be concluded that salinization is perceived as a large threat to agriculture in the low-lying regions of the Netherlands that should be addressed sooner rather than later, but that the urgency to do so is lacking due to low recognition and awareness of the issue amongst policy-makers and farmers, as well as the priority of other issues. Moreover, it can be concluded that current dominant land- and water-management practices like the focus on disposing water instead of retaining it, the lacking efficiency in the use and supply of freshwater, and the paradigm of 'water follows function' in spatial planning and chloride norms are largely stimulating salinization and/or are expected to become unsustainable in light of salinization. Furthermore, the opportunities and barriers of different mitigation and adaptation measures like anti-salinization drainage, a higher water level, soil conservation, more efficient water use and supply are that they often have other positive side-effects and/or are not too different from current practices, thereby relatively attractive to implement, but that these are not expected to be effective on the long-term and that their (cost-) effectiveness is still rather unknown. For salt-tolerant agriculture, the opportunities lie with the selection of more salt-tolerant cash crops, and although this could be a more long-term solution, the salt-tolerance of these crops and common soils is currently under-researched. For spatial differentiation of water prices and in functions, the opportunities are that this can respectively increase efficiency in use and supply, as well as secure enough freshwater for certain areas. Nevertheless, such measures are politically challenging and might take a long time to become the standard. A lock-in of the status quo is that the diffuse burdens and responsibilities between farmers and water managers are resulting in virtually no one having a strong incentive nor power to change the status quo. Moreover, the lacking long-term perspective amongst stakeholders and the uncertainty about the effectiveness and efficiency of different measures are preventing a transition to salinization-resilience. On the other hand, especially more communication and collaboration between the stakeholders can create opportunities for such a transition. Furthermore, it is strongly advised that more research is done on the effectiveness of different mitigation and adaptation solutions, as this is currently lacking and thereby limiting the action-perspectives for both farmers as well as water managers. Finally, it is advised that such research be supplemented by a societal cost-benefit analysis to identify the societal cost-effectiveness of different measures and the distributions of the costs and benefits, thereby informing decision-making on a preferred strategy.

ENDNOTES

1. In this research, salinization-resilience entails the ability to sustainably cope with, prevent or limit salinization-related stressors.
2. Note that the classes are skewed to the left, which is to account for incompleteness of answers on all the semi-structured questions, as not all stakeholders (e.g. innovative farmers) were able to answer every question.

REFERENCES

de Louw, P. G. B. 2013. "Saline seepage in deltaic areas: Preferential groundwater discharge through boils and interactions between thin rainwater lenses and upward saline seepage". Doctoral dissertation. Amsterdam: Vrije Universiteit.

de Louw, P. G. B., and Bogaart, P. 2014. *Rainwater lenses.* Amersfoort: STOWA. https://www.stowa.nl/deltafacts/zoetwatervoorziening/delta-facts-english-versions/rainwater-lenses (Accessed May 28).

de Louw, P. G. B., Essink, G. O., Stuyfzand, P. J., and Van der Zee, S. E. A. T. M. 2010. "Upward groundwater flow in boils as the dominant mechanism of salinization in deep polders, The Netherlands". *Journal of hydrology* 394, no. 3-4: 494–506.

de Vos, A., Bruning, B., van Straten, G., Oosterbaan, R., Rozema, J., and van Bodegom, P. 2016. *Crop salt tolerance under controlled field conditions in the Netherlands, based on trials conducted at Salt Farm Texel.* Texel: Salt Farm Foundation.

Delsman, J. R. 2015. "Saline groundwater-Surface water interaction in coastal lowlands". Doctoral dissertation. Amsterdam: Vrije Universiteit.

Delta Commissioner. 2018a. Leren van de droogteperiode 2018. *Deltanieuws, 5.* https://magazines.deltacommissaris.nl/deltanieuws/2018/05/zoet-water

Delta Commissioner. 2018b. *Delta programme 2019. Continuing the work on the delta: Adapting the Netherlands to climate change in time.* The Hague: Rijksoverheid.

Elzen, B., Barbier, M., Cerf, M., and Grin, J. 2012. Stimulating transitions towards sustainable farming systems. *Farming Systems Research into the 21st century: The new dynamic* (pp. 431–455). Dordrecht: Springer.

Friocourt, Y., Kuijper, K., and Leung, N. 2014. *Salt intrusion.* Amersfoort: STOWA. https://www.stowa.nl/deltafacts/zoetwatervoorziening/delta-facts-english-versions/salt-intrusion (Accessed May 28, 2019)

Fünfschilling, L., and Truffer, B. 2014. "The structuration of socio-technical regimes—Conceptual foundations from institutional theory." *Research policy* 43, no. 4: 772–791.

Geels, F. W. 2002. "Technological transitions as evolutionary reconfiguration processes: a multi-level perspective and a case-study." *Research policy* 31, no. 8–9: 1257–1274.

Geels, F. W. 2011. "The multi-level perspective on sustainability transitions: Responses to seven criticisms." *Environmental innovation and societal transitions* 1, no. 1: 24–40.

Ghassemi, F., Jakeman, A. J., and Nix, H. A. 1995. *Salinisation of land and water resources: human causes, extent, management and case studies*. Wallingford: CAB international.

Holtz, G., Brugnach, M., and Pahl-Wostl, C. 2008. "Specifying "regime"—A framework for defining and describing regimes in transition research." *Technological forecasting and social change* 75, no. 5: 623–643.

Huisman, P., Cramer, W., Van Ee, G., Hooghart, J. C., Salz, H., and Zuidema, F. C. 1998. *Water in the Netherlands*. Rotterdam: Netherlands Hydrological Society.

Maas, E. V., and Hoffman, G. J. 1977. "Crop salt tolerance–current assessment." *Journal of the irrigation and drainage division* 103, no. 2: 115–134.

OECD. 2014. *Water governance in the Netherlands: Fit for the future? OECD studies on water*. Paris: OECD Publishing.

Oude Essink, G., and Louw, P. D. 2014. Brackish seepage. Amersfoort: STOWA. https://www.stowa.nl/deltafacts/zoetwatervoorziening/delta-facts-english-versions/brackish-seepage#1572 (Accessed May 28, 2019).

Oude Essink, G. H. P., Van Baaren, E. S., and De Louw, P. G. 2010. "Effects of climate change on coastal groundwater systems: A modelling study in the Netherlands." *Water resources research* 46, no. 10.

Qadir, M., and Oster, J. D. 2004. "Crop and irrigation management strategies for saline-sodic soils and waters aimed at environmentally sustainable agriculture." Science of the total environment, 323, no. 1-3: 1–19.

Qadir, M., Quillérou, E., Nangia, V., Murtaza, G., Singh, M., Thomas, R. J., et al. 2014. "Economics of salt-induced land degradation and restoration." *Natural resources forum* 38, no. 4: 282–295.

Reed, M. S., Graves, A., Dandy, N., Posthumus, H., Hubacek, K., Morris, J., et al. 2009. "Who's in and why? A typology of stakeholder analysis methods for natural resource management." *Journal of environmental management* 90, no. 5: 1933–1949.

Shahid, S. A., Zaman, M., & Heng, L. 2018. Soil salinity: historical perspectives and a world overview of the problem. In Guideline for salinity assessment, mitigation and adaptation using nuclear and related techniques (pp. 43–53). Cham: Springer.

Singh, A. 2015. "Soil salinization and waterlogging: A threat to environment and agricultural sustainability." *Ecological indicators* 57: 128–130.

Snellen, B., and van Hattum, T. 2012. *Soil as a buffer*. Amersfoort: STOWA. https://www.stowa.nl/deltafacts/zoetwatervoorziening/delta-facts-english-versions/soil-buffer (Accessed May 28, 2019).

Stuyt, L. C. P. M., Blom-Zandstra, M., and Kselik, R. A. L. 2016. *Inventarisatie en analyse zouttolerantie van landbouwgewassen op basis van bestaande gegevens*. Report no. 2739). Wageningen: Wageningen Environmental Research.

ter Maat, J., Haasnoot, M., Hunink, J., van der Vat, M. 2014. *Effecten van maatregelen voor de zoetwatervoorziening in Nederland in de 21^e eeuw*. Report no. 1209141-000. Delft: Deltares.

Voorde, J., and Velstra., J. 2009. *Leven met zout water: overzicht huidige kennis omtrent interne verzilting*. Gouda: Acacia Water, Leven met Water, and STOWA.

Velstra J., Hoogmoed M., and Groen K. 2009. Inventarisatie maatregelen omtrent interne verzilting. In *Leven met zout water*. Gouda: Acacia water.

Wigboldus, S., Klerkx, L., Leeuwis, C., Schut, M., Muilerman, S., and Jochemsen, H. 2016. "Systemic perspectives on scaling agricultural innovations. A review." *Agronomy for sustainable development* 36, no. 3: 46.

Wigboldus, S., Hammond, J., Xu, J., Yi, Z. F., He, J., Klerkx, L., and Leeuwis, C. 2017. "Scaling green rubber cultivation in Southwest China—An integrative analysis of stakeholder perspectives." *Science of the total environment* 580: 1475–1482.

14 Mitigating and Adapting Agriculture of Coastal Areas in the Netherlands Wadden Sea Region to Increasing Salinization

From a Vision towards a Practical Implementation

Mindert de Vries, Jouke Velstra, Johan Medenblik, Joca Jansen, Linda Smit, Aaltje Rispens, and Gualbert Oude Essink

CONTENTS

DOI: 10.1201/9781003112327-14

14.1 INTRODUCTION

Our study area is situated in the Netherlands province of Friesland (also known as Fryslân), especially the area bordering the Wadden Sea. This area is considered rural, with some towns and villages, with a dominance of intense agriculture on the clay rich soils close to the coastline and dairy farming on the clay-peat soils, more inland. Agriculture is dominated by seed potato cropping and grain production; these have great importance for the local economy and for export. The area is largely below sea-level and is protected by a dike system. In the last decade, more attention has been put to the issues related to land subsidence, water management, and salinization of the groundwater and surface water in agricultural areas (e.g. Pauw et al., 2012). Due to the rising ambient temperature, increasing evaporation, increasing precipitation caused by climate change, and occurrence of some severe droughts, notably in the summer of 2018, there has been an increased sense of urgency to adapt water management and control salinization. It is expected that in coming decades an acceleration of sea-level rise and continuation of land subsidence will further exacerbate the issue.

The aim of this chapter is to first provide an overview of the status and then outline a vision (similar to Speelman et al., 2009) and strategy towards managing the salinization issue. This approach is based on the joint efforts of a working group on salinization that produced a shared analysis and a shared perspective that was accepted by government, farmers collectives, knowledge institutes, NGOs of the region. A more detailed description is found in Mooi Werk and Mooi Wad (2020). In the next section, the baseline situation will be described, and this will be followed by a section elaborating the innovative approach to managing salinization in the area.

14.2 ESTABLISHING A BASELINE

The larger Wadden region nowadays consists of an area of diked salt marshes and reclaimed coastal peat bogs bordering a dynamic area of complex natural large-scale, intertidal ecosystems where natural processes continue to take place largely undisturbed (Vos, 2015). It is viewed as a landscape of exceptional cultural historical value. Human impact on this region has occurred in stages, with changes in the way of life, development of agriculture, and increasing abilities to use technology for flood protection and maintaining productivity at large scales playing key roles.

With each stage, the impact of people on the natural environment increased. From the end of the early middle ages, after many centuries of continual adaptation to the dynamics of the sea, the inhabitants of the Wadden Sea coastal region were increasingly successful at bending their unruly environment to their will. Controlling the external seawater, and later the internal polder and lake water, was the critical factor here (Schultz, 1992). Between roughly 800 and 1760, colonization of the salt marshes was completed, large parts of the coastal area were dyked and the coastal peat bogs behind the salt marsh zone were converted into prosperous farmland. By these means, a predominantly natural landscape was transformed in the space of just a few centuries into a vast and varied productive cultural landscape (Bazelmans et al., 2012).

14.2.1 Geography

Large parts of the former peat and salt marsh areas are now enclosed by dikes and under the influence of freshwater, but underground still carry the legacy of the past influence of the sea. During the centuries the diked areas have been compacting and subsiding. At the present time, the resulting geography is recognizable in the topography of the region as illustrated in Figure 14.1. The areas of the mainland close to the Wadden Sea are youngest and therefore most elevated. The areas more inland are gradually lower until the Pleistocene sand formations in the East are reached. Towns and villages are often somewhat elevated due to their origin as man-made mounds in the landscape. Clearly, the bands parallel to the shorelines can be seen, resulting from centuries of step by step land reclamation, where the saltmarshes were reclaimed along the Middelzee and the Wadden Sea.

14.2.2 Soil Types

The soil types of the region (shown in Figure 14.2) are dominated by clay, peat, and sand, in diverse compositions. Going from the Wadden Sea and former branches of the Middelzee and Lauwerszee, we find clay-dominated regions. Inland, marsh formation has created peat layers, sometimes covered by clay. Further eastward, the remains of glacial moraines surface, creating sandy and sometimes loamy soils.

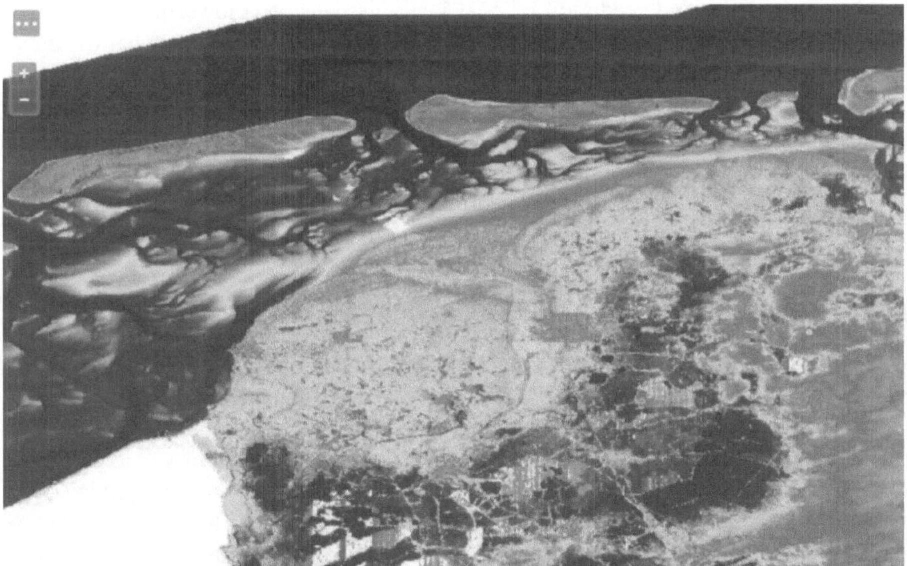

FIGURE 14.1 Elevation map of the Province of Fryslân and bathymetry of the Wadden Sea (projected from elevated viewpoint south of the province, color scheme: blue hues inland (approximate elevation below m.s.l.; green hues approx.0–2 m and yellow hues more than 2 m above m.s.l.). (Source: FAST Open Earth Geoserver.)

FIGURE 14.2 Soil types of the province of Fryslân (clay-dominated soils (greens), peat soils (purples), sandy soils (yellow-browns)). (Source: Province of Fryslan (2020).)

14.2.3 Land Subsidence

Land is subsiding gradually in the area. This is caused by two main processes. Firstly, compaction and oxidation of peat layers is causing a lowering of the land surface. This can be seen in Figure 14.3 (left) in the peat areas somewhat distance from the shoreline of the Wadden Sea (Haasnoot et al, 1999). However, in clay areas along the coastline, the dominant subsidence factor at present is the occurrence of gas and salt mining from underground (De Louw and Oude Essink, 2005; Verkaik and Oude Essink, 2008), seen as almost point sources surrounded with areas of influence on the map (caused by gas mining and salt mining activities). Figure 14.3 (right) extrapolates the impact of these mining activities into the near future.

14.2.4 Water Management and Groundwater

The province of Fryslân is situated for a large part below sea-level (Pauw et al, 2012). Rainwater and water flowing from the higher grounds in the east provide water to the central area of the province. A major freshwater source is Lake IJsselmeer that is connected to the Fryslân water system (the 'boezem') in times of water shortage. The surplus water from Fryslân is pumped out to Lake IJsselmeer, the Wadden Sea, and the Lake Lauwersmeer. From Figure 14.4 it becomes clear that the higher clay areas near the Wadden Sea coastline are quite detached from the freshwater source IJsselmeer. This is especially critical in periods of drought. In dry periods, the flow direction is reversed and Lake IJsselmeer water is pumped into the 'boezem' to provide water for irrigation. In an average year, more rainwater is pumped out of the province into the large receiving systems than is imported in dry periods in order to manage the accessibility of arable fields and meadows in the spring. This infrastructure

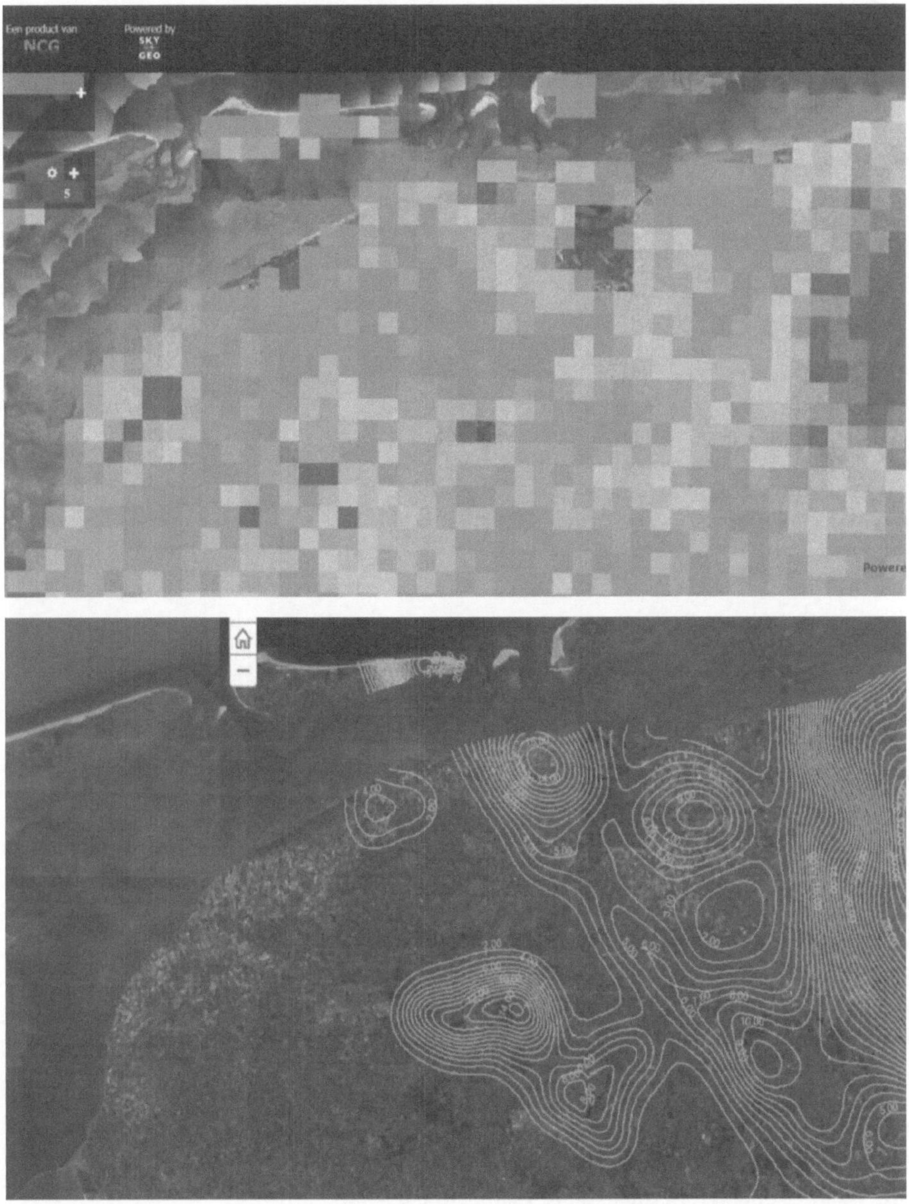

FIGURE 14.3 Subsidence rate observed from remote sensing sources in the province of Fryslân (left, mm/year, green colors approx. 0 mm/yr, red colors in the order of -5 mm/yr). (Source: map portal of the province of Fryslân.) The analysis is based on INSAR techniques. InSAR (Interferometric Synthetic Aperture Radar) maps millimeter-scale deformations of the earth's surface with radar satellite measurements. Total subsidence due to gas and salt mining predicted between 2010 and 2025 on the right. (Source: SWECO (2018).)

FIGURE 14.4 The Fryslân water system, depicting the larger channels and lakes. Note connections to the fresh lake IJsselmeer (South-West, blue arrows), the Wadden Sea and the exit points at Harlingen and lake Lauwersmeer (North, near Lauwersoog, blue arrows). Distance between Lemmer and Lauwersoog is about 70 km. (Adapted from Visplan Friese Boezem (2017).)

is not aiming at replenishing groundwater in wet periods and is therefore not optimal for combating the salinization trend observed in the groundwater. From Figure 14.5 it becomes clear that the average groundwater table in the province is quite close to the surface. In large areas, the distance is between 40 and 150 cm (green hues). It can be seen that in the coastal clay areas this distance is somewhat larger than in the low-lying central peaty areas, where average groundwater level is less than 40 cm below the surface (blue hues).

Depending on groundwater level and water pressures water will either flow upwards (seep) or flow downwards (infiltrate) with respect to the surface level (De Louw et al., 2011, 2013, 2015). Figure 14.5 (right) shows the distribution over the province (Grondwater Atlas van Fryslân, Boukes et al., 2019).

14.2.5 Salinization of the Groundwater and Surface Water

The extent of salinization of the phreatic groundwater is dependent on many factors. This is related to the geography and history of the area. In the case of Fryslân, it is to be expected that the diked and reclaimed areas of the former Middelzee and other branches will contain shallow saline groundwater (De Louw et al., 2009; Oude Essink and Van Baaren, 2009). As there is a fresh groundwater flow going from the higher East to the lower West a transition zone is found in the center of the province. Figure 14.6 (left) shows the salinity of the top of the saturated groundwater zone and illustrates this process. It is also shown in the figure (right) that in areas of diffuse upward groundwater seepage, the surface level will be exposed to a flux of salt that scales with the concentration in the groundwater times the strength of the seepage flux. From this, it becomes clear that the coastal agricultural areas in Fryslân (but likewise in other areas) are subject to upward fluxes in some places of more than 10,000 kilograms of chloride per hectare per year in the present situation. The risk to vegetation increases when the saline water reaches the root zone in dry periods with high evaporation as the fresh rainwater lens diminishes. Maintaining a salt balance in times of climate change (higher evaporation, more extreme dry periods) requires an adequate water management strategy.

The loading of salt to the surface as calculated by 3D groundwater salinity models is reflected in the measurement of chloride in the surface water found in ditches

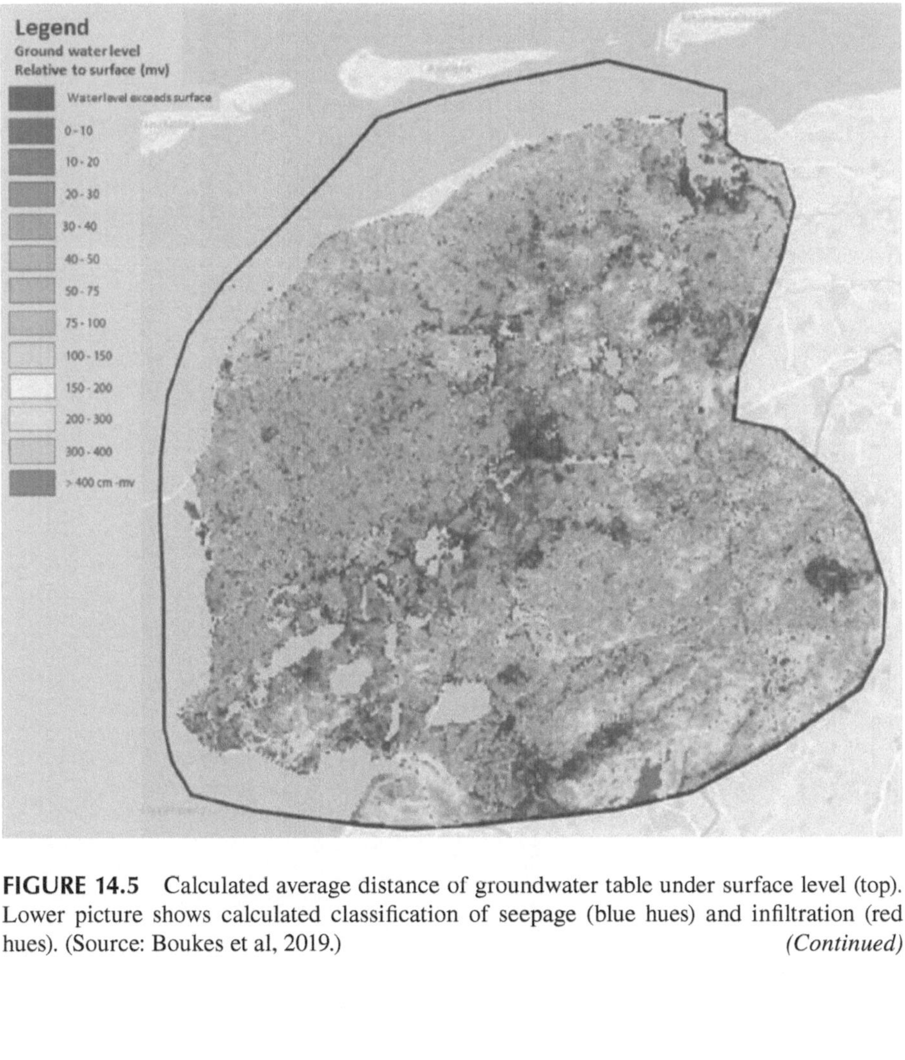

FIGURE 14.5 Calculated average distance of groundwater table under surface level (top). Lower picture shows calculated classification of seepage (blue hues) and infiltration (red hues). (Source: Boukes et al, 2019.) *(Continued)*

FIGURE 14.5 (Continued)

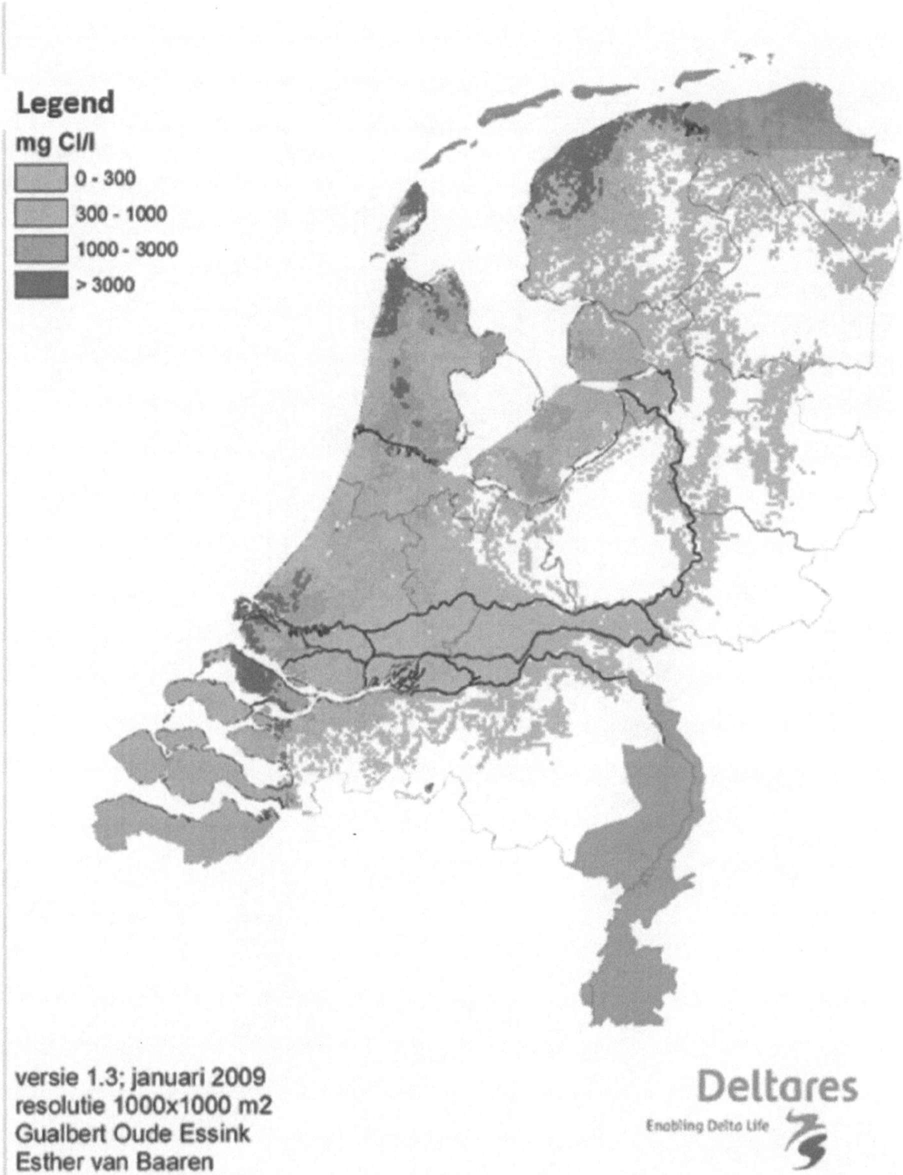

FIGURE 14.6 Calculated chloride concentration in shallow groundwater (mg Cl/l, left) and calculated yearly chloride load to the surface caused by seepage (kg/ha/year). (Source: Oude Essink and Van Baaren (2009).)

(Continued)

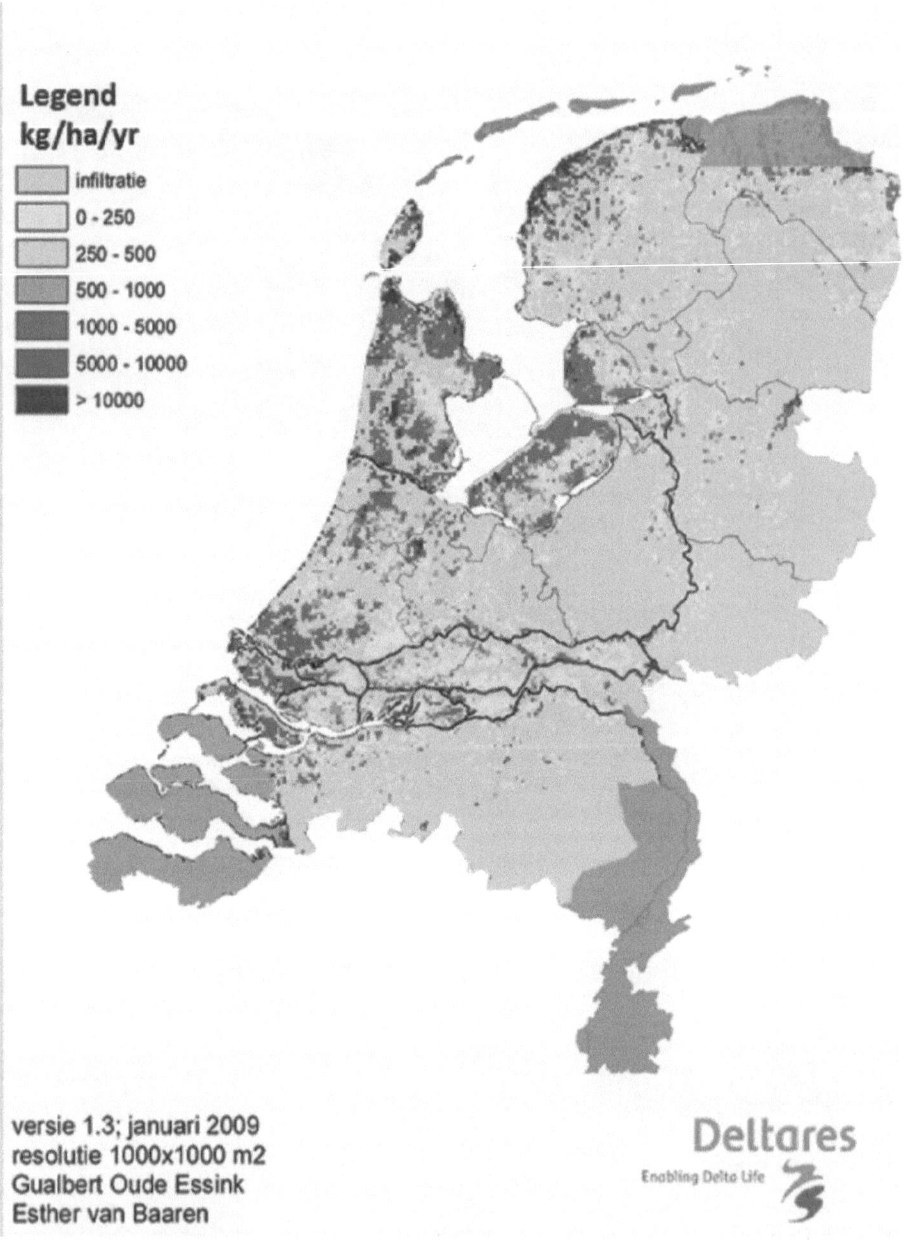

FIGURE 14.6 (Continued)

and canals in the coastal area of Fryslân. Figure 14.7 gives an impression of actual measurements. When Figures 14.6 and 14.7 are compared the patterns of the spatial distribution of chloride roughly align.

14.2.6 AGRICULTURE

The type of arable agriculture is always very specific to location and is mainly driven by soil type, water availability, and supporting technical and commercial infrastructure. In Figure 14.8 the spatial distribution of agriculture becomes very clear. The potato is the main cash crop of the region. Within a century of its introduction to the Netherlands, in the 1600s, the potato had become one of the country's most important food crops. The Netherlands is the world's major supplier of certified seed potatoes, with exports of some 700,000 tons a year, with a production of on average 45 tons per hectare (yearly value of export is approximately 250 million euro; source: potato-pro website, 2020). A considerable part is grown on the clay soils in Fryslân and Groningen (respectively 21,000 and 85,000 hectares of total cropped area with approximately 20% used for seed potatoes; source: CBS Statline website). In addition, grains and sugar beets are economically important crops of the area.

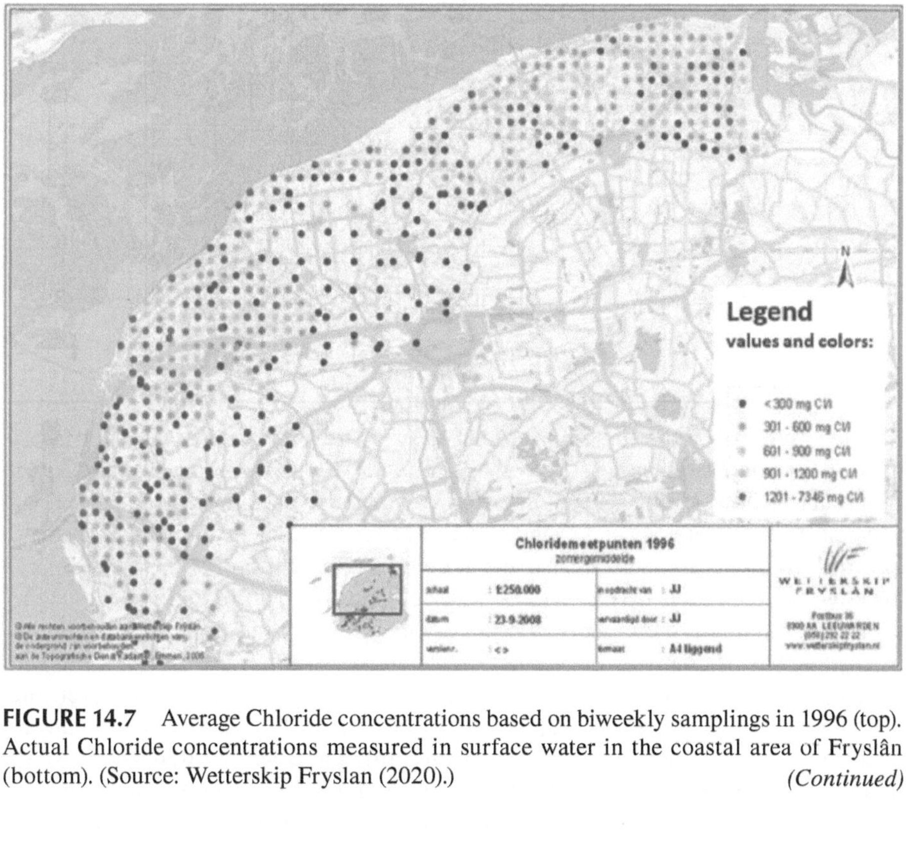

FIGURE 14.7 Average Chloride concentrations based on biweekly samplings in 1996 (top). Actual Chloride concentrations measured in surface water in the coastal area of Fryslân (bottom). (Source: Wetterskip Fryslan (2020).) *(Continued)*

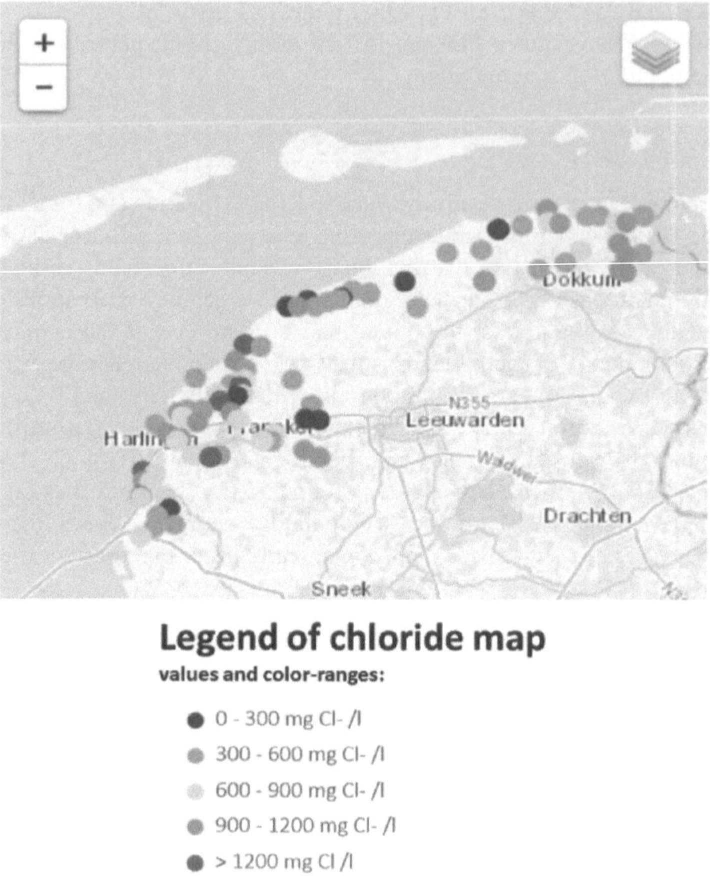

FIGURE 14.7 (Continued)

14.2.6.1 Sensitivity of Vegetation to Salinization

The sensitivity of crops to saline groundwater conditions can vary with a factor of 40 (Maas and Hoffman, 1977; Van Bakel and Stuyt, 2011; Stuyt et al., 2016). Crops such as flower bulbs and cut flowers sustain damage at chloride concentrations below 200 mg/L, potatoes and silage maize at concentrations from 700 to 800 mg/L upwards, grains, grass, and sugar beets at concentrations from 3600 to 4800 mg/L upwards. Sources used by the authors describe research within and outside of the Netherlands and will therefore represent various climatic conditions. Salinization is expected to have a relatively low impact on Dutch dairy farming. Both grassland and dairy cows easily tolerate chloride concentrations up to 2700 mg/L. However, the cultivation of silage maize, an important fodder crop, will already sustain damage at chloride concentrations from 800 mg/L upwards (de Boer and Radersma, 2011). Figure 14.7 gives an impression of chloride concentrations in surface water along the coast of the Wadden sea. Water with the salinities coded with the green and yellow

FIGURE 14.8 Rural land use map of Fryslân. The coastal areas show a dominance of cropping: potatoes (dark brown), grains (yellow) and beets/other crops (light purple); the central and eastern areas show the dominance of dairy farming (green). (Source: WUR LGN map (2020).)

dots are in the critical zone mentioned by de Boer and Radersma (2011) and could produce risks for crops when used for irrigation.

In addition to these large ranges, there is uncertainty on the coupling between the salinity of surface water, irrigation water, and the buildup of salinity in the groundwater. There is scarce data on this topic. Often it is assumed that the root systems are more tolerant to exposure to salinity than the above-ground biomass (van Dam et al, 2007).

Figure 14.9 shows the typical depths of the root systems of common herbs and of a potato plant. Their development is strongly influenced by soil type and water availability. For potato plants, the majority of roots develop in the area between 0 and 60 cm depending on soil conditions.

14.2.7 Risk of Salinization Effects on Agriculture

At present, occasional salt damage on crop production is reported (Figure 14.10) (Rozema and Flowers, 2008; Van Bakel and Stuyt, 2011; Stuyt et al., 2016). This was a reason for a consortium of farmers and researchers to start measuring salinization on fields that were suspected of possible salt damage (see also De Louw et al., 2006). It has become clear that impact of salinization is very local, and dependent on small-scale variations in elevation, the configuration of tile drainage, soil type, diffuse upward seepage conditions, and also on the type and sensitivity of crops (e.g., Stuyt et al, 2011; Stofberg et al., 2017).

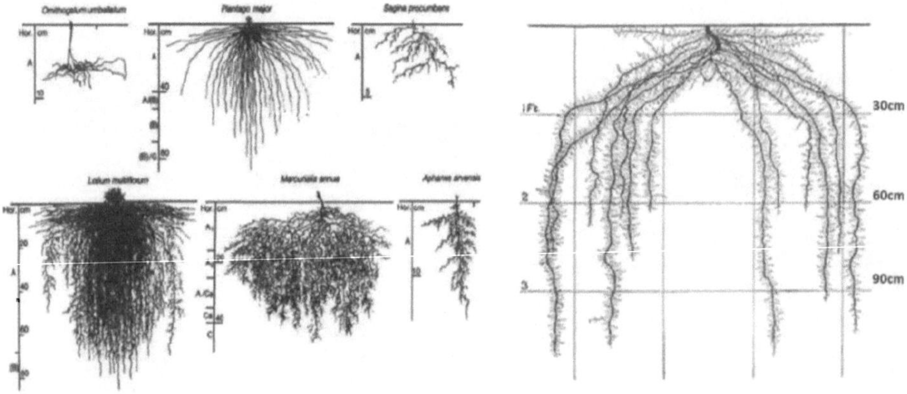

FIGURE 14.9 Typical root development of various common herbs (left in cm) and potato (in feet, 1 foot = 30,48cm). The potato plant was harvested after 94 days of growth. (Source: Soil and Health (2020).)

FIGURE 14.10 Monitoring ditches and groundwater at plot level (top). Damage of crops due to salinization at a site near Sexbierum (lower left, photo Jouke Velstra, Acacia Water). The graph on the lower right shows a measured cross section of the same field (ERT geophysical measurement) (blue, freshwater; red, salt water).

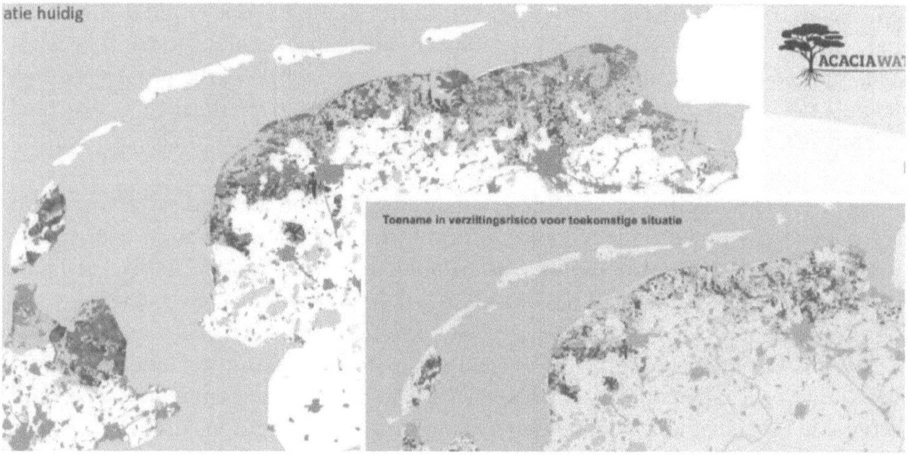

FIGURE 14.11 Risk of salinization of the root zone of vegetation in the coastal area of Fryslân (Source: Jouke Velstra, Acacia Water) for the present situation (2017) and 2050 (insert) based on average thickness of the freshwater lens above the saline groundwater table. Yellow (low risk, orange (medium risk), purple (high risk of thickness becoming less than 50 cm). (Source: Spaarwaterproject (2020).)

When geographical information on saline groundwater occurrence, seepage, groundwater table, evaporation, and water use by vegetation is combined, a risk of salinization of the root zone of plants can be estimated (see also Stofberg et al, 2017). In Figure 14.11 salinization risk is mapped in three classes based on average thickness of the freshwater lens above the saline groundwater table in relation to the depth of the root zone of vegetation. The calculation includes land subsidence, sea-level rise and a business-as-usual climate change scenario. In the climate change scenario, the risk of a dry year with a net surplus of evaporation of approximately 300 mm has increased in 2050 from 1 in 30 years to 1 in 15–25 years. At present some areas are at high risk; in 2050 these areas have extended due to the increasing risk of salinization in areas that are now classified at medium risk. More information can be found on the Spaarwater website (Spaarwater, 2020).

14.2.8 Monitoring Projects

Many data sources relevant to the salinization issues are available in the Netherlands and in the province of Fryslân. Very important generic information is derived from the monitoring services set up by government and knowledge institutes (for instance elevation maps or land use maps). Information on trends is increasingly derived from free to use high resolution spaceborne remote sensing sources (such as the subsidence data and land use data). Large-scale national and regional modeling and monitoring is based on projects such as FRESHEM (FREsh Salt groundwater distribution by Helicopter ElectroMagnetic survey; Delsman et al., 2018, Van Baaren et al., 2018).

In projects like GO-FRESH (www.go-fresh.info; Oude Essink et al., 2018; Pauw et al., 2015), and SPAARWATER (www.spaarwater.com), local-scale fresh groundwater storage and utilization solutions are investigated, while in the project 'boeren meten water', participative local-scale monitoring and citizen science is executed (Boeren meten Water, 2020, https://boerenmetenwater.nl/). Examples of monitoring and modeling are shown in Figure 14.12. Often the monitoring is supplemented with numerical modeling of variable groundwater flow and coupled salt transport to create a process-based understanding of the hydrogeological situation and to provide a tool 'to do' scenario analysis; an example of such a tool is The Netherlands Hydrological Instrument (De Lange et al., 2014). Monitoring is then crucial for validating model outputs. See for instance Deltares Zoetzout (2020) for an overview. In 2019 the Groundwater Atlas of Fryslân was published, based on monitoring and modeling work, in a cooperation between the provincial government, waterboard Wetterskip Fryslân and water company Vitens, providing lots of insights in the status of flows and salinization. More information can be found in Grondwater in Fryslân (2020), based on research funded and published by the Provincial government. Oude Essink and Forzoni (2017) provide an in-depth analysis of the groundwater processes of the province.

These projects are important to establish the groundwater baseline. A project that is focusing on creating awareness on the potential for adaptation to increasing salinity by introducing salt-tolerant crops and links to culinary markets is Interreg SalFar (2017–2020). The website of SalFar can provide more information (SalFar, 2017).

14.2.9 Learning from the Baseline

It becomes clear from this baseline analysis that salinization is a real issue in Fryslân. It can also be concluded that the situation is well understood on the large-scale. Model predictions and observations are in agreement. On the small-scale of farmers' plots, the situation is quite variable due to variations in farming practices, in soil type and structures, depth to groundwater, and seepage and bed elevation. The sensitivity of plants to salinization is also highly variable. Several cash crops are moderately sensitive; others are quite tolerant. The trend is negative; the risk of salinization effects is increasing in large part due to land subsidence and climate change (Pauw et al., 2010; see for the provinces South-Holland and Zeeland, Oude Essink et al., 2010 and Van Baaren et al., 2016, resp.). In addition, in a 10–30 km zone bordering the coastline, sea-level rise will also become a factor (Oude Essink, 2007; Oude Essink and Van Baaren, 2009; Voorde and Velstra, 2009). In general, it is clear that time scales for change are related to the spatial scale that is observed; changes on regional scale (50–100 km) with deep layers of groundwaters (100s of meters) are measured in decades and centuries. On the local scale of plots, 100s of meters, 2–3 m depth of groundwater, time scales are seasonal, and trends are observed in years to decades. From the baseline data in combination with climate and sea-level rise scenarios, it is clear that salinity levels will increase in more areas and closer to the surface. This will increase the pressure on mitigation and the move towards finding solutions for sustainable adaptation.

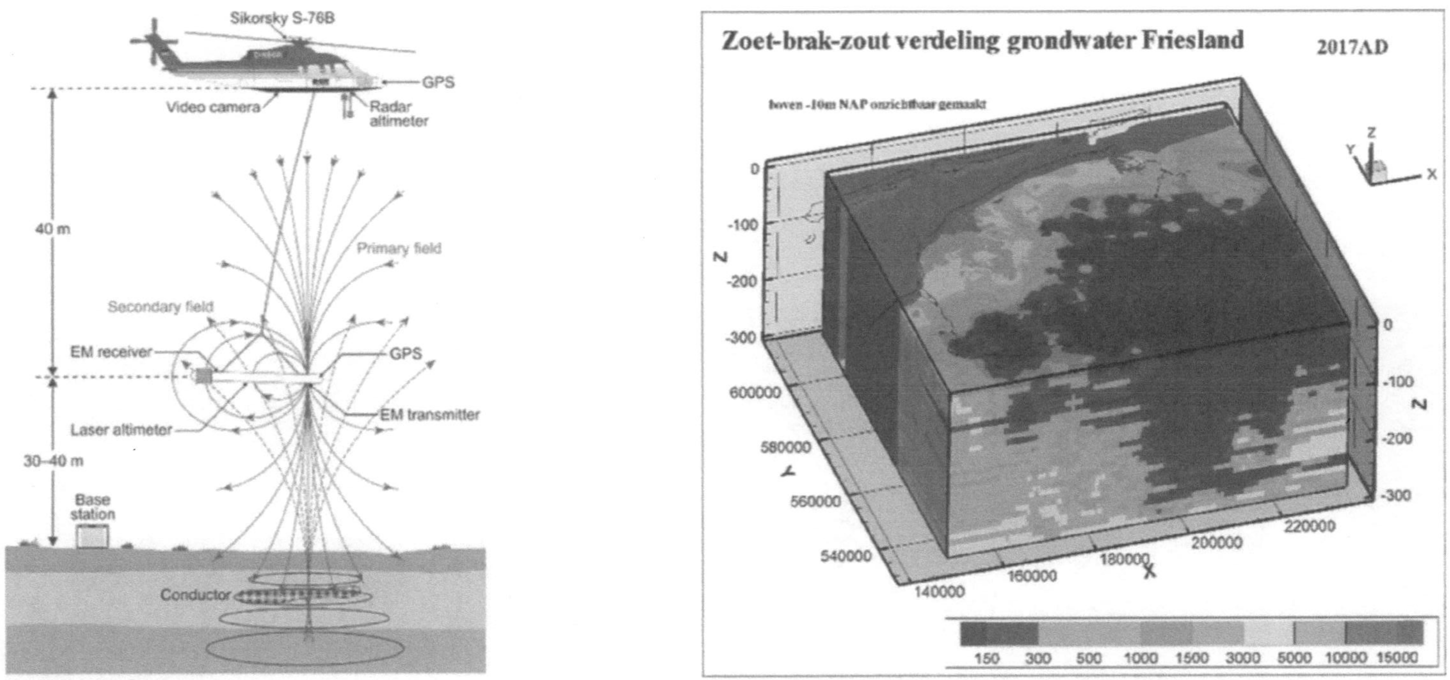

Source: Deltares

Source: Deltares, Oude Essink and Forzoni, 2017

(a)

FIGURE 14.12 Examples of regional scale monitoring and modeling of salinity and salinization (a), local-scale monitoring and modeling (b) and a participative monitoring facility on local-scale, see Hoogland et al, 2020 (c). *(Continued)*

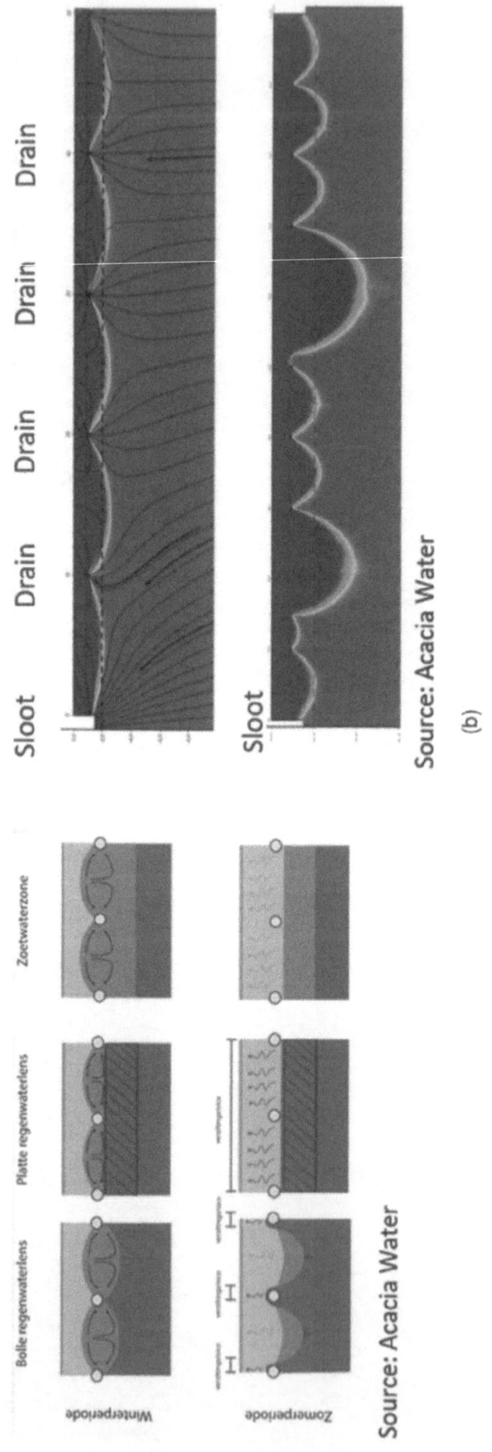

FIGURE 14.12 (Continued)

FIGURE 14.12 (Continued)

14.3 TOWARDS A STRATEGIC VISION AND A STRUCTURED IMPLEMENTATION PROCESS

In recent years, anecdotical information on crop damage caused by salinization has been shared by farmers and researchers. During the dry and hot summers of 2018–2019 questions were raised by many on the increasing impact of drought and limited availability of freshwater and the role of salinization as part of this impact. This was the starting point of a process leading to the building of a community of farmers, other stakeholders, and experts in relation to these issues and the setting up of a strategy and implementation process to further the systematic understanding of the issues and stimulate progress toward solutions.

In a bottom-up approach, a working group on salinization with stakeholders (agricultural representatives, entrepreneurs, government, nature management organizations) and salinization experts from various organizations, on request of the government, set up in 2018–2019 an approach towards establishing a vision and strategy to manage the salinization issues.

From the baseline analysis, it was clear that the coastal area of Fryslân (and of the Netherlands as a whole) is sensitive to salinization and that in the remainder of this century the risk of damage to the present type of agricultural production is increasing (Haasnoot et al., 2013, 2018, 2020). This will also impact on the characteristics of natural areas as they are situated in the same areas that will be affected. The baseline analysis shows that this impact will be variable in time and space. There is an increasing need to shift from a freshwater-dominated system towards a system that needs to accept and deal with more saline conditions (Speelman et al., 2009). The timing and location and necessity for such a shift will depend on the local situation.

From this perspective a distinction was made between mitigation and adaptation driven pathways to solutions (e.g., Stuyt et al., 2006; Kempenaer et al., 2007; Haasnoot et al., 2013; 2018).

- *Mitigation*: Taking measures (on small- and on larger-scale) to maintain the present system of cropping. This points to measures (optimized drainage, anti-salinization drainage, optimized irrigation, water management) that guarantee the long-term availability of sufficient freshwater resources.
- *Adaptation*: Adapting the land use to the changing environment. For example, switching from sensitive crops or varieties to tolerant crops or varieties, switching from agriculture to other land uses such as nature, or adapting the nature type of the area. This also means that local drainage and water management for a larger area should be adapted.

This has led towards formulating the following strategic objective for the region:

To aim at creating a climate robust area, with solutions implemented on the right scale that fit within the variable timing and location of increasing salinization, supported by the communities, and that are taking care of the local social, economic and natural values.

From this strategic objective, four operational objectives were formulated, that will be guiding the practical approach towards achieving the strategic objective:

1. *Increase the awareness of risk of salinization:* This will be made operational by implementing monitoring campaigns, sharing information, and overcoming taboos by initiating discussions on the topic. Awareness will help initiate a transformation process that is supported by stakeholders and citizens. The transformation process will help to find and accept location- and region-specific solutions that lead to a new socio-economic élan in the affected area.
2. *Increase understanding of the salinization impact:* Increasing knowledge needs to be translated into information that can be used for decision-making. This means that for each location and region a clear link between existing land use and functions needs to be established with salinization level and trends. This should lead towards the definition of tipping points that can help to decide if, where and when, decision-making and investments are needed for mitigation and adaptation solutions (Kwadijk et al., 2010).
3. *Clarify the effectiveness and scope of solutions:* Through field trials and learning from existing cases insight has been created on the various types of solutions that are effective from a technical, business, biological, ecological, and landscape perspective for the mitigation and adaptation pathways. Through cooperation and sharing of information this portfolio of solutions is shared and supported by stakeholders. Through experiments and pilots, new solutions will be researched leading toward innovations in technical, business, financial, and governance domains.
4. *Preparing financial arrangements and facilitating governance:* It is expected that mitigation and adaptation changes on local and regional scales will require large investments to adjust infrastructure on the scale of individual entrepreneurs and farmers (drainage, equipment, crops) and on the scale of regions (for instance in relation to water management). It is obvious that municipal, water board, provincial and national governments will be instrumental in facilitating this change as many regulations managing the present status quo will be challenged (for instance water levels and water quality related). It is very important to stimulate change by selecting and facilitating the frontrunners to develop pilots with the aim of creating convincing demonstrations fitting into the local and regional context.

The four operational objectives resemble four major hurdles that are to be overcome in order to realize transformations (awareness, understanding of the system, feasible solutions, operational financial and governance arrangements). Working along the lines of these operational objectives will result in a diversity of solutions, ensuing business models on a diversity of locations, and result in a coastal landscape and land uses that are probably more diverse in space and variable in time than today.

14.3.1 Implementation of a Structured Process-Based Approach

After formulating the strategic vision and the four operational objectives, a process is needed to be set in motion. For this process, the ABCD roadmap structure has been selected as an important building block (Figure 14.13). The roadmap encompasses an iterative process where the establishment of a baseline of system understanding (B, described in Section 1) and visioning (A, described as strategic and operational objectives in Section 2), are logical starting points and endpoints of a road (or in our case, the two pathways of mitigation and adaptation). Along the road, many solutions are proposed and studied (C in Figure 14.13) and a so-called 'Living Lab' process is implemented to evaluate solutions' success and decide on next steps fitting in the operational objectives and towards the strategic objective (D). In time, due to increased understanding, in an iterative manner, the baseline will move forward and the vision will be reiterated to better reflect what is known at that point in time.

14.3.2 Setup of the Fjildlab Living Lab

To make this ABCD Roadmap process work, people should work together towards the shared vision. In Fryslân emphasis by government is put on organizing and stimulating the 'Mienskip', the community, as supporter and host of transformational processes. As this process needs a home, the Fjildlab Living Lab (Figure 14.14 for an illustration) was set up in the region, involving representatives from the quadruple helix. In a Living Lab, the conditions are created to make progress as a group, welcome new ideas, support project teams, learn from each other and from projects

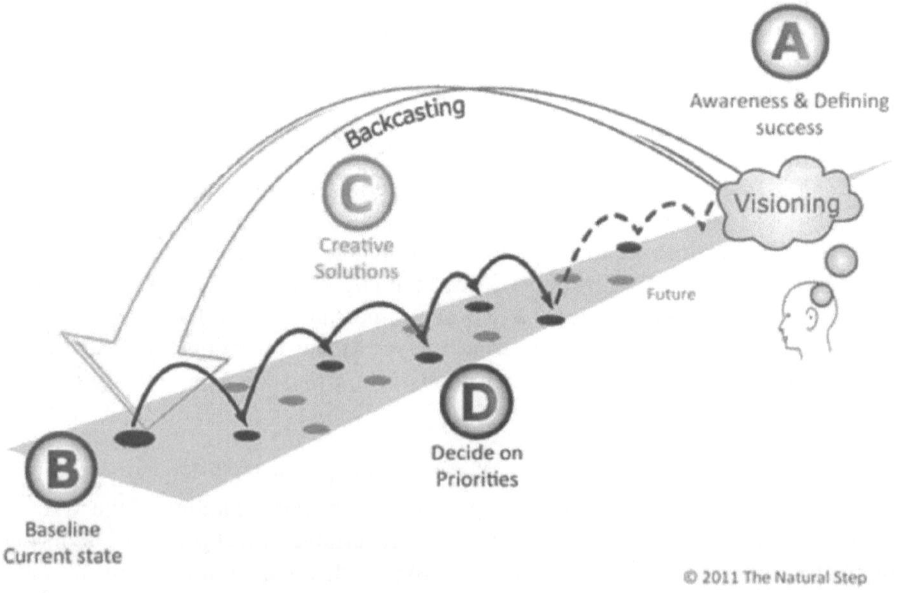

FIGURE 14.13 ABCD Roadmap backcasting and strategic planning approach to innovation. (Source: Holmberg and Robèrt, 2000.)

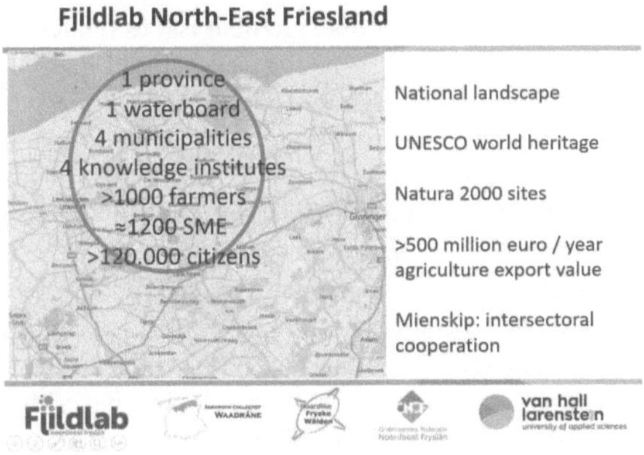

FIGURE 14.14 Fjildlab Living Lab as 'Mienskip' (Frisian word for 'Community') of many partners for learning, research and innovation is implemented in the Fjildlab Regiodeal program.

and pilots, consolidate knowledge and create support and awareness by involving stakeholders and citizens. A Living Lab is linked to a region; it is therefore culturally and socio-economically connected and sensitive and is open to new projects and contributors in order to allow adaptation of focus in time.

For the coming 5 years (2020–2025), the Fjildlab Living Lab (www.fjildlab.nl) is stimulating the interaction between the involved parties and project teams by providing a home with supportive infrastructure. In addition, the community is promoting the Agricultural Research Institute SPNA as a central location for experimenting and demonstrating salinization mitigation and adaptation solutions at a plot scale on clay-dominated soils (https://www.spna.nl/).

14.4 NEXT STEPS TOWARD SALINIZATION-PROOF AGRICULTURE

So far, a baseline has been established, a vision has been formulated, a process and hosting organization has been put in place. These are the major building blocks to achieve progress and assist in working toward the transformation that is required. At this point in time focus is put on working on the awareness raising objective and further increasing the understanding of the impact of increasing salinization. Many projects have been put forward and are now under consideration by funding organizations. The list of projects under execution and in planning (status September 2020) is impressive:

- Saline Farming, Interreg, 2016–2020, led by Province of Groningen;
- Implementation of salinization experimental farming facility on clay soil (2020–onwards), led by SPNA;
- Boeren meten Water awareness raising projects with implementation of monitoring on clay and peat soils, leads Acacia Water and LTO;
- Spaarwater projects, smart underground storage of freshwater, lead Acacia Water;

- Zoet op Zout program, overarching program structure integrating monitoring work, experiments and water management issues related to salinization, lead LTO;
- Zoute Verdienmodellen (investigation of business models for farmers in salinizing areas), Acacia Water for Programma Rijke Waddenzee;
- Fjildlab Living Lab, with an expert group on the theme of salinization and link to academic and professional education; Lead by farmers collectives, entrepreneurs and applied sciences university;
- Weerbare Waddenkust, A vision on creating a Wadden Sea coast that is economically vital and safe against flooding in times of sea-level rise, product of WUR, Deltares and VHL.
- Setting up a center of expertise on Salt-tolerant Farming, initiative of Province of Fryslân and The Potato Valley alliance.

Given this wealth of initiatives, a need arises to do a meta-level analysis to identify underdeveloped subjects that still need more focus to allow required innovation. This can be done in the Fjildlab Living Lab facility. For instance, viable business models, for instance on creating new products and developing new markets, are needed but are still rare or under-developed. A good mix of pilots linking the local- and regional-scale will be instrumental in developing new business models. All projects need to feed information on findings into a steady growing base of generic pre-competitive knowledge and know-how that is shared and can be used for training and education, and to facilitate the next generation of innovations, even beyond this Northern region. At this crucial stage of the process active support from government is needed to facilitate financial and regulation opportunities to frontrunners in exploring the mitigation and adaptation pathways, and in establishing a Center where the knowledge and know-how can be shared for the benefit of all.

From a governance perspective, this regional initiative needs to be embedded in a national perspective and programming. This is relevant for ministries of I&W and LNV in their task to manage and steer developments toward climate adaptation. In all coastal provinces of the Netherlands, salinization is recognized by politicians as a topic of importance and as a result it is getting embedded in supportive policies of the ruling parties. In 2020, the 'Zoet-Zout Knooppunt' initiative is taking off to facilitate this link between regional and national perspectives.

REFERENCES

Bazelmans, J., Meier, D., Nieuwhof, A., Spek, T., Vos, P., 2012. Understanding the cultural historical value of the Wadden Sea region. The co-evolution of environment and society in the Wadden Sea area in the Holocene up until early modern times (11,700 BCe1800 AD): An outline. Ocean Coast. Manag. 68 (2012) p 114–126.

Boeren meten Water, 2020. https://boerenmetenwater.nl/

de Boer, H., Radersma, S., 2011. Rapport 531, Verzilting in Nederland: oorzaken en perspectieven, ISSN 1570 - 8616)

Boukes, H., Haan, A. de, Medenblik, J., Rijen, J. van, Waaijenberg, J., Bonnema, F., Oosterhof, A., 2019. Grondwater Atlas van Fryslân. Book. ISBN: 90-80243-2-3

Dam, A. van, M., Clevering, O.A., Voogt, W., Aendekerk, T.G.L., van der Maas, M.P. 2007. *Zouttolerantie van landbouwgewassen: deelrapport Leven met zout water.* (PPO publicatie; No. 3234019400). Lisse: PPO Bloembollen en Bomen.

De Lange, W.J., Prinsen, G.F., Hoogewoud, J.C., Veldhuizen, A.A., Verkaik, J., Oude Essink, G.H.P., van Walsum, P.E.V., Delsman, J.R., Hunink, J.C., Massop, H.T.L., Kroon, T., 2014. An operational, multi-scale, multi-model system for consensus-based, integrated water management and policy analysis: The Netherlands Hydrological Instrument. Environ. Model. Softw. 59, 98–108.

De Louw, P.G.B., Oude Essink, G.H.P., 2005. Verzilting grondwatersysteem Wetterskip Fryslan, TNO rapport 2006-U-R0152/A.

De Louw, P.G.B., Oude Essink, G.H.P., Maljaars, P., Wils, R., 2006. Monitoring Verzilting Dongeradeel, TNO rapport 2006-U-R0177/B.

De Louw, P.G.B., Borren, W., Oude Essink, G.H.P., 2009. Orienterende studie verzilting Dongeradeel, TNO rapport 0906-0206.

De Louw, P.G.B., Eeman, S., Siemon, B., Voortman, B.R., Gunnink, J.L., Van Baaren, E.S., Oude Essink, G.H.P., 2011. Shallow rainwater lenses in deltaic areas with saline seepage. Hydrol. Earth Syst. Sci. 15, 3659–3678.

De Louw, P.G.B., Eeman, S., Oude Essink, G.H.P., Vermue, E., Post, V.E.A., 2013. Rainwater lens dynamics and mixing between infiltrating rainwater and upward saline groundwater seepage beneath a tile-drained agricultural field. J. Hydrol. 501, 133–145.

De Louw, P.G.B., Oude Essink, G.H.P., Eeman, S., van Baaren, E.S., Vermue, E., Delsman, J.R., Pauw, P.S., Siemon, B., Gunnink, J.L., Post, V.E.A., 2015. Dunne regenwaterlenzen in zoute kwelgebieden. Landschap 32, 5–15.

Delsman, J.R., Van Baaren, E.S., Siemon, B., Dabekaussen, W., Karaoulis, M.C., Pauw, P.S., Vermaas, T., Bootsma, H., De Louw, P.G.B., Gunnink, J.L., Dubelaar, W., Menkovic, A., Steuer, A., Meyer, U., Revil, A., Oude Essink, G.H.P., 2018. Large-scale, probabilistic salinity mapping using airborne electromagnetics for groundwater management in Zeeland, the Netherlands. Environ. Res. Lett. 13, 13.

Deltares FRESHEM, 2020. FRESHEM - Zoetzout - Deltares Public Wiki. See also https://www.deltares.nl/en/news/worldwide-treasure-hunt-for-scarce-fresh-groundwater-with-fresh-saline-mapping-technology-from-the-netherlands/

Deltares Zoetzout, 2020. https://publicwiki.deltares.nl/display/ZOETZOUT/Projecten

FAST Open Earth Geoserver, 2016. HOME - Project FAST EU FP7 Space - Deltares Public Wiki

Haasnoot, M., Vermulst, J.A.P.H., Middelkoop, H., 1999. Impacts of climate change and land subsidence on the water systems in the Netherlands. Report. ISBN 9036952786.

Haasnoot, M., Kwakkel, J.H., Walker, W.E., ter Maat, J., 2013. Dynamic adaptive policy pathways: A method for crafting robust decisions for a deeply uncertain world. Glob. Environ. Chang. 23, 485–498.

Haasnoot, M., Bouwer, L., Diermanse, F., Kwadijk, J.C.J., Van der Spek, A.J.F., Oude Essink, G.H.P., Delsman, J.R., Weiler, O., Mens, M.J.P., Ter Maat, J., Huismans, Y., Sloff, K., Mosselman, E., 2018. Mogelijke gevolgen van versnelde zeespiegelstijging voor het Deltaprogramma. Een verkenning. Deltares rapport 11202230-005-0002.

Haasnoot, M., Kwadijk, J.C.J., Van Alphen, J., Le Bars, D., Van den Hurk, B., Van der Spek, A., Oude Essink, G.H.P., Delsman, J.R., Mens, M.J.P., 2020. Adaptation to uncertain sea-level rise; how uncertainty in Antarctic mass-loss impacts the coastal adaptation strategy of the Netherlands. Environ. Res. Lett. 15, 1–24.

Hoogland, F., Roelandse, A., Burger, S., Feltmann, M., Velstra, J., 2020. Participatieve monitoring: samen werken aan een betere waterkwaliteit. Stromingen (26–1)

Holmberg, J. and Robèrt, K-H. (2000). Backcasting from non-overlapping sustainability principles – a framework for strategic planning. Int. J. Sustainable Dev. World Ecol. 7,291–308.

Kampenaer, J.G. de, Brandenburg, W.A., Hoof, L.J.W. van, 2007. Het zout en de pap; Een verkenning bij marktexperts naar langeretermijnmogelijkheden voor zilte landbouw. IMARES Research Report. InnovatieNetwerk Rapport nr. 07.2.154

Kwadijk, J.C.J.J., Haasnoot, M., Mulder, J.P.M., Hoogvliet, M., Jeuken, A.B.M., Van Der Krogt, R.A.A., Van Oostrom, N.G.C., Schelfhout, H.A., Van Velzen, E.H., Van Waveren, H., de Wit, M.J.M., 2010. Using adaptation tipping points to prepare for climate change and sea level rise: A case study in the Netherlands. Wiley Interdiscip. Rev. Clim. Chang. 1, 729–740.

Maas, E.V., Hoffman, G.J., 1977. Crop salt tolerance - current assessment. *ASCE J Irrig Drain Div* 103, 115–134.

Mooi Werk, Mooi Wad, 2020. Plan van Aanpak regie verzilting Noord-Nederland. Report in Dutch. Final, 14 februari 2019. https://www.thepotatovalley.nl/files/bijlagen/PvA%20verzilting%20Noord%20Nederland%20definitief%20rapport%2014%20feb%202019.pdf

Oude Essink, G.H.P., 2007. Effect zeespiegelstijging op het grondwatersysteem in het kustgebied. H_2O 19, 60–64.

Oude Essink, G.H.P., Van Baaren, E.S., 2009. Verzilting van het Nederlandse Grondwatersysteem Model versie 1.3 -2009-U-R91001. Deltares, Technical report of project: Effects of sea-level rise on groundwater: Case-study The Netherlands

Oude Essink, G.H.P., van Baaren, E.S., De Louw, P.G.B., 2010. Effects of climate change on coastal groundwater systems: A modeling study in the Netherlands. Water Resour. Res. 46, 1–16.

Oude Essink, G.H.P., Pauw, P.S., Van Baaren, E.S., Zuurbier, K., De Louw, P.G.B., Veraart, J., MacAteer, E., Van Der Schoot, M., Groot, N., Cappon, H., Waterloo, M., Hu-a-ng, K., Groen, M.M.A., 2018. GO-FRESH: Valorisatie kansrijke oplossingen voor een robuuste zoetwatervoorziening; Rendabel en duurzaam watergebruik in een zilte omgeving.

Oude Essink, G.H.P., Forzoni, A., 2017. Zoet-zout grensvlakkaarten grondwater in de Provincie Fryslân, Deltares rapport 11201095-000.

Pauw, P.S., De Louw, P.G.B., Oude Essink, G.H.P., 2012. Groundwater salinisation in the Wadden Sea area of the Netherlands: Quantifying the effects of climate change, sea-level rise and anthropogenic interferences. Netherlands J. Geosci. - Geol. en Mijnb. 91–3, 373–383.

Pauw, P.S., Van Baaren, E.S., Visser, M., De Louw, P.G.B., Oude Essink, G.H.P., 2015. Increasing a freshwater lens below a creek ridge using a controlled artificial recharge and drainage system: a case study in the Netherlands. Hydrogeol. J. 1–21.

Potato-pro, 2020. https://www.potatopro.com/nl/nederland/potato-statistics

Province of Fryslan, 2020. Map portal of province of Fryslân. Subsidence, https://bodemdalingskaart.nl/portal/index.

Province of Fryslan, 2020. Soil types of the province of Friesland. https://fryslan.maps.arcgis.com/home/webmap/viewer.html?webmap=913ffee88faa4978b65b0fdbd07a4177&extent=120990,533762,221778,617075,28992

Rozema, J., Flowers, T., 2008. Crops for a Salinized World. Science (80). 322, 1478–1480.

Schultz, B., 1992. Waterbeheersing van de Nederlandse droogmakerijen.

SalFar, 2014. SalFar is a project co-funded by the Interreg VB North Sea Region Programme 2014 - 2020.. https://northsearegion.eu/salfar/

Soil and Health, 2020. https://soilandhealth.org/wp-content/uploads/01aglibrary/010139fieldcroproots/010139ch15.html

Spaarwater, 2020. http://www.spaarwater.com/pg-27227-7-101924/pagina/spaarwater.html

Spaarwaterproject, 2020. http://www.spaarwater.com/content/27227/download/clnt/87235_Spaarwater_Verziltingsproblematiek_Jouke_Velstra.pdf

Speelman, H., Oost, A.P., Verweij, J.M., Wang, Z.B., 2009. De ontwikkeling van het Waddengebied in tijd en ruimte.

Stofberg, S.F., Oude Essink, G.H.P., Pauw, P.S., De Louw, P.G.B., Leijnse, A., Van der Zee, S.E.A.T.M., 2017. Fresh water lens persistence and root zone salinization hazard under temperate climate. Water Resour. Manag. 31, 689–702.

Stuyt, L.C.P.M., Van Bakel, P.J.T., Kroes, J.G., Bos, E.J., Elst, M. van der, Pronk, B., Rijk, P.J., Clevering, O.A., Dekking, A.J.G., Van der Voort, M.P.J., Wolf, M. de, Brandenburg, W.A., 2006. Transitie en toekomst van Deltalandbouw 434.

Stuyt, L.C.P.M., Van Bakel, P.J.T., Massop, L., 2011. Basic Survey Zout en Joint Fact Finding effecten van zout, Alterra-rapport 2200.

Stuyt, L.C.P.M., Blom-Zandstra, M., Kselik, R.A.L., 2016. Inventarisatie en analyse zouttolerantie van landbouwgewassen op basis van bestaande gegevens, Rapport 2739.

SWECO, 2018. Map on HVHL ARCGIS Map server. https://hsvhl.maps.arcgis.com/apps/MapSeries/index.html?appid=3ac65a88c8514c1dac80a8a93a82fcfc

Voorde, T, Velstra, J., 2009. Leven met zout water: overzicht huidige kennis omtrent interne verzilting. Acacia Water. Rapport STOWA (2009 45). ISBN 9789057734557

Van Baaren, E.S., Oude Essink, G.H.P., Janssen, G.M.C.M., De Louw, P.G.B., Heerdink, R., Goes, B.J.M., 2016. Verzoeting en verzilting freatisch grondwater in de Provincie Zeeland, Rapportage 3D regionaal zoet-zout grondwater model, Deltares report 1220185.

Van Baaren, E.S., Delsman, J.R., Karaoulis, M., Pauw, P.S., Vermaas, T., Bootsma, H., De Louw, P.G.B., Oude Essink, G.H.P., Dabekaussen, W., Gunnink, J.L., Dubelaar, W., Menkovic, A., Siemon, B., Steuer, A., Meyer, U., 2018. FRESHEM Zeeland - FREsh Salt groundwater distribution by Helicopter ElectroMagnetic survey in the Province of Zeeland, Deltares report 1209220. Utrecht, Netherlands.

Van Bakel, P.J.T., Stuyt, L.C.P.M., 2011. Actualisering van de kennis van de zouttolerantie van landbouwgewassen, Alterra-rapport 2201.

Verkaik, J., Oude Essink, G.H.P., 2008. Analyse van het verziltingsproces in relatie met bodemdaling door Frisia zoutwinlocatie gebied Oost, rapport 2008-U-R1282/B 19.

Visplan Friese Boezem, 2017. www.visstandbeheercommissie.nl

Wetterskip Fryslan, 2020. Actual chloride monitoring map. https://www.wetterskipfryslan.nl/kaarten/chloride-kaart

Vos, P.C., 2015. Origin of the Dutch coastal landscapes; Long-term landscape evolution of the Netherlands during the Holocene, described and visualized in national, regional and local palaeogeographical map series. Dissertation, ISBN: 9789491431821

Publisher: Utrecht UniversityWUR LGN map, 2020. https://www.wur.nl/nl/Onderzoek-Resultaten/Onderzoeksinstituten/Environmental-Research/Faciliteiten-Producten/Kaarten-en-GIS-bestanden/Landelijk-Grondgebruik-Nederland/lgn_viewer.htm

15 Saline Farming in the Wadden Sea Region of the Netherlands

Promising Initiatives for Salt-Tolerant Crops and Saline Aquaculture

Tine te Winkel, Jouke Velstra, Marc van Rijsselberghe, Klaas Laansma, and Titian Oterdoom

CONTENTS

15.1 INTRODUCTION

The coastal zone of the Wadden Sea region in the Netherlands is facing many transitions caused by salinization, climate change, and land subsidence (van de Meij & Minnema 1999; Velstra et al. 2011; Pauw et al. 2012; van Staveren & Velstra 2012). Regional land use is dominated by arable farming which is threatened by these external factors that directly or indirectly increase the risk of salinization. In this case study, based on market research conducted by consulting company Acacia Water B.V. and partners, we analyze in a systematic manner options for adaptation to salinization using different crop varieties and possible future development. A variety is classified as feasible when it has high scores regarding practical application, financial feasibility, and opportunities for large-scale production. We present examples of three main types of varieties: (1) conventional crops with some salt-tolerance, (2) saline crops like samphire, and (3) saline aquaculture. We present an overview of the potential of these three crop types for the more saline future of arable land in the Wadden region. As a part of the program 'Towards a rich Wadden-Sea' ('Programma

DOI: 10.1201/9781003112327-15

naar een Rijke Waddenzee', PRW), this research aimed to provide an overview of the current developmental status of different salinization adaptation measures to assess the feasibility of implementation in the short, medium and long term.

15.2 METHODS: PRACTICAL EXPERIENCES

In recent years, there have been various initiatives for salt-tolerant and saline crops, or saline aquaculture. The projects are generally initiated by applied research groups or are the local initiatives of individual entrepreneurs. One challenge lies in the collection and diffusion of the obtained results: these are often either confidential or remain unwritten and locked-up in the mind of the entrepreneur out of fear from competition. Therefore, it can be hard to determine why an initiative or pilot has ended, even though this could be very important information referring to project feasibility and the feasibility of future investments. The creation of future chances for saline agriculture will require entrepreneurs to step out of the spiral of non-investment because results are not written down. Our study has gathered practical experience through interviews with 10 entrepreneurs who have been working in saline conditions. A complementary desktop study including a review of > 20 publications was performed in combination with verification by three entrepreneurs and four researchers.

Our results in terms of the current status and feasibility of adaptation measures are listed per criterion. The conclusion illustrates the developmental status per criteria and areas where additional investments are needed. Results are plotted on a greyscale bar (Figure 15.1), where light grey means investments are necessary to further develop cultivation methods or the market supply chain, and dark grey indicates that the relevant criterion is well developed for further implementation. The timeline shows the future potential. The timeline is not meant as a summary of the criterion but indicates when it is reasonable to start development and investment.

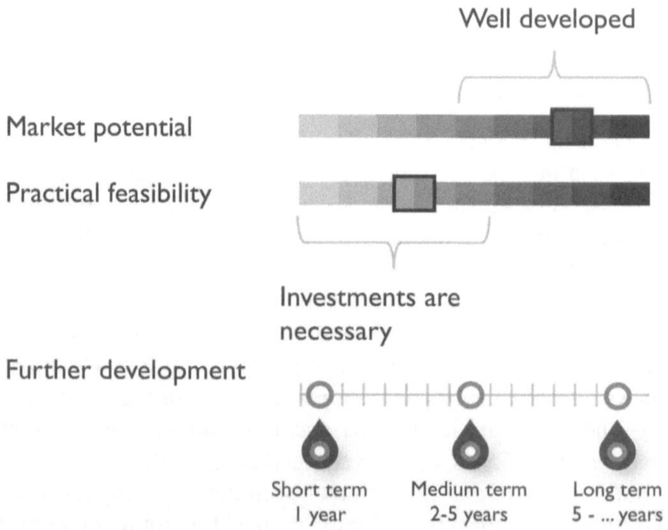

FIGURE 15.1 Methodology used to present results per variety.

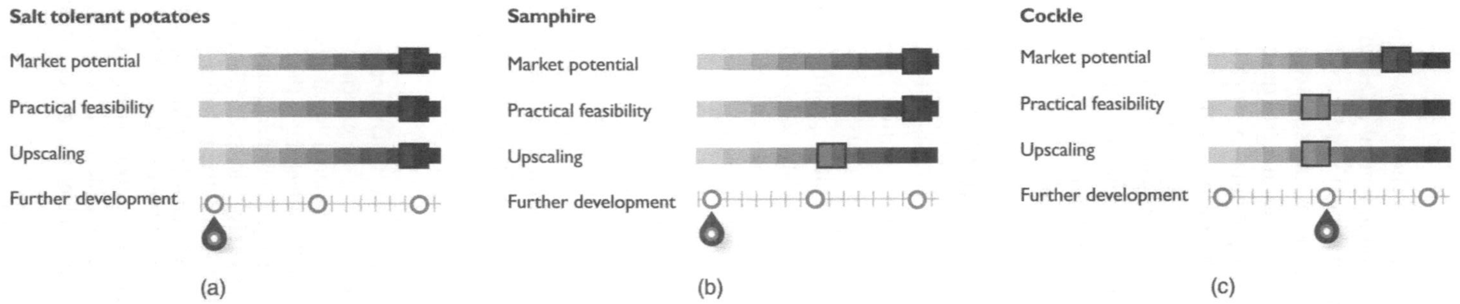

FIGURE 15.2 Potential in terms of different criteria for salt-tolerant potatoes (a), samphire (b), and cockles (c).

15.3 RESULTS

Based on the analysis, we recommend to choose three routes to pilot in the Wadden Sea region in the short term: (a) development of a salt-tolerant crop rotation plan for potatoes, (b) development of on-land cockle and seaweed cultivation, and (c) development of saline crops such as samphire and sea lavender. Minor saline crops (ice spice, sea kale, and sea fennel) have medium-term development potential.

We have determined the market potential per crop in two interdependent manners: by examining actual demand and marketing (Figure 15.2). For salt-tolerant crops and aquaculture, the emphasis in the early stages of development has to be on marketing to create sufficient demand. Marketing is an important part of the crops mentioned in this study because they are not conventional crops. Our study finds that there are often no sales channels near the farms focusing on saline agriculture, and the necessary infrastructure is often absent.

15.4 CONCLUSION AND RECOMMENDATIONS

Many initiatives in the Dutch province of Zeeland have sprouted because plot conditions for conventional crops became too saline. This has led to promising and successful initiatives. In a similar way, expansion could also be addressed in the Wadden Sea region. Policymakers can identify promising areas for adaptation measures using the salinization risk map (Velstra 2019) after which entrepreneurs are to be approached and opportunities will be discussed. It is notable that the majority of crops and varieties require new cultivation techniques. This requires large investments, a longer transition period, craftsmanship, and an intrinsic motivation for the entrepreneur. Physical preconditions also play an important role in this, not every crop and variety can be grown everywhere, giving the pre-feasibility phase an important place in the process of a farmer switching from one industry to another. The next step will be to take an in-depth look to connect and integrate the above factors and to make choices per crop and geographical location. This step must be taken in collaboration with agricultural entrepreneurs.

REFERENCES

Pauw, P., De Louw, P. G., & Essink, G. O. (2012). Groundwater salinisation in the Wadden Sea area of the Netherlands: quantifying the effects of climate change, sea-level rise and anthropogenic interferences. Netherlands Journal of Geosciences, *91*(3), 373–383.

van der Meij, J. L., & Minnema, B. (1999). Modelling of the effect of a sea-level rise and land subsidence on the evolution of the groundwater density in the subsoil of the northern part of the Netherlands. Journal of Hydrology, *226*(3-4), 152–166.

van Staveren, G., & Velstra, J. (2012) Klimaatverandering, toenemende verzilting en landbouw in Noord- Nederland (Climate change, increasing salinization and agriculture in the northern part of the Netherlands). Acacia Water, *333*.

Velstra, J. (2019). Spaarwater: Rendabel en duurzaam agrarisch watergebruik en waterbeheer in de verziltende waddenregio (Sustainable and profitable usage of water in the agricultural brackish environment of the Wadden Sea region). Acacia Water, *1*, 79.

16 Viability of the Saline Farming of Quinoa and Seed Potatoes in the Netherlands

An Assessment Supported by a Value Chain Analysis of Both Products

Mare Anne de Wit, Pier Vellinga, and Katarzyna Negacz

CONTENTS

DOI: 10.1201/9781003112327-16

16.1 INTRODUCTION

16.1.1 Salinization: A Growing Threat

Globally, salinization leads to the loss of the contained resources, goods and services of soil, resulting in land degradation and decreasing crop productivity (Daliakopolos et al. 2016). Climate change impacts such as increased droughts and a rising sea level are expected to further increase the salinization problem (Oude Essink et al. 2010). The main challenge will be to ensure sustainable cultivation in agricultural production without compromising the environment and natural resources. Alternative strategies and agricultural practices that adapt to salinization can thus represent a valid help for meeting the rising food demand, preserving the already overexploited freshwater and prohibiting land from severe degradation (Atzori et al. 2019).

16.1.2 Saline Farming: A Transition to Adaptation

Current academic research focuses on the identification and exploration of strategies that mitigate or adapt to salinization rather than on the socio-economic feasibility and viability of these identified strategies (Stuyt et al. 2011). One potential strategy identified in research is saline farming, i.e. the cultivation of salt-tolerant crops on marginal, saline soils while using salt containing water (De Vos, 2011). The strategy is often mentioned as bring a promising means of supporting future food demand as water scarcity increases (Stuyt, Kselik & Blom-Zandstra 2016), with approximately 400 million hectares of saline soil that could be utilized with the use of saline farming (Wicke et al. 2011). This study examines the opportunities and constraints for the scaling-up of two potentially salt-tolerant crops in the Netherlands: certain seed potato varieties and quinoa (De Vos 2011). Both are important crops because of their high nutritional value and therefore their increasing demand (Bazile et al. 2016). We choose the Netherlands as the case study area as it is inherently prone to salinization, with a quarter of the country located below sea level and over 65% at risk of flooding without the mitigative invention of dykes and polders (Huisman et al. 1998).

Within this research, two frameworks were used that have not been combined before in the field of saline agriculture; the local value chain development framework (LVCD) (Herr 2007) and the multi-level perspective framework (MLP) (Geels 2002). LVCD aids in the provisioning of information regarding the sector as a whole and the market requirements of a certain product, thereby supporting the identification of opportunities for increasing competitiveness and scaling (Herr 2007). LVCD has not yet been utilized for salt-tolerant cultivars.

MLP supports the examination of the opportunities and barriers to upscaling (i.e. increasing) and outscaling (i.e. expanding) of both products (Wigboldus et al. 2016). The framework assumes innovation to result from interactions at three different levels: (1) the fast-changing niche (micro-level), (2) the stable socio-technical regime (meso-level), and (3) the slow-changing socio-technical landscape (macro-level) (Geels 2002). The socio-technical regime consists of six regime-dimensions that potentially slow down or accelerate innovation: (1) industry, (2) technology, (3) policy, (4) science, (5) market, and (6) culture. The conditions at each level as well as these six dimensions affect the performance of an innovation (Geels 2002); for example, a shock at the landscape level could result in the regime level opening up and the niche level to take over, leading to the breakthrough of a particular innovation (Hermans et al. 2013). Supported by examination of the six dimensions, we mapped the various factors related to the scaling of saline farming of quinoa and seed potatoes. In addition, this research tends to identify potential lock-ins and windows of opportunity, thereby obtaining insights in the interaction of a variety of dynamics involved in scaling that could stimulate change but also lock-in current practice in its unsustainable mode (Geels 2002, 2011).

16.1.3 Research Question

This research aims to bring a novel, social-economical scope into the examination of saline farming by answering the following research question: *Is saline farming of quinoa and seed potatoes a viable option in the Netherlands, based upon an analysis of the opportunities and constraints for the upscaling and outscaling of both crops?*

16.2 DATA COLLECTION

We conducted semi-structured interviews, allowing the interviewer or interviewee the flexibility to deviate regarding a certain topic (Gill et al. 2008). The use of data triangulation ensured the validity of this study (Bryman 2012). Data triangulation is considered a suitable validation method that entails the combination of various sources to answer the research question: (1) scientific articles, (2) policy documents and (3) interviews with both stakeholders and experts (Turner & Turner 2009). We aimed to include different actors concerning the value chain as well as experts, including a balance between interviewees from the regime- and niche-levels. Guided by the snowballing method (Bryman 2012), we conducted 32 interviews in total, including eleven with experts, seven with farmers and four with policymakers. To broaden the scope of this research, three interviews were conducted with foreign experts, i.e. two from Belgium (Instituut voor Landbouw- Visserij- en Voedings-onderzoek) and one from Dubai (International Center for Biosaline Agriculture).

The interviews where recorded and summarized. Next, overarching or differentiating outcomes were analyzed. The results are presented with the support of a rating that identifies the quantity of interviewees that agreed on a certain topic; i.e. *few; <4 (−/+), multiple; 5-20 (−/++) and many; >20 (−/+++)*. An overview of the interviewees is found in the Appendix.

16.3 RESULTS

16.3.1 The Value Chain of Quinoa and Seed Potato Production in the Netherlands

16.3.1.1 Quinoa

Based on the interviews, the value chain of quinoa production in the Netherlands is as summarized in Figure 16.1.

16.3.1.1.1 Market Entrance

Approximately 30 years ago, Wageningen University and Research (WUR) pioneered quinoa breeding, aiming to develop a variety suitable for European (and preferably Dutch) conditions. WUR started to market their developed varieties under the auspices of The Dutch Quinoa Group, currently re-branded as The Quinoa Company (TQC). Another company, GreenFood50 (GF50), in turn distributes the quinoa to the retail sector. They sell the quinoa to processors, wholesalers, supermarkets, online stores and restaurants. Farmers produce quinoa for GF50 by contracted cultivation, which means that a pre-agreed amount is purchased at a fixed price.

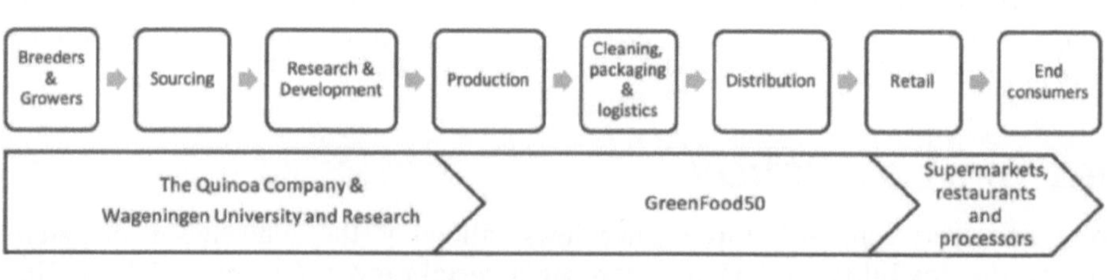

FIGURE 16.1 The value chain of Dutch quinoa. (Adapted from Herr, 2007.)

16.3.1.1.2 Market Requirements

The most common quinoa variety, "Royal Quinoa", is a protected Andean variety. Seed from this variety is white, large, contains saponins, and therefore needs to be washed and polished thoroughly. The first market requirement that was identified by *all* quinoa experts is therefore the common characteristic of Royal Quinoa. Quinoa cultivated in the Netherlands differs from this variety: *"the quinoa is less white [...] you can think of it as whole grain quinoa"* (quinoa expert 4). Secondly, the country of origin is identified to be important by *a few*: *"people are used to quinoa coming from Peru or Bolivia. Maybe they think that Dutch quinoa is less good"* (quinoa expert 3). However, *multiple* quinoa experts stated that the Dutch variety can be used for exactly the same things as the Andean. Thirdly, *multiple* people identify more demand for organic quinoa then imported conventional quinoa. To illustrate: *"the prices of organic crops are twice as high as those of conventional crops"* (quinoa expert 4). Therefore, it is of importance to gain or obtain organic certificates. This is possible, however expensive. Lastly, *multiple* interviewees identify local quinoa to be more expensive than conventional, imported quinoa. That makes the retail of local varieties difficult as *"quinoa is already an expensive product; consumers are expected to choose the product with the lowest price"* (quinoa expert 3). According to the interviewees, the higher price results from higher labor costs, labor-intensive production processes and the high investments regarding certificates.

16.3.1.2 Seed Potatoes

Based on the interviews, the value chain of seed potato production in the Netherlands is as summarized in Figure 16.2.

16.3.1.2.1 Market Entrance

There are multiple different trading houses with their own breeding programs, aimed to develop new varieties to be licensed and taken to the market. As an interviewee from trading house 1 explains *"we want to register new varieties for Plant Breeding Rights, which is similar to a patent. If it meets the requirements you get a license; the Plant Breeder's Rights Protection. This lasts 30 years for potatoes"*. Farmers produce seed potatoes for the trading houses through contracted cultivation. The amount to be produced is agreed to beforehand and thereby the farmers are secured revenue and take-off. What kind of seed potatoes farmers cultivate depends, as *all* trading houses agree that supply follows demand. This also applies to the (more) salt-tolerant varieties; as long as there is no (high) demand for salt-tolerant varieties, trading houses will wait with development and subsequently cultivation.

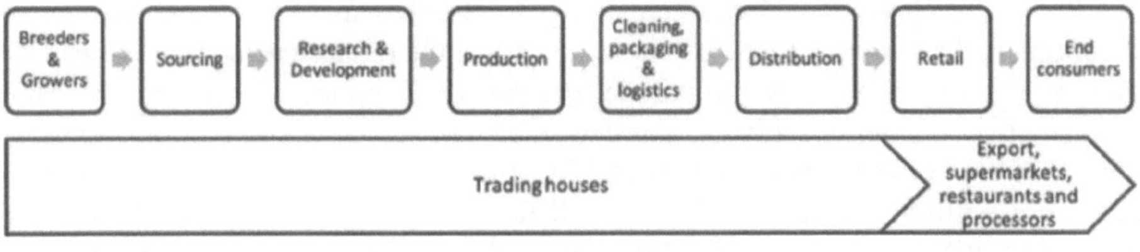

FIGURE 16.2 The value chain of seed potatoes. (Adapted from Herr, 2007.)

16.3.1.2.2 Market Requirements

There are a couple of overarching requirements identified during the interviews. Firstly, *many* agree that it needs to be determined whether the potatoes grown from seed potatoes need to be consumed freshly as a table potato or need to be processed for fries, chips or starch. Currently *"50% of the world market demands fresh potatoes, i.e. unprocessed"* (trading house 2). For western countries, however, specifically processed products are identified more important. This poses problems, as *"salt-tolerant potatoes produce more sugars, when fried or baked they turn brown"* (expert 2). Thus, they are unsuitable for the chips and fries market.

A few also identify the difference between salt-tolerant varieties and common varieties as an opportunity for the niche market. However, this is only a small market and certainly not bulk: *"we notice that there is a market in those special salt-tolerant products you don't need a lot of: purple crisps or purple potato, for decoration in a restaurant"* (farmer 3). In this market they prefer products that *"are actually only interesting if they are distinctive in taste or appearance"* (expert 1).

Additionally, the volumes a variety produces matters. *Multiple* people state that, currently, salt-tolerant varieties on average produce a lot less in terms of absolute volume. As sectoral representative 3 illustrates: *"Salt-tolerance itself says little; you also have to think about the volumes. A salt-tolerant variety can give less volumes under saline conditions than a highly productive, not salt-tolerant variety of which half fails."*

16.3.2 What Are the Opportunities and Constraints That Rise from the Dimensions of the Current Socio-Technical Regime of Quinoa?

The interviewees were questioned about the opportunities and constraints within the socio-technical landscape of quinoa, built upon six dimensions; industry, technology, policy, science, market and culture. Some of the findings are aligned with opportunities and constraints within the value chain.

16.3.2.1 Industry

Within the industry dimension of the socio-technical landscape of quinoa, there is no free market, as; *"the quinoa grown here* (i.e. in the Netherlands) *is developed by the WUR, owned by TQC and marketed by GF50"* (quinoa expert 4). *Multiple* interviewees identify the fact that there are scarcely any actors within the sector as a constraint: *"you may not secretly collect and grow our* (i.e. GF50) *seeds and sell them, because then you are violating the breeder* (i.e. TQC and WUR) *who has invested 20 years in it"* (quinoa expert 4). Besides, as GF50 manages supply according to the demand, there is no guarantee that the requested supply is constant. This dependency is identified to hamper the uptake of quinoa. Others are able and welcome to compete, but this has not happened yet. According to a quinoa expert, this is a result of the current numbers of TQC, that shows that the viability of quinoa cultivation within the Netherlands is very difficult: *"first, you need more demand to make cultivation profitable in the Netherlands, then a competitor would be of value"* (quinoa expert 5).

16.3.2.2 Technology

The cultivation of quinoa must be pesticide-free; thus, the whole cultivation plan must be pesticide-free due to legal jurisdictions. This is a prevailing issue within the technological dimension, highlighted by all quinoa farmers as well as *multiple* other interviewees. This means that the whole cultivation plan must be pesticide-free.

Dutch quinoa has a technological advantage as it is saponin-free. The absence of saponins in Dutch varieties is identified by *multiple* interviewees to smoothen the technological process of quinoa cultivation as it saves time and money for producers and consumers. As such, it is considered *"a huge advantage that makes it competitive to South-America. If it would contain saponins it would be at least 30% more expensive"* (quinoa expert 3).

The volumes of the quinoa are (too) low and unstable. A few interviewees identify a trade-off due to low productivity. As quinoa expert 3 illustrates, you can choose between *"more volume but lower price per product* (e.g. barley), *or less volume and a higher price per product* (quinoa)".

16.3.2.3 Policy

In 2013, the FAO initiated the "international year of quinoa (IYQ)" with the objective of increasing awareness of the benefits of quinoa. IYQ resulted in the establishment of TQC, as *"it was an opportunity to create a market for local quinoa because of the increasing awareness and demand"* (Quinoa expert 1).

The lack of authorization to use pesticides obstructs the scaling of quinoa cultivation. *Multiple* interviewees stated that lacking jurisdiction regarding pesticides negatively influenced quinoa cultivation (and thus consumption): *"in Europe pesticides have to be authorised per crop, per purpose. Nothing is allowed for quinoa. This makes it more difficult to grow quinoa here"* (quinoa expert 3 endorsed by *multiple* interviewees). This has various reasons: *"it is a matter of slow regulation, but also the fate of smaller crops because tests are expensive"* (quinoa expert 4). In addition, the difficult admission procedure plays a part in this issue: *"the investment for the pesticide admission process is too high, compared to the revenue it generates. This does constrain the scaling of quinoa cultivation in the Netherlands"* (quinoa expert 5).

16.3.2.4 Science

Multiple interviewees identify that Dutch quinoa research is relatively underdeveloped, especially in comparison to established cultivars and the timeframe of variety-development: *"the reason why we are working with salt-tolerant species today is that when you start breeding, this development can take fifteen years"* (quinoa expert 1).

Quinoa needs to become better protected from external threats and the production of quinoa has to become more stable in terms of volume. Currently, as pesticides are not allowed, *a few* endorse that: *"research needs to be done to find out what does not damage the quinoa but the weeds and threats"* (quinoa expert 5). When looking at the production, also the variability of the crop needs to be examined: *"we have 4 hectares; one year 2000 kg is produced; the other year 2500 kg. The exact reason is still unclear, there is not enough knowledge about that"* (farmer 2, endorsed by *multiple* interviewees). As the volume of the yields varies, constant revenue is not guaranteed.

According to *many* interviewees, there is a knowledge gap within Dutch consumer preferences regarding quinoa. Thus, consumer preferences and the quinoa market itself are under-researched. The reasoning of consumers to make certain choices is therefore currently subject to guessing.

16.3.2.5 Market

The demand is expected to grow, but likewise, the international supply is growing. *Multiple* interviewees stated that they expect increasing consumer demand as well as increasing supply: *"the market itself is growing very fast, but there is also more and more supply from Europe, Spain and France, but also from Canada and (South) America"* (quinoa expert 3). In general, the prevailing belief identified by *multiple* interviewees, is that for quinoa to be competitive on the (international) market, you need to do it in bulk: *"efforts only work if you can concentrate investments, and if you can give high volumes"* (quinoa expert 1).

There is a higher demand for organic quinoa; however, organic certificates demand high investments. *Multiple* interviewees identify the high investments in order to gain organic certificates (e.g. Skal) as a constraint. Imported quinoa often has such certificates, however, *a few* interviewees stated that is unclear to what extent these have been acquired fairly as *"there's a lot of corruption in the market"* (quinoa expert 4).

16.3.2.6 Culture

Quinoa research and development is mainly aiming to add value; *"if you invest in under-utilized plant species, you can actually add more diversity to food production systems"* (quinoa expert 1). As such, TQC is motivated to make quinoa a global staple crop; *"our objective is to accelerate the evolution of quinoa, so it can enter professional farming systems"* (quinoa expert 1). Increasing food demand is identified by *multiple* interviewees of importance as well; *"in 2050 we must produce 70% more food. That won't come from saline agriculture only, but we need to use the saline resources for unconventional crops like quinoa to reach that 70%"* (expert 2).

Curiosity (regarding business opportunities and the sustainability of quinoa) drives farmers to cultivate. In the interviews, farmers explained their motivation; *"I like to try new things and quinoa is a healthy crop, a crop to be proud of"* (farmer 5). Additionally, the sustainability issue is of value: *"for us, farmers, growing this product is good for biodiversity and keeps the soil healthy"* (farmer 4).

For retail and distribution, the two labels "traceability" and "sustainability" make the concept of local quinoa powerful, according to *multiple* interviewees. As quinoa expert 4 explains: *"if you want to do it right, besides organic, quinoa has to be local"*.

A few interviewees identified that the problem with quinoa is that people see it mostly as a specialty crop as consumption of quinoa is not embedded in the Dutch culture. To prevent this, *a few* argued that you should sell quinoa as an alternative to common products: *"when you introduce a new crop and you can offer it as an alternative to something very similar, that you can replace easily, the threshold for the consumer is lower"* (expert 2). This could entail e.g. rice or couscous. That quinoa is not culturally embedded also influences the use of pesticides: *"e.g. wheat and maize can be developed with a lot of pesticides* [...] *it is from the historical point of view that new crops are watched more closely than old ones*" (quinoa expert 3).

16.3.3 What Are the Opportunities and Constraints That Rise from the Dimensions of the Socio-Technical Regime of Salt-Tolerant Seed Potatoes?

16.3.3.1 Industry

Research and development regarding salt-tolerant seed potato varieties occurs in both the commercial and scientific fields and is supported by governmental agencies, e.g. ministries. However, "*researchers would like to cooperate more with the trading houses*" (expert 1). This lack of cooperation could result from the contradicting motivations of actors. There is cooperation within fundamental research, but the actual development of salt-tolerant varieties is separate. What could be of importance here is that the main revenue for trading houses comes from licensing varieties. Trading houses explain: "*closer to what you want to put on the market, there is simply no cooperation because of competition*" (trading house 1). As such, *multiple* interviewees identify this friction comes from the fact that trading houses focus on increasing revenue, whereas scientists generally do not have a commercial aim. *A few* interviewees state that r*esearch* is becoming more accessible. In general: "*research in the potato sector is becoming easier and cheaper* [...] *prices are dropping due to available technology. As a result, we start to understand complex things better and better*" (trading house 3).

16.3.3.2 Technology

Multiple interviewees identify the technological quest to develop a salt-tolerant potato variety that is close to existing, regular cultivation: "*it must remain a common, regular seed potato, but one that can withstand higher salt values*" (expert 4).

A few interviewees identify the subsequent requirement for salt-tolerant seed potatoes to be grown organically. However, this is considered difficult, as regular varieties are already hard to produce organically. Thus, the development of organic common varieties (non-salt-tolerant) is prioritized over salt-tolerance, as the demand for the first is bigger.

The quest for appropriate varieties will potentially be smoothened thanks to the emergence of hybrid breeding, as identified by *a few*. Hybrid breeding programs are faster and therefore cheaper. Solynta is a firm focused on this technique and expects to present their first seed potato seeds in the late 2020.

16.3.3.3 Policy

Many interviewees identify the need for official standardization of labeling for salt-tolerance. The absence of such a label is identified as a constraint because it leads to unsubstantiated claims. This poses problems for production as well as distribution and retail: "*if we* (i.e. trading house) *are asked if we have salt-tolerant varieties, we cannot and will not claim it*" (trading house 1 endorsed by trading house 2 and trading house 3). Therefore: "*we currently only tell which breeds are best and worst suited for saline soils*" (trading house 1).

Multiple interviewees identify the fact that some countries have import jurisdictions as a current constraint for the uptake of salt-tolerant seed potato varieties. As expert 4 argues: *"you cannot just grow any potato anywhere, and certainly not import it"*. There are numerous countries with breeding lists; *"often salt-tolerant breeds are not listed, making it difficult to cultivate them here"* (expert 2).

16.3.3.4 Science

A few interviewees identify that knowledge is currently not up to date. As water board 1 illustrates: *"a lot of knowledge is still based on old figures or outdated experiments. There are no recent norms or standards"*. However, *many* state that in recent years, a lot of important research has already been done in this domain, and that progress is made.

Multiple interviewees identify that the market is a research area that needs to be examined. There is need for information regarding the end use of salt-tolerant potato varieties (e.g. market research and consumer preferences).

Hybrid breeding is identified by *a few* as an opportunity. In addition, in this dimension, the faster, and potentially cheaper process of hybrid breeding could aid in accelerating research and development of new varieties.

16.3.3.5 Market

Many interviewees endorse the view that there is not yet an existing export market of salt-tolerant seed potatoes. As trading house 2 explains, *"No matter how good a variety performs in the Netherlands, it must also excel abroad"*. Currently, however, the demand from abroad is from developing countries where they have little budget. The need to target more prosperous countries is identified: *"focus on more wealthy European countries where salinization takes place: Spain, Italy and France"* (policymaker 2). The paradox is that *"The value will only increase as the target-area increases, thus as more area becomes saline"* (trading house 1 endorsed by *a few*).

Trading houses often have other priorities. Trading houses identified the need to rank the characteristics that future varieties need to obtain. This goes as follows: *"we ask our salespeople once a year what the most important aspects of a new variety are. They come up with a top 10, and salt tolerance is not yet in it"* (trading house 2 endorsed by trading house 3). Currently, organic cultivation is prioritized over salt-tolerant cultivation.

Multiple interviewees identify that farmers are highly dependent on seed potato cultivation as for many *"seed potato is a cash crop; if the yield and revenue would decrease it would mean end of business"* (sectoral representative 3). This could be interpreted as an opportunity for the development of salt-tolerant varieties.

16.3.3.6 Culture

Multiple interviewees highlight the leading position of the Dutch industry in the sector, but also its dependency on exports: *"we are dependent on 60-80% export. This has consequences that you see during such a crisis* (i.e. Covid-19), *that resulted in the collapse of the global fries' market"* (expert 1).

Multiple interviewees identify the cultivation of salt-tolerant varieties as means to maintain the leading position within the seed potato sector in the future. *A few* argue that overall system change is needed. However, trading house 2 disagrees *"you have to ask yourself; should we* (trading houses) *develop a variety that is more resistant to saline soils, or should people with salinized land start growing other crops?"*

There is ambiguity in how to market salt-tolerant seed potato varieties. For the retail sector *"there has to be some kind of marketing involved when you enter the product as a saline product in a certain segment. You have to try to make a product for which people are willing to pay a higher price, because you will be able to produce less"* (expert 3). Not everybody agrees with the idea to market the product as something special: *"we do not want to commercialise the product as a salty potato"* (trading house 1 endorsed by *multiple* interviewees).

16.3.4 Other Influences on the Scaling of Salt-Tolerant Cultivars

Certain themes were identified that were applicable to the socio-technical regimes of both quinoa and seed potatoes.

A few identify a change in attitude towards saline agriculture by governmental agents: *"at first, policy had the adage 'we must combat salinification'. This changed in 'we will prevent it as long as possible, but if it is no longer possible, we will invest in saline agriculture'"* (expert 5). This has results in the new adage of "mitigation where possible, adaptation where needed". As expert 4 explains: *"we strive for mitigation strategies with the aim of maintaining conventional agriculture as it is today. In some places however, it is better to let the saltwater take its course. In those areas, you could potentially apply saline agriculture"*. However, *multiple* interviewees identify that it also important to remember that saline farming is only suitable for parcels where the costs of keeping conventional agriculture 'sweet' outweigh the benefits of saline agriculture. As independent agent 1 argues; *"you shouldn't start recommending salt-tolerant crops to farmers unless they can't do otherwise"*.

Multiple interviewees identify the need for active policy to stimulate salt-tolerant cultivation as an adaptation strategy for salinization as well as freshwater shortages. Currently *"there is no active policy to stimulate salinization risk reduction [...] you have to pursue a targeted policy, with targeted strategies"* (expert 4). Policymaker 2 explains: *"active policy isn't on the schedule yet. The ministry prefers it tomorrow, but LTO* (i.e. Dutch farmers union) *is holding it back"*. In any case, this tardiness is not because of the Minister's personal involvement *"the motivation of her interest* (i.e. Ms. C. Schouten)*: food security. [...] We believe in growing salt-tolerant crops as one of the solutions to the salinization problem"* (policymaker 2). However, they cannot do it alone; *"we need Brussels to implement overarching policy at the European level"* (policymaker 2).

The "water distribution priority sequence" is identified by *a few* to possibly play a role in the uptake of saline farming: *"in case of extreme conditions like droughts, this* (i.e. leaching) *is no longer allowed, and we have to think about adaptation strategies"* (water board 2). Overall, *multiple* interviewees identify that the cooperation between the agricultural and the governmental sector should be smoothened, as the distance between practice and policy is currently too big. As farmer 6 states: *"the water boards are high and dry"*. Farmer 2 adds; *"we need scenarios from the water board. Everyone is unsure about what's going to happen, we're not familiar with possible scenarios."* However, water board 2 argues that the water boards choose to *"place the responsibility for choosing the right crops with the entrepreneur: 'This is the water you'll receive; it is your responsibility to choose a suitable crop. If you need better fresh water for your crop, then you have to come up with a trick yourself'"*. This can be seen as a "*passive invitation*" (water board 2) to start salt-tolerant cultivation in areas with salt-containing water supply. As such, *many* identify that, as long as the water boards continue to supply water with acceptable effort, there is no direct incentive to switch. In total, the cooperation within the agricultural industry *"is very difficult because different parties involved: policy-makers, landscape managers, environmental agents, water boards, soil experts et cetera. They all have an opinion and as a result, few decisions are taken because everyone keeps their hands off the risk as much as possible instead of looking at where to innovate"* (expert 3).

In the retail sector, *many* state that we should look for various ways to expose salt-tolerant products to the consumer market. It seems to be a matter of consciousness: *"we all have the mind-set of conventional agriculture; you use fresh water and it grows. You have to think differently to start saline agriculture on a large scale"* (expert 2). As expert 1 adds: *"the problem that everything is finite is slowly but surely becoming aware"*. Thus, the need for a different system is identified by *a few*: *"if different countries are supported to link their cultivars directly to water needs, water efficiency and primary protein supply for the local population, you'll see a difference"* (expert 1). For the farmers, also the long-term vision plays a role: *"many generations after us must be able to live off the land. Given the high demand for food for the ever-growing world population, this is a very important issue for us"* (farmer 4). Farmer 6 adds: *"I'm looking for salt-tolerant crops to keep the soil healthy for the next generation. My successor should be able to cultivate too"*.

16.3.5 What Locks in the Current Socio-Technical Regime and What Are Windows of Opportunities for the Scaling of Salt-Tolerant Cultivars?

16.3.5.1 Lock-Ins

The issue that there is no specific market for salt-tolerant products within the Netherlands, let alone for export, is identified as a lock-in. This lock-in can be explained because of: (1) no explicit demand for salt-tolerant cultivars from prosperous areas, therefore; (2) no direct priority/urgency for trading houses to develop varieties, therefore; (3) no ability to produce salt-tolerant products in bulk, therefore; (4) no direct incentive for farmers to cultivate salt-tolerant seed potatoes. Secondly, next to the lacking economic benefits there is also a lacking awareness regarding salinization and the threats it poses. As such, there is no awareness of the ability of salt-tolerant cultivars to give economic prospects on marginal land. Therefore, products obtained from saline land currently have no benefit on the consumer market as well as within the agricultural sector. This is primarily a result of short-term thinking and the prevailing idea of "mitigation first, adaptation later". Governmental agents and sectoral representatives see saline farming as a last resort thereby neglecting its possible potential. The third identified lock-in involves the fact that currently the societal demand for saline products stops at the niche market. The niche market might have benefits for current uptake; however, the question is whether this will help in the scaling of these crops. After all, a niche product is not a niche product anymore when produced in bulk. As trading houses focus on bulk, they are not interested in producing for the niche market.

16.3.5.2 Windows of Opportunity

The most frequently mentioned window of opportunity follows from the more frequent recent droughts. Many mention the summer of 2018 as a turning point where farmers and policymakers became increasingly aware of the problems regarding salinization. Besides, the Covid-19 crisis increased attention as the weakness in our current food system led to an interest in food security issues. Both led to a shock at the landscape-level and an increasing awareness regarding dependencies due to diminishing (export) markets. As a result, the interest in local value chain management could potentially increase.

16.4 DISCUSSION AND CONCLUSION

16.4.1 Discussion

When examining aspects considering the scaling of saline farming it is important to acknowledge the belief that a socio-technical regime is not deliberately shaped. It is the result of different (inter-) dependencies between actors and actions (Geels 2002). The acknowledgment that the Netherlands is new to water-scarcity issues can thus explain the prevailing motto "mitigation first, adaptation later". Involved actors are not so much denying or downplaying the possible issues that salinization entails, but rather not experiencing and recognizing the situation as urgent enough to take action. Based on the MLP there are certain pressures in place that may accelerate this. At the macro-level, there is a growing awareness regarding scarcity, vulnerability and interdependency due to the shocks of the recent droughts and Covid-19. At the micro-level, pressure is exerted in the form of further development of knowledge and pioneering, e.g. in the development of more salt-tolerant crops. Thus, according to Geels (2002), when saline farming goes through different niches, the market share will grow and even more pressure can be exerted on the existing regime. This brings up the perceived ambiguity of the MLP framework (Geels 2002). The different levels (i.e. niche, socio-technical regime and landscape) are rather broadly defined. As Berkhout states: "it is unclear how these conceptual levels should be applied empirically." (Berkhout 2004). As such, due to the difficulty to provide clear definitions and boundaries, the indication of saline farming as a niche, the sector as socio-technical regime and societal norms and values as landscape, are open for debate. A recommendation to overcome this ambiguity is to: (1) incorporate a degree of flexibility when utilizing the different levels, (2) include a clear outlook of the desired socio-technical regime, and (3) be transparent in the assumptions made.

There are more limitations to the findings of this study. First of all, the snowball method used affects the representativeness as two very important groups were not included in the examination: the retail sector and average farmers. This is unfortunate, as the retail sector is expected to play a big role in the uptake and thereby the economic viability of the products. Besides, their view on the framing of such products (niche vs. common) would be of high relevance. Additionally, the interviewed farmers should be considered as pioneers and the sectoral representatives speak for the concerns of their affiliates solely, not for all farmers. Hence, the "average farmer" is not represented. Additionally, it is important to mention that the occurrence of Covid-19 made face-to-face interviews rather difficult. Therefore, all but two interviews took place online. This limits the results, as it was not possible to read facial expressions and indirect responses of interviewees.

16.4.2 Conclusion

This research aimed to answer the question: *"Is the saline farming of quinoa and seed potatoes considered a viable option in the Netherlands based upon the opportunities and constraints of the scaling of both products?"* The obtained results indicate that the lack of economic benefits is the main constraint for the scaling of quinoa and/or seed potatoes cultivation on saline soil. Most actors identify the potential of saline farming either when: (1) its saline origin offers added value, or (2) when the saline origin of the product does not play a role, but cultivation on saline soil offers added value. The latter is the case when the costs of keeping conventional agriculture "sweet" outweigh the benefits of saline agriculture. Both seem not yet to be applicable.

The lacking economic benefits could potentially emerge from lacking awareness regarding the salinization issue. Currently, the uptake of and interest in saline products takes place in niche markets, which primarily entail small-scale production with a regional image. However, trading houses are only interested in bulk markets with products that can be competitive with similar products from other production systems. Conflicting interests (e.g. trading houses versus research institutes) further delay acceleration. The recent droughts and emergence of Covid-19 could possibly change this. These two windows of opportunity are expected to lead to a growing awareness and interest in adaptation measures.

Regardless of its current economic viability, saline farming is perceived as an attractive novel opportunity that deserves suitable support for required innovations and transitions. Supporting saline farming offers the opportunity to invest in a forward-looking way: one can facilitate new production in saline areas, thus giving marginal land economic meaning. It is not expected to be economically viable in the short-term; however, if salinization increases at the current rate, maybe its viability will come faster than expected.

ACKNOWLEDGMENTS

This research was conducted with support of the Wadden Academy, Leeuwarden, as a contribution to the European Union Interreg VB North Sea Region SalFar project, co-funded by the North Sea Region Programme 2014–2020.

REFERENCES

Atzori, G., Mancuso, S., & Masi, E. 2019. Seawater potential use in soilless culture: A review. *Scientia horticulturae* 249: 199–207.

Bazile, D., Jacobsen, S. E., & Verniau, A. 2016. The global expansion of quinoa: Trends and limits. *Frontiers in plant science* 7: 622.

Berkhout, F., Smith, A., & Stirling, A. 2004. Socio-technological regimes and transition contexts. *System innovation and the transition to sustainability: Theory, evidence and policy* 44, no. 106: 48–75.

Bryman, A. 2012. Sampling in qualitative research. *Social research methods* 4: 415–29.

Daliakopoulos, I.N., Tsanis, I.K., Koutroulis et al. 2016. The threat of soil salinity: A European scale review. *Science of the total environment* 573: 727–39.

Geels, F. W. 2002. Technological transitions as evolutionary reconfiguration processes: A multi-level perspective and a case-study. *Research policy* 31, no. 8: 1257–74.

Geels, F. W. 2011. The multi-level perspective on sustainability transitions: Responses to seven criticisms. *Environmental innovation and societal transitions* 1, no. 1: 24–40.

Gill, P., Stewart, K., Treasure, E., & Chadwick, B. 2008. Methods of data collection in qualitative research: interviews and focus groups. *British dental journal*, 204, no. 6: 291–95.

Hermans, F., van Apeldoorn, D., Stuiver, M., & Kok, K. 2013. Niches and networks: Explaining network evolution through niche formation processes. *Research policy* 42: 613–23.

Herr, M. L. 2007. Local Value Chain Development. International Labour Organization. Enterprise for Pro-poor Growth (Enter-Growth). http://www.oit.org/wcmsp5/groups/public/@ed_emp/@emp_ent/@ifp_seed/documents/instructionalmaterial/wcms_101319.pdf (accessed May 12, 2020).

Huisman, P., Cramer, W., Van Ee, P. et al. 1998. Water in the Netherlands. Paper presented at the Netherlands Hydrological Society, Rotterdam.

Oude Essink, G. H. P., Baaren, E. S. Van, & De Louw, P. G. B. 2010. Effects of climate change on coastal groundwater systems: A modeling study in the Netherlands. *Water resources research* 46 (October): 1–16.

Stuyt, L., Snellen, B., Essen, E. et al. 2011. Effecten van aan klimaatsverandering gerelateerde verzilting op de bedrijfsvoering van landbouwbedrijven in Noord-Nederland. Report presented by Alterra and Aequator, Wageningen.

Stuyt, L.C.P.M., Kselik, R.A.L., Blom-Zandstra, M. 2016. Nadere Analyse Zoutschade op basis van bestaande gegevens; Inventarisatie van eerder gerapporteerde zouttolerantiedrempels van beregeningswater met verhoogd zoutgehalte. Report by Wageningen University & Research. https://library.wur.nl/WebQuery/wurpubs/fulltext/391931 (accessed April 11, 2020)

Turner, P., & Turner, S. 2009. Triangulation in practice. *Virtual reality* 13, no. 03: 171–81.

de Vos, A. C. 2011. Sustainable exploitation of saline resources. Ecology, ecophysiology and cultivation of potential halophyte crops. PhD diss., Vrije Universiteit Amsterdam.

Wicke, B., Smeets, E., Dornburg, V. et al. 2011. The global technical and economic potential of bioenergy from salt-affected soils. *Energy & environmental science* 4, no. 8: 2669–81.

Wigboldus, S., Klerkx, L., Leeuwis, C. et al. 2016. Systemic perspectives on scaling agricultural innovations, a review. *Agronomy for sustainable development* 36, no. 3: 46.

APPENDIX: LIST OF INTERVIEWEES

Sector	Seed Potatoes	Quinoa
Trading houses	C.C. Meijer B.V.	
	HZPC	
	Solynta	
	Agrico	
Farmers	Texel	
		Noord-Holland
		Brabant
	Groningen	-
	Friesland	
		Flevoland
		Zeeland
Retail	Marc. Foods	
Policymakers	Wetterskip Frysland (2)	-
	Hoogheemraadschap Rijnland	-
	Ministry of Agriculture, nature and food quality	-
Independent agents	STOWA	-
	PBL	-
Sectoral representatives	Boerenverstand	-
	Potato Valley	
	LTO-Noord	
Experts	Salt Doctors	-
	ILVO	-
	Acacia water	-
	Rijksuniversiteit Groningen	
	ILVO	-
	SPNA	
Quinoa experts		ICBA
		The Quinoa Company
		Mercadero
		Wageningen University and Research
		GreenFood50

17 Dynamics of Soil Salinity in Denmark

Laurids Siig Christensen

CONTENTS

17.1 INTRODUCTION

Denmark is a nation of islands, generally with a low altitude, less than 200 m above sea level. The landscape is rich in marshland and meadows, and arable land constitutes 62% of the country (Christensen 2019). Because of its geography and because agriculture and food production in Denmark play a prominent economic role, Denmark is among the countries in the world that are expected to have to pay the highest costs of accommodating to climate changes. Denmark is also a relatively small country of 42,900 km^2 with an exceptionally long coastline of 7,300 km. No spot is more than 50 km from the sea, and the country is located between a brackish sea, the Baltic Sea, and the saline North Sea. This means that coastal areas in Denmark are exposed to sea salinities ranging from approximately 0.6% (wt/vol) on the island of Bornholm to 3.2% (wt/vol) on the North Sea coast of Jutland and the Danish islands of the Wadden Sea. The combination of the national importance of food production and this gradient of salt in the sea surrounding Danish islands also means that Denmark, among European countries, has optimal conditions for experimenting with halo-tolerant and halophilic plant production under various conditions in order to develop a contingency in food production against climate changes.

The state of salinity in soil on a national scale in Denmark is not known, nor have the dynamics of soil salinity and the mechanisms affecting this on a national scale been subject to research in Denmark. During the present study, an approach was established of sampling soil and water sources and measuring electrical conductivity (EC) in soil of highly variable nature, as well as in various water sources. Baselines of conductivity were defined in soil, as well as in water streams, as distant from the sea as possible in Denmark. In addition, coastal landscapes were chosen for monitoring because of records of flooding from the sea or because of other indications of salinity. A special focus was on smaller islands in Denmark, since on these islands, farming, in many cases, is practiced close to the sea.

DOI: 10.1201/9781003112327-17

17.2 MATERIALS AND METHODS

Top-soil samples of 15–30 grams were collected with a cylindrical soil spear sampling device to a depth of 30 cm. In some cases, surface soil (named as such in Table 17.1) samples were collected by scraping the soil surface with a spoon. Soil samples were dried for >8 hours in an incubator at 35–40°C. The soil samples were sieved through a mesh of 2 mm, and a total of 15 grams was diluted with 5 parts (wt/wt) of demineralized water and left for >1 hour. Finally, the samples were centrifuged at 2,850 g for 5 minutes in a Rotofix 32A Centrifuge. Water samples were collected in 50 mL PP tubes.

Prior to measurement, samples were incubated in a water bath at 25°C until a sample temperature of 20 – 25°C was reached. The EC was measured with a Thermo Scientific™ Orion Star™ A122 Conductivity Portable Meter. For soil samples, the conductivity measured, referred to as $EC_{1:1}$, was multiplied by the dilution factor (5 as a standard). Conductivity was measured >3 times in every sample and a mean value estimated.

A total of >140 top-soil samples and >120 water samples originating from streams, drainpipes, lakes, wells, etc. were collected in Denmark and analyzed. Locations of sampling were widely dispersed to get an overview, on a national scale, of variation in soil salinity. Multiple samples were collected from central regions of Jutland and Zealand, as well as on every island studied, to estimate a baseline soil conductivity in areas where the effects of the sea were expected to be minimal. Other locations, expected to be affected by the sea, were chosen, from which multiple samples were collected in order to increase the reliability of measurements and possibly reveal dynamics in variation of soil salinity. Samples were predominantly collected in March in both years in order to minimize the impacts on EC of fertilizer added by farmers and the effects of increased biological activity in the root zone.

17.3 RESULTS

A map of Denmark, Figure 17.1, shows three regions of Jutland and two islands in the northern Baltic Sea (Kattegat) where soil salinity was expected to be affected by the sea or by inlet seas. In addition to these areas, samples were collected from the Wadden Sea marsh in the south-west region of Jutland and from North Sea coastal areas north of the Wadden Sea marsh. The Varde River drains a considerable part of South-West Jutland and is the only river in Denmark that leads into the Wadden Sea via Ho Bay, without a sluice to regulate water level and flood. The river valley is subject to flooding by the North Sea almost every winter season. The Skjern River is the water-richest river in Denmark, draining an even larger area in Mid-West Jutland. It leads into Ringkoebing Fjord, which is regulated by a sluice to the North Sea. The fjord, Limfjorden, in northern Jutland, is open without sluices to the sea on both "sides" of Jutland and is subject to a considerable influx of water from the North Sea. Limfjorden is incidentally flooding arable land. Laesoe is an island, where a major marsh area (Rønnerne) is frequently flooded by the sea. Due to the combination of flooding in the winter season, a water-impermeable sub-terrain soil layer at a depth of 1–2 m at "Rønnerne" and evaporation of flooded water during the summer season,

TABLE 17.1
Electrical Conductivity Measured in Soil Samples

Sample Collection	Sample Size	Range of $EC_{1:1}$ mS/cm	Mean $EC_{1:1}$ mS/cm	SD
Baseline soil, central Jutland, 2019	8	0.11–0.75	0.38	0.21
Baseline soil, central Jutland, 2020	5	0.07–0.28	0.15	0.07
Baseline soil, central Zealand, 2019	5	0.49–0.67	0.57	0.05
Baseline soil, central Zealand, 2020	7	0.16–0.31	0.27	0.08
Soil of uncultivated forest, Allindelille Fredsskov, 2020	4	0.4–0.9	0.72	0.2
Sejeroe locations				
Soil of central island, baseline, 2020	6	0.2–0.49	0.29	0.06
Soil of rape field subject to spray from waves, 2019*	5	0.6–1.24	0.96	0.23
Wheat affected by sea spray, line 1, 25 m off coast, 2020	1		0.79	
Wheat affected by sea spray, line 1, 75 m off coast, 2020	1		0.51	
Wheat affected by sea spray, line 1, 150 m off coast, 2020	1		0.35	
Wheat affected by sea spray, line 2, 25 m off coast, 2020	1		0.48	
Wheat affected by sea spray, line 2, 75 m off coast, 2020	1		0.44	
Wheat affected by sea spray, line 2, 150 m off coast, 2020	1		0.33	
Laesoe locations				
Soil of central island, baseline, 2019	9	0.26–0.81	0.56	0.2
Rønnerne, soil above line of flood, 2020	1		1.07	
Rønnerne, flooded marsh (area of salt production), 2020	1		4.96	
Soil of flooded arable land, 2020	2	1.47–1.78	1.63	0.15
Other islands				
Fur, soil of high meadow, 2019	1		0.71	
Fur, soil of low meadow, 2019 (record of flood)	1		1.34	
Mors, soil of plowed field, 2020	1		0.28	
Areas near Limfjorden				
Sandy soil not flooded by the sea, 2020	3	0.11–0.32	0.24	0.06
Soil of meadow not flooded by the sea, 2020	1		0.4	
Soil of flooded meadows, 2020	3	1.87–7.15	3.87	2.35
Areas near Ringkøbing Fjord				
Soil baseline, 2019	5	0.52–0.92	0.71	0.15
Soil of plowed field, 2020	1		0.11	
Areas near the North Sea coast				
Soil samples collected 3–10 km off North Sea coast, 2019	3	1.2–2.8	1.79	0.7
Soil samples collected 3–10 km off North Sea coast, 2020	3	0.14–0.41	0.29	0.11
Varde Å (river) meadow				
River valley, 1 km from outlet to the Wadden Sea, 2020	3	2.33–5.3	3.66	1.76
River valley, 3 km from outlet to the Wadden Sea, 2020	2	2.28–2.96	2.62	0.34
River valley, 5 km from outlet to the Wadden Sea, 2020	1		0.5	
River valey, 8 km from outlet to the Wadden Sea, 2020	1		0.61	
The Wadden Sea marsh				
Tønder area, 2019	4	0.93–1.28	1.03	0.07

*See Figure 17.2.

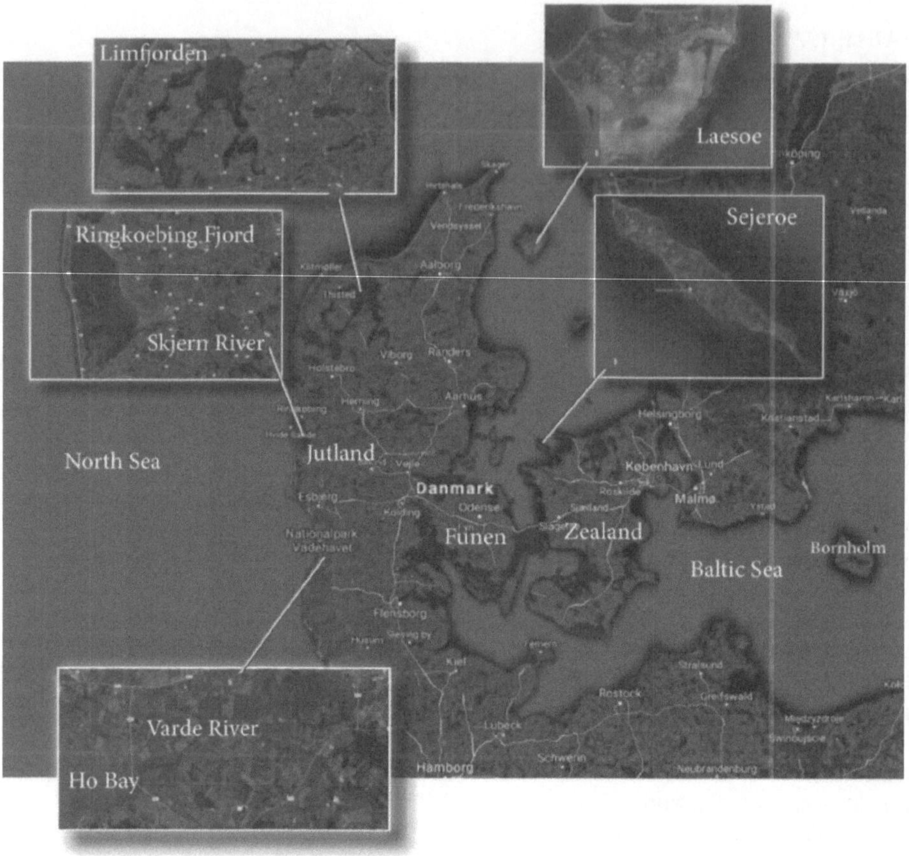

FIGURE 17.1 Map of Denmark showing five locations, potentially affected by marine salt, where multiple soil and water samples were collected.

salinity in wells may increase up to 17% (wt/vol) (Jørgensen 2002). These wells during the 12th–16th century AD served as substrate for a production of salt flakes, a production now re-established for archeological research (Laesoe salt). Sejeroe is an island <2 km wide without flooding during the past 70 years. The central axis of the island was covered by a glacier during the latest Ice Age and is now flanked by moraine hills toward the coast. The central axis of the island was sea floor, until the land was elevated approximately 6000 years ago.

Data on measurements of EC in soil and in water samples are compiled in Table 17.1 and Table 17.2, respectively. These tables do not include all data collected during the present study, but are representative of the sample collections and allow conclusions on causal relationships to be suggested. Other data comply with these causal relationships, but do not offer anything substantial that can be added to these conclusions.

Table 17.1 shows data on soil samples. Samples were stratified in order to define, tentatively, baseline values in soil in central areas of Jutland and Zealand, respectively,

TABLE 17.2
Electrical Conductivity Measured in Water Samples

Sample Collection	Sample Size	Range of EC_w mS/cm	Mean EC_w mS/cm	SD
Baseline water streams**, central Jutland, 2020	3	0.45–0.51	0.48	0.03
Baseline water streams, central Zealand, 2020	8	0.54–0.65	0.62	0.05
Baseline lakes and ponds, central Zealand, 2020	5	0.31–0.62	0.5	0.11
Rain puddles, central Jutland & central Zealand, 2020	8	0.15–0.64	0.41	0.19
Ponds of uncultivated forest, Allindelille Fredsskov, 2020	5	0.2–0.75	0.55	0.2
Sejeroe locations				
Rain puddles on well-drained cultivated land, 2020	7	0.27–0.62	0.45	0.12
Rain puddles on pasture not drained, nor cultivated, 2020	8	0.83–1.34	1.02	0.14
Creek/Canal water (effluent of drain pipes), 2020	3	1.64–1.85	1.71	0.1
Sub terrain (60 cm) water at Horsekaer, 2019	1		2.23	
Well at Horsekaer, 2019	1		1.89	
Lakes and ponds, 2020	3	0.95–2.02	1.32	0.75
Laesoe locations				
Creek/canal water, 2020	2	0.21–0.84	0.52	0.32
Puddle of remains of flood, 2020	1		47.2	
Other islands				
Fur, drain canal, 2019	1		1.63	
Mors, drain creek, 2020	1		2.63	
Areas near Limfjorden				
Influx sources to Limfjorden, 2020	3	0.45–0.51	0.48	0.03
Areas near Ringkøbing Fjord				
Influx sources to Ringkøbing Fjord, 2019	2	0.3	0.3	0
Influx sources to Ringkøbing Fjord, 2020	6	0.19–0.32	0.27	0.04
Ringkøbing Fjord, 2019	2	5.0–5.04	5.02	0.02
Ringkøbing Fjord, 2020	2	1.73–2.13	1.93	0.2
Areas near the North Sea Coast				
Henne Mølleå, 2019	1		0.27	
Varde Å (river) meadow				
Influx sources to Ho Bay, 2020	3	0.22–0.34	0.29	0.05
The Waddensea marsh				
Water of canal, 2019	1		2.62	

**Drain pipes, creeks and rivers.

suggested to be less affected by the sea than coastal areas. The samples showed little variation, as estimated by Standard Deviation (SD), within each of the sample collections. However, baseline $EC_{1:1}$ values in soil were significantly lower in Jutland than in Zealand in both years. In addition, in Jutland as well as in Zealand, baseline values were significantly lower in 2020 than in 2019, respectively. Samples from soil in the uncultivated forest (Allindelille Fredsskov), located in central Zealand, represent the variation in a unique ecosystem of forest, meadow, and ponds, known to be untouched by man for many centuries and without any drainage systems. EC in soil of this ecosystem was found to be significantly higher than outside the ecosystem in the same region and the same time of sampling (March 2020).

From locations where soil salinity is expected to be affected by the sea, most samples were collected on the two islands Sejeroe and Laesoe, both located in Kattegat with sea salinity expected to vary within a range of 2.0–2.6% wt/vol. In both cases, a tentative baseline soil salinity was defined in central regions of the island. The baseline soil $EC_{1:1}$ value was significantly higher on Laesoe in 2019, than on Sejeroe in 2020. However, this difference might reflect differences in precipitation between the two years (shown in Figure 17.2), rather than any difference related to location or geology of the two islands. It is a paradox that Laesoe, although renowned as the "island of salt" due to salt flake production on Rønnerne, has a salinity in the root zone of the sandy top-soil comparable to baseline values in other parts of Denmark.

On the south-west-oriented coast of Sejeroe, a significant effect was observed in a rape field, presumably due to saline spray caused by waves from two storms in late November and early December 2018 (Figure 17.3). Soil samples were collected from this field in March 2019 and an increase in top-soil salinity depending on distance to the coast and apparently correlating with rape growth performance was observed as shown (Figure 17.2). A similar effect, consistent with saline spray from recent storms, was observed in a wheat field on Sejeroe in early 2020. In order to establish an effect depending on distance to the coast, samples were collected in two lines in the field, and a decrease in $EC_{1:1}$ value with distance from the coast was observed

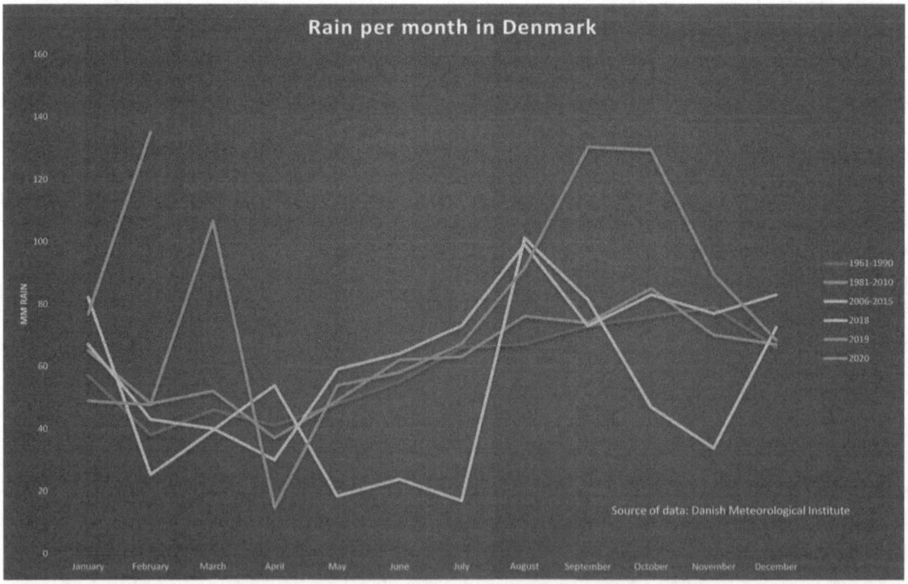

FIGURE 17.2 Precipitation of rain per month in Denmark. Mean values of rain are shown for three reference periods, 1961–1990, 1981–2010 and 2006–2016, In addition, precipitation of rain per month is shown for 2018, 2019 as well as for January and February 2020.

FIGURE 17.3 Rape field on the south-west oriented coast of the island of Sejeroe affected by saline spray from waves breaking on the beach.

in both lines. These values were all higher than baseline values during the same sampling period.

Laesoe is one of several islands that experienced flooding from the sea during the winter of 2019–20. Elevated values of $EC_{1:1}$ were observed in soil samples from flooded areas, while soil samples above the line of flood did not exhibit a significantly increased $EC_{1:1}$ value.

Results of $EC_{1:1}$ measures in samples from the islands of Fur and Mors are also shown in Table 17.1. The islands are both located in the western part of Limfjorden, and Fur has a record of incidental flooding of arable land. A significant difference in $EC_{1:1}$ in top-soil is seen between non-flooded meadow and land with a record of flooding on Fur, while $EC_{1:1}$ of soil samples on the island of Mors were within the national baseline levels (data not shown).

Soil samples from areas east of Ringkoebing Fjord in 2019 revealed slightly elevated values of salinity, while soil samples collected in 2020 did not indicate elevated values of $EC_{1:1}$ when compared to national baseline levels for 2019 and 2020, respectively. Soil samples collected near the North Sea coast in both 2019 and 2020 revealed higher $EC_{1:1}$ values than baseline soil in Jutland in 2019 and 2020, respectively.

Soil samples from the Varde River Valley – flooded almost every year and repeatedly flooded by the Wadden Sea in the winter of 2019–20 – showed significantly increased $EC_{1:1}$ values near the gate, Ho Bay, to the Wadden Sea. These levels decreased to values only slightly higher than national baseline levels at distances of >5 km from Ho Bay.

The marshland of the Danish Wadden Sea revealed slightly elevated levels of soil salinity compared to baseline values.

Table 17.2 shows representative data on water samples. Unfortunately, few water sample data were collected in 2019, and comparison of baseline EC_w values between 2019 and 2020 is not possible. However, as in the soil samples, baseline EC_w values in water streams were lower in Jutland than in Zealand.

EC_w values in terrestrial sources of water for Limfjorden are higher than in terrestrial sources of water for Ringkoebing Fjord and Ho Bay. Ringkoebing Fjord water exhibited a significantly higher EC_w in 2019 than in 2020.

EC in samples from ponds in the uncultivated forest (Allindelille Fredsskov) was within baseline values of central Zealand and these are suggested to be due to soluble components of microbial activity and oxidative and other forms of degradation of a relatively rich source of biological material.

On Sejeroe, significant differences were observed in EC_w values in puddles of rainwater depending on conditions of draining and cultivation of the location. Higher EC_w values were observed in puddles on land not drained, nor recently cultivated. Water samples from effluent of drainpipes, wells, lakes, and ponds, as well as in water 60 cm sub-terrain on Sejeroe, showed significantly higher values of EC_w than baseline values in central Zealand and central Jutland. On Laesoe, creeks and canals of drain effluent from central parts of the island do not exhibit EC_w values significantly above baseline values. A puddle of remains from the floods reveals the salinity of seawater affecting the soil salinity in such areas of flooding on this island. Results of EC_w values in water samples from the islands of Fur and Mors are also shown (Table 17.2). On both islands, EC_w values in drain effluent, significantly above baseline values, indicate reservoirs of higher salinity in deeper layers of soil. Values of EC_w in water streams leading to Ringkoebing Fjord were below baseline values and, hence, did not indicate any reservoirs of salinity in deeper layers of soil in this area. EC_w values in terrestrial sources of water leading to Ho Bay (Table 17.2) also did not indicate reservoirs of higher salinity in any soil layers east of the study locations in the Varde River Valley. The EC_w value, significantly above baseline EC_w values, found in canal water of the Wadden Sea marsh revealed sources of layers of soil with salinity significantly above baseline values.

17.4 DISCUSSION

The present study has been the first attempt to map soil salinity and parameters affecting the dynamics of soil salinity in Denmark. Defining a methodological approach to estimate soil salinity was the first obstacle – accentuated by the finding of only low to moderate salinity levels in soil and terrestrial water sources during the study period. Measurements of EC in pore water of the root zone soil is a standard procedure for a cost-efficient and rapid estimation of soil salinity potentially affecting the growth performance of cultured plants. It was found to be also an extremely sensitive analysis. NaCl is the salt of primary concern when addressing a problem of increased salinity in the North Sea region due to climate changes. However, it should also be stressed that other salts such as those in fertilizers, as well as charged components of soil resulting from microbial activity or oxidative and other forms of degradation of biological material, will contribute to the EC of a soil sample. This was illustrated by the $EC_{1:1}$ measurements of soil samples collected in the ecosystem reservation Allindelille Fredsskov, revealing a variation and values of $EC_{1:1}$ higher than those observed in arable land in the same central region of Zealand and same period of sampling. Rock salt is not present in Denmark and salt plugs originating from ancient seas are – when present – at a depth of more than 300 m.

A variety of sampling and sample processing procedures already exists (e.g. De Vos 2018), but common to most of them is that they are suitable to study variation over time of salinity on one location and in one type of soil. Mapping salinity on a national scale involves the analysis of a large variety of soil types and, hence, a more complex sampling and sample processing to make results comparable is needed. And even then, storage under ambient conditions in some samples revealed significant differences when analyzed one year later, probably as a result of microbial activity and the degradation of organic material. The sampling and sample processing steps prior to EC measurement adopted during the present study will be validated and published elsewhere. However, to overcome variation in $EC_{1:1}$ values as a result of sampling and sample processing procedures in soils of great variability it is recommended that a network of sampling sites be developed to be sampled repeatedly in the area studied. During the present study period, the number of sampling sites in Denmark were expanded continuously and as many sites as possible were tested in both years of testing. In those cases, variation in results confirmed the causal relationships of dynamics in salinity suggested here. In other cases, the lack of repetition in sampling in terms of suggesting causal relationships were evident. For instance, the lower baseline soil salinity observed on the island of Sejeroe in 2020 compared to baseline soil salinity on the island of Laesoe in 2019 and, likewise, the higher $EC_{1:1}$ values found 3–10 km off the North Sea coast in 2019, compared to those in the same area in 2020, are suggested not to reflect differences in geology or marine impacts on these locations. Instead, they are most likely due to differences in precipitation of rain between the two winters in combination with the low ionic retention ability of the sandy soil on the two islands and in West Jutland.

Combining top-soil sampling with sampling of water sources was found useful in trying to reveal the dynamics of soil salinity in Denmark. Since there is no given proportion between drain water volume and pore volume of the soil drained, values of EC of the two sources are not inter-comparable, and baseline values should be tentatively defined for both water sources and soil samples, respectively, to allow intra-comparisons. Doing so, water samples with significantly higher EC_w than baseline values are suggested as a useful source to screen for presence of salt deposits in larger areas of land and in sub-terrain layers of soil. Likewise, EC_w values significantly below baseline values as well as annual or seasonal variation of these values might reflect the wash-out of salt in sub-surface layers of the basins of West Jutland possibly depending on the annual and seasonal variation of precipitation. Monitoring in years to come will reveal if these suggestions are valid.

Soil salinization is not a new phenomenon, but one which has led to the disintegration of ancient civilizations, e.g. in Mesopotamia, as reviewed by Shahid et al. (2018). Sources of soil salinization are manifold on a global scale as reviewed by Manca et al. (2015), Daliakopoulos et al. (2016), and most recently by De Waegemaeker (2019) for the North Sea Region. Some of these sources are anthropogenic, such as sea-level rise related to climate changes (in part due to human activity), irrigation in some arid regions of the world, and intrusion of seawater in sub-terrain soil layers due to increased use of groundwater for drinking, irrigation and production processes in our part of the world. Others are natural and seasonally variable such as caused by (1) flooding by the sea, (2) spray from waves on the coast, (3) the deposition of aerosols from the sea, (4) seawater intrusion by seepage into coastal areas, and (5) capillary transportation of water from deeper layers to the root zone, as well as the counteracting effects of (6) rain. In the present study, we focused on the effects subject to seasonal variation. We saw examples of the impact of all these effects.

The locations chosen for this initial study on dynamics of soil salinity in Denmark represent the major geologies found in Denmark: (1) the moraine landscapes of central Jutland and Zealand, (2) the coastal areas of the major peninsula of Jutland, (3) the small islands, (4) the basin of Mid-West and South-West Jutland, and (5) the Wadden Sea marsh. Each location in Denmark, such as an island, has its own dynamic of soil salinization and may warrant many samples being collected at various locations and times of sampling. Doing this and understanding in depth the dynamics at any location in Denmark is beyond the scope of the present study.

No evidence of accumulation and deposition of salt in Danish soil was seen in our study. This suggests continuous downward transportation of salt from upper layers of soil due to the fact, that precipitation of rain exceeds evapotranspiration under our climatic conditions. Our data suggest that precipitation of rain in relation to time of sampling has a most prominent impact on top-soilEC. Mean values of rain per month (Danish Meteorological Institute) are shown in Figure 17.3 for three reference periods, 1961–1990, 1981–2010, 2006–2015, and for every month since January 2018. Mean values of annual rain for the three reference periods are increasing rapidly in Denmark from 712 mm during the period of 1961–1990 over 746 mm during 1981–2010 to 792 mm during the period of 2006–2015 – an increase of 11% in average annual precipitation during a period of 60 years. In addition, seasonal and annual variation in meteorological conditions seem to be increasing as well, with extremes in weather conditions being experienced more frequently. The year 2018 was characterized by a very low annual precipitation of 595 mm and a serious drought starting May 2018. The year 2019, on the other hand, was characterized by a very high annual precipitation of 905 mm. In particular, the winter season of 2019/20 (September 2019–February 2020) had an extremely high precipitation of 629 mm rain. The sampling in spring 2019 took place immediately after a peak precipitation in March, which assumably would reduce $EC_{1:1}$ values found at that time, but the extreme amount of rain during the winter of 2019–20 is reflected in the significant and consistently lower baseline values found in 2020 compared to those of 2019.

Denmark's landscape is without visible bedrock – except on the island of Bornholm in the Baltic Sea. It is resting on a pillow of chalk from ancient marine deposits. Geologically, Denmark can be divided into two parts due to the glaciers of the latest Ice Age stopping along the longitudinal central axis of Jutland. Thus, the land in Mid- and South-Jutland west of the longitudinal axis was a basin of water streaming from melting glaciers, while Northern Jutland, Funen, and Zealand are rich in moraine landscapes. The soil of Mid- and South-West Jutland has a high content of sand and gravel, while soil in the rest of the country is more variable and often has a higher content of clay. This is reflected in higher mean baseline EC values of soil and drain water on central Zealand compared to those of central Jutland, suggested to be due to the higher ionic retention ability of clay compared to that of sand. The soil, rich in sand and gravel, of the basin of Mid- and South-West Jutland also has good vertical drainage capacity, and this is reflected in the invariably low EC_w

values in terrestrial streams of water leading to Ringkoebing Fjord and the Wadden Sea via Ho Bay, respectively, compared to those of water streams in the rest of the country (Table 17.2). The EC_w values significantly higher than baseline values found in drain water in the Wadden Sea marsh and on islands not flooded (Table 17.2) are suggested to reflect seepage of seawater into these coastal areas. This suggestion is consistent with experiences in the past 70 years, that an increase in use of freshwater reservoirs for drinking or other purposes often leads to an increase in salinity in these reservoirs.

Airborne transmission of marine salt has been generally ignored in the North Sea region although acknowledged as the major source of soil salinity in other parts of the world such as the coastal perimeter of Australia (Hingston and Gailitis 1976). This discrepancy may be due to differences in the ratio of precipitation of rain and evapotranspiration. Yet, through epidemiological research comprehensive experience has been established on long-distance airborne transmission of infectious animal diseases in Denmark (Christensen et al. 1993, Christensen et al. 2005) and it is suggested that an airborne contribution to soil salinity in the North Sea region should not be ignored. Transport of salt in marine aerosols is also revealed by the salt deposited on windows and windscreens of cars in coastal areas subject to the mist of the North Sea. Transportation of salt by spray created by waves breaking on the coast was clearly demonstrated during the present study on the island of Sejeroe (Table 17.1), but this phenomenon is expected to be relevant for only a few hundred meters from the coast. Transportation of salt by aerosols developed at sea and selected for by their buoyant density and ability to float on ascending warm air, on the other hand, might take place over several hundred kilometers. Top-soil samples collected 3–10 km off the North Sea coast both in 2019 and in 2020 (Table 17.1) revealing higher EC values than baseline values found in Jutland in 2019 and 2020, respectively (Table 17.1), is suggested to reflect aerosol transmission of salt from the sea.

In both winters of 2018/19 and 2019/20 extensive flooding of arable land was seen in Denmark. In the winter of 2019/20, the effects were significantly counteracted by an extreme precipitation of rain. The impact of this precipitation of rain was seen in the variation of EC_w values of Ringkoebing Fjord. The regulation of the water level by a sluice towards the North Sea occasionally allows influx of seawater resulting in brackish water of the fjord. However, the extreme supply from terrestrial water sources in 2020 resulted in a drop in EC_w values in the fjord from 5.02 mS/cm in 2019 to 1.93 mS/cm in 2020 (Table 17.2). The balance between impacts of flood versus precipitation is also seen in the Varde River Valley in 2020. A significant drop in $EC_{1:1}$ values in soil samples over a few kilometers distance to the river outlet to Ho Bay is suggested to reflect the locations of the interface between effluent terrestrial water and influx of North Sea water during that particular flooding period. The location of this interface is expected to vary from year to year due to an annual balance between rain and flooding by the North Sea.

The EC values found during the present study do not indicate any presence of high soil salinity in Denmark – suggested to be due to a prevailing impact of high precipitation of rain as observed, in particular, during the winter of 2019/20. Due to the high precipitation of rain and the nature of soil in Denmark, elevated salinity in the root zone caused by any mechanism of supply of salt is expected to be of a temporary nature. Yet, increased salinity in soil compared to baseline values in Denmark was found. Coastal areas and some smaller islands have moderate salinity levels in sub-terrain soil layers, suggested to be due to seepage from the coast. This sub-terrain salinity could possibly affect growth of plants in very dry seasons due to upward capillary transportation from deeper layers of saline water. Also, farmers do report that the flooding of arable land is seen with increasing frequency in Denmark, in particular on the smaller islands. Experiences from such incidences – yet in an anecdotal form – suggest that the impact on performance of conventional agriculture is evident and only reverts to former performance over a period of several years (5–7 years have been mentioned by farmers). This may be due not only to an increased salinity in the root zone, but also due to waterlogging increasing the uptake of Na^+ and Cl^- and decreasing the uptake of K^+ into plant shoots (Barrett-Lennard 2003, Barrett-Lennard and Shabala 2013).

Thus, with extremes in meteorological conditions becoming more frequent as result of climate changes, our data do suggest that under the conditions of precipitation of rain significantly under average annual quantities, a root zone salinization of arable land might occur in some areas of Denmark due to flooding from the sea or marine aerosol transportation, and the impact on conventional agriculture might be significant in following years.

ACKNOWLEDGMENT

This study was co-funded by The European Union Development Fund of the North Sea Region and by the Danish Business Authority. Assistance in preparing the illustrations by Laila Dam and Margrethe Nielsen as well as editing the text by Stephen Valentine is appreciated.

REFERENCES

Barrett-Lennard, E.G. 2003. The interaction between waterlogging and salinity in higher plants: causes, consequences and implications. *Plant and Soil* 253: 35–54.

Barrett-Lennard, E.G. and Shabala, S.N. 2013. The waterlogging/salinity interaction in higher plants revisited – focusing on the hypoxia-induced disturbance to K^+ homeostasis. *Functional Plant Biology* 40: 872–882.

Christensen, L.S. 2019. Some structural aspects of food production, food retail markets and procurement in Denmark – implications for national strategies of the REFRAME approach. https://northsearegion.eu/media/8786/structural-aspects-of-food-production-in-denmark-v3.pdf

Christensen, L.S., Mortensen, S., Boetner, A. et al. 1993. Further evidence of long distance airborne transmission of Aujeszky's disease (pseudorabies) virus. *The Veterinary Record* 132: 317–321.

Christensen, L.S., Normann, P., Thykier-Nielsen. S. et al. 2005. Analysis of the epidemiological dynamics during the 1982-1983 epidemic of foot-and-mouth disease (FMD) in Denmark based on molecular high-resolution strain identification. *Journal of General Virology* 86: 2577–2584.

Daliakopoulos, I.N., Tsanis, I.K., Koutroulis, A. et al. 2016. The threat of soil salinity: a European scale review. *Science of the Total Environment* 573: 727–739.

Danish Meteorological Institute 2020. Archives of meteorological data. https://www.dmi.dk/vejrarkiv/

De Waegemaeker, J. 2019. SalFar framework on salinization processes. A comparison of salinization processes across the North Sea Region. A report by ILVO for the Interreg VB North Sea Region project Saline Farming (SalFar). https://www.researchgate.net/publication/333356242_SalFar_framework_on_salinization_processes_A_comparison_of_salinization_processes_across_the_North_Sea_Region

De Vos, A. 2018. Training video: How to take a soil sample and measure soil salinity. https://northsearegion.eu/salfar/news/taking-soil-samples-and-measuring-salinity/

Hingston, F.J. and Gailitis, V. 1976. The geographic variation of salt precipitated over Western Australia. *Australian Journal of Soil Research* 14(3): 319–335.

Jørgensen, N.O. 2002. Origin of shallow saline groundwater on the Island of Læsø, Denmark. *Chemical Geology* 184. 359–370. 10.1016/S0009-2541(01)00392-8.

Manca, F., Capelli, G. and Tuccemiei. P. 2015. Sea salt aerosol groundwater salinization in the Litorale Romano Natural Reserve (Rome, Central Italy). *Environmental Earth Science* 73: 4179–4190.

Shahid, S.A., Zaman, M. and Heng, L. 2018. Soil Salinity: Historical Perspectives and a World overview of the Problem. In M. Zaman, S.A. Shahid and L. Heng (Eds.) *Guideline for Salinity Assessment, Mitigation and Adaptation Using Nuclear and Related Techniques*. Springer, Cham. doi: https://doi.org/10.1007/978-3-319-96190-3_2.

18 Climate-Resilient Agricultural Practices in the Saline-Prone Areas of Bangladesh

Muhammad Abdur Rahaman, Md. Sahadat Hossain, and Md. Iqbal Hossain

CONTENTS

18.1 INTRODUCTION

Saline water intrusion is a common problem in the coastal areas of Bangladesh. Climate change-induced hazards including sea level rise, cyclones, storm surges and tidal inundation are contributing to this problem causing salinity ingression into water and land (Baten et al. 2015). The southwestern coastal region of Bangladesh is a food deficit area where net food production and the diversity of food production have declined significantly over recent decades due to the salinity problem. Regular devastating cyclonic storm surges, sea level rise and tidal inundation have

DOI: 10.1201/9781003112327-18

changed the level of salinity and these have increased the risks associated with normal crop production. The impacts of climate change on coastal regions include inundation from sea level rise, damage from storm surges and loss of water bodies and increased salinity of land from saltwater ingression. Including the coastal region of Bangladesh, worldwide about 600 million people currently inhabit low-elevation coastal zones that will be affected by progressive salinization (Wheeler 2011). One study predicts that the sea level may rise 1 m or more in the 21st century, which would increase the vulnerability of about 1 billion people by 2050 (Brecht et al. 2012, Dasgupta et al. 2015, Hansen and Sato 2012, Veermer and Rahmstorf 2009, Pfeffer et al. 2008). Normal agricultural land use practices are becoming more restricted due to the increasing degree of salinity and expansion of affected areas (Karim 1990). As a consequence, crop yields, cropping intensity and production levels have decreased more than in any other part of the country (Rahman and Ahsan 2001). SRDI (2010) notes that the affected areas of Bangladesh are still increasing rapidly. In the last four decades, the total salinity-affected area has increased from 0.833 Mha to 1.056 Mha. The worst salinity conditions are reported to be in the Khulna, Bagerhat, Satkhira and Patuakhali districts (SRDI 2010). Our research focused on the Shyamnagar Upazila which is in Satkhira district, adjacent to the Bay of Bengal and the mangrove forest of the Sundarbans. Being close to the Bay of Bengal, the area is highly vulnerable to salinity intrusion into agricultural land from cyclones, storm surges and tidal surges, impacts from sea level rise, drainage congestion and flooding (Figure 18.1).

Many coastal districts, including Satkhira, are facing increased levels of salinity in agricultural fields (Islam et al. 2015). Data from the Soil Resource Development Institute (SRDI) shows that in the top soil (upper 15 cm) of cultivated areas in Shyamnagar Upazilla, 71% are affected by high-level salinity (EC in a saturated extract (EC_e) above 12 dS/m). About 500 ha of agricultural land become saline in each year. In 2000, the average EC_e was about 23.9 dS/m, but by 2009 this had increased to about 28.6 dS/m in Shyamnagar Upazila (SRDI 2010). In 2018, average EC_e values of 32.0 dS/m were reported in the Barokupat village of Shyamnagar (Rahaman et al. 2018).

The area of land suited to the cultivation of rice in Syamnagar has also changed with time. According to Kibria (2016), the area cultivated to rice was 21,350 ha in 1996; this had decreased to 14.925 ha by 2008 and had fallen to 5,020 ha by 2013; these declines were all due to salinity intrusion into paddy fields.

It is expected that sea levels along the coast of Bangladesh will rise by 14 cm, 32 cm and 88 cm by 2030, 2050 and 2100 respectively (Baten et al. 2015). Sea Level Rise (SLR) will push saline water further inland and affect not only rice production but also other agricultural practices in the future. One study has estimated that Bangladesh may lose 0.2 million tons of crops due to saline intrusion in a moderate climate scenario, but this decline might more than double with a severe climate scenario (Huq and Ayers 2008).

Our study was conducted to explore the salinity-tolerant developed indigenous technologies and planned interventions adopted by farmers. Though several studies had been conducted in terms of climate-resilient farming practices, no integrated study had been conducted to explore the indigenous and planned technologies in the study area.

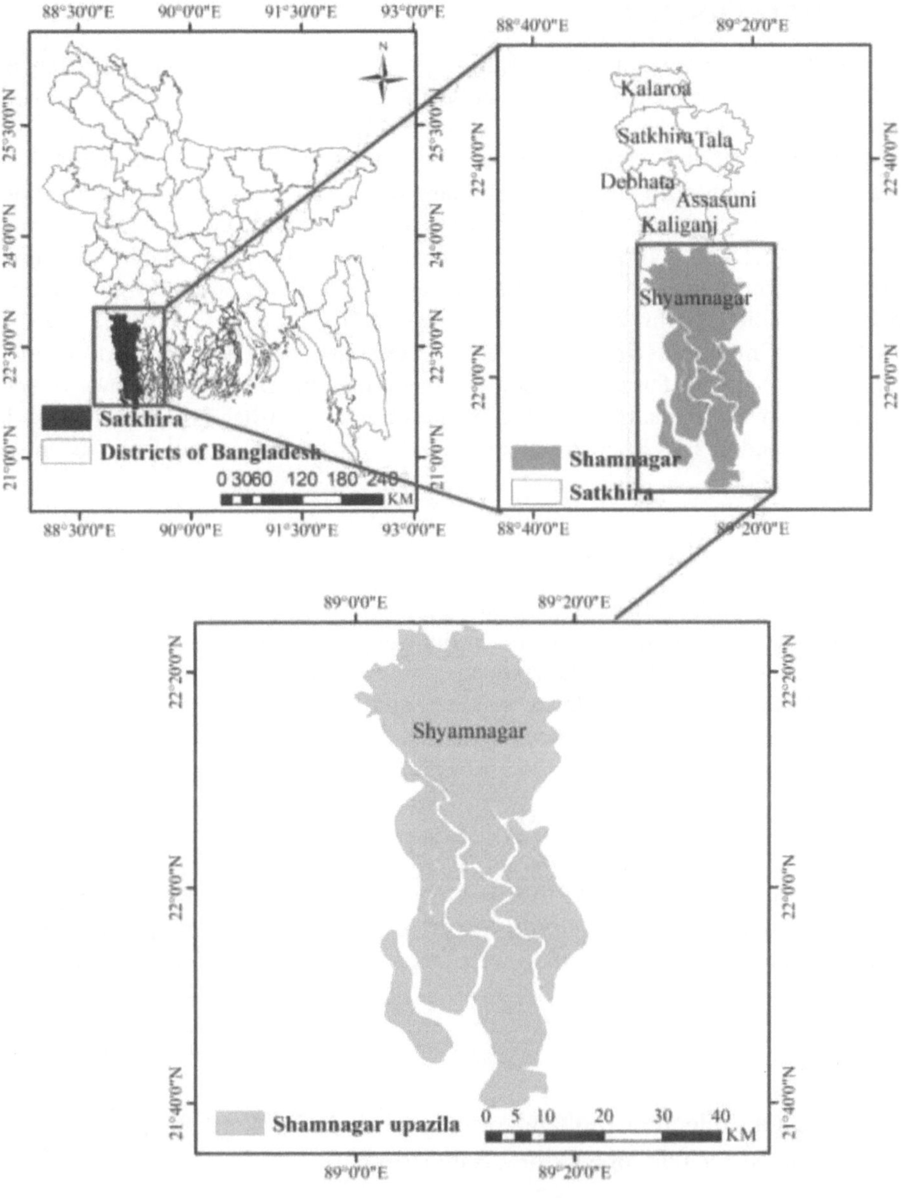

FIGURE 18.1 Map of the Shyamnagar Upazila, showing its location in Satkhira District within southern Bangladesh.

18.2 METHODOLOGY

The study aimed to explore the salinity-tolerant agricultural practices developed and adopted by farmers through the collection of field data and data from other secondary sources. The available scientific literatures related to the impact of salinity on agriculture were reviewed. The study incorporated both qualitative and quantitative information from secondary sources including the Department of Agricultural Extension (DAE), the Department of Fisheries (DoF), the Department of Livestock (DLS), the SRDI and through the reviewing of available literature related to salinity-tolerant agricultural practices in the study area. Primary information was collected through household surveys (HHS), Focus Group Discussions (FGD), Key Informants Interviews (KII), case studies and field observation and consultation with different stakeholders including the DAE, local government institutions and farmers, the Livestock Department, the Fisheries Department and NGOs involved in salinity-tolerant agricultural promotional activities in the study area (Table 18.1).

18.2.1 Household Survey

The HSS was conducted among climate-vulnerable people including smallholders, women, fishermen, livestock farmers. In the study, 60 respondents were male and 40 respondents were female. The respondents were selected randomly using multi-criteria like land size, household income level and whether they conducted livestock rearing, were fisher folk, or practiced homestead farming.

18.2.2 Focus Group Discussion

FGDs were conducted among different farming communities including crop farmers, paddy farmers, fisher folk, livestock farmers and women farmers to explore the salinity-tolerant practices in the study area. In each focus group, 10–12 participants from the respective group were selected based on homogenous farming practice.

TABLE 18.1
Primary Data Collection Tool and Sample Size for the Target People in Shyamnagar Upazila of Satkhira, Bangladesh

Data Collection Tool	Target People	Sample Size
Households survey	Climate-vulnerable people	100
Focus group discussion	Climate-vulnerable people	10
Key informant interviews	Department of Agricultural Extension, local government institutions, Department of Livestock, Department of Fisheries, NGOs	8
Case studies	Best practice documentation on salinity-tolerant interventions related to agriculture, livestock, fisheries	3

18.2.3 Key Informant Interviews

Key informant interviews were conducted at Upazila level with DLS, DAE, LGI, DoF and NGOs who are involved with agriculture in the study area.

18.3 RESULTS AND DISCUSSION

18.3.1 Household Types and Respondents

One hundred respondents (including women) who are involved with farming (crop, fish, livestock, poultry and homestead farming) were interviewed as part of this study. Among these respondents, 69% were males and 31% were females. Within households, 63% were headed by males and rest of the households were headed by females. Regarding religion, 83% were Muslim, 11% were Hindu and 6% were Christian.

18.3.2 Family Size

The average household size in the surveyed population was approximately five members. A majority of households (74%) had medium size families with four–six members while 11% had small families (one–three members) and another 15% had comparatively large families comprising seven or more members.

18.3.3 Education Levels

Of the households surveyed, 97% had someone who had obtained some level of education. Among the educated members, 17% had achieved primary education, 51% reached secondary education and 29% had obtained post-secondary education.

18.3.4 Community Perception of Salinity

Members of the community were invited to reflect on whether their land had no salinity or low, moderate or high salinity in 1998, 2008 and 2018. Most of the respondents of the study claimed that salinity intrusion had increased over the last decade (Figure 18.2). In 1998, 13% of respondents observed no salinity in the study area but over course of time salinity had increased, and in 2008 it stood at 9% and in 2018 it stood at only 5%. In 2018, 33% respondents claimed that they were affected by high salinity but 83% of respondents argued that rice production was restricted in the paddy field due to salinity.

18.3.5 Sources of Livelihoods

The sources of livelihoods of the respondents are summarized in Figure 18.3. Due to the loss in agricultural production, most of the respondents depended on fishing and extracting resources from the forest for their livelihoods; these included the collection of honey, firewood, fish, goalpata and crabs. Some of the respondents depended on small farming and fish farming as well as shrimp farming on their land. About

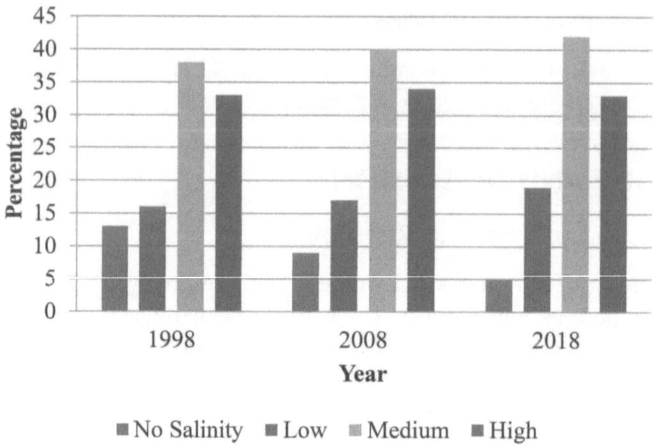

FIGURE 18.2 Perception of changes in salinity trends in the study area. Salinity classes were: no salinity (EC_e values below 2 dS/m), low salinity-affected soil (EC_e values 2–8 dS/m), medium salinity-affected soil (EC_e values 8.1–16 dS/m) and high salinity-affected soil (EC_e values above 16 dS/m).

2% of respondents earned wages as day laborers. Small trading, rickshaw pulling, day labor, small cottage industries and boating are wage earning sources of income of the respondents. Most (65%) households depended on natural resources for their livelihoods. A small proportion of households (5.5%) produced small livestock and poultry (like goats, duck and chicken) and livestock products. 84% of them produced fuel wood for household consumption. They also caught fish (57%) as well as collecting honey (30.1% of households).

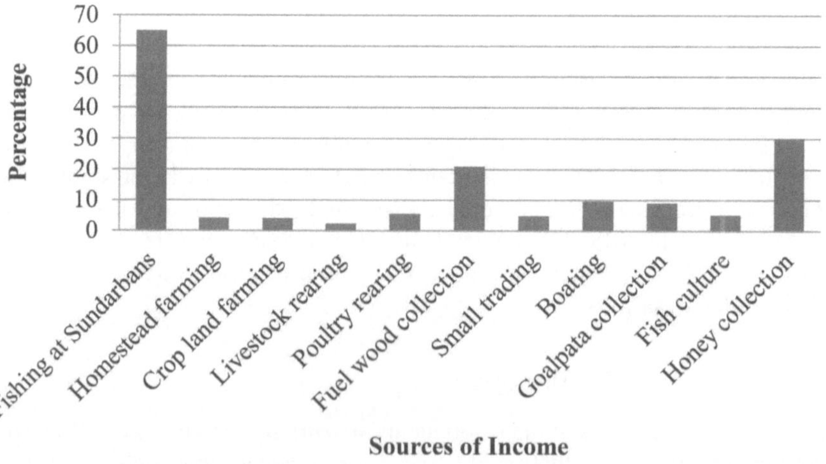

FIGURE 18.3 Source of livelihoods of the respondents in 2018. The number of households surveyed was 100, but many households had more than one source of income so the total of all sources exceeds 100.

18.3.6 Food Security

The monthly sources of food for the household were investigated and classified as to whether they came from their own farm, from off-farm sources or from the market for each month (on an average year). The survey found that only 2.1% of households consumed their own farm products for about three months a year, an average of 16% of households consumed food from off-farm sources throughout the year, and the rest of households depended on the market for their food products. Overall, only 31% of households were able to achieve food security. Respondents were also asked during which months of the year they struggled to have enough food to feed their household, from any source. Figures 18.4 and 18.5 indicate the sources of food and the months when the majority of households suffered from food insecurity in the study area.

Figure 18.4 shows the percentage of households who consume food from their own sources in a year. It is clear that the majority of households in Shyamnagar take food from market sources for six–eight months and they have to fully depend on the market during August–November. During this time, many of them also depend on off-farm sources such as relatives, friends and public food. The survey data also show that many households suffer from food shortages during July–November in this locality. Figure 18.5 shows a trend of food shortage (unable to meet the daily

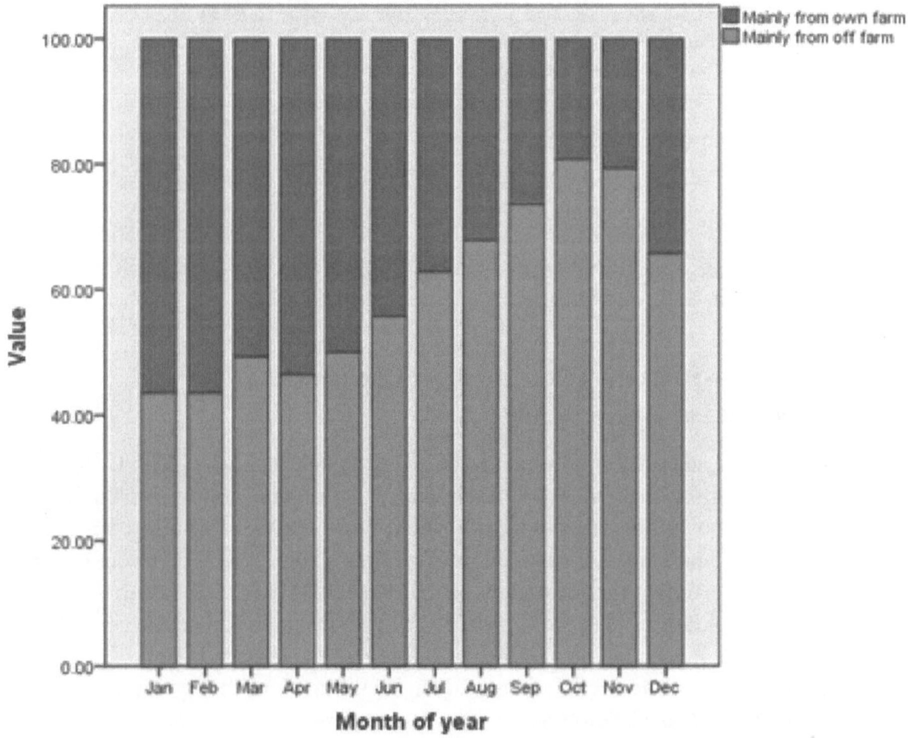

FIGURE 18.4 Effect of month of year on the main source of food for the household.

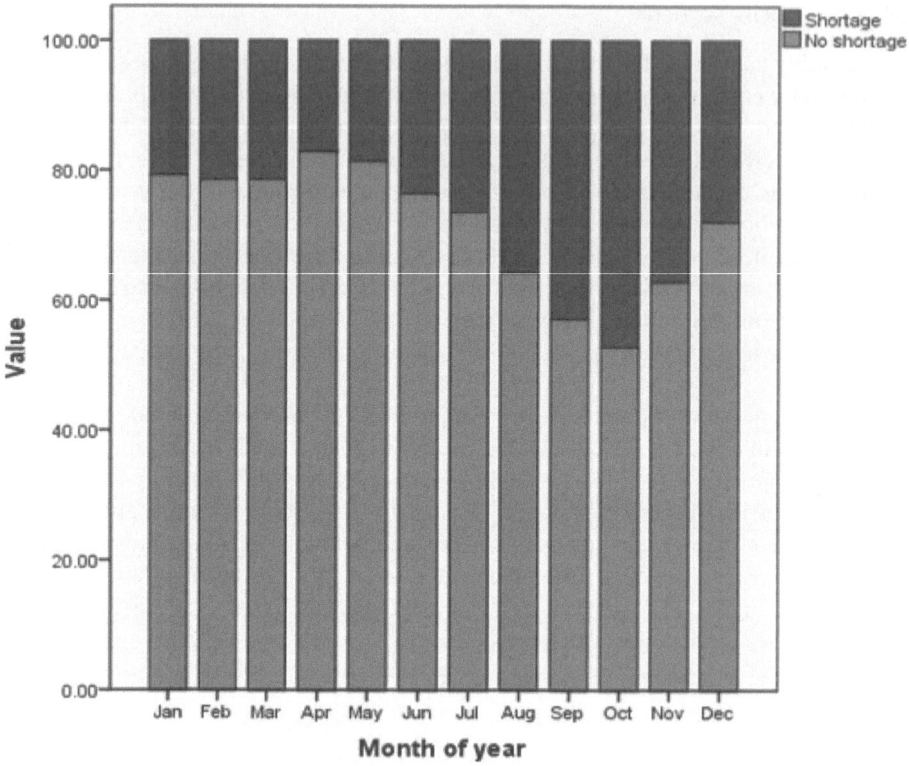

FIGURE 18.5 Food shortage months.

needs) in the study villages in Satkhira. It is to be noted that all households have some food shortage at all months of the year, but the shortages are most severe in August–December, and food shortage was greatest in October.

18.3.7 Salinity-Tolerant Livelihoods and Agricultural Practice in Shyamnagar

Many agricultural adaptation options are being practiced in Bangladesh to adapt to climate induced agricultural disasters like salinity, flood, waterlogging, drought, etc. Some of these innovations are devised by local communities through their indigenous knowledge, and some are planned interventions promoted by different government and non-government organizations (Rahaman et al. 2018). According to Nelson et al. (2007) and Alam et al. (2013), floating bed farming in the south-central and southern areas, plant bed raising and dyke cropping on the shrimp gher (water bodies which are artificially generated through raising dykes around the lowland) is an old but effective practice used nowadays.

The following are the key resilient agricultural and livelihood practices which were found in the study area to cope with salinity intrusion and ensure resilient

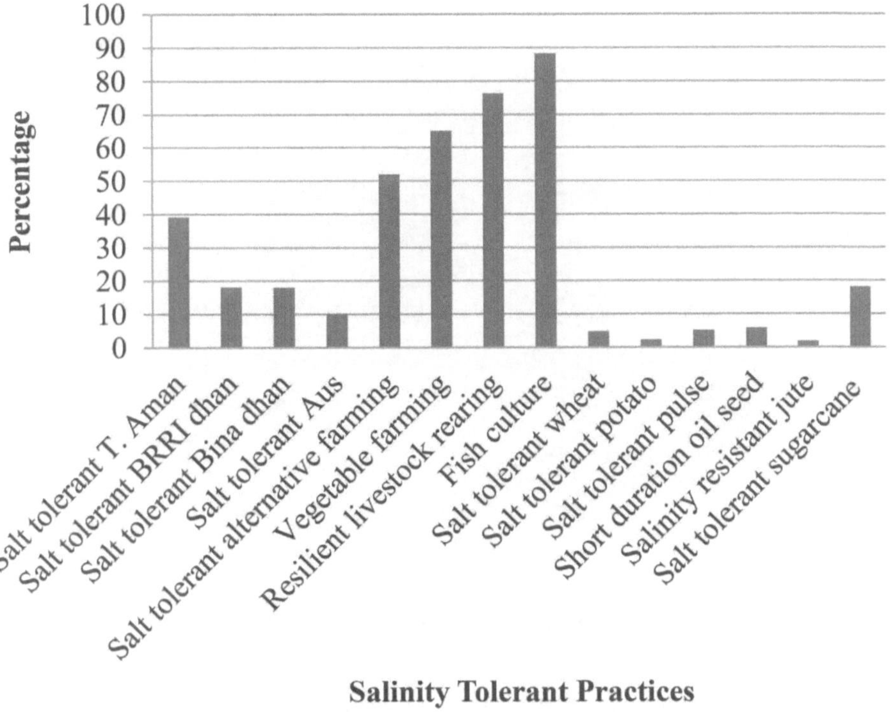

FIGURE 18.6 Salinity-tolerant practices used at Shyamnagar Upazila. (Source: Rahaman et al. 2018 and Field Study, 2019.)

livelihoods. The research indicates (Figure 18.6) that 31.3% of the farmers of the Shyamnagar Upazila are presently interested to adopt alternative land use practices like shrimp farming instead of crop production. Though the use of land for crop farming is decreasing gradually, there are some alternative options for agriculture emerging. The major salinity-tolerant practices of the study area which were identified by the different respondents (including farmers, agricultural professionals, government and non-government organizations who are involved with resilient livelihoods) include interventions like: (a) the introduction of salinity-tolerant rice varieties (such as T. Aman: BR-22 and BR-23; Bina shail; BRRI dhan 33, 40, 41, 46, 49, 53, 54, 56, 57, 62 and 65; Bina dhan 7, 8, 10 and 16), (b) cage fishing, (c) mele (reed) cultivation, (d) floating dhap cultivation (the practice of growing vegetable seedlings on water beds), (e) shifting the planting time of crops, (f) growing short duration rice varieties, (g) integrated farming, (h) crab farming, (i) semi-scavenger housing for goat, duck and hen rearing, (j) net fishing, (k) dyke farming, (l) the growth of salt-tolerant wheat (like Bijoy, BARI Gom-25, BAU-1059), (m) the growth of salt-tolerant potato (e.g. BARI Alo-22, CIP Clone 88, 163), (n) the growth of salt-tolerant sweet potato (e.g. BARI Mishti Alo-8,9), (o) the growth of salt-tolerant pulses (e.g. BARI Mug-2, 3, 4, 5, 6, BM-01, BM-08; BARI Falon-1, BARI Sola-9), (p) the growth of short duration and salt-tolerant oilseeds (e.g. BARI Sharisha-14,15; BARI Chinabadam-9, BINA

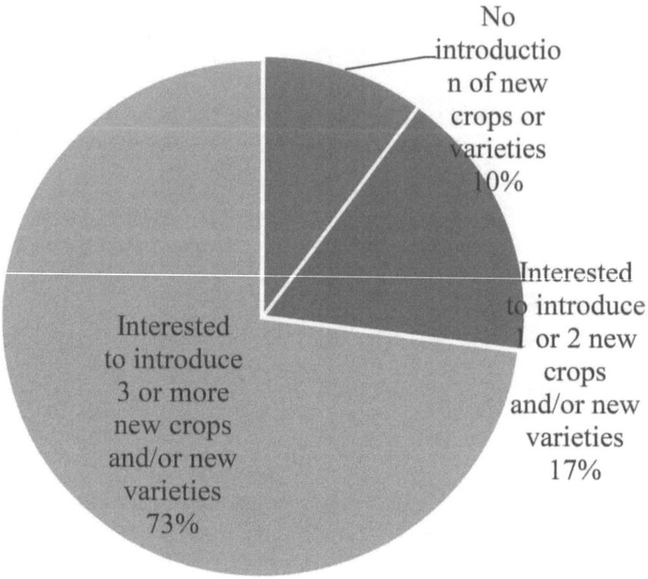

FIGURE 18.7 Development and adoption of saline tolerant crops/varieties.

China badam-1, BINA China badam-2, BARI Soyabean-6 BARI Til-2, 3, 4), (q) the growth of salt-resistant jute varieties (e.g. Bangladesh Jute Research Institute (BJRI): (i) HC-2, (ii) HC 95, (iii) CVL 1); and (iv) the growth of salt-tolerant sugarcane varieties from the Bangladesh Sugarcane Research Institute (BSRI) (eg. ISWARDI-40). Figure 18.6 shows that 39.2% of farmers use salt-tolerant T. Aman rice varieties BR-22, BR 23, BR 22 and Bina shail and 18.1% of farmers use salt-tolerant BRRI dhan (Rahaman et al. 2018 & Field Study 2019).

18.3.8 Peoples' Interest in Adopting Saline Tolerant Agricultural Practices

Due to food insecurity, most of the respondents have developed and adopted saline tolerant agricultural practices in the study area. 74% of respondents were interested to incorporate three or more new crops or varieties into their farming systems if they could get seed and capacity building support, while 17% were interested to introduce one or two new crops or varieties and 9% of households had no interest in any new crops or varieties because they do not own their own land for cultivation (Figure 18.7).

18.4 CONCLUSION

Our study has provided a unique portrait of a community at major risk of salinity from climate change. Most of the people of Shyamnagar are severely affected by salinity intrusion onto their land. They are trying to live with salinity in different ways.

Our study found that 39% farmers have developed saline tolerant technologies for home gardening, livestock rearing, fish farming, crab farming by their own knowledge, and 61% of the farmers have adopted saline tolerant verities for crop farming in their crop provided by government and non-government organizations. The government is mainly providing different varieties of rice, wheat and vegetables, but each year, people have a decreased capacity to adapt to the progressive spread of salinity. It is therefore time to think about long-term adaptation technologies in the salinity-affected areas and areas that could potentially be affected. Along with the innovation of agricultural technology and saline tolerant varieties, agricultural land needs to be protected from salinity intrusion and context-specific practices, including saline tolerant rice cultivation (seeds and methods), must be researched and implemented. In this work, it will be important to concentrate on the problems to coastal communities of saline intrusion caused by extreme events (e.g. cyclones Sidr and Aila).

REFERENCES

Alam, M., R. Ahammad, P. Nandy, and S. Rahman. 2013. "Coastal livelihood adaptation in changing climate: Bangladesh experience of NAPA priority project implementation." In *Climate change adaptation actions in Bangladesh*, 253–276. Springer, Switzerland.

Baten, M. A., L. Seal, and K. S. Lisa. 2015. "Salinity intrusion in interior coast of Bangladesh: challenges to agriculture in south-central coastal zone." *American Journal of Climate Change* no. 4 (03):248.

Brecht, H., S. Dasgupta, B. Laplante, S. Murray, and D. Wheeler. 2012. "Sea-level rise and storm surges: High stakes for a small number of developing countries." *The Journal of Environment & Development* no. 21 (1):120–138.

Dasgupta, S., M. M. Hossain, M. Huq, and D. Wheeler. 2015. "Climate change and soil salinity: The case of coastal Bangladesh." *Ambio* no. 44 (8):815–826.

Hansen, J. E., and M. Sato. 2012. "Paleoclimate implications for human-made climate change." In *Climate change*, 21–47. Springer, Switzerland.

Huq, S., and J. Ayers. 2008. *Climate change impacts and responses in Bangladesh.* Policy Department Economy and Science, European Parliament, Brussels, Belgium.

Islam, M. A., P. K. Shitangsu, and M. Z. Hassan. 2015. "Agricultural vulnerability in Bangladesh to climate change induced sea level rise and options for adaptation: a study of a coastal Upazila." *Journal of Agriculture and Environment for International Development (JAEID)* no. 109 (1):19–39.

Karim, Z. 1990. *Salinity problems and crop intensification in the coastal regions of Bangladesh. Soil Publication No. 33.* Dhaka, Bangladesh.

Kibria, G. 2016. "Storm surge propagation and crop damage assessment in a coastal polder of Bangladesh. Post graduate Dissertation, Bangladesh University of Engineering Technology (BUET), Dhaka, Banaagladesh"

Nelson, D. R., W. N. Adger, and K. Brown. 2007. "Adaptation to environmental change: contributions of a resilience framework." *Annual Review of Environment and Resources*.32:395–419.

Pfeffer, W. T., J. T. Harper, and S. O'Neel. 2008. "Kinematic constraints on glacier contributions to 21st-century sea-level rise." *Science* no. 321 (5894):1340–1343.

Rahaman, M. A., M. Rahman, and S. Hossain. 2018. "Climate-resilient agricultural practices in different agro-ecological zones of Bangladesh." In *Handbook of climate change resilience*, edited by Walter Leal Filho. Springer Nature, Switzerland.

Rahman, M., and M. Ahsan. 2001. "Salinity constraints and agricultural productivity in coastal saline area of Bangladesh." *Soil Resources in Bangladesh: Assessment and Utilization* no. 1:1–14.

SRDI. 2010. Saline Soils of Bangladesh. edited by SRMAF Project Soil Resource Development Institute. Dhaka: Ministry of Agriculture, Mrittika Bhaban, Krishikhamar Sarak Farmgate, Dhaka-1215.

Veermer, M., and S. Rahmstorf. 2009. "Global sea-level linked to global temperatura." *Proceeding National Academy of Science United States* no. 106 (51):21521–21532.

Wheeler, D. 2011. "Quantifying vulnerability to climate change: implications for adaptation assistance." *Center for Global Development Working Paper* (240).

19 Salinity Dynamics and Water Availability in Water Bodies over a Dry Season in the Ganges Delta

Implications for Cropping Systems Intensification

Afrin Jahan Mila, Richard W. Bell,
Edward G. Barrett-Lennard, and Enamul Kabir

CONTENTS

DOI: 10.1201/9781003112327-19

19.1 INTRODUCTION

The Ganges Delta is the world's largest delta containing the outlets of three major rivers, the Ganges, Brahmaputra and Meghna (Chowdhury 2010); it covers ~200,000 km^2 in Bangladesh and the Indian state of West Bengal (Alam et al. 2003; Mukherjee et al. 2007). A major tributary of the Ganges River is the Gorai-Madhumati River, which is the main source of fresh water in south-western Bangladesh (Mondal 2016). Since 1975, freshwater flow during the winter season (November to March) has decreased in the Gorai-Madhumati River due to the construction of the Farakka Barrage in India on the Ganges river (Mirza 1998), which is one of the reasons for elevated river salinity and increased salinity in the surrounding land. In addition, tidal salt water intrusion into rivers and aquifers is increasing salinity thereby making freshwater unavailable in the hydrologically connected canals and ponds (Mahmuduzzaman et al. 2014). In the south-central part of the Ganges Delta, the Meghna River remains fresh throughout the year. This study focuses on the water resources on agricultural land near Dacope, Khulna in south-western Bangladesh.

The south-western region, which comprises ~15% of the total area of Bangladesh, has a limited area of *Rabi* season cropping due to the perception that severe soil and water salinity would otherwise result in poor yields (BBS 2018). However, salinity (expressed as the electrical conductivity of the water – EC_w) changes seasonally, being highest in the late *Rabi* season (*Rabi* season covers the period from November to March; Payo et al. 2017) when the weather is dry and hot (mean maximum and minimum temperatures increase by 7.1°C and 5.8°C from December to March, Figure 19.1B), and early in the *Kharif*-1 season (March–June, Payo et al. 2017) during the hot and humid summer. In the main season for wetland rice, the *Kharif*-2 or monsoon season (June–November, Payo et al. 2017), the EC_w of all water sources, including surface water and groundwater, decreases because of seasonal rainfall. About 78% of the total yearly rainfall (~1,730 mm per annum) falls during the *Kharif*-2 season. Smaller but significant amounts of rainfall can fall between December and March (Yu et al. 2019). Water from canals, ponds or groundwater is used for irrigation of dry season (*Rabi* season) crops (e.g. *Boro* rice, sunflower, wheat, maize, mustard, potato, watermelon and spinach) and *Kharif*-1 crops (e.g. mung bean, sesame and *Aus* rice). A review of the limitations and prospects associated with the cropping system intensification in this region (Bell et al. 2019) suggested that changing the cropping season by early establishment of *Rabi* crops would reduce crop dependence on water for irrigation especially late in the *Rabi* season

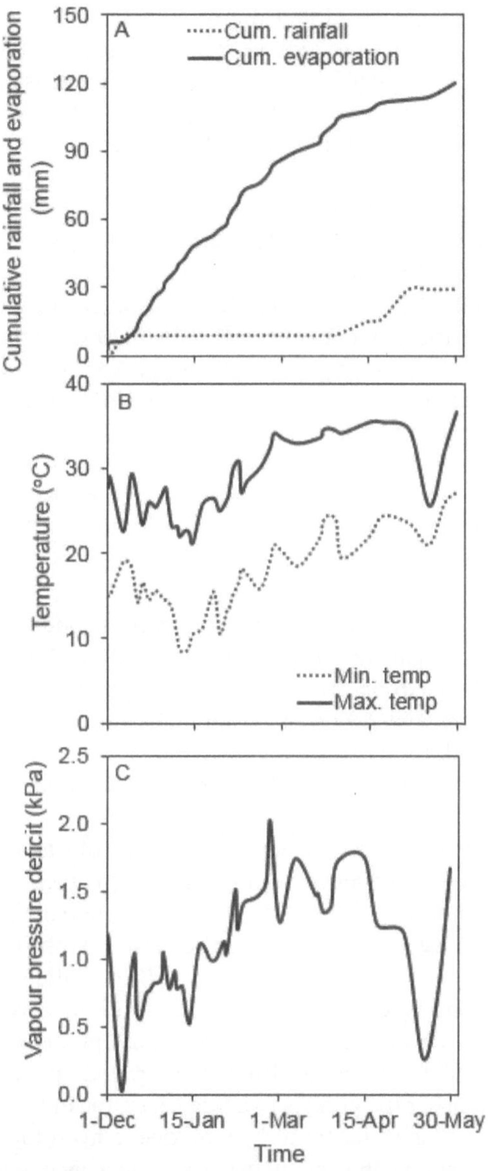

FIGURE 19.1 Weather variables as a function of time during the study period: (A) cumulative rainfall and evaporation, (B) maximum and minimum temperature and (C) vapour pressure deficit (VPD). Values are from the Khulna meteorological station 17.3 km north of Dacope.

when water is in short supply and its salinity is highest. Early crop establishment after the monsoon season needs to remove standing water but retain soil water storage after early harvested *Kharif*-2 rice so that crops can utilise fresh soil water for critical growth stages and minimise exposure to soil and water salinity until later growth stages (Bell et al. 2019; Kabir et al. 2019). Therefore, water availability and its quality are important factors enabling crops to reach their potential yield during the dry season in these saline areas.

Previous reports (Akanda et al. 2017; Hossain et al. 2019; Kabir et al. 2019; Mondal et al. 2006; Murad et al. 2018) on water resources in south-western Bangladesh have generally focused on the EC_w rather than the availability of water as constraints to cropping. For example, the mean monthly EC_w (1997–2004) of the Kazibachha River ranged from ~1.5 dS m^{-1} in January to ~20 dS m^{-1} in April or May at Kismat Fultola village, Khulna District (Mondal et al. 2006). The EC_w of canal water increased from 4.3 dS m^{-1} on 5 December 2016 to 9.8 dS m^{-1} on 18 April 2017 during the maize growing period at Benarpota Upazila, Satkhira, Khulna (Murad et al. 2018). And, of particular relevance to Dacope, the case study area of the present investigation, the EC_w of river water increased from 12 dS m^{-1} on 10 January 2017 to 13.9 dS m^{-1} on 30 April 2017, but the EC_w of the canal and pond water varied between 0.7–0.9 and 0.3–0.6 dS m^{-1}, respectively, between 30 December 2016 and 30 April 2017 (Akanda et al. 2017). In a sowing date study on wheat at Dacope, bunded canal and pond water were used for irrigation and the salinity remained below 3 and 2 dS m^{-1}, respectively (Kabir et al. 2019). Groundwater, the remaining water source, was reported to vary in EC_w between 1.5 and 2.3 dS m^{-1} from 5 December 2016 to 18 April 2017 at Benarpota Upazila, Satkhira District, Khulna, and between 1.1 and 1.3 dS m^{-1} from 30 December 2016 to 30 April 2017 at Dacope, Khulna (Akanda et al. 2017; Murad et al. 2018).

We hypothesise that EC_w is regulated by the volume of water stored in controlled canals or ponds, due to the influence of three factors: weather variables (cumulative evaporation, maximum temperature and vapour pressure deficit), hydrological connectivity to a river and elevation above mean sea level. We further hypothesise that lower elevation ponds or canals have a higher risk of shallow saline groundwater intrusion since the pond water level is more likely to be lower than the phreatic groundwater level. In addition, these ponds are more prone to receiving overland flow due to rainfall, which washes salt from the surrounding land surfaces. Another hypothesis is that the closer the pond is to a tidal river, the more susceptible its water will be to seepage of saline water from the river due to hydrological connectivity.

In this study, we explore the hypothesis that water availability rather than its EC_w is the main constraint to cropping systems intensification in this environment because changes in the planting date of *Rabi* season crops and crop choice, together with best-practice agronomy and irrigation, can be designed to manage, minimise or avoid the damaging effect of elevated EC_w on crop yields (Bell et al. 2019). Also, the relationships between EC_w and water availability with elevation above mean sea level and weather variables are explored. We examine the possibility of the conjunctive use of non-saline and slightly to moderately saline irrigation water from bunded canals and ponds for intensifying dry season cropping. Finally, we synthesise knowledge about the trends of EC_w and water availability in the *Rabi* and *Kharif*-1 season to propose novel cropping patterns to expand the area of *Rabi* season crop production. Our study focused on the salinity dynamics and water availability in water bodies in the *Rabi* and *Kharif*-1 season at a case study area at Dacope, Khulna in southwest Bangladesh.

19.2 MATERIALS AND METHODS

19.2.1 Study Location and Weather Condition

Our study was conducted on water bodies (canal, pond, river and groundwater) in Dacope, Khulna (Figure 19.2) to characterise salinity dynamics and water availability during the *Rabi* and *Kharif*-1 seasons from December 2017 to May 2018. Three canals (locations 1, 2 and 3) were selected (Figure 19.2B), and each canal was divided into 3–5 sections. Each location also included three ponds and tube wells, with positions illustrated in Figure 19.2B. In the case of the Sundarban River, three measuring points were selected shown in Figure 19.2B. The sampling was conducted on one (March–May), two (January–February) or three (December) days per week.

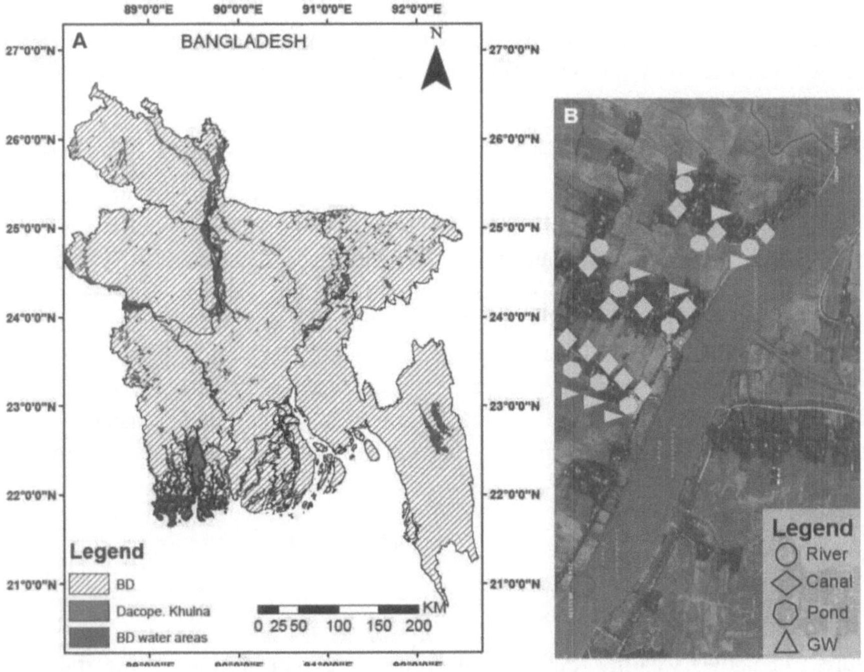

FIGURE 19.2 Location map of the study area and survey points at Dacope, Khulna, Bangladesh. (A). Location map of Bangladesh showing the large rivers (labelled BD water areas) and the position of Dacope, Khulna at the south-west of Bangladesh, (B). Survey points at Dacope on the river (circles), and three different locations for canal (diamonds), pond (heptagon) and groundwater (triangle) samples.

19.2.2 Water Salinity

The EC_w of all water bodies was recorded at high and low tides using a portable EC meter (Model: HI 8733).

19.2.3 Influence of Elevation and Weather Variables on Water Salinity and Water Availability

Data on the weather factors tested for their influence on EC_w (cumulative evaporation, maximum temperature and vapour pressure deficit) were collected from Khulna weather station while elevation above mean sea level was determined by mobile phone-operated elevation software (altimeter) at each survey point. Rainfall and evaporation were measured by 8-inch diameter ordinary rain gauge and USA class A Pan according to the World Meteorological Organization.

19.2.4 Calculation of Volume of Canal and Pond Water

The length and width of each canal were calculated in Google Earth. The length and width of ponds were determined with a measuring tape. Canals and ponds were considered to be rectangular in shape and in cross-section with depth. Water depth was measured one (March–May), two (January–February) or three (December) days per week, depending on the rate of change in EC_w. The volume of water at different times was calculated by multiplying the area of the water body by the water depth.

19.2.5 Cultivation Practice

Farmers of this region cultivate a range of dry season crops with both high- and low-water requirements including *Boro* rice (~1–2% of current cultivated land), sunflower, wheat, maize, potato, watermelon and bottle gourd. More than 90% of total cultivable land is kept fallow (personal communication). Among the crops, *Boro* rice needs ~870 mm of applied water (data from Jessore, Khulna) of good quality from crop establishment until maturity (Hossain et al. 2018), bottle gourd and watermelon need weekly irrigation, whereas sunflower, wheat, maize and potato need only two–four irrigations in total to reach their yield potential. The usual planting time (farmer's practice) for *Rabi* crops is from the last week of January to the first week of February.

19.3 RESULTS

19.3.1 Weather Conditions

Figure 19.1 shows weather variables at Khulna weather station during the study period (http://www.bmd.gov.bd). During the study period, the total rainfall was 29 mm, of which 20 mm fell from April to the first week of May (Figure 19.1A). The mean daily evaporation was 2.8 mm (calculated from Figure 19.1A). Temperature (maximum and minimum) and vapour pressure deficit increased between January and March (Figures 19.1B,C).

19.3.2 Variation of River, Canal, Pond, and Groundwater Salinity over Time

The EC_w of river, canal, pond and groundwater significantly increased during the progression of the *Rabi* season until the middle of the *Kharif*-1 season (Figure 19.3). The EC_w of the river and of canals 2 and 3 increased significantly ($P < 0.001$;

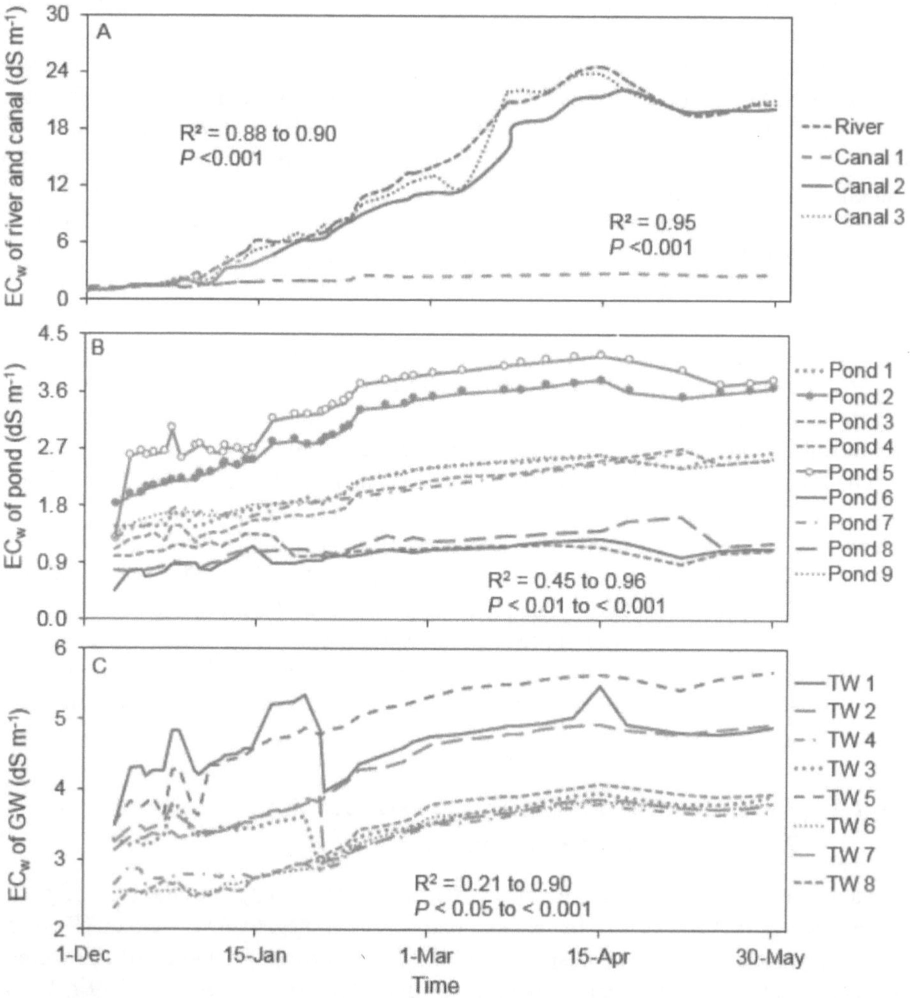

FIGURE 19.3 Scatter diagrams showing the relationship between EC_w of water bodies and time: (A) mean river and canal water, (B) pond water and (C) groundwater from tubewells (TW). For linear lines of best fit (not shown), the *P*- and R^2 values are indicated. In case of water quality, canal 1: slightly (green) to moderately saline (orange), canal 2 and 3 and river: slightly to severely saline (green, orange, red and pink), pond 6: non-saline to slightly saline (blue), pond 3 and 8: slightly saline (green), pond 1, 2, 4, 5, 7 and 9: slightly to moderately saline (orange), TW 2, 3, 4, 6, 7 and 8: moderately saline (orange) and TW 1 and 5: highly saline (red).

r^2 = 0.88 to 0.90) with time with the highest value recorded in mid-April (Figure 19.3A). Among the three canals, the canal at location 1 was not directly linked with the river due to the construction of a major bund; consequently, the EC_w was ~90% lower than that of the river during the peak time but still increased significantly (*P* <0.001; r^2 = 0.95) with time (Figure 19.3A). The EC_w of this canal increased from 2 dS m^{-1} in February to a maximum of 2.9 dS m^{-1} in mid-April. The EC_w values of canals 2 and 3 were positively (*P* <0.001) correlated with mean river EC_w with r^2 values of 0.98 and 0.99. The river EC_w increased from 2 dS m^{-1} on 26 December to a maximum of 24.6 dS m^{-1} in mid-April.

The mean pond EC_w at the three locations increased significantly (*P* <0.01–<0.001) with time, with r^2 values for linear correlations of 0.45–0.96 (Figure 19.3B). The EC_w of three ponds (3, 6 and 8) remained below 2 dS m^{-1} over the dry season, while for four ponds (1, 4, 7 and 9) the EC_w increased from 2 dS m^{-1} in February to a maximum of 2.6 dS m^{-1} between mid-April and late May and for two ponds (2 and 5) the EC_w reached a maximum of 4.2 dS m^{-1} in mid-April. In terms of water quality criteria[1], pond 6 was non-saline to slightly saline, and ponds 8 and 3 were slightly saline, and these would have been suitable for supplying water for human consumption as well as for irrigation[2]. Ponds 1, 4, 7 and 9 were slightly to moderately saline with maximum EC_w values below 3 dS m^{-1} and these could have been suitable as water sources for late irrigation. Ponds 2 and 5 had higher EC_w values but were still slightly to moderately saline and water could have been used for late irrigation in grain filling stages (i.e., wheat, sorghum, cowpea and maize)[3].

The EC_w of the groundwater at three locations increased significantly (*P* <0.05 to <0.001) with time, with r^2 values for linear correlations of 0.21 to 0.90 (Figure 19.3C). Among the tube wells (TW), three (TW 4, 6 and 8) had EC_w values that increased from 2.3 dS m^{-1} on 9 December to a maximum of 4.1 dS m^{-1} on mid-April, and five (TW 1, 2, 3, 5 and 7) had EC_w values that increased from 3.1 dS m^{-1} on 9 December to a maximum of 5.5–5.7 dS m^{-1} between mid-April and the end of May. In the case of water quality, TW 2, 3, 4, 6, 7 and 8 were moderately saline; this water could be used for later-stage irrigation (grain filling), whereas TW 1 and 5 were highly saline and this water was not suitable for irrigation.

19.3.3 Relationship between Elevation and Weather Parameters with Water Body Salinity

Elevation and weather criteria were tested for their relationships with the EC_w of pond water (Figures 19.4 and 19.5). The EC_w of pond water was negatively correlated (*P* <0.05 to <0.01) with elevation above mean sea level (MSL) (Figures 19.4A and 19.4B). However, for each elevation, the EC_w increased with time until April (Figure 19.4A and 19.4B) and decreased after April (not shown) due to rainfall (Figure 19.1A).

A strong correlation was found between cumulative evaporation and EC_w in the river, canals, ponds and groundwater; exponential lines of best fit were significant (*P* < 0.001 with r^2 values of 0.89–0.98; Figures 19.5A and 19.5D). Maximum temperature and vapour pressure deficit were of lower importance; exponential lines of best fit were significant (*P* < 0.001) but r^2 values were 0.48–0.65 (Figures 19.5B and 19.5E) and 0.57–0.76 (Figures 19.5C and 19.5F), respectively.

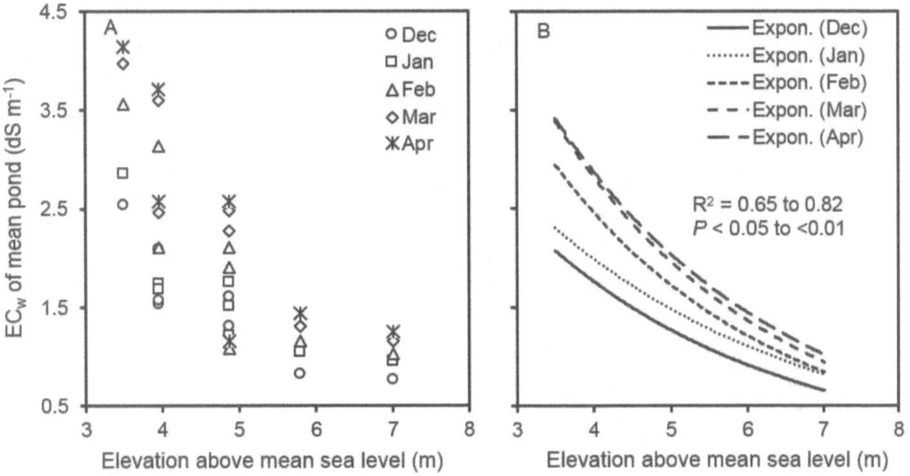

FIGURE 19.4 Relationships between EC_w of pond water and elevation above mean sea level for sampling times between December and April: (A) scatter points showing mean monthly EC_w values and elevation above mean sea level for months from December to April (B) and exponential lines of best fit to these points showing the *P*- and R^2 value ranges.

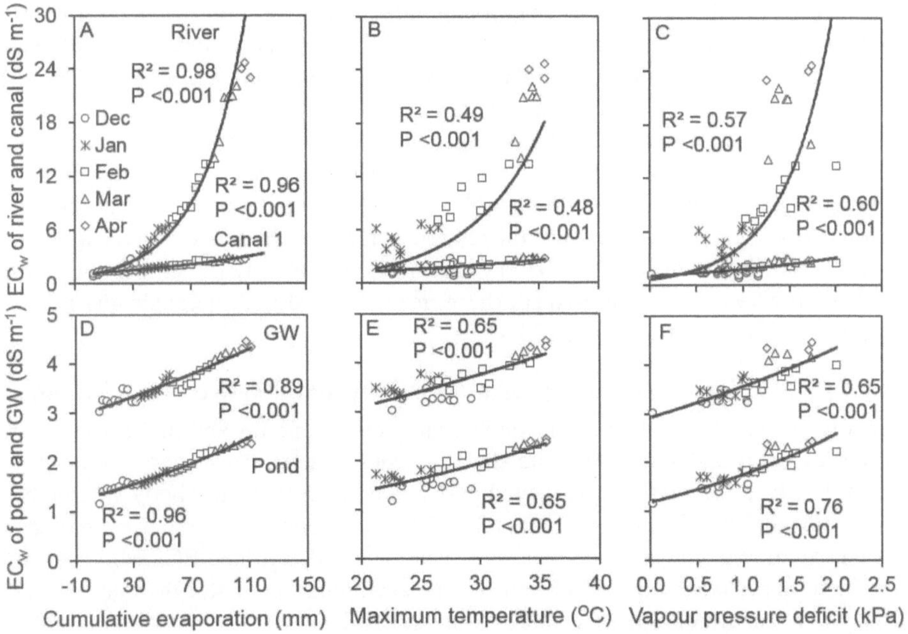

FIGURE 19.5 Relationships between mean river, canal, pond and groundwater EC_w and weather variables during the *Rabi* and early *Kharif*-1 season: (A, D) effect of cumulative evaporation, (B, E) effect of maximum temperature and (C, F) effect of vapour pressure deficit (VPD). Parts A, B and C show river and canal EC_w values. Parts D, E and F show pond and groundwater EC_w values. The points have been fitted to exponential lines of best fit with the *P*- and R^2 values indicated.

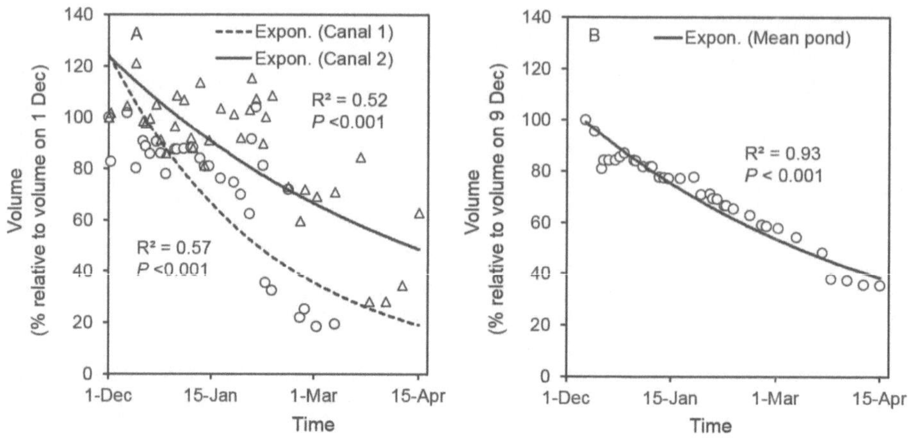

FIGURE 19.6 Relationships between relative water volume (V_r) in canals and ponds over time: (A) canal at location 1 and location 2 and (B) the mean for all ponds. The points were fitted to exponential lines with the *P*- and R^2 values indicated.

19.3.4 Variation of Water Availability in Canals and Ponds over Time

Ponds and canals differed by orders of magnitude in the maximum volumes of water that these contained. The canals at locations 1 and 2 contained 3,089 m^3 and 16,961 m^3 respectively, whereas the measured ponds contained a maximum of 244 to 1,078 m^3 of water. Because of this variation, we transformed the volume of water in canals and ponds to values relative to 'maximum water content' – defined as the water content of canals on 1 December and ponds on 9 December, respectively. Figure 19.6 shows change in relative water volume (V_r) in canals (A) and ponds (B) over the dry season period. For all water bodies, the relative volume of water decreased exponentially (P <0.001; r^2 values 0.52 to 0.93) with time between early December and mid-April (Figure 19.6). After mid-April, the water volume stored in canals and ponds increased due to rainfall (Figure 19.1A).

19.3.4.1 Relationship between Relative Salinity and Relative Water Volume

One way to test whether there was connectivity between ponds and the surrounding sources of water and salt (rivers and shallow groundwater) was to test whether the EC_w of the water increased as expected with a decline in the water body volume by evaporation and irrigation water extraction. That is, a doubling of EC_w for a halving of water volume would indicate that only evaporation determined EC_w while deviation from this relationship would indicate the accumulation of salt in water bodies from external sources, irrigation water extraction, or leakage of water from the water body. To determine this, we built scattergrams of the volume of water in canals and ponds relative to 'maximum water content' against the EC_w of water in canals and ponds relative to EC_w when canals and ponds were at maximum water content (Figure 19.7). These scattergrams included the theoretical line showing the expected increase in relative EC_w associated with a decline in volume relative to maximum. Values above this line (i.e. relative EC_w values greater than expected assuming that

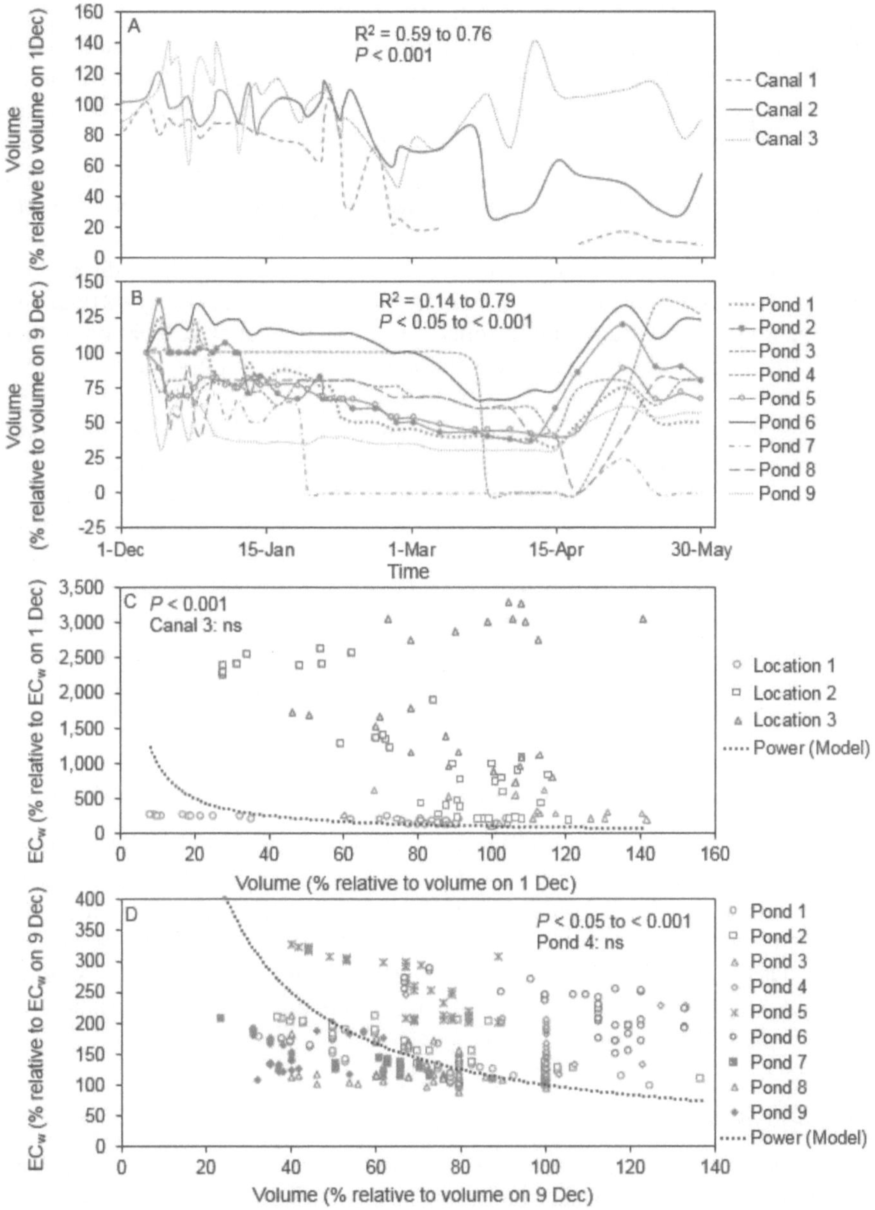

FIGURE 19.7 Relationships between water body (canal and pond) relative water volume (V_r) and relative EC_w during December to May: (A) change in V_r of canals over time, (B) change in V_r of ponds over time, (C) scattergram for canals showing the effect of V_r (as a % of volume on 1 December) on relative EC_w (% EC_w on 1 December) and (D) scattergram for ponds showing the effect of V_r (% volume on 9 December) on relative EC_w (as a % of EC_w on 9 December). In case of water quality, canal 1: slightly (green) to moderately saline (orange), canal 2 and 3: slightly to severely saline (green, orange, red, and pink), pond 6: non-saline to slightly saline (blue), pond 3 and 8: slightly saline (green) and pond 1, 2, 4, 5, 7 and 9: slightly to moderately saline (orange).

increases in EC_w were due to evaporation and irrigation water extraction) might suggest that salt was entering water bodies from the surrounding land, groundwater or rivers. It was found that the decline in a relative water volume (V_r) of canals and ponds was significantly correlated ($P < 0.05$ to < 0.001) with the increase in their relative salinity over time (Figures 19.7C and 19.7D). Changes in EC_w with V_r for the canal at location 1 (Figure 19.7C) was below the theoretical line suggesting that there was a little intrusion of salt through the bund protecting the canal from the river while irrigation water extraction decreased V_r. However, points for the canal at locations 2 and 3 (Figure 19.7C), were substantially above the hypothetical line indicating that factors in addition to evaporation, direct connectivity to tidal river and irrigation water extraction controlled EC_w. For the canal at location 3, there was a little change at all in relative volume but the EC_w values increased about 30-fold.

Among the ponds, ponds 4, 5 and 6 were above and ponds 1, 2 and 8 were partially above the hypothetical line, while ponds 3, 7 and 9 were below the hypothetical line (Figure 19.7D). The ponds that had substantial seepage of salt into the water were ponds 4, 5 and 6. The other ponds had little leakage salt or entry salt from outside the pond. A major change in salinity at 100% volume for pond 4, 80% volume for pond 8 and 113% volume for pond 6 might be explained by the runoff of rainfall as overland flow into the pond, or by seepage into the pond from hydrological connectivity to river or from shallow groundwater. However, the relative increase in salt into the leaky ponds was far less (2.5-fold) than into the uncontrolled canals connected to the river (about 30-fold) (Figures 19.7C and 19.7D).

19.3.5 Relationship between Weather Parameters and Water Availability

Figure 19.8 shows correlation between the relative water volumes of canals and ponds with the weather variables during the *Rabi* and up to the middle of the *Kharif-1* season. Cumulative evaporation, maximum temperature and vapour pressure deficit were negatively correlated ($P < 0.001$) with the relative water volume (V_r) of the canal with r^2 values of 0.75, 0.63 and 0.23, respectively (Figures 19.8A–19.8C). A similar relationship ($P < 0.001$) was found between the V_r of pond and weather variables with r^2 values of 0.71, 0.50 and 0.58 (Figures 19.8D–19.8F).

19.4 DISCUSSION

Previous work on water resources in saline coastal areas of the Ganges Delta has generally focused on EC_w in surface water and groundwater as the major constraint to the irrigation of dry season crops (Akanda et al. 2017; Hossain et al. 2019; Kabir et al. 2019; Mondal et al. 2006; Murad et al. 2018). However, this study has also examined the quantity of water in bunded canals and ponds because EC_w can be regulated by changes in water volume, hydrological connectivity to landscape and weather variables. This discussion is focused around three themes. We discuss the changes in EC_w and water availability in the bunded canal and ponds, the causes of change in EC_w and the availability of water in canals and ponds, and the use of these water resources for dry season cropping.

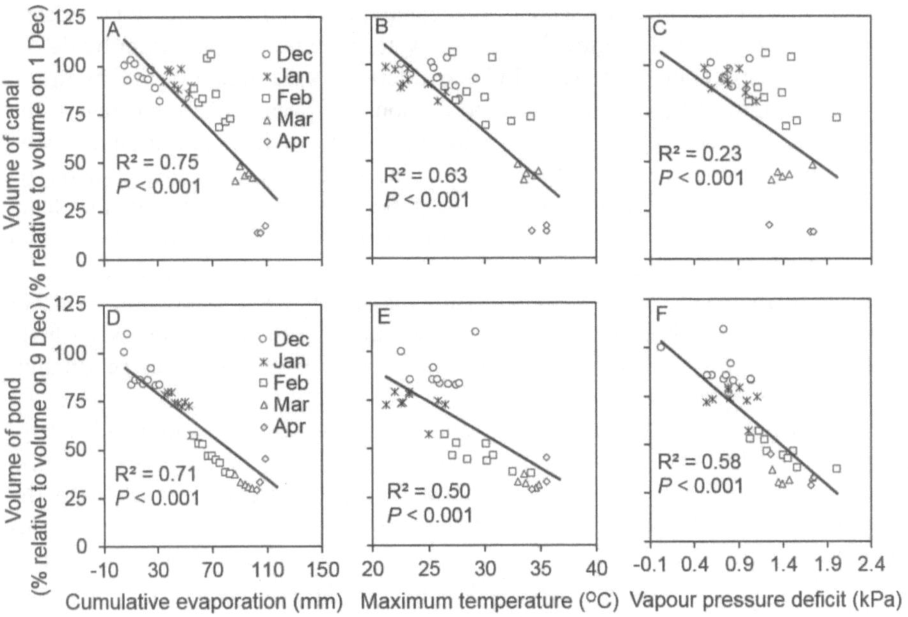

FIGURE 19.8 Relationships between relative water volume of canals and ponds (V_r) with the weather variables during the *Rabi* and up to the middle of the *Kharif*-1 season: (A, D) effect of cumulative evaporation, (B, E) effect of maximum temperature and (C, F) effect of vapour pressure deficit (VPD). Parts A–C are canal V_r values and parts D–F are pond V_r values. The points have been fitted to linear lines of best fit with the *P*- and R^2 values indicated.

19.4.1 Salinity and Water Availability Dynamics in the Bunded Canal and Ponds

In the present study, the EC_w of a controlled or bunded canal during the late *Rabi* and early *Kharif-1* (mid-April) seasons was ~90% lower than that of the uncontrolled canal directly linked with the tidal saline river (Figure 19.3A). Similar EC_w values in controlled canals (1.3 to 3.0 dS m^{-1} in the same study area and 1.1 to 2.3 dS m^{-1} in Amtali, Barisal) have been previously recorded during the dry season (Hossain et al. 2019; Kabir et al. 2019). The best time for placing the bund in canals was when the river EC_w was below 4 dS m^{-1} on 10–15 December (Hossain et al. 2019). For a protected canal, the EC_w increased 2.3 fold from the first week of December 2016 until mid of April 2017 in Benarpota, Satkhira, Khulna (Murad et al. 2018). By contrast, river EC_w increased 13 fold from January to April or May at Kismat Fultala village, Khulna (Mondal et al. 2006).

In the present study, the EC_w of ponds fell in the range 0.5–4.2 dS m^{-1} and ponds could be considered a promising source of irrigation for dry season cropping (Figure 19.3B). In the same study area, in the dry season of 2016–2017, pond water remained non-saline (Akanda et al. 2017; Kabir et al. 2019). However, the current storage capacity of ponds was not sufficient to meet the irrigation requirements. To solve this issue, the involvement of government agencies for promoting pond development is important (Bell et al. 2019).

Relative water volume of the controlled canal and of ponds decreased over time during the study period (Figure 19.6). This water volume change might be due to cumulative evaporation (increasing EC_w), extraction for irrigation or loss by drainage (not changing EC_w), groundwater infiltration (increasing EC_w) into ponds and later rainfall (decreasing EC_w). However, groundwater level in one location remained lower than the pond water depth from November until May 2017–2018. Therefore, there was no chance of inflow towards the pond from groundwater for that location but there was a chance of outflow from the canal towards the groundwater (not measured during the study). In this study, volume change for canal and ponds over time was attributed mostly to water extraction for irrigation and evaporation. Further water balance research is needed involving measuring groundwater and water body water elevations together with cumulative evaporation, extraction/drainage, groundwater infiltration and rainfall over time.

In the current study, the increase in relative EC_w of a bunded canal and several ponds (3, 7 and 9) with the decrease in relative water volume due to evaporation and irrigation water extraction was below the hypothetical line (doubling the EC_w for a halving of water volume due to evaporation and irrigation water extraction) (Figures 19.7C and 19.7D). This water would have been suitable as a source of irrigation for dry season cropping. By contrast, the relationship between relative EC_w and relative water volume of uncontrolled canals and other ponds (4, 5 and 6) was substantially above the theoretical line. This suggesting that these ponds might have high connectivity to the adjacent tidal river, or that there had been overland flow from rainfall, which allowed additional salt to enter the water body (Figures 19.7C and 19.7D). Groundwater EC_w remained in the range of 2.3–5.7 dS m^{-1} over the dry period (Figure 19.3C). Groundwater can, therefore, be considered a supplementary water resource if controlled canal and pond water volumes are insufficient.

19.4.2 Causes of Change in Water Salinity and Water Availability

A further cause suggesting hydrological connectivity of stored water to the landscape was the highly significant negative relationship between elevation above mean sea level and pond EC_w (Figures 19.4A and 19.4B). At present, we do not know for sure the cause of this relationship. Perhaps, the intrusion of shallow saline groundwater into ponds is affected by the depth to water-table and the periodic lowering of pond level during extraction for irrigation, and this is related to the elevation above sea level since the water-table will be deeper in the elevated parts of the landscape and shallower in the less elevated parts of the landscape: there may also be differences in groundwater salinity with elevation. Given this, we would expect more seepage (and therefore saline intrusion) from shallow groundwater in ponds with low elevation. The app that we have used does not specify vertical resolution of the elevation measurement but we note there was a strong correlation between elevation and EC_w which suggests there was a physical relationship between elevation and EC_w. This could be tested by measurement of depth to water-table around ponds using high-resolution digital elevation surveying or differential global positioning system devices together with shallow groundwater EC_w at differing elevations in the landscape.

In the current study, among the weather variables, cumulative evaporation was the most important cause for increasing salinity (Figures 19.5A and 19.5D). We propose that the mechanism for this effect was by decreasing relative water volume along with irrigation water extraction (Figures 19.8A and 19.8D). Cumulative evaporation was also positively ($P < 0.001$) correlated with maximum temperature and vapour pressure deficit. Analysis of 40 years (1970–2017) of climate data in the coastal zone of Bangladesh showed that maximum temperature increased to 0.04°C per year (Yu et al. 2019). Further increases in temperature may exacerbate the increases in EC_w of stored water by decreasing water volume due to increased evaporation and extraction of irrigation water. Precipitation after April increased the water storage in canals and ponds (Figures 19.7A and 19.7B) and decreased their EC_w (Figures 19.3A and 19.3B). This water can be used for seedbed preparation of *Kharif*-1 rice. For example, in Amtali, Barishal, in 2018–2019, *Kharif*-1 rice cropping area was reduced drastically due to lack of fresh surface water (no rainfall) during seedbed preparation (Bell et al. 2019).

19.4.3 Utilisation of Irrigation Water for Dry Season Cropping

In this study, water quality measurements suggested three types of pond water (non-saline to slightly saline, slightly saline and slightly to moderately saline) during the *Rabi* season and into the middle of the *Kharif*-1 season (Figure 19.3B). The EC_w of the bunded canal was also non-saline to slightly saline throughout the season (Figure 19.3A). Therefore, considering the water availability, importance should be given on alternative crops rather than *Boro* rice because of its requirement for substantial volumes of water of low salinity (Bell et al. 2019) and on more water-efficient irrigation methods such as drip irrigation (Mahanta et al. 2019) and also to employing mulches and tillage to decrease soil salinity and conserve soil water (Paul et al. 2020a,b; Sarangi et al. 2018a, b). We suggest that the cropping area could be increased by early sowing of *Rabi* season crops including waterlogging-tolerant, salt-sensitive crops (grass pea and mustard) along with existing salt-tolerant non-rice crops (sunflower, maize, wheat, barley and watermelon) as well as through the conjunctive use of non-saline and slightly to moderately saline water during the dry season (Kabir et al. 2019). Fitting a non-rice crop (mustard, garden pea, spinach and potato) into the *Rabi* season after *Kharif*-2 rice increased the overall benefit-cost ratio relative to that from a single *Kharif*-2 season rice in the same study area (Saha et al. 2019). Also, the focus should be given on the development of alternative crops which are of short duration, and tolerant to early waterlogging, and to salinity and drought at the end of the growing season (Bell et al. 2019). Drainage strategies to facilitate earlier sowing and harvest of *Rabi* season crops might enable these crops to escape waterlogging at the start of the growing season and ripen before the adverse effects of salinity and drought at the end of the season curtail yield (Bell et al. 2019; Kabir et al. 2019). Another opportunity is to increase the storage capacity of existing water bodies (canal and pond) (Bell et al. 2019) so that water can be stored during the monsoon season from rainfall as well as from river sources when water salinity is below 2 dS m^{-1}. Rainfall from February to April may be beneficial for increasing storage of freshwater in ponds and bunded canals. Indeed, the possibility of 20 mm and 50 mm rainfall events has increased by 25–65% and 5–30% in the last 30 years during February to April (Yu et al. 2019). At the same time, a more effective weather forecasting system could enable surface water storage decisions during infrequent heavy rainfall during the dry season which may also alleviate damage from flooding of crop fields. Finally, cutting down on direct evaporation from the surface of the ponds and bunded canals might be possible by cultivating lightweight vine-type vegetables on top of the water bodies supported by bamboo and rope trellises. However, ponds and canals in lower-lying land need to be protected from runoff of saline surface water from higher land.

19.5 CONCLUSION

In the coastal zone of the Ganges Delta, an increase in soil and water salinity and drying of soil profiles during the dry season impedes the production of *Rabi* crops by farmers. Current practices involving a delay in crop establishment increases the exposure of non-rice crop to increasing soil and water salinity, decreasing surface water availability and increasing dependency on groundwater. The salinity of canals and ponds started to increase after February and reached a maximum during mid-April. Pond and groundwater salinity was comparatively lower than the river and canal (without bund) water salinity. The water salinity of the canals isolated from the river remained comparatively low (1.1–2.9 dS m^{-1}). Cumulative evaporation, water extraction and saline water intrusion were the most important factors for increasing water salinity and decreasing the volume of stored water. Therefore, if crop irrigation can be mostly completed by February, stored water would remain sufficiently low in EC_w to be used for irrigation (0.5–3.9 dS m^{-1}).

The present study relates to a particular part of the Ganges, Brahmaputra and Meghna Delta where river salinity is high in the dry season. Further study is required at other locations in the delta with either continuous fresh or saline river water flow, and with differences in elevation above mean sea level and distance from the river to determine how the present case study results can be extrapolated more widely.

Therefore, it is recommended that the early establishment of non-rice crops, increasing the existing storage capacity of reservoirs (canal and pond), construction of temporary bunds in canals and ponds before the *Rabi* season, cultivation of lightweight leafy vegetables to shade the canals and pond and strengthening the weather forecasting system can increase agricultural productivity and food security for low-lying deltas.

ACKNOWLEDGEMENT

The authors are grateful to the Australian Centre for International Agricultural Research (Project LWR/2014/073) for financial support and for a John Allwright Fellowship to the senior author. The senior author is also thankful for travel grants from Murdoch University and the Saline Futures Conference that allowed the oral presentation of the research from which this paper is derived. We are grateful to the people of Dacope for cooperation during the survey.

ENDNOTES

1. Irrigation water was classified for salinity (EC_w; dS m^{-1}) as follows (modified after Rhoades et al. (1992) : 0–0.7 (non-saline), 0.7–2 (slightly saline), 2–5 (moderately saline), 5–10 (highly saline), 10–25 (severely saline), 25–45 (extremely saline) and >45 dS m^{-1} (brine).
2. According to Rhoades et al. (1992), non-saline (<0.7 dS m^{-1}) water can be used for drinking and irrigation; slightly saline (0.7–2 dS m^{-1}) water can be used for irrigation.
3. Studies of crops (wheat, sorghum, cowpea and maize) suggest that plants are more sensitive to saline water irrigation at their initial growth stages and are less sensitive at the later stages of growth (Maas et al. 1983; Maas and Poss 1989a,b; Maas et al. 1986).

REFERENCES

Akanda, M. A. R., K. K. Sarker, S. A. Kamar, and M. Mainuddin. 2017. *Conjunctive use of fresh and saline water irrigation for sunflower in coastal areas of Bangladesh*, 188–202. Gazipur: Annual research report, Irrigation and Water Management Division, Bangladesh Agricultural Research Institute.

Alam, M., M. M. Alam, J. R. Curray, M. L. R. Chowdhury, and M. R. Gani. 2003. An overview of the sedimentary geology of the Bengal Basin in relation to the regional tectonic framework and basin-fill history. *Sedimentary Geology* 155:179–208.

Bangladesh Bureau of Statistics (BBS) 2018. *Population and housing census 2011*. Dhaka, Bangladesh: Statistics Division, Bangladesh Bureau of Statistics, Government of the People's Republic of Bangladesh..

Bell, R. W., M. Mainuddin, E. G. Barrett-Lennard, S. K. Sarangi, M, Maniruzzaman, K. Brahmachari, K. K. Sarker, D. Burman, D. S. Gaydon, and J. M. Kirby. 2019. Cropping systems intensification in the coastal zone of the Ganges Delta: Opportunities and risks. *J Indian Soc Coastal Agric Res* 37:153–161.

Chowdhury, N. T. 2010. Water management in Bangladesh: An analytical review. *Water Policy* 12:32–51.

Hossain, M. B., M. Maniruzzaman, J. C. Biswas, and N. Kalra. 2018. Irrigation water requirement of major cropping systems in Bangladesh: A model study. In: *Project completion report on modeling climate change impact on agriculture and developing mitigation and adaptation strategies for sustaining agricultural production in Bangladesh*, 78–87. KGF, Farmgate, Dhaka pp.

Hossain, M. B., M. Maniruzzaman, M. S. Yesmin, A. B. M. Mostafizur, P. K. Kundu, M. J. Kabir, J. C. Bishwas, and M. Mainuddin. 2019. Water and soil salinity dynamics and dry season crop cultivation in coastal region of Bangladesh. *J Indian Soc Coastal Agric Res* 37:24–31.

Kabir, M. E., B. C. Sarker, A. K. Ghosh, M. Mainuddin, and R. W. Bell. 2019. Effect of sowing dates on yield of wheat grown in excess water and salt affected soils in southwestern coastal Bangladesh. *J Indian Soc Coastal Agric Res* 37:51–59.

Maas, E. V., G. J. Hoffman, G. D. Chaba, J. A. Poss, and M. C. Shannon. 1983. Salt sensitivity of corn at various growth stages. *Irrig Sci* 4:45–57.

Maas, E. V., and J. A. Poss. 1989a. Salt sensitivity of cowpea at various growth stages. *Irrig Sci* 10:313–320.

Maas, E. V., J. A. Poss, and G. J. Hoffman. 1986. Salinity sensitivity of sorghum at three growth stages. *Irrig Sci* 7:1–11.

Maas, E. V., and J. A. Poss. 1989b. Salt sensitivity of wheat at various growth stages. *Irrig Sci* 10:29–40. doi:10.1007/BF00266155.

Mahanta, K. K., D. Burman, S. K. Sarangi, U. K. Mandal, B. Maji, S. Mandal, S. Digar, and M. Mainuddin. 2019. Drip irrigation for reducing soil salinity and increasing cropping intensity: Case studies in Indian Sundarbans. *J Indian Soc Coastal Agric Res* 37:64–71.

Mahmuduzzaman, M., Z. U. Ahmed, A. K. M. Nuruzzaman, and F. R. S. Ahmed. 2014. Causes of salinity intrusion in coastal belt of Bangladesh. *Int J Plant Res* 4:8–13.

Mirza, M. M. Q. 1998. Diversion of the Ganges water at Farakka and its effects on salinity in Bangladesh. *Environ Manage* 22:711–722.

Mondal, M. 2016. The Keys to unlocking production potentials of the coastal zone. The Daily Star. Accessed May 27, 2020. https://www.thedailystar.net/supplements/25th-anniversary-special-part-4/the-keys-unlocking-production-potentials-the-coastal

Mondal, M. K., T. P. Tuong, S. P. Ritu, M. H. K. Choudhury, A. M. Chasi, P. K. Majumder, M. M. Islam, and S. K. Adhikary. 2006. Coastal water resource use for higher productivity: participatory research for increasing cropping intensity in Bangladesh. In *Environment and livelihoods in tropical coastal zones: Managing agriculture-fishery-aquaculture conflicts*, ed. C. T. Hoanh, T. P. Tuong, J. W. Gowing, and B. Hardy, 72–85. UK: CABI Publishing.

Mukherjee, A., A. E. Fryar, and P. D. Howell. 2007. Regional hydrostratigraphy and groundwater flow modeling in the arsenic-affected areas of the western Bengal basin, West Bengal, India, *Hydrogeol J* 15:1397–1418.

Murad, K. F. I., A. Hossain, O. A. Fakir, S. K. Biswas, K. K. Sarker, R. P. Rannu, and J. Timsina. 2018. Conjunctive use of saline and freshwater increases the productivity of maize in saline coastal region of Bangladesh. *Agric Water Manage* 204:262–270.

Paul, P. L. C., R. W. Bell, E. G. Barrett-Lennard, and E. Kabir. 2020a. Variation in the yield of sunflower (Helianthus annuus L.) due to differing tillage systems is associated with variation in solute potential of the soil solution in a salt-affected coastal region of the Ganges Delta. *Soil Till Res* 197:104489.

Paul, P. L. C., R. W. Bell, E. G. Barrett-Lennard, and E. Kabir. 2020b. Straw mulch and irrigation affect solute potential and sunflower yield in a heavy textured soil in the Ganges Delta. *Agric Water Manage* 239:106211.

Payo, A., A. N. Lázár, D. Clarke, R. J. Nicholls, L. Bricheno, S. Mashfiqus, and A. Haque. 2017. Modeling daily soil salinity dynamics in response to agricultural and environmental changes in coastal Bangladesh. *Earth's Future* 5:495–514. doi:10.1002/2016EF000530.

Rhoades, J. D., A. Kandiah, and A. M. Mashali. 1992. *The use of saline waters for crop production* vol 48. Rome: FAO.

Saha, R. R., M. A. Rahman, M. H. Rahman, M. Mainuddin, R. W. Bell, and D. S. Gaydon. 2019. Cropping system intensification under rice based system for increasing crop productivity in salt-affected coastal zones of Bangladesh. *J Indian Soc Coastal Agric Res* 37:72–81.

Sarangi, S. K., B. Maji, S. Digar, K. K. Mahanta, U. K. Mandal, D. Burman, S. Mandal, P. C. Sharma, and M. Mainuddin. 2018a. Zero tillage potato cultivation with paddy straw mulching increase yield, water productivity and income in the coastal saline soils. In *Extended Summaries XXI Biennial National Symposium on Doubling farmers' income through agronomic interventions under changing scenario*, 24–26. India: Indian Soc Agron.

Sarangi, S. K., B. Maji, S. Digar, K. K. Mahanta, P. C. Sharma, and M. Mainuddin. 2018b. Zero tillage potato cultivation. *Indian Farming* 68:23–26.

Yu, Y., M. Mainuddin, M. Maniruzzaman, U. K. Mandal, and S. K. Sarangi. 2019. Rainfall and temperature characteristics in the coastal zones of Bangladesh and West Bengal, *India. J Indian Soc Coastal Agric Res* 37:12–23.

20 The International Farmers' Café on Salinization and Saline Agriculture

A Test Case for Participatory Research on Saline Agriculture

Jeroen De Waegemaeker and Elke Rogge

CONTENTS

20.1 THE RISE OF PARTICIPATION IN POLICY AND RESEARCH

In the twentieth century, state policies were largely implemented through centralized planning and framed within the vision of the "provider-state". Today, however, many policy spheres are moving towards more engagement with stakeholders for the development and implementation of their government objectives, thus shifting from government to governance (Curry 2001). Instead of top-down, unilateral decision-making in government, governance reconciles politics and citizens by consulting and involving people and organizations in the shaping and monitoring of policy-making (Jansen et al. 2006). In other words, *"political decisions are being discussed and negotiated between state actors and private actors"*, resulting in a co-creation of policy (Böcher 2008, p. 373).

The framework of participatory governance, a subset of governance theory, aims to involve citizens in public decision-making in a more direct and meaningful way

DOI: 10.1201/9781003112327-20

(Fischer 2012). Participatory governance aspires to deepen the ways in which citizens can effectively participate in and influence policies that directly affect their lives (Fung and Wright 2001). Benefits of such an approach include greater responsiveness to complex situations and more deliberation than traditional governance processes (Leach 2006).

Similar to policy work, co-creation and participation are on the rise in academic research, including in agricultural research. Opposing unilateral knowledge transfer and linear innovation processes, there is increasing acknowledgment of the importance of knowledge co-creation processes that recognize science and society as equal co-producers of knowledge (Moschitz and Home 2014; Sumane et al. 2017). Hence, the farming community is not merely a consumer of research but adds to research. The premise of participatory research on agriculture is to bring that farming community into the research process and to facilitate collaboration between professional researchers and farmers in order to achieve better research results (Hoffmann et al. 2007).

20.2 A CALL FOR PARTICIPATORY RESEARCH ON SALINE AGRICULTURE

In this chapter, we argue that the development of saline agriculture requires a participatory governance perspective and, by extension, that research on saline agriculture must be embedded in a participatory process. Four distinct arguments, found in the literature on participatory research, underpin this call for participatory research on saline agriculture.

Firstly, there is the **substantive** argument. There are various types of knowledge on agriculture. Besides researchers' expertise that is acquired through experiments, there is farmers' tacit knowledge that builds on years of hands-on experience. Researchers and farmers work in different ways and have diverging epistemologies, nevertheless, it is vital to agricultural innovation that both actors collaborate and create synergies (Hoffmann et al. 2007). Likewise, we argue that both researchers and farmers each add essential pieces to the proverbial jigsaw that is saline agriculture.

Secondly, there is the **methodological** argument. By connecting different actors and by sharing experiences one often acquires crucial information much faster. For example, setting up a test site for saline agriculture and getting to credible (scientific) results can take months or even years, yet farmers might point out important opportunities and pitfalls for field trials at an early stage. Consequently, a broader participation in the research process can save time and resources.

Thirdly, there is the **moral** argument. We need to involve various actors in policy making and research processes in order to increase their legitimacy. In short, research on saline agriculture needs to include those stakeholders whom they will affect: farmers and water managers.

Finally, there is the **social** argument. Different actors with little or no connection meet throughout the participatory process and discover mutual interests and build bridges. Alternative modes of farming, e.g. saline farming, are often based on new linkages between farming, local resources and the local community (Renting et al. 2003, Kirwan 2004). In this alternative mode of farming, the formerly independent processes such as food production, food processing, distribution and consumption constitute a singular, all-encompassing process. Hence, the social interaction in a participatory process is paramount for the creation of the multi-actor partnerships on saline agriculture.

20.3 AN INTERNATIONAL FARMERS' CAFÉ AT THE SALINE FUTURES CONFERENCE

In accordance with our call to embed the research on saline agriculture in a participatory process, the Saline Futures Conference (September 2019, Leeuwarden, the Netherlands) was not limited to scholarly discussions amongst academic experts on saline agriculture. In parallel to this conference – *which eventually gave rise to this book* – the Interreg VB North Sea Region project Saline Farming (SalFar) organized the International Farmers' Café on Salinization and Saline Agriculture. To clarify, a farmers' café is a farming-oriented version of the better-known World Café methodology: a structured conversational process that aims for open, meaningful discussions in an informal, "café-like" setting (Gordijn et al. 2018). Both the academic conference and the farmers' café took place at the same location, thus facilitating interaction between the researchers and the farmers.

The International Farmers' Café on Salinization and Saline Agriculture was open to farmers, agricultural advisors, consultants, water managers, policy workers, NGOs and other relevant stakeholders in the North Sea Region. A total of 32 practitioners from Germany (12), the Netherlands (9), Belgium (7) and Norway (4) participated in the two-day event. The international farmers' café had multiple objectives: inform about salinization and salt-tolerant crops, visit the SalFar test site for saline agriculture on the island of Texel, discuss the future of (saline) agriculture in the North Sea Region, exchange saline strategies across borders and receive feedback into the research of the SalFar project. An extensive report on the event and its outcome can be found in De Waegemaeker et al. (2020). This chapter first elucidates the architecture of the international farmers' café, and next addresses its results.

20.4 THE ARCHITECTURE OF AN INTERNATIONAL FARMERS' CAFÉ

Despite growing academic attention for participatory research, knowledge on the practical implementation of participatory governance in policy work and research remains limited. How do we structure and conduct a multi-actor, multi-level and multi-sector process? Participatory governance processes require customization and adaptability and, as a consequence, there is no fixed "blueprint" for participation (Ostrom 2007; Rogge et al. 2013). In this section, we discuss the "architecture" of the International Farmers' Café on Salinization and Saline Agriculture to provide a source of inspiration for future participatory research on saline agriculture.

First and foremost, it must be stressed that the topic of the international farmers' café was not limited to saline agriculture but included salinization. On the one hand, broadening the topic helped to draw more participants to the event. Since there was little awareness on salinization in the North Sea Region, many participants wanted to get a better understanding of the problem (salinization) rather than the proposed solution (saline agriculture). Broadening the topic in this way facilitated an open and constructive debate even though the need for saline agriculture is contested by many farmers in the North Sea Region. Throughout the international farmers' café there were multiple opportunities to debate climate adaptation strategies that mitigate salinization, e.g. the flushing of surface waters and level-controlled drainage. In other words, the event put forward the strategy of saline agriculture, yet not as the sole solution to salinization. Leeuwis and van den Ban (2004, p. 45–46) define a code of conduct for the organizers of learning processes on agricultural innovation and stress the importance of respectfulness, which includes a genuine willingness to see things from other people's perspectives. In line with this code of conduct, we stress that future participatory research on saline agriculture in the North Sea Region must provide opportunities to address all possible saline strategies, both the mitigation of as well as the adaptation to salinization.

Secondly, the farmers' café started with an informative plenary session that provided participants with knowledge about salinization and saline agriculture. It included presentations about the different types of salinization processes in the North Sea Region, the impact of salinization on soils, the experience with saline agriculture at the Salt Farm Foundation in Texel, and the array of climate adaptation strategies for the (Dutch) agricultural sector. The goal of the plenary session was to create a shared understanding about and a common language for salinization and saline agriculture. The need for such shared understanding and common language was particularly high because the participants in the international farmers' café had diverging professional backgrounds; in agriculture, water management and rural development. Moreover, the participants live and work in different parts of the North Sea Region, an area where the extent of salinization and its causes vary strongly (see chapter 5). The plenary session was needed to familiarize the participants with the terminology on salinization and saline agriculture, especially since none of them were native English speakers.

After the plenary session, there were two consecutive rounds of two parallel workshops: a workshop on the impact of salinization on soils and crop production (Figure 20.1), and a workshop on the potential for saline agriculture in the North Sea Region (Figure 20.2). Fixed and thoroughly researched methodologies for both of these workshops were developed to ensure that all voices would be heard. In the workshop on the impact of salinization on soils and crop production, for example, the participants were divided into small groups of eight. In this rather intimate setting, each practitioner was asked to present their local "saline context"; the location of the farm, the crops that they grew, and the experiences with salinization in their practice. An aerial view (via Google Maps) of the participant's working area was projected as a visual support. In the workshop on saline agriculture, the participants were first asked which crops they wanted to discuss. The participants indicated their interest on a series of posters with markers and post-its, a method known as

FIGURE 20.1 Workshop on the impact of salinization on soils and crop production. (Courtesy of Wim Van Isacker.)

FIGURE 20.2 Workshop on the potential of saline agriculture in the North Sea Region. (Courtesy of Wim Van Isacker.)

FIGURE 20.3 Results of the "dotmocracy" exercise in the workshop on saline agriculture.

"dotmocracy" (Figure 20.3). In this way, we prioritized those agricultural crops that multiple participants found interesting and that, as a consequence, could benefit from a transnational discussion. Furthermore, it helped to engage the entire group in the discussion rather than just the loudest voices.

The international farmers' café deliberately attributed a lot of time to the lunch and coffee breaks, since farmers highly appreciate and value interactions with their peers. In other words, farmers enjoy teaching to and learning from other farmers (Franz et al. 2010a). Long breaks throughout the international farmers' café offered participants the opportunity to exchange their experiences in one-to-one conversations. In the ex-post online evaluation of the event, the participants highlighted the added value of these interactions with their peers (e.g. quote).

"It was insightful to talk to people from another region/country about their specific situation. For me this is of great added value. It is good to be able to talk about specific problems on a practical level."

Finally, the international farmers' café included an excursion to the island of Texel (the Netherlands) on the second day of the two-day event. In this way, we did not limit the international farmers' café to indoor, oral knowledge exchange but offered opportunities for on-farm interactions. These types of interactions, e.g. hands-on teaching, demonstrations and farm visits, are highly preferred by farmers (Franz et al. 2010b). During the excursion, the participants visited the demonstration sites on saline agriculture of the Salt Farm Foundation. Here, the participants could see the saline irrigation system as well as touch, smell and taste the test crops (Figure 20.4). What is more, many eagerly measured the salinity of the surface waters throughout the island (Figure 20.5).

FIGURE 20.4 Impressions from the visit to the demonstration site of Salt Farm Foundation. (Courtesy of Wim Van Isacker.)

FIGURE 20.5 Participants eagerly measured salinity levels in surface water by using an EC measuring tool. (Courtesy of Wim Van Isacker.)

20.5 THE ADDED VALUE OF THE INTERNATIONAL FARMERS' CAFÉ

The results of the International Farmers' Café on Salinization and Saline Agriculture are extensively described in De Waegemaeker et al. (2020). This section addresses only some of the insights that were acquired via this participatory event. In line with the scope of this book, we focus on those insights that are particularly relevant to future research on saline agriculture in the North Sea Region.

Firstly, the participants of the international farmers' café discussed which "saline crops" they favored and, as a result, the event identified key research questions on saline agriculture in the North Sea Region. First and foremost, the international farmers' café highlighted an enormous demand for salt-tolerant pastures in the North Sea Region. Far more than any other agricultural crop, participants were looking for information on salt-tolerant grass varieties. This interest for grass resulted from the fact that salinization currently occurs predominantly in the participants' pastures rather than on their arable land. The debate on grass clarified the farmers' expectations in terms of salt-tolerant grass varieties: saline grass varieties must be able to grow in saline conditions and, at the same time, need to maintain a high level of productivity and digestibility. Furthermore, there was a lot of interest for the salt-tolerance of conventional arable farming crops such as potato, wheat, barley, maize, rapeseed, oat and onion. As such, the international farmers' café indicated that farmers in the North Sea Region conceptualize saline agriculture as an incremental substitution of the current agricultural production by salt-tolerant cultivars rather than the cultivation of new, unknown crops such as halophytes (see quote 2).

> *"Why we are interested in the salt-tolerance of onions? We are growing them! And we would like to keep them. They are already in our system."*

Secondly, the practice-oriented discussions at the international farmers' café uncovered important barriers and opportunities for saline agriculture in the North Sea Region. Based on these discussions, De Waegemaeker et al. (2020) list guidelines for future research on saline agriculture. For example, the international farmers' café highlighted the need for research on "saline crop rotations". The participants stressed that the development of a sequence of alternating saline crops rather than one saline crop is a prerequisite to put saline agriculture into practice in the North Sea Region. Moreover, they pointed towards the history of the North Sea Region in order to find salt-tolerant varieties for the region's saline future. The participants often argued that historically farmed crops provide a useful gene pool for research on saline agriculture.

Finally, the international farmers' café clarified why it is necessary to research saline agriculture in the North Sea Region. Most of the participants stressed that they currently experience only a minor level of salinization. This observation, however, should not be interpreted as an argument to postpone local research on saline agriculture. The participants indicated that the road from the laboratory to the field is long. Moreover, they stressed that it takes a long time for the seed industry to commercialize the results of fundamental research. Hence, the research on the salt-tolerance of crops is urgent even though the salinization in the North Sea Region is not yet acute.

> "[A salt-tolerant crop] *takes a lot of time to develop. If we don't ask it now, we don't have it when we need it.*"

20.6 A FARMERS' CAFÉ, A FIRST STEP

As saline agriculture grows to a field of research at the global scale, this chapter advocates embedding this research within a participatory research perspective. We hope that our description of the International Farmers' Café on Salinization and Saline Agriculture may inspire the organization of similar events in future research. We need to clarify, however, that a farmers' café is merely a first step in participatory research on saline agriculture. Since a farmers' café focuses on informing and consulting the farming community, the methodology is situated at the lower end of the participation ladder (Arnstein 1969; Pretty 1995). Hence, there is much room for growth in terms of participation. Besides the organization of the International Farmers' Café on Salinization and Saline Agriculture at the Saline Futures conference, the Interreg VB North Sea Region SalFar project currently experiments with other forms of participatory research on saline agriculture in the North Sea Region. In the United Kingdom, for example, the farming community is involved in the selection of crops and varieties for field trails on saline agriculture at the University of Lincoln. In Sweden, researchers from the University of Gothenburg are testing the salt-tolerance of wheat varieties on a private farm. In Norway, NMBU researchers and farmers are working together on the development of an irrigation system for saline agriculture. Future research on these participatory research processes is needed to define important parameters for these processes, and to clarify their added value to the development of saline agriculture.

ACKNOWLEDGMENT

The authors would like to thank Lies Debruyne (ILVO) and Anne M Asselin de Williencourt (NMBU) for all their contributions to the farmers' café and their input on this book chapter. We would like to thank the SalFar consortium, Pier Vellinga and Katarzyna Negacz from Waddenacademie in particular, for their help in organizing the international farmers' café.

REFERENCES

Arnstein, S. 1969. A ladder of citizen participation. *Journal of the American Planning Association* 35, no. 4 (November): 216–224 https://doi.org/10.1080/01944366908977225

Böcher, M. 2008. Regional governance and rural development in Germany: The implementation of LEADER+. *Sociologia Ruralis* 48, no. 4 (October): 372–388 https://doi.org/10.1111/j.1467-9523.2008.00468.x

Curry, N. 2001. Community participation and rural policy: Representativeness in the development of millennium greens. *Journal of Environmental Planning and Management* 44, no.4 (August): 561–576 https://doi.org/10.1080/09640560120060966

De Waegemaeker, J., Asselin de Williencourt, A., and L. Debruyne. 2020. The International Farmers' Café on Saline Agriculture, a report by ILVO and NMBU for the Interreg VB North Sea Region project Saline Farming (SalFar). Available at: https://northsearegion.eu/media/12556/salfar_report_on_international_farmers_cafe_final.pdf

Fischer, F. 2012. Participatory Governance: From Theory To Practice. In *The Oxford Handbook of Governance*, ed. Levi-Faur, D., Oxford: Oxford University Press https://doi/org/10.1093/oxfordhb/9780199560530.013.003

Franz, N., Piercy, F., Donaldson, J., Richard, R., and J. Westbrook. 2010a. How farmers learn: Implications for agricultural educators. *Journal of Rural Social Sciences* 25, no. 1: 37–59.

Franz, N., Piercy, F., Donaldson, J., Westbrook, J., and R. Richard. 2010b. Farmer, agent, and specialist perspectives on preferences for learning among today's farmers. *Journal of Extension* 48, no. 3: 1–10

Fung, A., and E. O.Wright. 2001. Deepening democracy: Innovations in empowered participatory governance. *Politics & Society* 29, no. 1 (March): 5–41 https://doi.org/10.1177%2F0032329201029001002

Gordijn, F., Eernstman, N., Helder, J., and H. Brouwer. 2018. *Reflection Methods: Practical Guide for Trainers and Facilitators, Tools to Make Learning More Meaningful.* Wageningen Centre for Development Innovation, Wageningen University and Research, Available at: https://doi.org/10.18174/439461

Hoffmann, V., Probost, K., and A. Christinck. 2007. Farmers and researchers: How can collaborative advantages be created in participatory research and technology development. *Agriculture and Human Values* 24, 355–368 https://doi.org/10.1007/s10460-007-9072-2

Jansen, T., Chioncel, N., and H. Dekkers. 2006. Social cohesion and integration: Learning active citizenship. *British Journal of Sociology of Education* 27, no. 2 (August): 189–205 https://doi.org/10.1080/01425690600556305

Kirwan, J. 2004. Alternative strategies in the UK agro-food system: Interrogating the alterity of farmers' markets. *Sociologia Ruralis* 44, no. 4 (October): 395–415 https://doi.org/10.1111/j.1467-9523.2004.00283.x

Leach, W.D. 2006. Collaborative public management and democray: Evidence from western watershed partnerships. *Public Administration Review* 66, no. 1 (November): 100–110 https://doi.org/10.1111/j.1540-6210.2006.00670.x

Leeuwis, C., and A. van den Ban. 2004. *Communication for Rural Innovation: Rethinking Agricultural Extension.* Oxford: Blackwell Publishing.

Moschitz, H., and R. Home. 2014. The challenges of innovation for sustainable agriculture and rural development: Integrating local actions into European policies with the Reflective Learning Methodology. *Action Research* 12, no. 4 (June): 392–409 https://doi.org/10.1177%2F1476750314539356

Renting, H., Marsden, T.K., and J. Banks. 2003. Understanding alternative food networks: Exploring the role of short food supply chains in rural development. *Environment and Planning A* 35, no. 3 (March), 393–411 https://doi.org/10.1068%2Fa3510

Rogge, E., Dessein, J., and A. Verhoeve. 2013. The organization of complexity: A toolbox to organize the interface of rural policy making. *Land Use Policy* 35, 329–340 https://doi.org/10.1016/j.landusepol.2013.06.006

Ostrom, E. 2007. A diagnostic approach for going beyond panaceas. *Proceedings of the National Academy of Sciences of the United States of America* 104, no. 39 (September): 15181–15187 https://doi.org/10.1073/pnas.0702288104

Pretty, J. 1995. Participatory learning for sustainable agriculture. *World Development* 23, no. 8 (August): 1247–1263 https://doi.org/10.1016/0305-750X(95)00046-F

Sumane, S., Kunda, I., Knickel, K., Strauss, A., Tisenkopfs, T., des Ios Rios, I., Rivera, M., Chebach T., and A. Ashkenazy. 2017. Local and farmers' knowledge matters! How integrating informal and formal knowledge enhances sustainable and resilient agriculture. *Journal of Rural Studies* 59, 232–241 https://doi.org/10.1016/j.jrurstud.2017.01.020

21 Putting Saline Agriculture into Practice

A Case Study from Bangladesh

Arjen De Vos, Andrés Parra González, and Bas Bruning

CONTENTS

21.1 INTRODUCTION

Coastal Bangladesh is severely affected by salinity. In 2010, over 1 million hectares of land were salt-affected. This is a 26.7% increase since 1973. The Soil Resource Development Institute estimates that an additional 36,440 hectares of new land have become affected by salinity during the past nine years (SRDI, 2010). In Bangladesh, there are three distinct seasons: a hot, humid but dry summer (March-June), a cool, wet monsoon season (June–October) and a cool, dry winter (October–March). Salinity issues mostly occur during the two dry seasons (October–March and March–June). A recent publication (Chen and Mueller, 2018) shows that soil salinity is one of the main forces driving migration in coastal Bangladesh. It is estimated that this could affect up to 27 million people by 2050. At present, around 44% of the salt-affected area is moderately saline (EC_e values between 4 and 12 dS/m – SRDI, 2010).

The Dutch social enterprise "Salt Farm Texel" has worked on salt-tolerant crops for the past 13 years and has identified salt-tolerant varieties of common crops

DOI: 10.1201/9781003112327-21

including, among others, potato, cabbage, cauliflower, carrot and beets that can be cultivated successfully under these moderately saline conditions (De Vos et al. 2016; Van Straten et al. 2020). These crops have now been introduced to coastal Bangladesh as part of the Salt Solution project. The first implementation focused on the October–March cropping season.

Here, we describe a project on saline agriculture aimed at improving the livelihoods of farmers living in salt-affected areas. We present the current situation of the project in terms of soil salinity levels and which types of cultivation strategies have been recommended. Additionally, the impact of this project is discussed according to the Sustainable Development Goals (SDGs), to assess how effective (or not) the introduction of saline farming practices can be in farming communities in coastal Bangladesh. The improvements to the livelihoods realized by the project have been quantified through the independent evaluation by a third party, Grameen Bikash Foundation (GBF), Bangladesh (commissioned by ICCO), using the SDGs as an overarching theme, and these results are also reported here.

21.2 THE PROJECT

The Salt Solution project was funded by the Dutch Postcode Lottery and implemented by a consortium led by the ICCO Group BV; ICCO is a leading non-governmental organisation with headquarters in the Netherlands and regional offices in several countries, such as Bangladesh. Salt Farm Texel (specializing in saline agriculture in the Netherlands), CODEC (a Community Development Centre from Bangladesh), Acacia Water (specializing in water in the Netherlands) and Lal Teer Seed (a seed company from Bangladesh) were also part of the consortium. The Salt Solution project was an innovative climate-smart agriculture-based project and throughout its 3-year duration, it trained 5,000 farmers directly on saline agriculture in four coastal districts (Khulna, Bagerhat, Barguna and Patuakhali). This indirectly benefited 25,000 household members and some aspects of the project continue to this day, increasing the number of people reached. The project had four main outputs:

- salt-tolerant crop production on salt-affected land
- increased nutritious food consumption
- increased participation and decision-making by women in crop production and water management
- creation of a network of farmers, extension officers, policy makers and scientists to create solutions aimed at adapting to salinity

The project involved government personnel with the project intervention designed to sensitize them towards the intended project production technologies of salt-tolerant crop varieties. Key interventions of the project were: setting up a field station and the development of best practices, demonstration and promotion, scaling up, linking with input suppliers and with markets for the sale of produce, research, water management and the influencing of government policies.

The project aimed to ensure that farmers in the salt-affected area in coastal Bangladesh were empowered to improve their yields and livelihoods. At present, the majority of the farmers in the coastal area grows only one crop per year commercially (rice during the monsoon season); while in the north of the country, three crop cycles per year are standard. By making use of salt-tolerant crops and smart soil and water management, it became possible to introduce two additional crop cycles per year to the coastal area. By growing different high yielding, nutritious crops with good market value, smallholder farmers were able to adapt to the increasing salinities and improve their livelihoods, so that ultimately migration out of the area could be stopped. However, before this climate-smart, resilient form of agriculture could be introduced on a large scale, several limiting factors had to be addressed.

21.3 APPROACH AND LIMITING FACTORS

Part of the approach of the project is summarized in Figure 21.1. First of all, the project should be embedded at an institutional level. Extension programs have to be developed as well as "best practices" for crop cultivation that includes crop, soil and water management and an agro-service for farmers need to be established to assist farmers upon request. Salt Farm Texel has tested the yield potential of several salt-tolerant vegetable crops and crop varieties (including potato, cabbage, cauliflower, carrot, beetroot and kohlrabi) at the Salt Farm Texel Research and Training Centre in the Netherlands (for crop species and varieties see Bruning et al. 2015; de Vos et al. 2016). Given the evidence from these field experiments, Salt Farm Texel had begun to pilot the feasibility of introducing some varieties into the production portfolios of lead farmers in coastal Bangladesh. A Saline Agriculture Research and Training Centre (referred to as the Training Centre in the rest of this chapter) has been set up in coastal Bangladesh in collaboration with Lal Teer Seeds, as part of the ongoing project. At the start, the project focused on the validation of the crop performance under local conditions and the development of the "best practices" for crop

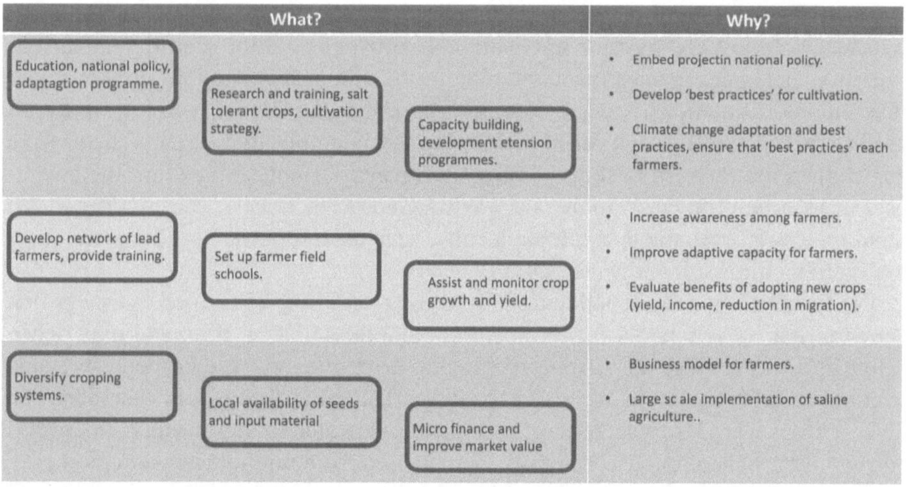

FIGURE 21.1 Overview of the approach of the Salt Solution project to ensure large-scale implementation of saline agriculture in coastal Bangladesh.

cultivation under saline conditions in coastal Bangladesh during the dry seasons. In total, six different crops have been introduced at the farm level so far: cabbage, cauliflower, kohlrabi, carrot, beetroot and potato.

21.4 ROLE OF THE TRAINING CENTER

A cultivation strategy for the year-round production of rice, potato and other vegetables under saline conditions was developed in the Training Center during the Salt Solution project. Smart soil and water management strategies were developed and tested at the Training Center. The introduction of new technologies such as (underground) rainwater harvesting (in partnership with Acacia Water) and drip irrigation was implemented. Soil monitoring (soil analysis and frequent salinity measurements) and management focused on crop rotation, raised bed cultivation, mulching and the use of organic inputs to improve the structure and fertility of the soil are conventional practices that might play an important role in fighting salinization. An additional added value of the training center is that the Bangladesh Agricultural Research Institute is performing important work in developing and selecting suitable varieties of crops such as rice to be cultivated under saline conditions. All these crops and varieties continue to be tested and demonstrated at the Training Centre developed in The Salt Solution Project.

21.5 STRUCTURE OF THE PROJECT

As the Salt Solution project matured, the structure was such that once the best strategy had been determined in the Training Center, this was implemented by a network of lead farmers who had been trained and assisted closely. According to the "train-the-trainer" principle, the knowledge and know-how of "Salt Farm Texel" were passed on to the staff of ICCO and CODEC, who subsequently trained the lead farmers. The training focused on pre-sowing activities (formation of raised beds, fertilizer, compost and gypsum application, improvement of the seedling nurseries), soil, crop and water management and monitoring and data collection. Protocols and illustrations, animations, videos, community theater and farmer field days were all used to inform and instruct farmers. The lead farmers became trainers themselves and trained the farmers in their community (group farmers), spreading the awareness of tolerant crop cultivation as a way to adapt to increasing salinity levels. All input materials were made available locally, such as appropriate seeds as well as a toolkit to monitor soil and water salinity levels.

For farmers, "seeing is believing". By setting up a network of lead farmers that act as farmer field schools for the local community, all farmers could experience saline agriculture close up. Also, through crop diversification, the farmers were able to produce different crops to ensure good market value throughout the year. The diverse crops were also selected for their nutritional value to improve nutritious food consumption since around 20% of the yield is used for household consumption.

Soil salinity was closely monitored at the Training Center and at an additional 50 different farms across the whole project. Lead farmers and, subsequently, the farmers' groups were trained in determining the salinity of their soil using various methods, as well as the salinity of the irrigation water. Results of these measurements were converted to the international standard of EC_e values to be able to link crop performance to soil salinity values.

The best way to convince a farmer to change their way of farming is by showing the financial return on investment. For this reason, special attention was focused on farm business models, also aimed at Shifting traditional agriculture to a commercial basis. Close monitoring took place by ICCO and CODEC, collecting data of the different crops at a minimum of 30 different farms per crop.

21.6 RESULTS

21.6.1 Soil Characteristics

For 11 locations, soil samples were collected and analyzed in detail. The majority of the analyzed soils were silt loam and silt clay loam soils. The pH and organic matter percentage ranged from 6.2 to 7.5 and 1.1 to 2.5%, respectively. The saturation of the Cation Exchange Capacity (CEC) was, on average, 81, 13, 4.8 and 1.1% for Ca, Mg, K and Na, respectively. These results indicate that few problems occur with soil structural issues in relation to salinity, and soils appear to be non-sodic (low Na saturation in CEC, pH<8). As mentioned above, the soil salinity was monitored at 50 different locations distributed over 4 coastal districts where the project implementation took place. The average soil salinity level (EC_e) of all 50 locations was 3.6 ± 2.0 (s.d.) dS/m at the first sampling event and 5.6±3.3 (s.d.) dS/m at the last sampling event, with a seasonal average of 4.7 dS/m. Of the 50 locations, 5 locations were in the 0–2 dS/m range, 21 locations in the 2–4 dS/m range, 18 locations in the 4–8 dS/m range, 4 locations in the 8–12 dS/m range and 2 were in the >12 dS/m range (based on the seasonal average EC_e).

21.6.2 Cost Effectiveness

Data were collected and analyzed regarding the input costs (costs for fertilizer use, crop protection, labor, seeds, irrigation equipment) as well as the yield and market value. Although the analyzes are ongoing, and the second year of data will be required to obtain a robust and reliable data set, the first trends do show some interesting results. The introduced crop varieties showed no yield reductions even at the higher salinity levels; however, there was considerable variation in crop yield within salinity classes suggesting that other factors besides salinity have a great effect on crop yield, i.e. greater than the effect of salinity. This effect was also observed under controlled field conditions (Van Straten et al. 2020), and additional research is needed to determine which other factors affect the yield in which manner and how this can be improved.

On average, the input costs for fertilizers and labor were more than twice as high as the costs for crop protection and irrigation (irrigation costs are mostly based on pump renting and fuel costs). The input cost for seeds vary greatly depending on the crop, and in the case of carrot, beetroot and potato, the input cost for the seeds was the largest investment for the farmer. However, the market value of beetroot and carrot was the highest (no market value for potato has been obtained so far) and in all cases (for all five crops that were evaluated) the cultivation was, on average, profitable for the farmers. The profit for the farmers varied, on average, between 70 and 240 euros per decimal (commonly used unit of area in Bangladesh, equals around 40 m^2), with beetroot, cabbage and cauliflower showing the highest profit. These results are based on data sets collected at around 30 farms for each crop. Again, these first results are only based on one dry season and should be validated in a second year, especially since these profits appear to be very high.

21.6.3 Project Accomplishments according to the SDGs

Two years after the start of the project, directly after the season in which the above-mentioned inputs costs and potential profits were obtained, an independent third-party evaluation, executed by the GBF, took place to determine the effect of the project (mid-term results, see Table 21.1). The SDGs were used to determine the impact of the project.

First, results showed that 75% of the farmers now use the salt-affected land during both dry seasons as described in Section 21.1, which addresses SDG 2, Zero

TABLE 21.1
The Mid-Term Results of "The Salt Solution" Project, Based on an Independent Evaluation for Which 260 Farmers from a Group of 2,000 Farmers Were Interviewed. Project Outcomes Are Linked to the Relevant SDG

SDG Number	SDG Description	Accomplishment in the Project after 2 Years	Before Start of Project	2 Years after Start Project
1	No poverty	**Average household income increased**		34%
		Households with more than €100, - monthly increase:		
		- *Lead farmers*		55%
		- *Group farmers*		4%
		Employment increased		
		- *Lead Farmers*		10%
		- *Group farmers*		41%
2	Zero hunger	**Food security increased**[a]	15%	65%
		Use of salt affected fallow land increased[b]	0%	76%
3	Good health and well being	**Vegetable consumption increased**[c]	26%	74%
		Households improved dietary diversity	75%	100%
5	Gender equality	**Skills on sustainable food production in women increased**	9%	79%
		Access to land for women increased	4%	87%

[a] Food security is based on household food insecurity access scale-0 (full food security).
[b] During the first part of the dry season (December–February).
[c] Defined as the consumption of a minimum of 150 g/day, during at least 10 months/year.

Hunger. At the start of the project, none of the farmers used this land. Vegetable consumption increased from 26% to 74% which addresses SDG 3, Good Health and Well-Being, and food security increased from 15% to 65% also addressing SDG 3. Average household income increased by 34% which falls under SDG 1, No Poverty. The project also deals with gender inequality (SDG 5), providing training for women, resulting in an increase of the number of women with improved skills for sustainable food production from 9% to 79% and increasing access to land for women from 4% to 87%.

21.7 NEXT STEPS

The results shown in Table 21.1 are based on one year (one season). The second year of similar results would validate the possibilities and the profitability of crop cultivation under saline conditions. It is the ambition of the team to collect and analyze this second year of data but is beyond the scope of this case study.

21.7.1 Conditions for Further Testing and Training

The second crop cycle in the dry season (March–June) will be more challenging, as salinity levels increase until the next monsoon rains, in combination with high temperatures and low freshwater availability. Crops should be both salt and heat tolerant, and water availability should be ensured by (rain) water harvesting in times of surplus. Underground freshwater can be stored using novel methods in making optimal use of the existing soil profile. Rainwater is stored in coarse sand surrounded by natural clay layers that form a barrier. Later, this water can be pumped up and used for irrigation.

The newly built test facility has a special focus on the summer crops (such as okra, Indian spinach, bitter gourd and eggplant) and underground storage of water. More testing is needed to determine which varieties are most suitable for cultivation under saline conditions. Before new crop varieties can be introduced more broadly at the farm level, the proof of concept should expand, demonstrating a relatively low risk for the farmers involved. This can be centered around the Training Center in Bangladesh and the first pilots have begun. Breeding for salt tolerance to introduce even better varieties is also taking place at the Training Center but this is a time-consuming effort and is planned for the coming years. At present, the economic viability of underground water storage is being evaluated. This includes the price of equipment, installation costs and the amount of water that can be stored and the market value of the crops that can be cultivated off season.

Once the input materials are locally available and the farmers are able to acquire these materials and know-how to implement the best practice cultivation strategy, then large-scale implementation of saline agriculture is feasible. Currently, vegetable cultivation takes place on the land around the farmhouses; much greater impacts might be achieved when the rice paddies are used for vegetable cultivation in the dry season. This will also involve mechanization, large-scale water harvesting, improved market access and improved export potential.

TABLE 21.2
Overview of Limiting Factors for Agriculture during the Dry Season and the Proposed Solutions

Limiting Factor	Proposed Solution
Scarcity of quality irrigation water	Rainwater harvesting during monsoon, (underground) storage
Unstable rainfall	Rainwater harvesting during monsoon, (underground) storage
Limited salt-tolerant crop choices	Introduction of salt-tolerant varieties by Salt Farm Texel, Lal Teer Seed
Restricted sowing/planting time	Develop cultivation strategies to increase the window of planting
Polder structure of areas	Develop a strategy to shift from rice paddy to vegetable field and back
Soil salinity	Lowering soil salinity levels by making use of monsoon rains for leaching
Low soil fertility	Introduction of cultivation strategy with organic inputs, crop rotation
Shallow saline groundwater	Minimize capillary rise of saline groundwater into rootzone
Heavy soil that requires tillage	Cultivation strategy of raised beds with minimum tillage
Perennial water logging	Low tech drainage system, raised cultivation beds
Lack of extension programs	Capacity building of extensions services
Insufficient training for saline agriculture	Intensive training programs for (lead) farmers
Difficult communication, marketing	Develop a network of lead farmers, farmer field schools, training center, engage local communities
Availability of input materials	Ensure local availability seeds, (organic) fertilizers, irrigation equipment
Financial means	Availability of micro-finance
Proven minimum risk for farmers	Demonstration of crops and cultivation strategies, business model for farmers
Low market value	Diversify cropping systems, improve "going-to-market window", improve farmer organization, improve market access, create more export opportunities
Limited success after the end of the project	Several social indicators will be identified and quantified to measure factors of success and these factors will be used to stimulate the continuation of the success after the project ends

21.7.2 Limiting Factors

Several limiting factors need to be addressed before scalable solutions for crop cultivation under saline conditions can be introduced successfully. The limiting factors, as presently identified, are listed in Table 21.2. The proposed solutions for the identified limiting factors are also listed.

21.8 CONCLUSION

Although most data presented in this case study are based on a single season, the results do indicate that crop cultivation under (moderate) saline conditions in coastal Bangladesh is possible and profitable. The vast majority of the farmers previously did use the salt-affected land in the dry season, but now with the help of this project, three out of four farmers in the local area are empowered to use salt-affected land. The results in Table 21.1 clearly show the cascading effect of this and the impact of the project in addressing the SDGs.

These accomplishments closely match some of the SDGs. Now, "new" land can be used for crop cultivation (SDG 2: Zero Hunger) which improves the vegetable consumption (SDG 3: Good Health and Well-Being) and diversifies the diet, increases the food security and the income for the households (SDG 1: No Poverty, 2: Zero Hunger and 3: Good Health and Well-Being). Since women were actively involved, their skills and participation increased greatly (SDG 5: Gender Equality). The project is a clear example of Climate Action (SDG 13) and setting up (public-private) partnerships (SDG 17) to reach the goals. Although this project is already training 5,000 farmers, many more farmers need to be reached in coastal Bangladesh and beyond. Most farmers started this project dedicating only a small piece of land for demonstration in the project but have turned their whole land into the tailor-made adaptive farming system developed at the Training Center. This shows that the chosen approach does work and farmers are willing to adopt the new farming strategies in order to better adapt to climate change.

Salt-affected lands are often considered to be unsuitable for crop production, but, in fact, the saline resources of the world have the potential to help improve the livelihoods of millions of farmers and contribute to global food security.

ACKNOWLEDGMENTS

The successful implementation of the salt-tolerant crop varieties from the Netherlands in Bangladesh could not have taken place without the contribution of the ICCO staff in Bangladesh. This was really a team effort and especially the contribution of Tessa Schmelzer, Abul Azad, Arun Ganguly and Masud Rana should be mentioned here. Also, the team of CODEC did a great job with the farmer training and data collection in the field, which is gratefully acknowledged.

REFERENCES

Bruning, B., van Logtenstijn, R., Broekman, R., de Vos, A. C., Parra González, A., Rozema, J. 2015. Growth and nitrogen fixation of legumes at increased salinity under field conditions: implications for the use of green manures in saline environments. *AOB Plants*, 7, plv010.

Chen, J., and V. Mueller. 2018. Coastal climate change, soil salinity and human migration in Bangladesh. *Nature Climate Change* 8, 981–985.

De Vos, A.C., Bruning B., van Straten, G., Oosterbaan, R., Rozema, J., van Bodegom, P. 2016. Crop salt tolerance under controlled field conditions in The Netherlands, based on field trials conducted by Salt Farm Texel. https://edepot.wur.nl/409817 (Accessed September 2020).

Soil Resource Development Institute (SRDI). 2010. Saline soils of Bangladesh. http://srdi.portal.gov.bd/sites/default/files/files/srdi.portal.gov.bd/publications/bc598e7a_df21_49ee_882e_0302c974015f/Soil%20salinity%20report-Nov%202010.pdf. (Accessed September 2020).

Van Straten, G., Bruning, B., De Vos, A.C., Parra González, A., Rozema, J., van Bodegom, P.M. 2020. Distinguishing potato varieties by salt tolerance through novel analysis of multiple-year field tests. *Submitted*.

22 Case Study – Stichting De Zilte Smaak

'Discovering Saline Farming Potential on Terschelling'

Jacqueline Wijbenga and Stichting De Zilte Smaak

CONTENTS

22.1 INTRODUCTION

Stichting De Zilte Smaak (The Saline Taste Foundation) was founded in 2017. Three entrepreneurs on Terschelling initiated the foundation: Flang Cupido (Cooking Studio Flang in de Pan), Ria Laanstra (Catering Service 't Lokaal) and Hans Wilmink (owner of holiday apartments De Zeeboer). Being active in the hospitality sector themselves, they saw opportunities for saline products in menus of local restaurants and for developing other specific island food products.

Tourists associate the island Terschelling and its surroundings with sun, sand, sea and salt. Logically, these features are translated into the commercial touristic

DOI: 10.1201/9781003112327-22

products on offer at the island. Saline vegetables and herbs might fit perfectly into the existing range of these products. With this in mind, the idea arose to explore the possibilities of saline farming on the island. It fits perfectly in the growing attention for producing, tasting and buying local products as well.

22.2 SALINITY

Salinity is not new to the island. Being situated in the UNESCO world heritage region Waddenzee, the island is surrounded by salt sea water. For its fresh drinking-water supply, Terschelling relies on a fresh groundwater reservoir in the dunes of the island. In addition, the water company provides freshwater through a pipeline from the mainland.

The freshwater available for farming purposes on the island itself is plentifully available during autumn and winter and to a lesser extend also in spring. During these months, the water used for field irrigation in the summer months is replenished by rainfall. However, during the past few sunny, warm and extremely dry summers, the demand has been greater than the replenishment. The availability of freshwater for farming purposes, therefore, has become a major concern for the island.

22.3 CLIMATE CHANGE

Also, with the rise of the sea-level due to climate change, it is expected that the brackish or saline water pushing in from the sea towards the freshwater system on the island will affect farming more in the years to come. The plots closest to the Waddenzee dike on the south of the island have already experienced the effect of the saline seawater pushing in: waterways close to the dike mainly have brackish water. The salinity becomes a problem for farmers, as with increasing salt levels in the water it cannot be used as drinking water for the cattle anymore. All farmers on the island of Terschelling nowadays are dairy farmers. Land close to the Waddenzee dike is used for the harvest of roughage for cattle, in some cases for the grazing of sheep and if applicable as a resting area for the many geese and other migratory birds.

22.4 NEW OPPORTUNITIES

Saline crops are normally found on soils that are regularly flooded by salty or brackish water and tend to occur in clay-rich soils. As the land close to the coastline becomes more saline, growing saline arable crops might have potential on the soils of Terschelling. However, the soil conditions vary depending on the location on the island. The same goes for the saline conditions of the available water. The entrepreneurs of De Zilte Smaak looked for – and found – a piece of land on the island on which they could grow the saline crops. They started farming with crops such as samphire, sea lavender and ice plant (Figure 22.1). Their aim was to make positive use of the salinization of water and/or soil to learn how to grow these unknown crops and to share their knowledge and experiences.

FIGURE 22.1 Ice plant.

22.5 SALFAR

Besides gathering information on how to grow the crops as well as possible, the initiators have focused on product development based on the harvested produce. At first instance, the aim was to combine the fresh vegetables into a durable, shelf-stable product such as pesto or dried herbs. These kinds of products are not new to the market; however, they are new to Terschelling. In order to make use of already available knowledge and to benefit from ongoing research, the Stichting decided to participate in the international Interreg VB North Sea Region SalFar project.

22.6 EXPERIENCE

Based on experiences in the first year, the initiators learned that growing saline crops requires focus, attention and crop knowledge. In order to professionalise the growth of the crops and the work in the field, more information on arable farming was required. This was obtained by recruiting an experienced arable crop farmer from the mainland on a freelance basis. His external expertise proved beneficial to the success of the field design, crop development and harvest of ready produce.

22.7 PRODUCT INTRODUCTION

First steps into the market proved that there certainly is an interest in saline products. The freshly harvested saline crops were offered for taste sessions to consumers through local markets on the island. Consumers were enthusiastic about the possibilities to incorporate freshly grown saline products in an everyday (vacation) dinner to add some extra island experience to their daily meal. Initiators Ria Laanstra and Flang Cupido successfully developed pesto based on the saline crops. When offered at the local market, the public enthusiastically welcomed this pesto as well.

The warm welcome of the products by the consumers at the local Terschelling markets proved that the initiators were on the right track with product development. However, the fresh produce offers a major challenge with regard to market development. Contrary to the ideal situation, the fresh product is not always available to meet the demand.

22.8 MORE VARIETIES

On the other hand, the recipe for pesto, although it can change based on the availability of the crops, has certain restrictions as well. Not all of the surplus of the crops can be processed into a balanced pesto. This experience in the second year of saline crop growing led to the conclusion that even more control of the field experiment was needed. Furthermore, a greater variety of crops – such as sea aster, beach beet, sea banana and oyster leaf – would be worth exploring as it was clear that the harvested produce could not all be sold as the fresh product (Figures 22.2 and 22.3).

FIGURE 22.2 Sea banana.

FIGURE 22.3 Oyster leaf.

22.9 DURABLE PRODUCTS

Development of more durable consumer products such as saline herb mixtures should, therefore, be explored further. Mixtures cannot be based on only a few different ingredients; therefore, a wider range of saline crops should be incorporated in the field design. Every crop has its own unique saline taste experience. The herb mixtures might be used for different applications and offer another opportunity to produce a product with an extended shelf life.

22.10 NEW CHALLENGES

The year 2019 offered new challenges for the test field of De Zilte Smaak. Unfavourable growth conditions in spring, an extraordinary dry hot summer and a very wet autumn hindered the development of many of the crops. The production of fresh produce was very irregular and neither suitable for regular delivery of products to the local farmers market nor for offering to consumers or being used by cooks in their restaurant menus. Most of the harvest, therefore, was used for making pesto.

22.11 NEW AMBITIONS

However, there were some small successes in the test field over the past few years that have given the team of De Zilte Smaak new ambitions (Figure 22.4). Some saline crops turned out to be very resilient and survived the unfavourable growing conditions. And some of the potato breeds that were tested in 2019 have been offering yield, even though much less than the commercially grown crop on the mainland.

FIGURE 22.4 Test field.

22.12 NEW FIELD DESIGN

The experience gained by the team resulted in yet another field design for 2020. The design is more robust and is based on more regulated growth conditions in the field. Furthermore, it was found that some of the saline crops have a troublesome start when sown directly in the open field. Therefore, it was decided to sow these particular crops in the future in a greenhouse and plant the seedlings later on as soil temperatures and growing conditions are more favourable for these plants.

Some of the crops need to be restricted as these have a tendency to overgrow other crops. Growing them in containers can prevent this. It enables the team to easily keep the soil free from weeds and prevents the crops from expanding uncontrollably. Growing the different herbs in restricted spaces also offers an opportunity to produce the product more in accordance with market demand.

22.13 IRRIGATION

Under dry conditions, the soil on Terschelling has difficulties retaining moisture. The team invested in drip irrigation for the open field as well as for the containerised crops. This enabled the team to irrigate crops in periods of drought, and also offered the opportunity to irrigate seeds to ensure germination when moisture levels were low or when crops needed less saline water to stimulate germination. The irrigation should also guarantee stable growth of the crops, giving a more secure base for the

FIGURE 22.5 Drip irrigation for rucola.

development of the crops during the season. This would benefit the market position of the fresh produce towards restaurants on the island by providing a more stable delivery of saline vegetables (Figure 22.5).

22.14 MARKET DEMAND

Market demand in the past years has been largely based on the development and production of pesto. The team of De Zilte Smaak would like to explore further which products could be developed based both on the potential of field crops and the market opportunities on the island. In order to do so, cooperation will be sought with other hospitality entrepreneurs who have shown interest in using the fresh produce in their restaurants.

22.15 COOPERATION

In the summer of 2020, the first successful cooperation with local restaurants was started. Two restaurants on the island signed an agreement with Stichting De Zilte Smaak. Following the agreement, restaurants can harvest crops during the season to meet the demands of the menu their cooks have designed. In return, the restaurants sponsor the initiative financially. As the summer of 2020 again was sunny, hot and dry, the initiators were careful in their approach and selected two restaurants for the first experience. The insecurity of crop availability was the main reason for a somewhat hesitant market introduction. During the summer, this turned out to be the right choice. As the summer progressed into autumn, some crops had an extra growth boost from the rainwater that fell in August.

22.16 PRODUCT DEVELOPMENT

Some of the crops turned out to deliver far more produce than could be used as fresh product in the restaurant menus. Therefore, the initiator will invest in making pesto from the surplus this year as well, knowing it will sell as an interesting local product in specialised deli shops on the island. The team will also continue to experiment with products from the test field and develop durable products such as herb mixtures, as they can be sold to tourists to let the memory of their stay on the island linger on. Obviously, the team will keep searching for other products in which the saline herbs can be used.

22.17 THE PEOPLE OF STICHTING DE ZILTE SMAAK

Stichting De Zilte Smaak (The Saline Taste Foundation) is an initiative of the people of Terschelling. Islanders do the fieldwork and product development as well. By exploring and developing saline farming on the island and creating a market for the saline products, the team of De Zilte Smaak aims to contribute to more sustainable crop production and offer local entrepreneurs an opportunity to differentiate themselves with unique products. External experts from the island of Terschelling support the work of the foundation, and in some cases, expertise is gained from the mainland (Figure 22.6).

FIGURE 22.6 Field visit.

Section III

Crop Salt Tolerance and Microbiological Associations

23 Developments in Adaptation to Salinity at the Crop Level

Theo Elzenga, Edward G. Barrett-Lennard, and Redouane Choukr-Allah

In the ten chapters of this section, we focus on alternative crops adapted to saline environments and on the responses of plants to salinity. The first chapter (Jovanović and Radović 2021) emphasizes the mechanisms that enable the salt-tolerant plant to survive and even thrive and produce in saline environments that are fatal or severely inhibitory to salt-sensitive plants. The presence of high levels of NaCl in the root zone has two major effects on plants. A direct osmotic effect, or physiological drought, reducing the uptake of water by the plants, and a toxic effect caused mainly by the replacement of K^+ by Na^+ in the cytoplasm resulting in inhibition of a multitude of cellular processes (Munns 2002; Cuin et al. 2009). Typical adaptations to soil salinity in specialized, halophytic plant species are the more efficient use of water and the accumulation of NaCl in the vacuole, making use of Na^+ as a 'cheap' osmoticum, balancing the osmotic potential in the soil with NaCl, rather than by synthesizing metabolically expensive small molecular weight organic compounds (Greenway and Munns 1980); also the compartmentalization of NaCl keeps it away from the cytoplasm and sensitive enzyme systems (Flowers 1972; Greenway and Osmond 1972). Plant species from saline environments are often also capable of getting rid of excess NaCl, using salt glands (Marcum 1999).

The domestication of halophytic species and their use in agriculture has been tested by several authors to replace or supplement the vegetation of saline rangelands (Norman et al. 2013). For example, pioneering research by Clive Malcolm and colleagues in Western Australia showed that the revegetation of saline habitats with halophytic forage shrubs (*Atriplex* spp.) was profitable and provided many additional benefits (see citations in Barrett-Lennard and Norman 2021); the Western Australian technology is now being applied in other parts of Australia and Pakistan (Barson and Barrett-Lennard 1995; Qureshi and Barrett-Lennard 1998).

In non-salt tolerant ('glycophytic') plant species, exposure to salinity often induces a reduction in Na^+ uptake in the roots, activating 'reflux' transporter systems in the root, thereby decreasing the amount of Na^+ reaching the leaves (Roy and Tester, 2013); the salt that does enter the xylem stream accumulates in the older leaves and the early senescence of these leaves can help to protect the younger still actively growing parts of the plant and the decrease in leaf area reduces water uptake. True salinity-adapted plant species (halophytes) typically show a growth rate increase at

DOI: 10.1201/9781003112327-23

moderate salinity levels, whereas glycophytic species tend to have a strong reduction in growth at similar salt concentrations (Flowers and Colmer 2008).

Why is agriculture under saline conditions so difficult? The answer lies principally in the origin of most of our crop plants. In adopting plants that were originally gathered in nature for controlled cultivation under farmers' care, the selection favored plant species that were fast-growing, performed well in the absence of other competition and disturbed soil, did not have extensive defense traits, such as thorns or toxins, and produced many seeds. Although all these traits would have been attractive to early farming communities, they can also be considered the root cause of many of the problems that modern agriculture is now facing. The traits selected for 'tick all the boxes' of plants are categorized as ruderals (Grime 2001), the typical pioneer plants, often annuals or short-lived perennials of disturbed ecosystems. This category of plants invests relatively little resources into pathogen resistance and, as recent studies show, also may have antagonistic relations with soil and rhizosphere micro-organisms (Karol et al. 2010). In contrast, most perennial plant species invest in defense and competitiveness and in a mutualistic relation with the microbiome of the root zone. Many of the traits that enable these plants to tolerate stress, optimize resource utilization, and resist pathogens can be partly attributed to their symbiotic relationship with micro-organisms. On the other hand, the species that have been selected as our crop plants encourage antagonistic micro-organisms to thrive in the root zone, leading to the accumulation of pathogens and competing soil micro-organisms. Breeding programs have, unknowingly, further reduced the potential of plants to attract so-called plant growth-stimulating bacteria or fungi (Pérez- Jaramillo et al. 2015).

In the chapters of this section (Chapters 25, 26, 27, 28, 29), we have also reviewed the successes and difficulties that breeders and plant physiologists have encountered in trying to select for and use salt--tolerant plants in breeding programs. Most selection programs have consciously or unconsciously, been biased toward plant traits that made our agriculture more dependent on farming practices that aimed to control and optimize growth conditions (e.g. use of irrigation, fertilizers, herbicides, and pesticides). They have also reduced the capacity of our crop plants to use the services of natural allies; this has triggered new research programs on the role of symbiotic micro-organisms in stress resistance (Chapters 32, 33). The predicted increase in extreme weather events, short periods of heavy precipitation and prolonged periods of drought during the growing season, has made us aware that relying on full control of growth conditions is untenable and that our crops should be based on plant varieties that are more robust, that can fend for themselves better, and depend less on chemical means for keeping them healthy (Chapters 30, 31).

Let's look at a number of potential answers to the question how to make our crops more salt-resistant. The first is the breeding of varieties of crops that are considered glycophytic, still possessing some of the traits that, in combination, could yield very hardy varieties (Zaki 2011). For other glycophytic species, the presence of wild sister species with desirable salinity tolerance traits could be exploited by backcrossing and selecting for combinations of stress-tolerance and elite yield traits (Brozynska et al. 2015; Mickelbart et al. 2015). This approach has been adopted by researchers looking for genes for salt tolerance in grasses allied to commercial wheat (Colmer et al. 2005).

We could also turn to nature and adopt plant species that are naturally growing in a saline environment but have not been adapted at a large scale for agricultural purposes. Most of the saline agriculture crops have only recently been entered into breeding and selection programs, although several studies have been done on what we can consider landraces, many of the traits associated with elite varieties have not been systematically studied yet. In our traditional crop plants, the elite varieties have been optimized for growth parameters such as water and nutrient use efficiency, pathogen resistance and yield, market traits, such as ability to keep in storage, and consumer traits, such as nutritional value, taste, smell, and visual appeal. For the novel saline farming crops, most of the optimization in extensive breeding programs still has to be done. Studies aimed at filling this gap can be found in Chapters 28, 29, 30, and 31.

The last two chapters in this section of the book (Jovanović and Radović 2021; Munikumar et al. 2021) explore the potential for using plant growth-promoting bacteria of fungi to improve stress tolerance. These studies are based on the hypothesis that the symbiosis between a plant and specific micro-organisms can strongly affect a plant's resistance to abiotic stress. Isolating microbial symbionts from plants that grow naturally under particular abiotic stress conditions and exposing (or infecting) crop plants with these isolates may improve the stress resistance of the crop plant that hosts the micro-organism.

REFERENCES

Barrett-Lennard EG and Norman H (2021) Agriculture in salinizing landscapes in southern Australia: Selected research 'snapshots'. In: *Future of Sustainable Agriculture in Saline Environments*. Routledge.

Barson M and Barrett-Lennard EG (1995) Productive use and rehabilitation of Australia's saline lands. Australian Journal of Soil and Water Conservation 8, 33–37.

Brozynska M, Furtado A, and Henry RJ (2015) Genomics of crop wild relatives: Expanding the gene pool for crop improvement. Plant Biotechnology Journal 14, 1070–1085. doi.org/10.1111/pbi.12454

Colmer TD, Munns R, and Flowers TJ (2005) Improving salt tolerance of wheat and barley: Future prospects. Australian Journal of Experimental Agriculture 45, 1425–1443.

Cuin TA, Tian Y, Betts SA, Chalmandrier R, and Shabala S (2009) Ionic relations and osmotic adjustment in durum and bread wheat under saline conditions. Functional Plant Biology 36, 1110–1119.

Flowers TJ (1972) The effect of sodium chloride on enzyme activities from four halophyte species of Chenopodiaceae. Phytochemistry 11, 1881–1886.

Flowers TJ and Colmer TD (2008) Salinity tolerance in halophytes. New Phytologist 179, 945–963.

Greenway H and Munns R (1980) Mechanisms of salt tolerance in nonhalophytes. Annual Review of Plant Physiology 31, 149–190.

Greenway H and Osmond CB (1972) Salt responses of enzymes from species differing in salt tolerance. Plant Physiology 49, 256–259.

Grime, JP (2001) *Plant strategies, vegetation processes and ecosystem properties.* (2nd edn) Wiley and Sons. p 417.

Jovanović Z and Radović S (2021) Plant growth promoting bacteria as an alternative strategy for the amelioration of salt stress effects in plants. In *Future of Sustainable Agriculture in Saline Environments.* Routledge.

Kardol P, Bezemer TM, and van der Putten WH (2006) Temporal variation in plant-soil feedback controls succession. Ecology Letters 9 doi: org/10.1111/j.1461-0248.2006.00953.x

Marcum KB (1999) Salinity tolerance mechanisms of grasses in the subfamily Chloridoideae. Crop Science 39, 1153–1160.

Mickelbart MV, Hasegawa PM, and Bailey- Serres J (2015) Genetic mechanisms of abiotic stress tolerance that translate to crop yield stability. Nature Reviews Genetics 16, 237–251.

Munns R (2002) Comparative physiology of salt and water stress. Plant, Cell and Environment 25, 239–250.

Munikumar S, Nataraja KN, and Elzenga JTM (2021) Tolerance to environmental stresses: Do fungal endophytes mediate plasticity in *Solanum dulcamara*? In *Future of Sustainable Agriculture in Saline Environments*. Routledge.

Norman HC, Masters DG, and Barrett-Lennard EG (2013) Halophytes as forages in saline landscapes: Interactions between plant genotype and environment change their feeding value to ruminants. *Environmental and Experimental Botany* 92, 96–109.

Pérez- Jaramillo JE, Mendes R, and Raaijmakers, JM (2015) Impact of plant domestication on rhizosphere microbiome assembly and functions. *Plant Molecular Biology* 90, 635–644. DOI: 10.1007/s11103-015-0337-7

Qureshi RH and Barrett-Lennard EG (1998) *Saline Agriculture for Irrigated Land in Pakistan: A Handbook.* Monograph No. 50, Australian Centre for International Agricultural Research, Canberra, p 142.

Roy SJ and Tester M (2013) *Increasing slat tolerance of crops.* In: *Sustainable Food Production.* Christou P, Savin R, Costa-Pierce BA, Misztal I, Whitelaw CBA Eds. Springer. DOI: https://doi-org.proxy-ub.rug.nl/10.1007/978-1-4614-5797-8_429

Zaki F (2011) The determinants of salinity tolerance in maize (Zea mays L.) PhD Thesis University of Groningen, p 226. http://hdl.handle.net/11370/ae616bc6-de6d-442c-8f20-0318fcaa63c0

24 Salt Effects on Plants
An Overview

Živko Jovanović and Svetlana Radović

CONTENTS

24.1 SALINITY STRESS

Plants as sessile organisms are constantly exposed to different constraining environmental conditions, such as drought, salinity, extreme temperatures, UV radiation, heavy metals and hypoxia. These factors, collectively referred to as abiotic stresses, limit plant growth, and in agricultural crops, decrease productivity and yield. Resistance or sensitivity to these factors is very important, given that they can affect different stages of plant growth and development, and may act synergistically with other abiotic stresses (Chinnusamy et al. 2004). Therefore, the discovery of mechanisms underlying tolerance to abiotic stresses and adaptations to these are a major focus of contemporary research.

In parallel with climate change, which will affect crop production worldwide, there is an increasing need to maintain food security for a growing global population. The quality of land and the availability of water will be critical for agriculture in the future. The Food and Agriculture Organization estimates that salinity, as one of the major abiotic stress factors of plants, currently affects more than 6% of the world's land area (Ilangumaran and Smith 2019). However, much of this land is not used for cultivation; according to the FAO, 450 million ha of irrigated land and 32 million ha under dryland agriculture have been affected by salinity (Munns and Tester 2008). Some reports suggest that the area of salinized land will increase with climate change as the scarcity of water will lead to increased salt concentrations.

DOI: 10.1201/9781003112327-24

24.1.1 Soil Salinity

According to the USDA salinity laboratory, a soil is defined as saline when its electrical conductivity of the saturation extract (EC_e) exceeds 4 dSm^{-1}, which is approximately 40 mM sodium chloride (NaCl). Salt-affected lands have been present for thousands of years; the oldest written records date back to ~2400 BCE when salinity was recorded in the alluvial plains of Iraq (Russel et al. 1965). Salinization of soil can be caused by different mechanisms (natural or human activities) that can increase the concentration of dissolved salts. The salts present in soils are predominant NaCl, but may also include sodium nitrate ($NaNO_3$), sodium sulphate (Na_2SO_4), calcium sulphate ($CaSO_4$), potassium sulphate (K_2SO_4), magnesium chloride ($MgCl_2$) and magnesium sulphate ($MgSO_4$). Salt-affected lands are widespread and occur in all climatic regions, at different altitudes, from below sea level (e.g. the Dead Sea) to mountains above 5000 meters. Natural processes, which lead to the accumulation of the salt in the soil and groundwater, during extended periods of time, result in the formation of salt lakes, marine sediments and salt scalds (Ilangumaran and Smith 2017). However, the process of soil salinization is aggravated and accelerated by cultivation, mainly by crop irrigation, but also by land clearing and inadequate drainage. In general, irrigated lands are more affected by salinity than drylands; this is because the deposition of salt is an inescapable part of the irrigation process. The water used for irrigation can contain calcium (Ca^{2+}), magnesium (Mg^{2+}) and sodium (Na^+). In alkaline soils, after evaporation, Ca^{2+} and Mg^{2+} ions can precipitate as carbonates, leaving sodium as a dominant ion in the soil (Serrano et al. 1999). In addition, irrigation of crops with more water than they can utilize can lead to an increase in the water table, which can mobilize salts from depth in the soil profile towards plant roots.

The capacity of soil to store salt depends on soil type. Sands are more readily leached than clays. Also, sandy soils have low cation exchange capacity compared to clay; negatively charged clay particles are, therefore, able to adsorb Na^+. If Na^+ is adsorbed to the cation exchange complexes of clays, the soils can become dispersive causing soil aggregates to break down, increasing bulk density and, thus, changing soil porosity. This can affect soil and root aeration. Because of these factors, plants in saline soils can suffer from high levels of sodium ions, water deficits (similar to drought conditions) as well as hypoxia (Carillo et al. 2011).

24.2 EFFECT OF SOIL SALINIZATION ON PLANTS

Soil salinity is a major factor limiting the yield of agriculturally important plants and may well jeopardize the capacity of agriculture to cope with a projected food demand for 10 billion people by 2050 (Flowers 2004; Parida and Das 2005; Munns and Tester 2008; GAP Report 2018).

Plants can be divided into two different groups depending on their ability to survive under saline conditions: glycophytes and halophytes (Munns and Termaat 1986). Most plants, including most crop plants, belong to the first group; their growth is inhibited and they may even be killed by 100–200 mM NaCl. These plants have evolved under conditions of low soil salinity. On the other hand, halophytes can survive high concentrations of salt, up to 400 mM NaCl. Depending on their salt-tolerance, they can be further subdivided into two groups – obligate and facultative halophytes. Facultative halophytes thrive in less saline habitats, mainly marginal soils representing the border between saline and non-saline soils. On the other hand, obligate halophytes are found in landscapes of continual high salinity.

24.2.1 Plant Response to Salinity Stress

Salinity is a complex stress for plants, affecting growth and development in different ways. First, soil salinity disturbs the capacity of roots to extract water because of an osmotic (water) stress. A second stress is cytotoxicity due to the excessive uptake of ions (predominantly sodium and chloride). Third, there can be stress caused by nutritional imbalance. All these effects are typically accompanied by oxidative stress due to the generation of reactive oxygen species (ROS) (Tsugane et al. 1999; Hernández et al. 2001; Isayenkov 2012).

A two-phase model describes plant responses to salinity. During the first phase, osmotic shock causes stomatal closure and inhibition of cell expansion. This phase is immediate, taking place within minutes to hours (Munns and Passioura 1984; Munns and Termaat 1986; Rajendran et al. 2009). A second phase of response takes longer, days to weeks; in this phase, the cytotoxicity of ions slows down metabolism, causing senescence and cell death (Roy et al. 2014; Isayenkov and Maathuis 2019). Plants can adapt to salinity through three mechanisms: control of ion uptake, osmotic adjustment and detoxification.

24.2.1.1 Control of Ion Uptake

Overall, the most important strategy for plants is to control the uptake of ions rather than dissipating energy on detoxification and subsequent damage repair.

Ion uptake into plants can be through two different pathways – the symplastic and the apoplastic pathway (Gao et al. 2007; Maathuis et al. 2014; Negrao et al. 2017; Isayenkov and Maathuis 2019). The apoplastic pathway involves the direct flow of ions in cell walls and the extracellular space between the outside of the root and the root xylem (Krishnamurthy et al. 2009). In some monocots, apoplastic flux can be responsible for up to 50% of total Na^+ uptake, as well as up to 50% of Cl^- translocation to shoots (Shi et al. 2013; Yeo et al. 1987; Kronzucker and Britto 2011). On the other hand, the symplastic pathway requires ions to cross a cell membrane and then move from cell to cell in the symplasm through plasmadesmata. The uptake of Na^+ and Cl^- into roots into the symplastic pathway is catalyzed by different transporters. The most investigated amongst these belong to the group of nonselective cation channels. Another subclass of transporters is HKT (Isayenkov and Maathuis 2019). Arabidopsis possess only the subclass 1, Na^+ selective AtHK1 isoform, but monocotyledonous plants have multiple HKT isoforms. In addition, PIP isoforms as well as low-affinity cation transporter LCT1 and Na^+/H^+ antiporters could be involved in Na^+ transport (Schachtman et al. 1997; Kronzucker and Britto 2011; Huang et al. 2012; Byrt et al. 2017). In contrast to sodium, chloride is an essential nutrient for plants. It can be transported into plant cells by an H^+/Cl^- symport, but the precise mechanism is still unknown. The cation-chloride cotransporters represent another class of potential Cl^- transporters (Zhang et al. 2011).

Although there is a lot of data about mechanisms of salt entering into plants, how Na^+ and/or Cl^- are sensed by plants still remains unknown. In animals, primary Na^+ sensors are specific Na^+ selective ion channels, with Na^+ binding sites. These sites modulate the gate and may act as the reporters of Na^+ concentrations in body fluids. These binding sites can also modulate transporter activity. However, similar mechanisms have not yet been identified in plants (Isayenkov and Maathuis 2019). It is speculated that Ca^{2+} signaling could play an important role in early signaling events, but all upstream components are still unknown (Choi et al. 2014; 2016). It seems that the SOS pathway is a key regulator of Na^+ homeostasis in plant cells, but other components are ROS as well as cyclic nucleotides, such as cGMP (Kiegle et al. 2000; Donaldson et al. 2004). It has been shown that glycosyl inositol phosphorylceramide (GIPC) – sphingolipid in plant plasma membrane – could sense salts (Jiang et al. 2019). MOCA1 as a glucuronosyltransferase for GIPCs in the plasma membrane is required for salt-induced depolarization of the cell-surface potential, Ca^{2+} spikes and waves, Na^+/H^+ antiporter activation, and regulation of growth. Na^+ binds to GIPCs to gate Ca^{2+} influx channels.

24.2.1.2 Osmotic Adjustment

One of consequences of increased salt concentration in the soil is osmotic shock. Salt decreases the water potential of the soil; if this becomes more negative than the water potential of the roots, there can be tissue dehydration. Osmotic shock can decrease the rate of plant growth and development, even causing cell and plant death. One way of estimating the osmotic component of salt stress is necessary to compare the effects of salt exposure to exposure to an equi-osmolar quantity of an inert osmoticum such as polyethylene glycol. Many studies have shown that osmotic stress has more influence on plant growth than ion toxicity (Castillo et al. 2007; Zhao et al. 2010; Isayenkov and Maathuis 2019).

Plants can regulate their osmotic potential through the process of osmotic adjustment, which prevents macromolecules from denaturation. One of the most common mechanisms of osmotic adjustment is the synthesis and accumulation of compatible organic solutes – organic acids (e.g. malate, aspartate and oxalate), amino acids and their derivates (e.g. glutamate, proline and glycinebetaine), carbohydrates (sucrose) and polyols (e.g. sorbitol, mannitol and pinitol). These molecules act as the protectors of macromolecules under conditions of increasing ionic strength. Apart from their role as osmoprotectants, some of those molecules can also serve as sources of nitrogen. Thus, proline accumulated as a free amino acid for osmotic adjustment can also regulate the availability of nitrogen. In addition, some of these osmolites can stabilize cell membranes, preventing and reducing the damage caused by high salt levels (Iqbal et al. 2014; Vaishnav et al. 2016).

24.2.1.3 Detoxification

One secondary effect of salt stress and other abiotic stress is the occurrence of oxidative stress. Oxidative stress occurs because of an imbalance between the production and scavenging of ROS. These include hydrogen peroxide (H_2O_2), the hydroxyl radical ($OH\cdot$), synglet oxygen ($^1O_2$) and the superoxide anion radical (O^{2-}) (Sharma et al. 2012). Apart from ROS, the production of reactive nitrogen species can also cause oxidative stress. Many ROS are able to propagate themselves, as well as producing other ROS, leading to the damage of macromolecules in the cell; indeed, cell death can be caused by lipid peroxidation, protein oxidation and the oxidation of nucleotids in nucleic acids. Plants have developed mechanisms to protect them from ROS that are both non-enzymatic (e.g. the production of glutathione, ascorbate, tocopherols, carotenoids, phenolics) and enzymatic. Enzymatic defense mechanisms include the synthesis of enzymes that regenerate the reduced forms of antioxidants, such as ascorbate peroxidase (APX) and glutathione reductase, and ROS-interacting enzymes (ROS scavengers), such as superoxide dismutase, catalase (CAT), non-specific peroxidase (POD) and APX (Jovanović et al. 2011).

24.3 HOW TO PRODUCE SALT-TOLERANT PLANTS

A variety of strategies can be used to improve the tolerance of plants to different abiotic stresses, and all of them have been extensively studied because of the economic importance of plant tolerance to agriculture. These vary from traditional breeding and genetic engineering to the use of chemical (priming agents) and biological strategies (biofertilizer).

The traditional approach, i.e. breeding and selection for higher yields, is time consuming and labor intensive (Vaishnav et al. 2016). However, the development of new molecular techniques, such as the use of QTLs and molecular markers, gene mapping etc. have accelerated the development of marker-assisted breeding (Agarwal et al. 2013).

The development of molecular biology and genetic engineering provide the most progressive approach for producing salt tolerant plants. Halophytes (salt-tolerant plants) and glycophytes (salt-sensitive plants) can be used to identify genes associated with the salt response. Once identified and characterized, these genes can be transferred into different plants, in order to improve their tolerance against salt stress. There have been many attempts to obtain transgenic plants tolerant to salinity. Zhang et al. (2019) developed transgenic salinity-tolerant rice varieties using CRISPR/Cas9 targeted mutagenesis. Experimental data have shown that such genes are involved in different activities during salt stress, such as ROS scavenging, and the biosynthesis of compatible osmolytes and protective proteins (Ashraf and Akram 2009). Although the use of transgenic approaches has great potential for the development of salt tolerant varieties, a lot of ethical issues will need to be solved (Vaishnav et al. 2016).

Another method for improving plant growth in saline environments may be to prime plants with exogenous chemicals; in this case, chemical products of plant metabolism produced at low concentration in plants could be applied to enhance the salt tolerance of plants. These compounds, such as nitric oxide, H_2O_2, salycilic acid, jasmonates, proline and glycine betaine, could be applied to plants to control downstream processes. One major problem with the use of these compounds is that their use is not yet cost-effective. Furthermore, it is possible that these chemicals might cause a number of other environmental problems.

One of the most promising approaches for the enhancement of plant tolerance to salinity is the use of biocontrol agents such as plant growth promoting bacteria

(PGPB) (Russo et al. 2010; Vejan et al. 2016; Numan et al. 2018; Abbas et al. 2019; Egamberdieva et al. 2019).

24.4 FUTURE PERSPECTIVES

Salinity as a major limiting factor for crop productivity affects almost 30% of the irrigated land of the world. Every year this area increases by 1–2% (FAO 2014). Climate change, with an increase in temperature as well as with a decrease in average annual rainfall, will worsen the situation. Studies on plant responses to salinity have to be done not only with model plants (such as Arabidopsis), but also with other plant species, especially with agriculturally important and woody plants. Although many studies on salinity tolerance have focused at the agronomic, physiological and biochemical level, this focus is now increasingly moving towards the molecular level providing valuable information about candidate genes that could be used in salt resistant plants. New molecular methods will accelerate the development of transgenic salt-tolerant plants, with desirable traits. However, the use of transgenic plants will bring new problems, not only with consumers but also with legal issues. Another feasible strategy to mitigate salinity impacts on crop production could be the use of halotolerant PGPB, which may help plants thrive in more extreme environmental conditions.

REFERENCES

Abbas R, Rasul, Aslam K, Baber M, Shahid M, Mubeen F, Naqqash, T. 2019. Halotolerant PGPR: A hope for cultivation of saline soils. *Journal of King Saud University Science* 31:1195–1201.

Agarwal P, Bhatt V, Singh R, Das M, Sopory SK, Chikara J. 2013. Pathogenesis-related gene, JcPR-10a from *Jatropha curcas* exhibit RNase and antifungal activity. *Molecular Biotechnology* 54:412–425.

Ashraf M, Akram NA. 2009. Improving salinity tolerance of plants through conventional breeding and genetic engineering: An analytic comparison. *Biotechnology Advances* 6:744–752.

Byrt CS. Zhao M, Kourghi M, Bose J, Henderson SW, Qiu J et al. 2017. Non-selective cation channel activity of aquaporin At PIP2; 1 regulated by Ca^{2+} and pH. *Plant, Cell and Environment* 40:802–815.

Carillo P, Annunziata M G, Pontecorvo G, Fuggi A, Woodrow P. 2011. Salinity stress and salt tolerance. In *Abiotic Stress in Plants – Mechanisms and Adaptations*, ed. Sanker and Venkateswaralu, 21–38. Rijeka, In Tech.

Castillo EG, Tuong TPh, Ismail AM, Inubushi K. 2007. Response to salinity in rice: comparative effects of osmotic and ionic stresses. *Plant Production Science* 10:159–170.

Chinnusamy V, Shumaker K, Zhu, JK. 2004. Molecular genetic perspectives on cross-tolk and specificity in abiotic stress signaling in plants. *Journal of Experimental Botany* 55:225–236.

Choi WG, Hilleary R, Swanson SJ, Kim SH, Gilroy S. 2016. Rapid, long distance electrical and calcium signaling in plants. *Annual Review of Plant Biology* 67:287–307.

Choi WG, Toyota M, Kim SH, Hilleary R, Gilroy S. 2014. Salt stress-induced Ca^{2+} waves are associated with rapid, long-distance root-to-shoot signaling in plants. *Proceedings of the National Acadamy of Science of USA* 111:6497–6502.

Donaldson L, Ludidi N, Knight MR, Gehring C, Denby K. 2004. Salt and osmotic stress cause rapid increases in *Arabidopsis thaliana* cGMP levels. *FEBS Letters* 569:317–320.

Egamberdieva D, Wirth S, Bellingrath-Kimura SD, Mishra J and Arora NK. 2019. Salt-tolerant plant growth promoting Rhizobacteria for enhancing crop productivity of saline soils. *Frontiers in Microbiology* 10:2791.

FAO Report.2014. The state of food and agriculture. Innovation in family farming. Rome, http://www.fao.org/3/a-i4040e.pdf

Flowers TJ. 2004. Improving crop salt tolerance. *Journal of Experimental Botany* 55:307–319.

Gao JP, Chao DY,Lin HX. 2007. Understanding abiotic stress tolerance mechanisms: Recent studies on stress response in rice. *Journal of Integrative Plant Biology* 49:742–750.

GAP Report. 2018. Global Agricultural Productivity Reports (GAP Reports). Global Harvest Initiative, Washington. https://globalagriculturalproductivity.org/wp-content/uploads/2019/01/GHI_2018-GAP-Report-FINAL-10.03.pdf (assessed April 15, 2019)

Hernández JA, Ferrer MA, Jiménez A, Barceló AR, Sevilla F. 2001. Antioxidant systems and $O^{(2)(.-)}/H_{(2)}O_{(2)}$ production in the apoplast of pea leaves. Its relation with salt-induced necrotic lesions in minor veins. *Plant Physiology* 127:817–831.

Huang J, Lu X, Yan H, et al. 2012. Transcriptome characterization and sequencing-based identification of salt-responsive genes in *Millettia pinnata*, a semi-mangrove plant. *DNA Research* 19:195–207.

Ilangumaran G, Smith DL. 2017. Plant growth promoting rhizobacteria in amelioration of Salinity stress: A systems biology perspective. *Frontiers in Plant Science* 8:1-14

Isayenkov S. 2012. Physiological and molecular aspects of salt stress in plants. *Cytology and Genetics* 46:302–318

Isayenkov S, Maathuis, F. 2019. Plant salinity stress: Many unanswered questions remain. *Frontiers in Plant Science.* 10:1–18

Iqbal N, Umar S, Khan NA, Khan MIR. 2014. A new perspective of phytohormones in salinity tolerance: Regulation of proline metabolism. *Environnemental and Experimental Botany* 100:34–42.

Jiang, Z, Zhou, X, Tao, M. et al. 2019. Plant cell-surface GIPC sphingolipids sense salt to trigger Ca2+ influx. *Nature* 572:341–346.

Jovanović Ž, Rakić T, Stevanović B, Radović S. 2011. Characterization of oxidative and antioxidative events during dehydration and rehydration of resurrection plant *Ramonda nathaliae. Plant Growth Regulation* 64: 231–240.

Kiegle E, Moore CA, Haseloff J, Tester MA, Knight MR. 2000. Cell-type-specific calcium responses to drought, salt and cold in the Arabidopsis root. *Plant Journal* 23: 267–278.

Krishnamurthy P, Ranathunge K, Franke R, Prakash H, Schreiber L. Mathew M. 2009. The role of root apoplastic transport barriers in salt tolerance of rice (*Oryza sativa* L.). *Planta* 230:119–134.

Kronzucker H J, Britto DT. 2011. Sodium transport in plants: A critical review. *New Phytologist* 189: 54–81.

Maathuis FJM, Ahmad I, Patishtan J. 2014. Regulation of Na+ fluxes in plants. *Frontiers in Plant Science* 5:467.

Munns R, Passioura JB.1984. Effect of prolonged exposure to NaCl on the osmotic pressure of leaf xylem sap from intact, transpiring barley plants. *Functional Plant Biology* 11:497–507.

Munns R, Termaat A.1986. Whole-plant responses to salinity. *Functional Plant Biology* 13:143–160.

Munns R, Tester M. 2008. Mechanisms of salinity tolerance. *Annual Review of Plant Biology* 59:651–681.

Negrao S, Schmöckel S, Tester M. 2017. Evaluating physiological responses of plants to salinity stress. *Annals of Botany* 119:1–11.

Numan M, Bashir S, Khan Y, Mumtaz R, Shinwari ZK, Khan AL, Khan A, AL-Harrasi A. 2018 Plant growth promoting bacteria as an alternative strategy for salt tolerance in plants: A review. *Microbiology Research* 209:21–32.

Parida A, Das A. 2005. Salt tolerance and salinity effects on plants: A review. *Ecotoxicology and environment safety* 60:324–349.

Rajendran K, Tester M, Roy SJ. 2009. Quantifying the three main components of salinity tolerance in cereals. *Plant, Cell and Environment* 32:237–249.

Roy S, Negaro, S, Tester M. 2014. Salt resistant crop plants. *Current Opinion in Biotechnology* 26:115–124.

Russel JC, Kadry H, Hanna AB. 1965. Sodic soil in Iraq. *Agrokémia és Talajtan* 14:91–97

Russo A, Carrozza GP, Vettori L, Felici C, Cinelli F, Toffanin A. 2010. Plant beneficial microbes and their application in plant biotechnology. In *Innovations in biotechnology*, ed. Agbo E.D., 57–72, Rijeka: InTech.

Schachtman, DP, Kumar R, Schroeder JI, Marsh EL. 1997. Molecular and functional characterization of a novel low-affinity cation transporter (LCT1) in higher plants. *Proceedings of National Academy of Science of USA* 94:11079–11084.

Serrano R, Mulet JM, Rios G, Marquez JA, de Larrinoa IF, Leube MP, Mendizabal I, Pascual-Ahuir A, Proft M, Ros R,Montesinos C. 1999. A glimpse of the mechanisms of ionhomeostasis during salt stress. *Journal of Experimental Botany* 50:1023–1036.

Sharma P, Jha A, Dubey R, Pessarakli M. 2012. Reactive oxygen species, oxidative damage, and antioxidative defense mechanism in plants under stressful conditions. *Journal of Botany.* Article ID 217037.

Shi Y, Wang Y, Flowers TJ, Gong H. 2013. Silicon decreases chloride transport in rice *(Oryza sativa* L.) in saline conditions. *Journal of Plant Physiology* 170:847–853.

Tsugane K, Kobayashi K, Niwa Y, Ohba Y,Wada,K, and Kobayashi HA. 1999. Recessive arabidopsis mutant that grows photoautotrophically under salt stress shows enhanced active oxygen detoxification. *Plant Cell* 11:1195–1206.

Vaishnav A, Varma A, Tuteja N, Choudhary DK. 2016. PGPR-mediated amelioration of crops under salt stress. In *Plant Microbe Interaction: an approach to sustainable agriculture*, ed. D.K. Choudhary, A.Varma, N.Tuteja, 205–226, Singapore: Springer.

Vejan P, Abdullah R, Khadiran T, Ismail S, Nasrulhaq Boyce A. 2016. Role of plant growth promoting rhizobacteria in agricultural sustainability—A review. *Molecules* 21:573.

Yeo, A, Yeo, M, Flowers T. 1987. The contribution of an apoplastic pathway to sodium uptake by rice roots in saline conditions. *Journal of Experimental Botany* 38: 1141–1153.

Zhang, A, Liu, Y, Wang, F. et al. 2019. Enhanced rice salinity tolerance via CRISPR/Cas9-targeted mutagenesis of the OsRR22 gene. *Molecular Breeding* 39:47.

Zhang, J L, Wetson, A M, Wang, S M, Gurmani, AR, Bao, AK, Wang, CM. 2011. Factors associated with determination of root $^{22}Na^{(+)}$ influx in the salt accumulation halophyte *Suaeda maritima. Biological Trace Element Research* 139: 108–117.

Zhao, KF, Song J, Fan H, Zhou S, Zhao M. 2010. Growth response to ionic and osmotic stress of NaCl in salt-tolerant and salt-sensitive maize. *Journal of Integrative Plant Biology* 52:468–475.

25 Global Analysis of Differences in Plant Traits between Salt-Tolerant and Salt-Sensitive Plants

Bas Bruning, William K. Cornwell, and Jelte Rozema

CONTENTS

25.1 INTRODUCTION

The morphological, physiological and genetic mechanisms that convey salinity tolerance in flowering plants (Angiosperms) have received substantial scientific interest (e.g. Munns 2005; Flowers and Colmer 2008; Rozema and Flowers 2008; Rengasamy 2010; Munns et al. 2012). This research effort is, in part, motivated by the negative effect that soil salinity has on crop yields worldwide. Salinity stress in plants relates to two main issues. The first is related to the plants (in)ability to absorb water under saline conditions (Munns and Tester 2008), since to absorb water the plant water

DOI: 10.1201/9781003112327-25

potential has to be more negative than that of the soil moisture. Second, the ionic phase of salt stress relates to the accumulation of toxic levels of sodium (Na^+) and chloride (Cl^-) into the leaves and the capacity of the plant to maintain an appropriate K^+ status through competition between Na^+ with K^+ (Munns and Tester 2008; Shabala and Cuin 2008;) and with Ca^{2+} uptake (Lambers et al. 2008). Because Na^+ can have direct toxic effects on cellular metabolism, plants generally prevent Na^+ from entering the cell, and/or compartmentalize accumulated Na^+ inside the vacuole (Munns and Tester 2008).

Plant species show different levels of tolerance to salinity, varying from extremely sensitive, e.g. plant mortality of a salt sensitive variety of *Cicer arietinum* at 25 mM NaCl (Flowers et al. 2010b), to extremely salt tolerant, growing well at seawater salinities or higher (English and Comer 2013), and to 'obligate' halophytes (Ungar 1978). Obligate halophytes depend on the internal accumulation of Na^+ ions for unimpeded growth (for example, *Salicornia dolichostachya*, Katschnig et al. 2013; Rozema and Schat 2013). Those species that are able to survive and reproduce at salinities of the soil solution equivalent to 200 mM NaCl are called halophytes (Flowers and Colmer 2008).

To increase the salinity tolerance of crops, which would allow crop cultivation in salt-affected areas or use saline water for irrigation, the underlying mechanisms, both genetic and physiological, are being studied worldwide (Flowers 2004; Munns and Tester 2008; Bruning and Rozema 2012; Yuan et al. 2016; 2019). However, decades of research into the mechanisms of salinity tolerance and attempts to improve it, have so far achieved little success and resulted in minor increases in salinity tolerance in both conventional crops and scientific model species (Flowers 2004; Flowers and Flowers 2005; Munns et al. 2012).

A better understanding of the fundamentals of salinity tolerance among Angiosperms may be achieved by adopting an evolutionary approach (Flowers et al. 2010a; Bennett et al. 2013; Saslis-Lagoudakis et al. 2014; Cheeseman 2014). Phylogenetic analyses of salt tolerance and other plant traits across a large number of species may help to identify underlying plant traits that correlate with salinity tolerance, for example, the presence of salt glands in the Plumbaginaceae (Caperta et al. 2020). Moreover, it will allow us to determine whether such underlying plant traits are labile or conserved within plant clades (Donoghue 2008). This type of analysis has not been done until quite recently (Flowers et al.,2010a; Kadereit et al. 2012; Bennett et al. 2013; Saslis-Lagoudakis et al. 2014; Caperta et al. 2020) and has provided new insights. For example, Flowers et al. (2010a) have shown that among Angiosperms, halophytes (*sensu* Flowers and Colmer (2008)) are rare but widely spread across Angiosperm phylogeny, with a concentration in the Caryophyllales. Bennett et al. (2013) took a more in-depth approach and estimated that the number of times salt tolerance has evolved in the Poales, concluding that salt tolerance has independently evolved in that clade no less than 70 times. Saslis-Lagoudakis et al. (2014) took a similar approach and reconstructed the independent evolution of salt tolerance in the Angiosperms to have occurred at least 59 times, based on a database containing ~2600 salt tolerant species (Menzel and Lieth 2003).

There are two concepts that may help us understand these patterns. First, certain habitats may require a certain trait. If this trait is rare across all plant species but concentrated in a certain lineage, it has a much higher chance of colonizing this habitat and the trait, thus, represents a pre-requisite for pioneering this habitat. This phenomenon is called habitat filtering: the habitat represents a selective filter through which a small selection of the available total species pool can pass.

The other concept is local adaptation. This is just the evolutionary process of adaptation by gradual change in the genetic make-up of individuals and populations. This process starts after the initial colonization of the new habitat. If the habitat represents certain specific challenges, these adaptations are expected to be similar in nature but different in specifics, i.e. examples of convergent evolution.

We took the eHALOPH salt tolerance database (University of Sussex, http://www.sussex.ac.uk/affiliates/halophytes/) that finds its origin in Aronson (1989), and reconstructed Angiosperm phylogeny with salt tolerance as a binary character state on the tree. The minimum limit of salt tolerance in this database is 7–8 dS/m (precise definition below). We used the Angiosperm Phylogeny Group III phylogeny (APG III 2009) as the backbone of our tree to plot the number of salt-tolerant species in each Angiosperm order (Figure 25.1). The length of the blue bars represents the number of salt-tolerant species in each order of the Angiosperms. The width of the tip of each order is proportional to the log number of species in that order. The two orders with the highest incidence of salt tolerance are highlighted in light blue. Two conclusions can be drawn from the distribution of salinity tolerance across Angiosperms shown in Figure 25.1. First, many orders contain salt-tolerant representatives, which is in agreement with Flowers et al. (2010a) and Saslis-Lagoudakis et al. (2014). Second, salinity tolerance is not equally distributed across the Angiosperm orders, nor related to the number of species in the respective order. There are clear ‘hot spots’ and ‘cold spots’ for salt tolerance in the phylogenetic tree. For example, the Caryophyllales and the Poales both show a high number of salt-tolerant species (highlighted). For a more in-depth study on the distribution of salt tolerance across Angiosperms, we refer to Saslis-Lagoudakis et al. (2014).

However, our main objective in this analysis was to study traits that are associated with salt tolerance. This follows because the mechanisms of salt tolerance in those different orders may differ substantially (Flowers 2004). For example, some of the most tolerant Eudicotyledonous (Eudicots hereafter) species, such as those found in the Caryophyllales (*Salicornia*, *Suaeda* etc.) have a marked succulent appearance, i.e. low specific leaf area (SLA) measured as leaf area divided by leaf dry mass. In addition to elasticity of cell walls, this relates to ion (Na^+) accumulation in the vacuoles, which together with the production and accumulation of compatible solutes in the cytoplasm helps the plant to create a low osmotic potential. These compatible solutes often contain nitrogen, and as such nitrogen content in leaves may also be different in leaves from salt tolerant and salt sensitive species. By contrast, Monocotyledonous (Monocots hereafter) species (such as the Poales) do not show increased succulence to high salinity (except some species, e.g. the genus *Triglochin*), nor do they show growth stimulation upon salinity and their cell walls show rigidity (Rozema et al. 1987, 1991; Flowers and Colmer 2008; Munns and Tester 2008). These differences between two major plant groups within the Angiosperms suggest fundamentally different strategies in salinity tolerance between Monocots and Eudicots. To fully understand responses to salinity common to all Angiosperms and

FIGURE 25.1 Salt-tolerance data based on the eHALOPH database which finds its origin in Aronson (1989). The figure shows a phylogenetic tree (fan dendrogram) of Angiosperm orders (APG III). The width of the tips is proportional to the log number of species in each order. The number of known salt-tolerant species is shown as the blue bars around the tips. Higher level topology is based on the APG (2009) with branch lengths from the RAxML tree as described by Zanne *et al.* (2014).

those that are clade specific, a whole-plant physiological, biochemical and molecular approach must be adopted, and the uniqueness or ubiquity of traits associated with salinity tolerance should be identified.

Common to all species living under saline conditions are problems associated with water absorption due to the low water potential of the substrate. Hence, a tight control over water loss may be crucial to any plants' ability to live and reproduce (Yuan et al. 2019) in saline environments. Any genetic or physiological advantage that is already present in a species may facilitate the colonization of saline habitats. Indeed, a higher water use efficiency (WUE; units of carbon fixed per units of water lost) of certain cultivars over others is consistently related to a higher salinity tolerance and is, thus, called a pre-requisite for salinity tolerance (for example, Barbieri et al. 2012; Orsini et al. 2012). Existing leaf trait variation between species, such as variation in stomatal aperture size or stomatal density to minimize water loss, that convey higher WUE increase the likelihood of such species to enter saline or dry niches (Cheeseman 2014). The pattern apparent from the phylogeny, where certain clades have a high number of salt-tolerant species while other clades have none (Figure 25.1), is also suggestive of the existence of one or more pre-requisites or 'enablers' to evolve salt tolerance and consistent with trade conservation within plant lineages (Cornwell et al. 2014). Such a pre-requisite has been identified for the ability of grasses (Poaceae) to evolve C_4 photosynthesis (Christin et al. 2013). In that paper, the authors describe that the specific proportion of different cell types (outer sheath and inner sheath cells) in the bundle sheaths of leaves greatly affects the likelihood of the evolution of C_4 photosynthesis. This is because certain ratios of cell numbers and sizes favour the transition from C_3 to C_4 photosynthesis and also convey selective advantages to the intermediate states. Since all Angiosperms, i.e. also all C_3 species, have the basic machinery (different kinds of bundle sheaths cells) that is theoretically compatible with C_4 photosynthesis, it is cell-level characteristics such as cell size and ratios between cell types that determines if a plant species can evolve C_4 photosynthesis.

The photosynthetic metabolic pathway which a plant uses is related to the plants' WUE. Both C_4 and CAM photosynthesis convey higher WUE to plants and a link between C_4 photosynthesis and salt tolerance has been shown in the Poaceae (Bennett et al. 2013) and in the Chenopodiaceae (Kadereit et al. 2012).

The connection between C_4 photosynthesis and salinity tolerance is an example of the obvious connection that must exist between cell-level traits and salt tolerance. Differences in genetic make-up may underlie physiological differences. As such, there are two levels of explanation why some species are salt tolerant and others not: phenotype (physiology) and genotype. One aspect of genotype is C-value. C-value, representing the amount of DNA in an unreplicated haploid cell (Bennett and Leitch 2012), is extremely variable across living organisms. In Angiosperms, C-value varies over three orders of magnitude (Bennett and Leitch 2012). Most of the variation in genome size is not associated with differences in gene number, i.e. DNA sequences that are transcribed and translated (Bennetzen et al. 2005) but may reflect differences in regulatory DNA. Even though genome size does not affect phenotype through the conventional way, genome size correlates with a multitude of plant traits. For example, it strongly correlates with cell-level traits such as nuclear volume (Vanthof and Sparrow 1963), cell size (Cavalier-Smith 2005; Beaulieu et al. 2008) and cell division rate (Símova and Herben 2012). Phenotypic traits such as stomatal size and density also correlate with genome size (Beaulieu et al. 2008). Finally, life history related traits such as seed mass (Beaulieu et al. 2007) and absence or presence in extreme environments (Knight et al. 2005) also correlate with genome size. Some of the correlations between genome size and cell-level traits are so strong (Bennett 1987) that genome size has been suggested to serve as a proxy for cell- and tissue-level processes (Herben et al. 2012). Hence, genome size may act as an explanation both at the genetic and the phenotypic levels or may be considered to be one of the links between the two.

Cell size, rate of mitosis, seed mass and stomatal density, in turn, affect higher level plant physiological processes that ultimately relate to differences in species' phenologies (Knight and Ackerly 2002), dispersal rate (Cornelissen et al. 2003) and other plant physiological parameters (Knight et al. 2005). For example, the above-mentioned correlation between genome size and stomatal guard cell length and stomatal density (Beaulieu et al. 2008) are important traits for a plant's water relations and WUE. This, in turn, has a strong effect on the potential habitat of a species, such as the ability to colonize dry or saline habitats.

Because of the correlation between genome size and various plant traits that we assume to affect a species' tolerance to salinity and given the large variation in genome size across Angiosperms, we correlated genome size with salinity tolerance. Because of the positive correlation between genome size and stomatal guard cell length and the negative correlation with stomatal density (Beaulieu et al. 2008), we hypothesize that salt-tolerant species should have small genomes assuming that a larger number of smaller stomata should provide plants with an improved control over water loss. Furthermore, due to the higher WUE conveyed by C_4 and CAM photosynthesis, we expect a higher proportion of these photosynthetic metabolic pathways among salt-tolerant species, as already shown in some lower-level clades: Kadereit et al. (2012); Bennett et al. (2013). Finally, we correlated SLA and leaf nitrogen to salt tolerance to formally check the general notion of an association between SLA and salt tolerance in the Eudicots but not in the Monocots (SLA), and to see whether nitrogen-containing compatible solutes may appear as higher leaf N in any of the clades. Taken all together, we expect some of these traits to be common to all Angiosperms because they increase the likelihood of passing the initial filter that saline habitats represent, whereas other traits are likely lineage specific, representing adaptations that have evolved after the colonization of saline environments.

To conclude: salt tolerance emerges via a number of mechanisms across Angiosperms. Here, we ask: to what extent are there any general patterns in the genome size and (leaf) morphological and physiological traits among salt tolerant species among all Angiosperms, especially with respect to the traits that influence WUE? Are these patterns consistent across the two mayor clades in the Angiosperms, Monocots and Eudicots? Is there a correlation between the C_3, C_4, CAM type of photosynthesis and salinity tolerance of plants?

25.2 METHODS

25.2.1 Database Assembly

A dataset of 66,243 Monocotyledoneous and Dicotyledoneous plant species was compiled containing five different plant traits: SLA (leaf area divided by leaf mass), leaf nitrogen content (% N of leaf dry weight, and on an per area basis: SLA * leaf N), seed mass (1000 seed weight, g), genome size (pg) and type of photosynthesis (C_3, C_4 or CAM). Leaf nitrogen and SLA data was taken from the TRY database Kattge et al. (2011), additional data on SLA came from LEDA (Kleyer et al., 2008).

Data on type of photosynthesis also came from the TRY database (Kattge et al. 2011). Data on seed mass (available on http://data.kew.org/sid/, release 7.1, 2008) and genome size was obtained online from the KeW website (www.kew.org). The plant C-value database (8510 species; Bennett and Leitch 2012) and the eHALOPH salt tolerance database (University of Sussex, http://www.sussex.ac.uk/affiliates/halophytes/). The eHALOPH database is based on the database of Aronson (1989). To create a binary variable for salt tolerance, we used the definition from Aronson: HALOPH, a Data Base of Salt Tolerant Plants of the World (1984) and qualified the following species as salt tolerant:

> 'known or presumed tolerance to electrical conductivity [i.e. EC_e] measuring (or estimated to be) at least 7–8 dS/m, during significant periods or the plant's entire life'.

Note that this is a much more inclusive definition of salt tolerance than the definition given for halophytes by Flowers and Colmer (2008) using 200 mM NaCl (~ 40% seawater salinity) as a salinity criterion. Of the 1554 species in the eHALOPH database, a more precise estimate of the salinity tolerance (in dS/m) is reported for 292 species of which 221 are tolerant to seawater salinities (~500 mM NaCl) or more.

Higher plants of the Magnoliidae clade were excluded from the analyses since this clade contained very few salt-tolerant species and fewer still for which we also had data on other plant traits. Also, non-annotated clades were excluded, such that all species represent the sum of the Monocots and Eudicots.

The database we used, thus, includes species that are known to be salt tolerant. However, absence of a salt tolerance score is no proof for absence of salt tolerance in the species since there are undoubtedly species with unrecorded salt tolerance in our dataset. More importantly, salt tolerance is not a discrete trait of a species but rather a continuum of salt tolerance is present ranging from the very sensitive to the very tolerant. It is necessary to draw a line somewhere since exact salt-tolerance scores are available for only a very limited number of species. Notwithstanding those two issues, we believe the dataset we have assembled should provide us with a reasonable coverage of salt tolerant species across Angiosperm phylogeny.

25.2.2 Phylogenetic Distribution of Salt Tolerance

To test for the generality of traits associated with salinity tolerance across different levels of Angiosperm phylogeny, we performed our analysis repeatedly on various levels of phylogeny. First, all analyses were done on the complete dataset. We subsequently performed our analyses on the two mayor Angiosperm phylogenetic levels: Monocots and Eudicots. To look for the consistency of our findings in the clade Caryophyllales that contains the highest number of salt tolerant species, we also performed all our analysis on this clade, plus an analysis on all the Eudicots excluding the Caryophyllales. This allowed us to assess the robustness of our findings across the phylogeny and identify specific adaptations at the hot spots of salinity tolerance in the Angiosperms.

25.2.3 Statistical Analysis

We used logistic regression to test for the effect of salt tolerance (yes/no) on the natural log of C-value, ploidy level, seed mass, SLA and leaf nitrogen per leaf area and per weight. We used Ln transformed data to meet the assumptions of the statistical tests. A Chi-squared test was used to assess differences in photosynthetic pathways among salt-tolerant and salt-sensitive species. Chi square tests were performed on the same levels of phylogeny as described above.

Even though logistic regression does not incorporate phylogeny and, thus, may be confounded by phylogenetically conserved traits within lineages, our analyses on separate levels provided information on the distribution of the significant correlations we identified across Angiosperms. Our research questions specifically relate to the ubiquity of certain traits correlated with salt tolerance across Angiosperms, and thus an unequal distribution of correlations between salt tolerance and certain traits between different phylogenetic levels does not pose a specific problem. To identify traits that correlate with salinity tolerance across all Angiosperms, but also other traits that may be confined to lower-order phylogenetic clades, provides clues about the pre-requisites and/or subsequent adaptations to inhabit saline habitats. However, we are aware that there are limits to our methodology (Freckleton 2009).

25.3 RESULTS

We found a number of traits that significantly differed between salt-tolerant and non-salt-tolerant species. A trait that shows a negative correlation across all levels with salt tolerance is genome size (Table 25.1) (in contrast to the findings of Garcia et al. 2008). Natural log (Ln)-transformed seed mass was significantly lower in salt-tolerant species (Figure 25.3); however, average non-log-transformed seed mass was not significantly different between salt-tolerant and salt-sensitive species. The C_4 photosynthesis pathway was significantly more common among salt-tolerant species than among salt-sensitive species (Table 25.2). Those species with the C_4 photosynthesis pathway have a smaller genome size than species with the C_3 photosynthetic pathway (Figure 25.2).

25.3.1 Specific Leaf Area

SLA is significantly lower in salt-tolerant Eudicots than in non-salt-tolerant Eudicots, i.e. more succulent leaves, but significantly higher (although barely) in salt tolerant Monocots than in non-salt-tolerant Monocots. Within the Eudicots, the pattern did not change if the Caryophyllales are included or excluded (Table 25.1).

25.3.2 Leaf Nitrogen

Leaf nitrogen is significantly higher in salt-tolerant Eudicots than in salt-sensitive Eudicots. This is only true for leaf nitrogen on a per weight basis, not on a per area basis (leaf N x SLA) (Table 25.1). Within the Eudicots, the Caryophyllales did not

TABLE 25.1
The Average, Standard Deviation, Number of Observations and P-Values of C-Value, Seed Mass, SLA and Leaf Nitrogen of Salt Tolerant and Non-Salt Tolerant Plants on the Different Phylogenetic Levels

				Sla				Leaf N (per area)				Leaf N (per weight)			
				C-Value				Seed Mass				Ln Seed Mass			
				mean	sd	*N*	*p*	mean	sd	*n*	*P*	mean	Sd	*n*	*p*
All Plants			**Salt Tolerance**								N	p			
			yes	2.39	2.66	112	0.02	1303	19210	614	0.66	0.78	2.89	614	0.01
			no	5.90	8.93	2882		852	115845	31323		1.08	2.72	31277	
	Angiosperms	Monocots													
			yes	3.72	3.46	34	<0.001	4154	40677	131	0.41	0.09	3.28	131	0.98
			no	9.25	11.33	961		4331	293879	4866		0.33	2.49	4862	
		Eudicots													
			yes	1.69	1.65	67	0.07	573	4688	445	0.09	0.96	2.81	445	0.23
			no	2.71	4.12	1458		196	1946	23159		1.12	2.72	23121	
			Caryophyllales												
			yes	1.87	1.53	32	0.58	10	65	200	0.86	-0.05	1.87	200	0.86
			no	1.67	1.89	77		12	92	1683		-0.02	2.01	1683	
			All other Eudicots												
			yes	2.60	2.98	35	0.03	1927	23378	245	0.006	1.18	3.20	245	0.006
			no	6.02	9.02	1381		900	119095	21438		1.14	2.75	29594	
		Others	yes	2.59	3.44	11	0.001	15	31	38	<0.001	1.07	1.98	38	0.03
			no	8.98	10.56	463		327	2633	3298		1.91	2.81	3294	

(Continued)

TABLE 25.1 *(Continued)*
The Average, Standard Deviation, Number of Observations and P-Values of C-Value, Seed Mass, SLA and Leaf Nitrogen of Salt Tolerant and Non-Salt Tolerant Plants on the Different Phylogenetic Levels

				Sla				Leaf N (per area)				Leaf N (per weight)			
				C-Value				Seed Mass				Ln Seed Mass			
				mean	sd	*n*	*p*	mean	Sd	*n*	*p*	mean	sd	*n*	*p*
All Plants			**Salt Tolerance**								N	p			
			yes	148	125	189	0.006	2.161	0.845	108	0.003	0.025	0.017	62	0.010
			no	162	111	6679		1.958	0.875	4373		0.018	0.011	2874	
	Angiosperms	Monocots													
			yes	239	211	36	0.640	1.832	0.751	22	0.490	0.016	0.013	12	0.900
			no	190	125	842		1.969	0.829	521		0.016	0.014	307	
		Eudicots													
			yes	126	80	145	<0.001	2.226	0.847	82	0.002	0.027	0.018	47	0.003
			no	157	107	4889		1.983	0.887	3226		0.019	0.011	2179	
			Caryophyllales												
			yes	123	69	65	<0.001	2.580	0.747	37	0.220	0.031	0.022	18	0.080
			no	179	113	278		2.469	0.954	142		0.024	0.028	73	
			All other Eudicots												
			yes	128	88	80	0.010	1.936	0.820	45	0.860	0.025	0.015	29	0.070
			no	156	107	4611		1.961	0.878	3084		0.018	0.009	2106	
		Others	yes	131	99	8	0.280	2.161	0.845	108	0.170	0.025	0.008	3	0.120
			no	172	112	898		1.884	0.844	567		0.017	0.009	354	

TABLE 25.2
Number of Salt-Tolerant and Salt-Sensitive Species according to Their Photometabolic Pathways

				C_3	C_4	CAM	
				N	N	N	P
All Plants			**Salt Tolerance**				
			yes	305	170	5	< 0.001
			no	7026	801	96	
	Angiosperms	Monocots					
			yes	62	47	0	< 0.001
			no	1043	624	32	
		Eudicots					
			yes	230	123	5	< 0.001
			no	5169	172	63	
			Caryophyllales				
			yes	120	115	4	< 0.001
			no	372	136	38	
			All other Eudicots				
			yes	110	8	1	< 0.001
			no	4797	36	25	

Note: Number of observations are presented. p-values are derived from Chi square tests.

show a significant difference between salt-tolerant and salt-sensitive species because all of the species in this order have high leaf nitrogen values.

25.3.3 C-VALUE

As stated in Section 25.1, we use the terms genome size and C-value interchangeably. However, we have tested all the components related to genome size *sensu* Greilhuber et al. (2005), i.e. including ploidy level on correlations with salt tolerance (data not shown). Salt-tolerant species have a lower C-value than salt-sensitive species in all clades except the Caryophyllales (Table 25.1). The pattern changes slightly for genome size *sensu stricto* (i.e. C value * ploidy level) where the difference ceases to be significant ($p = 0.078$) in the Eudicots at all levels tested (data not shown). C-value is about 2.2 times smaller in salt-tolerant species than in salt-sensitive species (2.3 *versus* 5.2, respectively). Ploidy level was not significantly different for salt-tolerant and salt-sensitive species). The difference in C-value between salt-tolerant and salt-sensitive species is strongest in the Monocots, which have a higher average C-value than the Eudicots. Mean C-value in the Caryophyllales is small (Table 25.1; Leitch et al. 1998; Soltis et al. 2003; Bennett and Leitch 2012).

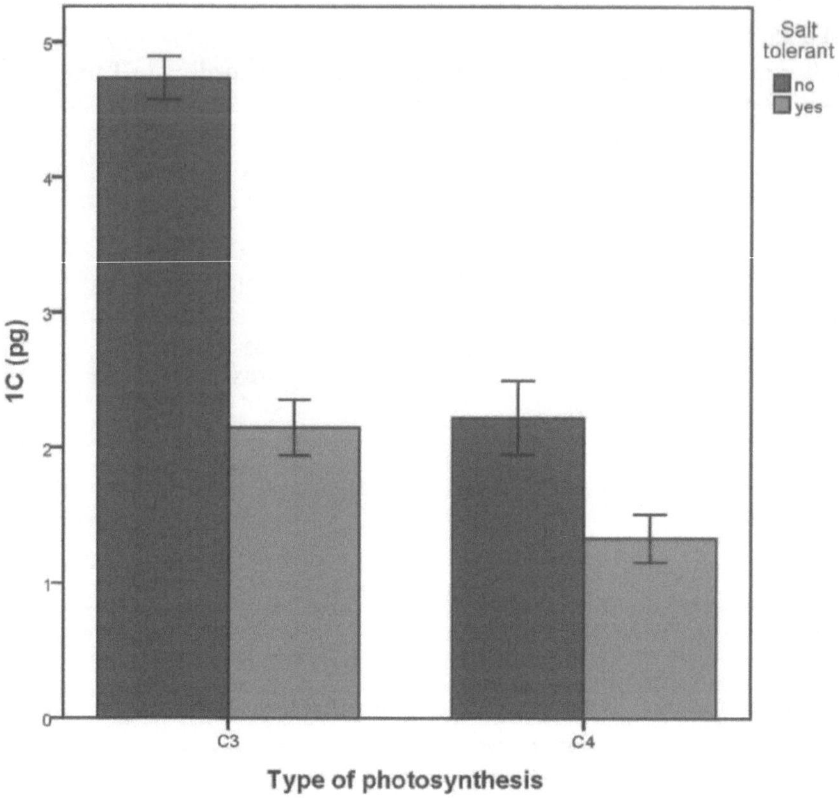

FIGURE 25.2 Mean 1C value (± 1SE) of C_3 and C_4 species. C_4 species have a significantly smaller genome size than C_3 species, and within both groups salt-tolerant plants have a significantly smaller genome that non-salt-tolerant species. There was no significant interaction between the two factors (Two-way ANOVA, $P < 0.0001$).

25.3.4 Seed Mass

Ln seed mass of the whole dataset was significantly lower ($p = 0.001$) in salt-tolerant species than in salt-sensitive species, but non-transformed data showed that seed mass in salt-tolerant species did not significantly differ from seed mass of salt-sensitive species ($p = 0.92$).

25.3.5 Photosynthetic Metabolic Pathway

Our comparison between salt-tolerant and salt-sensitive species of photosynthetic metabolic pathways revealed that salt-tolerant species had a significantly higher proportion of C_4 photosynthesis in all groups (Table 25.2). Furthermore, mean genome size was significantly smaller in species with C_4 photosynthesis (Figure 25.2). Data on CAM genome size were not shown due to small number of observations (Table 25.2).

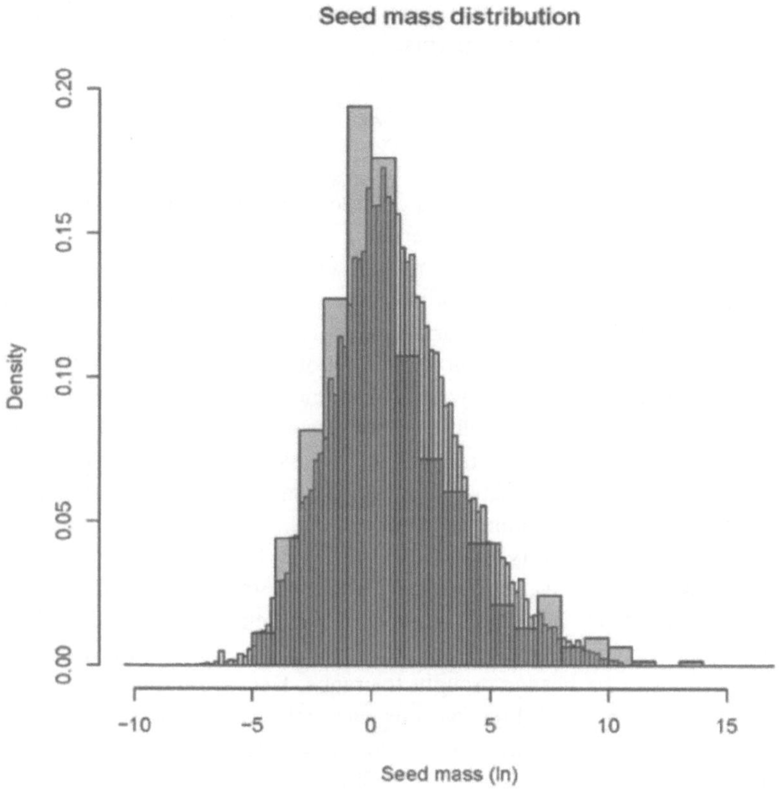

FIGURE 25.3 Density plot of Log10 transformed seed mass of salt-tolerant (blue) and non-salt-tolerant (green) species. Salt-tolerant species generally have smaller seeds but some species have very large seeds.

25.4 DISCUSSION

We present here a global analysis of differences in plant traits between salt-tolerant and salt-sensitive plant species. At the global scale, we found a significant negative correlation between salt tolerance and genome size across Angiosperms. We also found correlations between salt tolerance and SLA and leaf nitrogen content in the Eudicots and Ln transformed seed mass for all Angiosperms, plus a significantly different distribution of photosynthetic metabolic pathways between salt-tolerant and salt-sensitive species in various Angiosperm clades. In general, salt tolerance was associated with small genomes, small seeds, and a high incidence of C_4 photosynthesis. However, while there were highly significant differences between certain traits across all Angiosperms, there was also important lineage-specific complexity for some traits. Genome size and mode of photosynthesis were traits consistently differing between salt-tolerant and salt-sensitive species throughout the Angiosperms. SLA was significantly lower in the salt-tolerant Eudicots. Despite the differences between Monocots and Eudicots in their specific mechanisms to deal with a saline environment, our results indicate that there are fundamental traits underlying salinity tolerance common to all Angiosperms.

25.4.1 Genome Size and Salt Tolerance

Salt-tolerant species have, on average, a 2.2 times smaller genome than non-salt-tolerant species. A direct effect of genome size on salt tolerance seems unlikely since variation in genome size is most often caused by variation in non-coding regions of the genome (Bennetzen et al. 2005) and is, thus, not related with differences in gene products (i.e. proteins, RNA etc.). However, as mentioned above, genome size itself strongly correlates with a variety of cell-level characteristics (Bennett 1977) and those, in turn, affect several ecological, evolutionary and phenotypic traits (reviewed by Herben et al. 2012). The exact mechanism by which genome size exerts this effect on numerous traits and patterns still remains elusive, though some are undoubtedly linked to the effects of genome size on cell size and cell division rate. In case of salinity tolerance, the positive correlation between genome size and stomatal guard cell size (presumably via the correlation between genome size and cell size; Bennett 1987) and the negative correlation with stomatal density (Beaulieu et al. 2008; Hodgson et al. 2010) suggest a connection with the plants' water relations. This means that salt-tolerant species, having small genomes, tend to have small stomata in high densities. In fact, Beaulieu et al. (2008) almost predicted the genome size relationship presented in this chapter, albeit not for salinity tolerance but for drought tolerance. Here we hypothesize that small stomatal size allows for a rapid response to water stress and a close control of water loss, while a high density in stomata allows for optimal CO_2 uptake during favourable photosynthetic conditions (Beaulieu et al. 2008). This, in turn, suggests that plants in a saline environment are adapted to the optimal use of occasional exposure to water of lower salinity for the uptake of water. It is known that halophytes make optimum use of spatial heterogeneity in salinity levels, taking advantage of local less-saline conditions (Bazihizina et al. 2009) and it is possible that they also make optimal use of temporal variation in salinity levels (Ball and Farquhar 1984).

Recently, it has been shown that genome size evolution correlates well with speciation rates in Angiosperms. However, Puttick et al. (2015) have shown that it is the rate of genome size evolution, not genome size per se, which correlates well with speciation rate in the Angiosperms. Angiosperm evolutionary history is characterized by multiple genome duplications. However, genome duplications do not necessarily lead to larger genome sizes in Angiosperms, which is also why we have tested both C-value and genome size in the analysis. Even though our results did not change when correlating C-value or genome size (C-value * genome size) with salt tolerance, it is the rapid downsizing after whole genome duplications that may be the reason for fast speciation rates in the Angiosperms.

In this light, our result may reflect recent speciation events in salt-tolerant species. This would also explain the 'tippiness' of salt tolerance across Angiosperms; many Angiosperm clades contain salt-tolerant species but they are all of relatively recent evolutionary origin (Saslis-Lagoudakis et al. 2014; Caperta et al. 2020).

A small genome may represent a filter to colonize saline habitats. Angiosperms' abilities to rapidly evolve changes in genome size may, thus, explain a number of things such as: why there are so few salt-tolerant Gymnosperms which on average have large genomes, why many Angiosperm clades have some salt-tolerant species, and why salt tolerance is more common in those clades that already have small genome size, such as the Caryophyllales.

25.4.2 Leaf Traits and Salt Tolerance

Leaf nitrogen was higher in salt-tolerant species in the Eudicots but not in the Caryophyllales. SLA was significantly lower in salt-tolerant plants than in salt-sensitive plants in the Angiosperms, and higher (albeit not significantly) in salt-tolerant Monocots than in salt-sensitive Monocots (Table 25.1). In the case of the Eudicots, these traits can also be interpreted in relation to WUE. The correlation between salinity tolerance and leaf nitrogen and SLA in the Eudicots reflects the adaptations to saline environments; many species in this clade employ succulence or sclerophyllous-type leaves. Along with stomatal closure and reduced stomatal density, a low area to leaf mass ratio helps to reduce water loss (Wright et al. 2001). High WUE is generally associated with low leaf N. To compensate for this negative correlation, our results suggest that salt-tolerant species have increased leaf nitrogen levels. The Caryophyllales represent a good example, since all species in this clade have a high leaf N content, which explains why there is no significant difference in Leaf N between salt-tolerant and salt-sensitive Caryophyllales. This mechanism may thus be both an evolutionary adaptation to salinity, as well as an adaptive, plastic response to salinity stress. For example, Zhu and Meinzer (1999) reported increased leaf nitrogen content per leaf area with increasing salt stress in *Atriplex lentiformis*. Even though their findings could be explained by a decreased SLA under salinity while maintaining the same leaf nitrogen, our analyses on leaf nitrogen per leaf weight (Table 25.1) reveal that this is not the case. Leaf N per weight in salt-tolerant species deviated most from leaf N per weight in salt-sensitive species at the lower end of the SLA spectrum (data not shown).

An alternative explanation may be sought in need for salt-tolerant species to accumulate high concentrations of (nitrogenous) compatible solutes such as glycinebetaine and proline (Munns and Tester 2008). These compatible solutes are produced by plants as a mechanism to balance the intracellular osmotic potential with the osmotic potential in the environment, both internal (i.e. vacuole versus cytosol) and external (the root environment). Additionally, these compounds may aid in protein and membrane stability in cells exposed to high Na^+ concentrations, or function as Reactive Oxygen Species scavengers (Smirnoff and Cumbes 1989).

25.4.3 Seed Mass and Salt Tolerance

Seed mass was not significantly different between salt-tolerant and salt-sensitive species but this was caused by a small number of some very heavy seeded salt-tolerant species (Figure 25.3). Log transformed seed mass was significantly lower in salt-tolerant species than in salt-sensitive species, indicating that most salt-tolerant plants analyzed have small seeds. These results agree with the findings of Beaulieu et al. (2007), even though in our analyses no Gymnosperms were included, whereas

Gymnosperms were the main factor in the correlation reported between genome size and seed mass in Beaulieu et al. (2007). The correlation we find may, thus, be a confounding factor via the correlation between salinity tolerance and genome size and genome and seed sizes. We identified the genome sizes for the four salt-tolerant species in our dataset with a seed mass of over 1000 grams (i.e. *Cocos nucifera, Mauritia flexuosa, Barringtonia acutangula* and *Phoenix dactylifera*). Their average C-value was 2.6, which is a value that is close to the average value of salt tolerant species in this analysis, and much lower than the average C-value in Monocots, both salt tolerant and salt sensitive.

Seed mass relates to dispersal capacity (i.e. Cornelissen et al. 2003) and small seeds can be produced in larger numbers and be dispersed further either by wind or water. This suggests that saline habitats are ephemeral, i.e. existing relatively briefly, and that the optimal dispersal strategy is for most salt-tolerant species is to produce many small seeds. The exceptions, the large seeded species at the right of Figure 25.3, are some mangrove and coconut species. These species rely on the sea for their seed dispersal (i.e. hydrochorous dispersal). Hence, it seems that there are two distinct optima in the salt tolerance fitness landscape for seed dispersal strategies, either small seeds or large, floating seeds (Moles et al. 2005).

25.4.4 Photosynthetic Metabolic Pathway and Salt Tolerance

The photosynthetic metabolic pathway is another physiological trait that shows a different distribution among salt-tolerant species and salt-sensitive species. The proportion of the C_4 photosynthetic pathway is much larger in salt-tolerant species than in salt sensitive species, more than half of the salt-tolerant plants use the C4 pathway in all but one of the phylogenetic levels tested (Table 25.2). In addition of the skewed proportion of the photosynthetic metabolic pathway in salt-tolerant species, genome size is still smaller in salt-tolerant species in that use either C_3 or C_4 photosynthesis (Figure 25.2).

C_4 and CAM photosynthesis are both more WUE than C_3 photosynthesis ultimately due to a CO_2 concentrating mechanism (Sage 2004; Ghannoum 2009). This concentration of CO_2 allows the plant to partly escape the connection between water loss and carbon gain, which gives them an advantage in arid – and saline – environments. In fact, the C_4 photosynthetic pathway may even have evolved, in some cases, as a response to salinity stress (Sage 2004; Ghannoum 2009). It is perhaps no coincidence that some genera containing some of the most tolerant terrestrial Eudicot species to salinity, i.e. *Salsola, Atriplex* (some species in the genus) and *Sueada*, all independently evolved the C_4 photosynthetic pathway; the common ancestor of the Amaranthaceae, a family in the Caryophyllales most likely used the C_3 photosynthetic pathway (Sage 2004).

Christin et al. (2013) identified differences in leaf anatomy in the Poaceae that affect the likelihood of different clades of evolving the C_4 pathway. As mentioned in Section 25.1, the distance between bundle sheath cells seems critical for the ability of a species to evolve C_4 photosynthesis. A reduction in cell size in the BEP clade (subfamilies *Bambusoideae, Pooideae* and *Ehrhartoideae*) of the Poaceae prevented this clade of the ability of evolving C_4 photosynthesis. Even though this does not directly corroborate our findings, it is this type of cell-level processes identified by Christin et al. (2013) that affect the potential of a species to adapt to certain circumstances and that may be ultimately driven by differences in genome size.

25.4.5 Habitat Filtering

Previously, it has been shown that species with large genomes are excluded from a number of different extreme environments, including very dry places (Wakamiya et al. 1996; Knight and Ackerly 2002;). If salt tolerance is thought of as a type of extreme environments, the results we presented are consistent with those previously published results (Knight and Ackerly 2002). However, other extreme environments seem to favour large genomes (for example cold environments, Macgillivray and Grime 1995) and in many of the extreme environments, positive, negative and no significant correlations with genome size have been identified (Knight et al. 2005). The absence of large genomes in extreme environments is most prominent in the extreme cases and failing to detect negative correlation between genome size and environmental gradients may be, in many studies, due to a limited sample size and/or not including the full range of the gradient under study.

Our results can, thus, be interpreted to reflect two phenomena in salt tolerance: adaptations and pre-requisites. Having a small genome and C_4 photosynthesis may increase the likelihood of a species in colonizing a saline habitat and, thus, represent a possible example of habitat filtering (Cornwell et al. 2006). The other traits we have found to correlate with salinity tolerance (confined to the Eudicots) fit the traditional evolutionary framework in which an ecological niche selects for certain physiological traits. This is the classical directional selection that changes trait means through differential survival among the individuals in a population and occurs *after* the filter. Here we identify some leaf traits in the Eudicots that most likely represent these subsequent adaptations but unfortunately not for the Monocots. It will be interesting for future studies to identify traits that are common to all salt-tolerant Monocots.

The Caryophyllales support the hypothesis of pre-requisites. As a clade, their species have small genomes and high leaf nitrogen content and are thus suitable to colonize saline habitats. This is the only clade in which we did not find a significant difference in genome size and leaf nitrogen content between salt tolerant and salt sensitive.

25.5 CONCLUSION

The tight correlations between genome size and certain cell-level traits and processes such as nuclear volume, cell size, duration of mitosis and meiosis appear to show some very fundamental physics rules in cell biology (Bennett 1987) and seem to fit the concept of a habitat filter. Several mechanisms may underlie the correlation between small genomes and salt tolerance. Alternatively, small genome size is perhaps a mechanical requirement for plant physiology to function under salt stress.

These findings may have practical applications since both genome size (Bennett and Leitch 2012) and type of photosynthesis (C_3, C_4, CAM) show inter- and intraspecific variation and can, thus, be selected for by plant breeders. By selecting cultivars of species of interest on their small genome sizes as targets by which to improve salt tolerance, breeders can give themselves a head-start when breeding for salt tolerance. Our findings show that while some traits (genome size, seed mass) are associated with salt tolerance in all Angiosperms, other traits are lineage specific (SLA, leaf N content). This is to be expected in the event of multiple evolutionary origin of a complex trait such as salt tolerance.

REFERENCES

APG III. 2009. An update of the angiosperm phylogeny group classification for the orders and families of flowering plants: APG III. Bot J Linn Soc 161, 105–121.

Aronson, J.A., 1989. HALOPH a database of salt tolerant plants of the world. Office of Arid Land Studies, University of Arizona, Tucson, Arizona. Updated and available online at http://www.sussex.ac.uk/affiliates/halophytes/

Ball, M.C., Farquhar, G.D. 1984. Photosynthetic and stomatal responses of the grey mangrove, Avicennia marina, to transient salinity conditions. Plant Physiol 74, 7–11.

Barbieri, G., Vallone, S., Orsini, F., Paradiso, R., De Pascale, S., Negre-Zakharov, F., Maggio, A. 2012. Stomatal density and metabolic determinants mediate salt stress adaptation and water use efficiency in basil (Ocimum basilicum L.). J Plant Physiol 169, 1737–1746.

Bazihizina, N., Colmer, T.D., Barrett-Lennard, E.G. 2009. Response to non-uniform salinity in the root zone of the halophyte Atriplex nummularia: Growth, photosynthesis, water relations and tissue ion concentrations. Ann Bot-London 104, 737–745.

Beaulieu, J.M., Leitch, I.J., Patel, S., Pendharkar, A., Knight, C.A. 2008. Genome size is a strong predictor of cell size and stomatal density in Angiosperms. New Phytol 179, 975–986.

Beaulieu, J.M., Moles, A.T., Leitch, I.J., Bennett, M.D., Dickie, J.B., Knight, C.A. 2007. Correlated evolution of genome size and seed mass. New Phytol 173, 422–437.

Bennett, M.D. 1977. Time and duration of meiosis. Philos T Roy Soc B 277, 201–226.

Bennett, M.D. 1987. Variation in genomic form in plants and its ecological implications. New Phytol 106, 177–200.

Bennett, M.D., Leitch, I.J. 2012. Plant DNA C-values database (release 6·0, October 2012). http://www.kew.org/cval/homepage.html

Bennett, T.H., Flowers, T.J., Bromham, L. 2013. Repeated evolution of salt-tolerance in grasses. Biol Lett 9: 20130029

Bennetzen, J.L., Ma, J.X., Devos, K., 2005. Mechanisms of recent genome size variation in flowering plants. Ann Bot-London 95, 127–132.

Bruning, B., Rozema, J. 2012. Symbiotic nitrogen fixation in legumes: Perspectives for saline agriculture. J Environ Exp Botany 92, 134–143. doi:10.1016/j.envexpbot.2012.09.001.

Caperta, A.D., Róis, A.S., Teixeira, G., Garcia- Caparros, P., and Flowers, T.J. 2020. Secretory structures in plants: Lessons from the Plumbaginaceae on their origin, evolution and roles in stress tolerance. Plant, Cell Environ 43, 2912–2931.

Cavalier-Smith, T. 2005. Economy, speed and size matter: Evolutionary forces driving nuclear genome miniaturization and expansion. Ann Bot-London 95, 147–175.

Cheesman, J.M. 2014. The evolution of halophytes, glycophytes and crops, and its implications for food security under saline conditions. New Phytologist 206, 557–570. doi: 10.1111/nph.13217.

Cornelissen, J.H.C., Lavorel, S., Garnier, E., Diaz, S., Buchmann, N., Gurvich, D.E., Reich, P.B., ter Steege, H., Morgan, H.D., van der Heijden, M.G.A., Pausas, J.G., Poorter, H. 2003. A handbook of protocols for standardised and easy measurement of plant functional traits worldwide. Aust J Bot 51, 335–380.

Cornwell, W.K., Schwilk, D.W., Ackerly, D.D. 2006. A trait-based test for habitat filtering: Convex hull volume. Ecology 87, 1465–1471.
Cornwell, W.K., Westoby, M., Falster, D.S., Fitzjohn, R.G., O'Meara, B.C., Pennell, M.W., McGlinn, D.J., Eastman, J.M., Mole,s A.T., Reich, P.B. et al. 2014. Functional distinctiveness of major plant lineages. J Ecol 102, 345–356.
Christin, P.A., Osborne, C.P., Chatelet, D.S., Columbus, J.T., Besnard, G., Hodkinson, T.R., Garrison, L.M., Vorontsova, M.S., Edwards, E.J. 2013. Anatomical enablers and the evolution of C-4 photosynthesis in grasses. Proc Natl Acad Sci USA 110, 1381–1386.
Donoghue, J.M. 2008. A phylogenetic perspective on the distribution of plant diversity. Proc Nat Acad Sci 105, 11549–11555.
English, J.P., Colmer, T.D. 2013. Tolerance of extreme salinity in two stem-succulent halophytes (Tecticornia species). Funct Plant Biol 40, 897–912.
Flowers, T.J. 2004. Improving crop salt tolerance. J Exp Bot 55, 307–319.
Flowers, T.J., Flowers, S.A. 2005. Why does salinity pose such a difficult problem for plant breeders? Agr Water Manage 78, 15–24.
Flowers, T.J., Colmer, T.D. 2008. Salinity tolerance in halophytes. New Phytol 179, 945–963.
Flowers, T.J., Galal, H.K., Bromham, L. 2010a. Evolution of halophytes: Multiple origins of salt tolerance in land plants. Funct Plant Biol 37, 604–612.
Flowers, T.J., Gaur, P.M., Gowda, C.L.L., Krishnamurthy, L., Samineni, S., Siddique, K.H.M., Turner, N.C., Vadez, V., Varshney, R.K., Colmer, T.D., 2010b. Salt sensitivity in chickpea. Plant Cell Environ 33, 490–509.
Freckleton, R.P. 2009, The seven deadly sins of comparative analysis. J Evol Biol, 22: 1367–1375.
Garcia, S., Canela, M.A., Garnatje, T., McArthur, E.D., Pellicer, J., Sanderson, S.C., Valles, J. 2008. Evolutionary and ecological implications of genome size in the North American endemic sagebrushes and allies (Artemisia, Asteraceae). Biol J Linn Soc 94, 631–649.
Ghannoum, O. 2009. C-4 photosynthesis and water stress. Ann Bot-London 103, 635–644.
Greilhuber, J., Dolezel, J., Lysak, M.A., Bennett, M.D. 2005. The origin, evolution and proposed stabilization of the terms 'genome size' and 'C-value' to describe nuclear DNA contents. Ann Bot-London 95, 255–260.
Herben, T., Suda, J., Klimesova, J., Mihulka, S., Riha, P., Simova, I. 2012. Ecological effects of cell-level processes: Genome size, functional traits and regional abundance of herbaceous plant species. Ann Bot-London 110, 1357–1367.
Hodgson, J.G., Sharafi, M., Jalili, A., Diaz, S., Montserrat-Marti, G., Palmer, C., Cerabolini, B., Pierce, S., Hamzehee, B., Asri, Y., Jamzad, Z., Wilson, P., Raven, J.A., Band, S.R., Basconcelo, S., Bogard, A., Carter, G., Charles, M., Castro-Diez, P., Cornelissen, J.H.C., Funes, G., Jones, G., Khoshnevis, M., Perez-Harguindeguy, N., Perez-Rontome, M.C., Shirvany, F.A., Vendramini, F., Yazdani, S., Abbas-Azimi, R., Boustani, S., Dehghan, M., Guerrero-Campo, J., Hynd, A., Kowsary, E., Kazemi-Saeed, F., Siavash, B., Villar-Salvador, P., Craigie, R., Naqinezhad, A., Romo-Diez, A., Espuny, L.D., Simmons, E. 2010. Stomatal vs. genome size in Angiosperms: The somatic tail wagging the genomic dog? Ann Bot-London 105, 573–584.
Kadereit, G., Ackerly, D., Pirie, M.D. 2012. A broader model for C4 photosynthesis evolution in plants inferred from the goosefoot family (Chenopodiaceae s.s.). Proc Roy Soc B 279, 3304–3311.
Katschnig, D., Broekman, R.A., Rozema, J. 2013. Salt tolerance in the halophyte Salicornia dolichostachya Moss: Growth, morphology and physiology. J Environ Exp Botany 92, 32–42.
Kattge, J., Díaz, S., Lavorel, S., Prentice, I.C., Leadley, P., Bönisch, G., Garnier, E., Westoby, M., Reich, P.B., Wright, I.J., Cornelissen, J.H.C., Violle, C., et al. 2011. TRY – A global database of plant traits. Global Change Biol 17, 2905–2935.

Kleyer, M., Bekker, R.M., Knevel, I.C., Bakker, J.P., Thompson, K., Sonnenschein, M., Poschlod, P., Van Groenendael, J.M., Klimeš, L., Klimešová, J., et al. 2008. The LEDA Traitbase: A database of life-history traits of the Northwest European flora. J Ecol 96: 1266–1274. doi: 10.1111/j.1365-2745.2008.01430.

Knight, C.A., Ackerly, D.D. 2002. Variation in nuclear DNA content across environmental gradients: A quantile regression analysis. Ecol Lett 5, 66–76.

Knight, C.A., Molinari, N.A., Petrov, D.A. 2005. The large genome constraint hypothesis: Evolution, ecology and phenotype. Ann Bot-London 95, 177–190.

Lambers, H., Raven, J.A., Shaver, G.R., Smith, S.E. 2008. Plant nutrient-acquisition strategies change with soil age. Trends Ecol Evol 23(2), 95–103. doi:10.1016/j.tree.2007.10.008

Leitch, I.J., Chase, M.W., Bennett, M.D. 1998. Phylogenetic analysis of DNA C-values provides evidence for a small ancestral genome size in flowering plants. Ann Bot-London 82, 85–94.

Macgillivray, C.W., Grime, J.P. 1995. Genome size predicts frost-resistance in british herbaceous plants – implications for rates of vegetation response to global warming. Funct Ecol 9, 320–325.

Menzel, U., Lieth, H. 2003. Halophyte database version 2.0. In: H. Lieth and M. Mochtchenko (Eds.). Cash crop halophytes: Recent studies (pp. 221–250). Dordrecht: Kluwer Academic Publishers.

Moles, A.T., Ackerly, D.D., Webb, C.O., Tweddle, J.C., Dickie, J.B., Westoby, M. 2005. A brief history of seed size. Science 307, 576–580.

Munns, R., 2005. Genes and salt tolerance: Bringing them together. New Phytol 167, 645–663.

Munns, R., James, R.A., Xu, B., Athman, A., Conn, S.J., Jordans, C., Byrt, C.S., Hare, R.A., Tyerman, S.D., Tester, M., Plett, D., Gilliham, M. 2012. Wheat grain yield on saline soils is improved by an ancestral Na+ transporter gene. Nat Biotechnol 30, 360–U173.

Munns, R., Tester, M. 2008. Mechanisms of salinity tolerance. Annu Rev Plant Biol 59, 651–681.

Orsini, F., Alnayef, M., Bona, S., Maggio, A., Gianquinto, G. 2012. Low stomatal density and reduced transpiration facilitate strawberry adaptation to salinity. Environ Exp Bot 81, 1–10.

Puttick, M. N., Clark, J., Donoghue, D. C. J. 2015. Size is not everything: rates of genome size evolution, not C-value, correlate with speciation in angiosperms. Proc Roy Soc B 282: 20152289.

Rengasamy, P. 2010. Soil processes affecting crop production in salt-affected soils. Funct Plant Biol 37, 613–620.

Rozema, J. 1991. Growth, water and ion relationships of halophytic monocotyledonae and dicotyledonae – A unified concept. Aquat Bot 39, 17–33.

Rozema, J., Arp, W., Vandiggelen, J., Kok, E., Letschert, J. 1987. An ecophysiological comparison of measurements of the diurnal rhythm of the leaf elongation and changes of the leaf thickness of salt-resistant dicotyledonae and monocotyledonae. J Exp Bot 38, 442–453.

Rozema, J., Flowers, T.J. 2008. Crops for a salinized world. Science 322, 1478–1480.

Rozema, J., Schat, H. 2013. Salt tolerance of halophytes, research questions reviewed in the perspective of saline agriculture. J Environ Exp Botany 92, 83–95. http://dx.doi.org/10.1016/j.envexpbot.2012.08.004

Sage, R.F. 2004. The evolution of C-4 photosynthesis. New Phytol 161, 341–370.

Shabala, S., Cuin, T. 2008. Potassium transport and plant salt tolerance. Physiologia Plantarum 133, 651–69.

Saslis-Lagoudakis, C.H., Moray, C., Bromham, L. 2014. Evolution of salt tolerance in angiosperms: a phylogenetic approach. In: Rajakaruna N, Boyd RS, Harris TB, eds. Plant ecology and evolution in harsh environments (pp. 77–95). Hauppauge, New York: Nova Science Publishers.

Símova, I., Herben, T. 2012. Geometrical constraints in the scaling relationships between genome size, cell size and cell cycle length in herbaceous plants. P Roy Soc B-Biol Sci 279, 867–875.

Smirnoff, N., Cumbes, Q.J. 1989. Hydroxyl radical scavenging activity of compatible solutes. Phytochemistry 28, 1057–1060.

Soltis, D.E., Soltis, P.S., Bennett, M.D., Leitch, I.J. 2003. Evolution of genome size in the Angiosperms. Am J Bot 90, 1596–1603.

Ungar, I.A. 1978. Halophyte seed germination. Botanical Rev 44(2), 233–264.

Vanthof, J., Sparrow, A.H. 1963. A relationship between DNA content, nuclear volume, and minimum mitotic cycle time. P Natl Acad Sci USA 49, 897–902.

Wakamiya, I., Price, H.J., Messina, M.G., Newton, R.J. 1996. Pine genome size diversity and water relations. Physiol Plantarum 96, 13–20.

Wright, I.J., Reich, P.B., Westoby, M. 2001. Strategy shifts in leaf physiology, structure and nutrient content between species of high- and low-rainfall and high- and low-nutrient habitats. Funct Ecol 15, 423–434.

Yuan, F., Lyu, M. J. A., Leng, B. Y., Zhu, X. G., and Wang, B. S. 2016. The transcriptome of NaCl-treated Limonium bicolor leaves reveals the genes controlling salt secretion of salt gland. Plant Mol Biol, 91(3), 241–256.

Yuan, F., Guo, J., Shabala, S., and Wang, B. 2019. Reproductive physiology of halophytes: Current standing. Front Plant Sci, 9, 1954.

Zanne, A.E., Tank, D.C., Beaulieu, J.M. 2014. Three keys to the radiation of angiosperms into freezing environments. Nature 506, 89–92.

Zhu, J., Meinzer, F.C. 1999. Efficiency of C-4 photosynthesis in Atriplex lentiformis under salinity stress. Aust J Plant Physiol 26, 79–86.

26 Comparative Study on the Response of Several Tomato Rootstocks to Drought and Salinity Stresses

Abdelaziz Hirich, Abdelghani Chakhchar, and Redouane Choukr-Allah

CONTENTS

26.1 INTRODUCTION

Soil and water salinity are major problems affecting several parts of the world, and the affected areas are on the increase due to brackish groundwater overexploitation, sea water intrusion, misuse of fertilizers and climate change (Valipour 2014; Elhag 2016). However, there is a great potential in using salt-affected lands and waters by adopting several practices that alleviate the negative impact of salinity on crops, yield and farmer's income (Dasgupta et al. 2015; Wichelns and Qadir 2015). Increasing the salt tolerance of crops through plant breeding could improve the sustainable management of saline water by reducing the leaching requirement of irrigated soils and enabling the selection of tolerant varieties that can grow under saline conditions (Ashraf and Wu 1994; Flowers and Yeo 1995).

Tomato is the most cultivated vegetable crop in the world. According to FAOSTAT (2014), the cultivated area of tomato is 6 Mha, around 20% of total area cultivated for vegetables, while production is equal to 223 million tonnes, accounting for ~50% of the total amount of vegetable production. Due to the importance of tomato

DOI: 10.1201/9781003112327-26

production, the enhanced use of rootstocks under different growing conditions, both in the field and in the greenhouse, is essential.

Grafting is currently regarded as a rapid technique aimed at increasing the environmental stress tolerance of fruit vegetables as well as maximizing yield (Schwarz et al. 2010; Al-Harbi, 2017a; 2018). It is used to reduce infections by soil-borne pathogens as well as enhance tolerance against abiotic stresses (Louws et al. 2010). The selection of a commercial rootstock is often carried out either by experienced farmers based on the recommendations of extension advisors or by the seed companies, based on their breeding and development programs. The most adopted selection criterion is plant vigor, including stem diameter (SD) (Navarrete et al. 1997). Other criteria used to select the best rootstock include root weight, shoot weight, and sodium concentration in shoots or leaves (Zijlstra et al. 1994; Santa-Cruz et al. 2002). Under saline conditions, the most suitable rootstock for tomato is the one that best excludes sodium from the root system, as long-term damage caused by salinity in tomato has been mainly related to high accumulation of Na^+ and Cl^-. Therefore, salt-tolerant rootstocks should slow or prevent the accumulation of toxic levels of sodium and chloride in the leaves (Estañ et al. 2005).

Grafting has been used to increase the salinity tolerance of trees and recently grafting has been used for vegetables, in particular tomato (Santa-Cruz et al. 2002; Al-Harbi et al. 2017b,c). However, tomato rootstocks were initially used to increase resistance of tomato to soil diseases, as tomato was mainly grown with fresh water irrigation.

Taking into account facts that tomato is one of the most important horticultural crops in the world, and its production is concentrated in semi-arid regions, where salinity and drought are major problems, it is of great interest to know whether the grafting technique is a valid strategy for improving tomato salt tolerance. The objective of this study was to evaluate the responses to both water- and salinity-stress of twelve commercial rootstocks and one wild Solanum species.

26.2 MATERIALS AND METHODS

Twelve commercial tomato varieties (DR9011TV, Emperador, Empower, APIT 0919, King Kong, Maxifort, Optifort, Silex, Sousspro, Superpro, Unifort, Arazi) and two wild genotypes (*S. pimpinellifolium* and *S. chessmanii*) were tested in a field trial using four salinities in the irrigation water (EC_w values of 1.5, 4, 7 and 10 dS/m) and three irrigation levels (100, 50 and 25% of the full irrigation requirement) using low saline water (1.5 dS/m). The field trial was conducted between February and May 2016 in the experimental station of the International Center for Biosaline Agriculture, Dubai, UAE. Plants were first sown into a nursery and then transplanted into the field after 4 weeks. For plant fertilization, a modified Hoagland solution was supplied to the plants (Hoagland and Arnon 1950; Cooper 1988; Hochmuth and Hochmuth 2001).

Salinity treatments were applied using four tanks filled with saline groundwater mixed with fresh water to achieve the required level of salinity. Irrigation treatments were applied using drippers with different flows. Drippers with 2, 4 and 8 L/hr flow rates were used to irrigate treatments at 25, 50 and 100% of full fresh irrigation

TABLE 26.1
Fresh and Groundwater Analysis

Water Type	pH	Electrical Conductivity (dS/m)	Ca+Mg (mg/L)	Na (mg/L)
Fresh water	8.35	0.495	122	63
Groundwater	7.15	24.70	6462	4864

requirement (FI). Irrigation amounts were determined using the ET_0 approach, as described by Allen et al. (1998) for 100% full irrigation requirement. All salinity treatments received full irrigation. Analyses of fresh and saline groundwater are presented in Table 26.1.

SD, root, shoot and biomass weight, and sodium and potassium concentrations in leaves were determined. SD was measured using digital calipers at 5 cm above the ground. Fresh root weight (FRW), fresh shoot weight (FSW), fresh biomass weight (FBW), root to shoot ratio (R/S ratio) were measured at the end of the growth period. Plants were first weighed as a whole, then roots and shoots weights were determined separately. Sodium and potassium concentrations of leaves were analyzed using a flame photometer after digestion of dried tissue in 10 mM H_2NO_3, using the standard procedures (Negrão et al. 2017).

Differences in the response of the plants to the applied treatments were assessed with a general linear model in the StatSoft STATISTICA 8.0.550 software. Comparison of means was performed using analysis of variance with two factors (ANOVA2). A Tukey test was used to statistically identify the homogeneous groups. All statistical differences were considered significant at $\alpha = 0.05$ or lower.

26.3 RESULTS

26.3.1 Biomass Partitioning and Mineral Concentrations

The fourteen rootstocks were studied under drought or saline stress conditions in order to assess their tolerance levels. Growth and biomass as well as macronutrients were determined under different stress levels. The results of this study showed significant differences in the measured characteristics: rootstock, stress level and their interaction according to the factors ($P<0.05$) (Tables 26.2 and 26.3). Except for root/shoot ratio (R/S) under drought stress, both drought and saline stress treatments significantly reduced all growth and biomass parameters (FRW, FSW, R/S ratio, SD and FBW) for all rootstocks ($P<0.05$); this decrease was greater under the 25% FI and 10 dS/m treatments respectively, compared to the 100% FI non-saline control.

Under drought stress conditions, Na^+ concentrations increased significantly with increasing stress intensity (especially under 25% FI), while Na/K ratio decreased ($P<0.05$). However, no statistical difference was recorded for K (Table 26.2). These both ions and their ratio (Na/K) changed significantly depending on the saline stress level ($P<0.05$) (Table 26.3). In comparison with the control, Na/K ratio significantly decreased, while both Na and K increased, where the highest concentrations were found at 10 dS/m and 7 dS/m, respectively. Furthermore, all studied

TABLE 26.2
The Variation among the Monitored Parameters of Investigated Rootstocks under Drought Treatments and ANOVA Analysis Results (P-Values)

	FRW	FSW	FBW	R/S ratio	SD	Na	K	Na/K ratio
		(g/plant)		-	cm	mg/g of DM*		-
Rootstock								
AIPT0919	102.25a	381.42a	483.68a	0.27abc	1.14abc	0.58b	17.84a	36.43abc
ARAZI	94.83a	354.83ab	449.67ab	0.32abc	1.28a	0.52b	20.22a	45.22a
S. Cheesmanii	16.71c	51.13d	67.83e	0.33ab	0.51f	1.93a	20.70a	13.24c
DR9011TV	98.54a	302.29abc	400.83abc	0.33ab	1.18abc	0.58b	18.77a	34.46abc
EMPERADOR	65.50ab	325.50abc	391.00abcd	0.23abc	1.11abcd	0.42b	18.51a	47.94a
EMPOWER	62.92ab	355.50a	418.42abc	0.23abc	1.16abc	0.50b	17.81a	38.91ab
KING KONG	62.04ab	253.08abc	315.13bcd	0.26abc	1.01bcd	0.57b	17.28a	34.86abc
MAXIFORT	96.17a	385.08a	481.25a	0.26abc	1.29a	0.58b	18.25a	34.52abc
OPTIFORT	88.37a	292.25abc	380.63abcd	0.37a	1.16abc	0.49b	18.78a	40.93ab
S. Pimpinilifolium	94.13a	300.17abc	394.29abcd	0.32abc	0.90de	1.46a	19.54a	18.47bc
SILEX	64.46ab	217.25bc	281.71cd	0.31abc	0.97cde	0.40b	18.29a	50.71a
SOUSSPRO	73.75ab	347.54ab	421.28abc	0.21bc	1.19ab	0.44b	17.54a	41.56ab
SUPERPRO	62.46ab	363.33a	425.79abc	0.18bc	1.19ab	0.53b	16.97a	42.12ab
UNIFORT	32.87bc	198.71c	231.58d	0.17c	0.79e	0.70b	15.56a	26.72abc
Drought Treatment								
100 % FI	113.62a	434.65a	548.26a	0.29a	1.13a	0.55b	17.91a	42.68a
50 % FI	55.46b	261.96b	317.42b	0.25a	1.09a	0.66b	17.36a	34.36b
25 % FI	48.43b	187.98c	236.41c	0.28a	0.97b	0.87a	19.61a	31.41b
ANOVA (*P*-values)								
Rootstock	<.001	<.001	<.001	<.001	<.001	<.001	ns	<.001
Drought	<.001	<.001	<.001	ns	<.001	<.001	ns	0.003
Rootstock x Drought	0.042	0.030	0.024	0.018	ns	ns	ns	Ns

*: dry matter

Abbreviations: FRW: fresh root weight, FSW: fresh shoot weight, FBW: Fresh biomass weight, R/S: Root to shoot ratio, SD: Stem diameter, Na: Sodium concentration, K: Potassium concentration, Na/K: sodium to potassium ratio.

Note: Means within a column accompanied by the same letter are not significantly different (5% probability level, the Tukey test).

parameters had significant statistical differences between rootstocks, except K^+. Under drought stress, both AIPT0919 and MAXIFORT rootstocks showed high values of FRW, FSW and FBW, where MAXIFORT and ARAZI had the highest SD values among all rootstocks. We also noted that high values of RW were found in ARAZI, DR9011TV, OPTIFORT and *S. Pimpinilifolium*; while high values of SW were observed in EMPOWER and SUPERPRO. In addition to its high RW value, OPTIFORT had the highest R/S ratio.

Under saline conditions, the highest RW values were with DR9011T, whereas MAXIFORT, AIPT0919 and SUPERPRO had the highest SW values. ARAZI had

TABLE 26.3
The Variation among the Monitored Parameters of Investigated Rootstock under Salinity Treatments and ANOVA Analysis Results (P-Values)

	FRW	FSW	FBW	R/S ratio	SD	Na	K	Na/K ratio
		(g/plant)		-	cm	mg/g of DM*		-
Rootstock								
AIPT0919	113.69abcd	457.12a	570.86a	0.25abcde	1.12abc	0.98c	17.69a	25.45ab
ARAZI	116.69abc	394.72abc	511.41a	0.32a	1.29a	1.17c	20.34a	30.22a
S. Cheesmanii	10.97g	39.34d	50.30c	0.26abcd	0.47f	5.06a	15.85a	6.56c
DR9011TV	135.78a	425.91ab	561.69a	0.32a	1.21ab	1.17c	19.10a	21.94abc
EMPERADOR	78.75cdef	375.22abc	453.97ab	0.21bcde	1.10abcd	0.84c	19.45a	31.13a
EMPOWER	66.94ef	432.88ab	499.81ab	0.18cde	1.17abc	0.87c	16.82a	25.24ab
KING KONG	84.81bcdef	349.66abc	434.47ab	0.23abcde	1.08bcd	1.21c	18.84a	24.50ab
MAXIFORT	110.22abcde	459.88a	570.11a	0.23abcde	1.29a	0.98c	19.48a	24.99ab
OPTIFORT	116.88abc	404.09abc	520.95a	0.29ab	1.21ab	1.05c	18.92a	25.66ab
S. Pimpinilifolium	126.41ab	413.34ab	539.74a	0.32a	0.91de	2.64b	17.43a	10.46bc
SILEX	70.59def	260.56c	331.17b	0.28abc	0.98cde	0.75c	17.66a	31.77a
SOUSSPRO	70.53def	370.75abc	441.28ab	0.18bcde	1.14abc	0.89c	18.20a	26.71ab
SUPERPRO	68.94def	447.34a	516.28a	0.16de	1.15abc	0.85c	17.91a	27.89a
UNIFORT	42.16fg	293.97bc	336.13b	0.14e	0.84e	1.19c	15.44a	19.20abc
Salinity Treatment								
Control	113.62a	434.65a	548.26a	0.29a	1.13a	0.55c	17.91b	42.68a
4 dS/m	109.35a	463.71a	573.04a	0.26ab	1.09a	1.00c	18.12b	24.76b
7 dS/m	83.96b	343.66b	427.62b	0.24b	1.07a	1.76b	21.10a	17.77b
10 dS/m	39.75c	222.20c	261.97c	0.18c	0.98b	2.31a	15.19c	9.56c
ANOVA (*P*-values)								
Rootstock	<.001	<.001	<.001	<.001	<.001	<.001	Ns	<.001
Salinity	<.001	<.001	<.001	<.001	<.001	<.001	<.001	<.001
Rootstock x Salinity	ns	Ns	ns	Ns	ns	<.001	Ns	ns

*: dry matter

Abbreviations: FRW: fresh root weight, FSW: fresh shoot weight, FBW: Fresh biomass weight, R/S: Root to shoot ratio, SD: Stem diameter, Na: Sodium concentration, K: Potassium concentration, Na/K: sodium to potassium ratio.

Note: Means within a column flanked by the same letter are not significantly different at 5% (probability level, the Tukey test).

high values of both R/S ratio and SD. Similar values of R/S ratio were observed in both DR9011TV and *S. pimpinilifolium* and of SD in MAXIFORT. Among all studied rootstocks, seven rootstocks (AIPT0919, ARAZI, DR9011TV, MAXIFORT, OPTIFORT, *S. pimpinilifolium*, and SUPERPRO) had highest FBW values. However, under drought and saline conditions, the lowest values of FRW, FSW, SD and FBW were recorded for *S. cheesmanii*, whereas UNIFORT showed the lowest R/S ratio values. Regarding to the minerals under both drought and saline

TABLE 26.4
Correlation Coefficients between Studied Parameters under Drought Conditions for the 14 Rootstocks

	FRW	FSW	R/S Ratio	SD	FBW	Na	K	Na/K Ratio
FRW	1							
FSW	0.59	1						
R/S ratio	0.46	−0.23	1					
SD	0.18	0.47	−0.24	1				
FBW	0.76	0.97	−0.05	0.43	1			
Na	−0.24	−0.32	0.00	−0.46	−0.33	1		
K	0.12	−0.10	0.05	−0.04	−0.07	0.20	1	
Na/K ratio	0.23	0.24	0.02	0.31	0.26	−0.57	0.47	1

Note: All correlations were significant at $P = 0.05$.

conditions, *S. cheesmanii* had highest Na^+ and lowest Na^+/K^+ ratio. Nonetheless, the highest significant values of Na^+/K^+ were found in ARAZI, EMPERADOR and SILEX rootstocks ($P<0.05$). According the ANOVA analysis, a significant rootstock × drought interaction was observed for FRW, FSW, R/S ratio and FBW under drought conditions, while the rootstock × salinity interaction was only significant for Na^+ ($P<0.05$).

Significant positive correlations were recorded between FRW, R/S ratio and FBW, between RW and SW, and between K^+ and Na^+/K^+ ratio, under both stress conditions ($P<0.05$) (Tables 26.4 and 26.5). However, significant negative correlations were observed between Na^+ and the studied growth and biomass traits (FRW, FSW, SD and FBW) and between Na^+ and Na^+/K^+ ratio ($P<0.05$) (Tables 26.4 and 26.5).

TABLE 26.5
Correlation Coefficients between Studied Traits under Saline Conditions for the 14 Rootstocks

	FRW	FSW	R/S Ratio	SD	FBW	Na	K	Na/K Ratio
FRW	1							
FSW	0.68	1						
R/S ratio	0.60	−0.06	1					
SD	0.22	0.47	−0.15	1				
FBW	0.82	0.98	0.12	0.43	1			
Na	−0.23	−0.34	−0.06	−0.45	−0.33	1		
K	0.18	0.10	0.11	0.14	0.13	−0.02	1	
Na/K ratio	0.15	0.16	0.07	0.20	0.17	−0.38	0.41	1

Note: All correlations were significant at $P = 0.05$.

26.3.2 Canonical Discriminant Analysis

In order to discriminate between the fourteen studied rootstocks, we performed canonical discriminant analysis (CDA) using all studied traits as predictors of membership in the diagnostic group (fourteen rootstocks) under drought stress conditions (CDAd) and under saline stress conditions (CDAs). These traits allowed us to separate the best groups of individuals and provide graphical representations to highlight this separation. The results obtained from CDA affirmed the existence of differences in the global characteristics of the studied rootstocks. Wilks's lambda showed a high significance of the differences (Wilks's $\lambda = 0.169$ for CDAd and Wilks's $\lambda = 0.203$ for CDAs) and the calculated F value also indicated significant differences at $P \leq 0.001$. The null hypothesis of discriminant functions generated in this study is tested using χ^2 test. The latter demonstrated a significant discriminatory power for all the functions in both analyses ($P < 0.001$). For CDAd, the first two discriminant functions (DF1 and DF2), with high eigenvalues (1.43 and 0.57, respectively), accounted for most of the total variance (87.8%) and their canonical correlations were $r_1 = 0.77$ and $r_2 = 0.60$, respectively (Table 26.6). Regarding the CDAs, the high eigenvalues of the first two functions (1.65 and 0.55, respectively) allowed them to explain most of the total variance (84.7%) and the corresponding canonical correlations were $r_1 = 0.79$ and $r_2 = 0.60$ for DF1 and DF2, respectively (Table 26.6).

The 2D canonical plots (Figures 25.1 and 25.2) showed the distribution of the fourteen rootstocks spanned by the first two functions. According to the standardized coefficients of the canonical discriminant functions (Table 26.7), SD and R/S ratio were highly weighted in the positive part of DF1 (CDAd), whereas Na^+ and

TABLE 26.6
Statistical Characteristics of the Discriminant Functions Extracted from Canonical Discriminant Analysis (CDA)

	Discriminant Function	Eigenvalue	Variance (%)	Cumulative Variance (%)	Canonical Correlation
Drought Stress	1	1.43	62.7	62.7	0.77
	2	0.57	25.0	87.8	0.60
	3	0.19	8.4	96.2	0.40
	4	0.05	2.1	98.3	0.21
	5	0.02	0.9	99.2	0.14
	6	0.01	0.5	99.7	0.11
	7	0.01	0.3	100.0	0.08
Saline Stress	1	1.65	63.6	63.6	0.79
	2	0.55	21.1	84.7	0.60
	3	0.22	8.5	93.2	0.43
	4	0.11	4.1	97.3	0.31
	5	0.03	1.3	98.6	0.18
	6	0.03	1.1	99.7	0.17
	7	0.01	0.3	100.0	0.10

TABLE 26.7
Standardized Canonical Discriminant Function Coefficients of the Studied Parameters under Both Drought and Saline Stress, according to CDA

	Drought Stress		Salinity Stress	
	Function 1	Function 2	Function 1	Function 2
RW	–0.12	0.38	0.41	0.32
SW	0.15	0.10	–0.02	0.08
R/S ratio	0.37	0.47	–0.09	0.62
SD	0.70	0.59	0.67	0.37
TB	0.19	0.13	–0.02	0.10
Na	–0.63	0.34	–0.57	0.68
K	0.04	0.53	0.14	0.07
Na/K ratio	0.05	-0.65	–0.17	–0.32

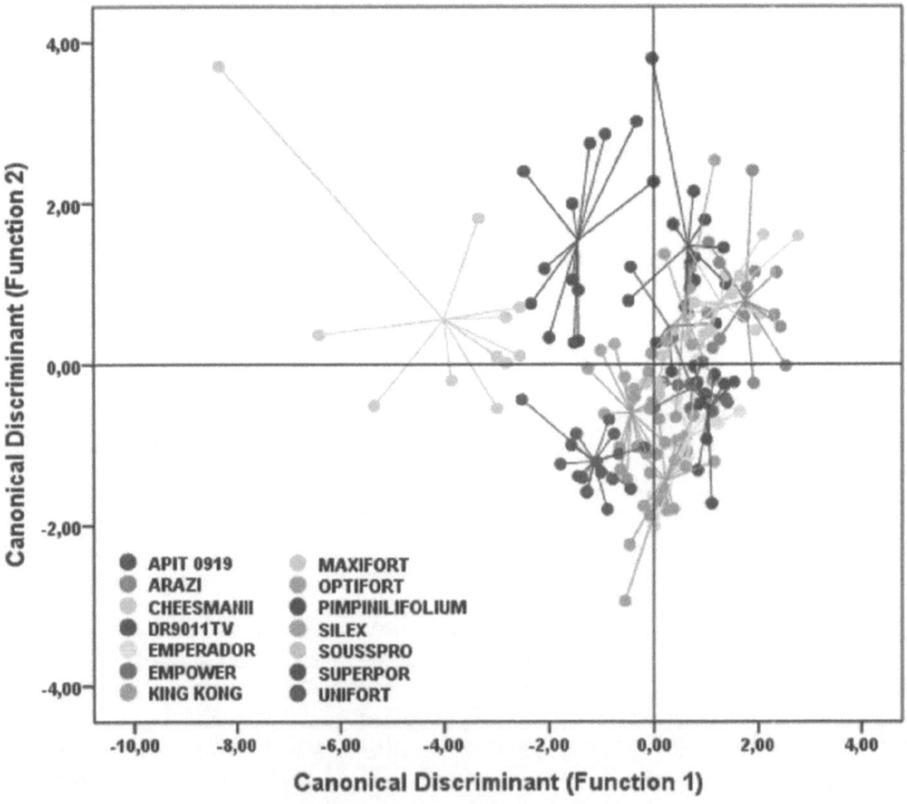

FIGURE 26.1 2D scatterplot showing the distribution of the fourteen rootstocks according to the two discriminant function gradients obtained by CDA for the studied traits under drought stress conditions (CDAd).

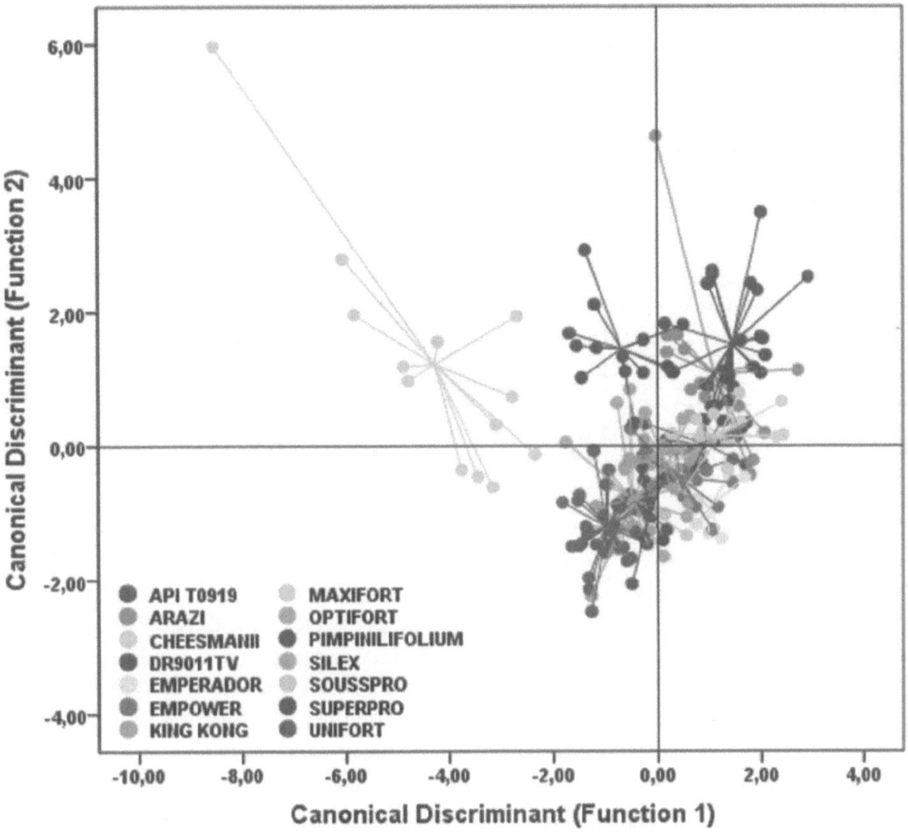

FIGURE 26.2 2D scatterplot showing the distribution of the fourteen rootstocks according to the two discriminant function gradients obtained by CDA for the studied traits under saline stress conditions (CDAs).

RW in the negative part. SD and K^+ were highly weighted in the positive part of DF2 (CDAd), while Na^+ was the only trait most weighted to the negative part of CDAd. Regarding DF1 of CDAs, SD and FRW exhibited the highest standardized coefficients in the positive part, whereas Na^+ and Na^+/K^+ ratio were highly weighted in the negative part. Na^+ and R/S ratio were highly weighted in the positive part of DF2 (CDAs), while Na^+/K^+ ratio was highly weighted in the negative part as for DF2 of CDAd. The DF1 of CDAd allowed a clear separation of DR9011TV, ARAZI, OPTIFORT, MAXIFORT, SUPERPRO, EMPERADOR, EMPOWER and SILEX from the other rootstocks, where ARAZI, MAXIFORT and DR9011TV were characterized by the highest SD and RW. According to the DF2, both *S. Cheesmanii* and *S. pimpinilifolium* were distinguished from other rootstocks especially by having highest Na^+ concentration and low levels of biomass. Compared to CDAd, DF1 of CDAs separated DR9011TV, ARAZI, OPTIFORT, MAXIFORT, EMPERADOR, EMPOWER, KING KONG and APIT0919 from others, where DR9011TV was distinguished by the highest RW. Similar to CDAd, both ARAZI and MAXIFORT were discriminated by the highest SD. Yet again, both *S. cheesmanii* and *S. pimpinilifolium* were clearly discriminated from other rootstocks by having highest Na^+ concentration and highest R/S ratio, respectively.

26.4 DISCUSSION

The study found a limited variation in salt tolerance within the cultivated species. However, there are several wild species within *Lycopersicon* that present a potential source of useful genes for breeding to salt tolerance (Mousa et al. 2013). Researchers have also identified salt-tolerant accessions within the wild tomato species *L. pimpinellifolium, L. peruvianum, L. cheesmanii, L. hirsutum. and L. pennellii* which can be of interest in breeding programs under salinity conditions (Rush and Epstein 1976; Cuartero and Fernández-Muñoz 1998; Foolad 2007; Rzepka-Plevnes et al. 2007).

The differential accumulation and re-distribution of dry matter biomass among the study tomato genotypes with respect to root and shoot tissues, under a range of water and salt treatments, suggested a negative linear relationship between biomass accumulation and increased stress. Increased salinity and stress led to reduction in root and shoot weight with more pronounced effects on shoots. *S. pimpinellifolium* has an extremely efficient system for the re-distribution of photosynthates from the source (shoot) to the sink (root), thus, enabling aggressive root production under salinity. Several reports indicate that *S. pimpinellifolium* is a salt-tolerant wild relative of tomato. Bolarin et al. (1991) found that *S. pimpinellifolium* was the most tolerant genotype to salinity based on slope value of the shoot weight vs salinity relationship curve compared to other wild tomato relatives *(L. pennellii, L. hirsutum and L. peruvianum).* Our results show that under high salinity stress (10 dS/m) *S. pimpinellifolium* had the highest root and shoot weight compared to commercial varieties. This was confirmed by Rao et al. (2013) who reported that shoot dry weight of *S. pimpinellifolium* accessions was less affected under salt stress than *S. lycopersicum.*

It has been shown that Arazi showed the thickest SD under both water and salt stress. Several studies have suggested that SD has positive effects on a rootstock's vigorous root system which helps absorb water and nutrients more efficiently (Oztekin and Tuzel 2011; Wahb-Allah et al. 2011) and may also serve as a supplier of endogenous plant hormones (Albacete et al. 2008; 2014). Rootstock diameter is, therefore, an important criterion for rootstock selection. Vigorous rootstocks have high water and nutrient uptake due to their well-developed xylem structure (Klepper et al. 1971). Bletsos et al. (2003) indicated that depending on the increase in plant development, the grafted plants showed a better vigor compared to non-grafted plants with regard to the plant length and SD.

It has been suggested that commercial crop yield should be the ultimate agronomic criterion for establishing the salt tolerance of crops. However, as salinity affects a wide range of physiological, biochemical and molecular functions and processes in plant growth and development, the enhancement of crop salt tolerance requires a combination of several physiological traits, not simply those directly influencing yield (Rao et al. 2013). There is considerable evidence that salt exclusion (specifically Na^+ exclusion) is the mechanism of survival for most species of agricultural importance when they are exposed to saline conditions (Walker 1986; Schachtman et al. 1991; Reimann 1992;

Fortmeier and Schubert 1995; Hauser and Horie 2010). This study clearly suggests that *S. pimpinellifolium* was the most suitable rootstocks under high saline conditions in terms of root and shoot development; however, this genotype accumulated more sodium in the leaves compared to commercial varities and thus had an increased Na^+/K^+ ratio. This was confirmed by Rush and Epstein (1981) who showed that, under high salt conditions (50–100 mM NaCl), the salt-tolerant tomato genotype freely accumulated Na^+ in the shoot, while the salt-sensitive cultivar excluded it from the leaves. It appears that Silex rootstock tends to exclude Na^+ from the leaf tissue, while *S. pimpinellifolium* freely accumulates Na in the leaf with no toxic effects. Potassium levels in Silex rootstock remained high in the leaves, indicating that K^+ was being selectively accumulated even at these high Na^+ solution levels where the Na^+/K^+ ratio was greater than 15. From this, it appears that the inability of *S. pimpinellifolium* to withstand salinity is linked to its limited efficiency in keeping Na^+ in the leaf tissue below toxic levels and compensating for the lower water potentials associated with salinity by increasing tissue levels of organic solutes (Rush and Epstein 1976). According to the ANOVA and CDA analysis of the studied traits, we can suggest a classification of the studied rootstocks according to their tolerance degree. DR9011TV, ARAZI, OPTIFORT, MAXIFORT seem to be more tolerant to drought stress, whereas both DR9011TV and ARAZI also perform better under saline stress. However, *S. cheesmanii* appears to be the most sensitive to both drought and saline stress.

26.5 CONCLUSION

With the rising of the groundwater salinization problem, there is an urgent need to explore alternative practices to use saline water to grow vegetable cash crops. Grafting use salt-tolerant rootstocks potential solution to producing good yields under saline conditions. However, the screening of suitable rootstocks is not an easy task as it depends on several parameters. Root development as well as sodium exclusion is the main criteria for the selection of suitable rootstocks. This study has shown that both water and salinity stress affected negatively rootstock biomass production with less effect observed for SD. Water stress slightly increased sodium accumulation in rootstock leaves, however, salinity stress greatly increased sodium accumulation and consequently increased the Na^+/K^+ ratio. Our results have shown that there is a great potential for using rootstocks in tomato grafting under saline conditions. Arazi and *S. pimpinellifolium* had highest root development; however, the latter accumulated more sodium in the shoot. Nevertheless, the use of *S. pimpinellifolium* for tomato grafting under saline conditions needs to be studied. Silex, Empower, Sousspro, Maxifort and Emperador rootstocks had the lowest Na^+ concentration in their leaves, which means that they exclude sodium in their root system and this makes them potentially useful rootstocks to be used for tomato grafting under saline conditions.

REFERENCES

Albacete, A., Cantero-Navarro, E., Balibrea, M.E., Großkinsky, D.K., De La Cruz González, M., Martínez-Andújar, C., Smigocki, A.C., Roitsch, T., Pérez-Alfocea, F. 2014. Hormonal and metabolic regulation of tomato fruit sink activity and yield under salinity. *Journal of Experimental Botany*, no. 65 (20): 6081–6095. https://pubmed.ncbi.nlm.nih.gov/25170099

Albacete, A., Ghanem, M.E., Martínez-Andújar, C., Acosta, M., Sánchez-Bravo, J., Martínez, V., Luttsn, S., Dodd, I.C., Pérez-Alfocea, F. 2008. Hormonal changes in relation to biomass partitioning and shoot growth impairment in salinized tomato (*Solanum lycopersicum* L.) plants. *Journal of Experimental Botany* no. 59 (15):4119–4131. https://pubmed.ncbi.nlm.nih.gov/19036841/

Al-Harbi, A., A.M. Al-Omran, K. AlHarbi. 2017a. Grafting improves cucumber water stress tolerance in Saudi Arabia. *Saudi Journal of Biological Sciences*, no. 25 (2): 298–304. https://doi.org/10.1016/j.sjbs.2017.10.025.

Al-Harbi, A., A.M. Al-Omran, T. Alqardaeai, H. Abdel-Razzak, K. AlHarbi, S. Montasir, A. Obiad. 2017b. Tomato grafting impacts on yield and fruit quality under water stress conditions. *Journal of Experimental Biology and Agricultural Sciences*, no. 5: 136–147. http://dx.doi.org/10.18006/2017.5(spl-1-SAFSAW)

Al-Harbi, A., A. Hejazi, A.M. Al-Omran. 2017c. Responses of grafted tomato (*Solanum lycopersiocon* L.) to abiotic stresses in Saudi Arabia. *Saudi Journal of Biological Sciences*, no. 24 (6):1274–1280. https://www.sciencedirect.com/science/article/pii/S1319562X16000073

Al-Harbi, A., A.M. Al-Omran, T. Alqardaeai, H. Abdel-Razzak, K. AlHarbi, S. Montasir, A. Obiad. 2018. Grafting affects tomato growth, productivity and water use efficiency under different water regimes. *Journal of Agricultural Science and Technology*, no. 20 (6): 1227–1241. https://jast.modares.ac.ir/article-23-20094-en.pdf

Allen, R.G., Pereira, L.S., Raes, D., Smith, M. 1998. Crop evapotranspiration–Guidelines for computing crop water requirements. *FAO Irrigation and Drainage Paper*, no. 56:326.

Ashraf, M., Wu, L. 1994. Breeding for salinity tolerance in plants. *Critical Reviews in Plant Sciences*, no. 13 (1):17–42. https://www.tandfonline.com/doi/abs/10.1080/07352689409701906

Bletsos, F., Thanassoulopoulos, C., Roupakias, D. 2003. Effect of grafting on growth, yield, and Verticillium wilt of eggplant. *HortScience*, no. 38 (2):183–186. https://journals.ashs.org/hortsci/view/journals/hortsci/38/2/article-p183.xml

Bolarin, M., Fernandez, F., Cruz, V., Cuartero, J. 1991. Salinity tolerance in four wild tomato species using vegetative yield-salinity response curves. *Journal of the American Society for Horticultural Science*, no. 116 (2):286–290. https://journals.ashs.org/jashs/view/journals/jashs/116/2/article-p286.xml

Cooper, A. 1988. 1. The system. 2. Operation of the system. *The ABC of NFT Nutrient Film Technique*, pp. 3–123. Grower Books. London, UK

Cuartero, J., Fernández-Muñoz, R. 1998. Tomato and salinity. *Scientia Horticulturae*, no.78 (1):83–125. https://www.sciencedirect.com/science/article/abs/pii/S0304423898001915

Dasgupta, S., Hossain, M.M., Huq, M., Wheeler, D. 2015. Climate change and soil salinity: The case of coastal Bangladesh. *Ambio*, no. 44 (8):815–826. https://link.springer.com/article/10.1007/s13280-015-0681-5

Elhag, M. 2016. Evaluation of different soil salinity mapping using remote sensing techniques in arid ecosystems, Saudi Arabia. *Journal of Sensors*, no. 2016: 8. https://doi.org/10.1155/2016/7596175

Estañ, M.T., Martinez-Rodriguez, M.M., Perez-Alfocea, F., Flowers, T.J., Bolarin, M.C. 2005. Grafting raises the salt tolerance of tomato through limiting the transport of sodium and chloride to the shoot. *Journal of Experimental Botany*, no. 56 (412):703–712. https://academic.oup.com/jxb/article/56/412/703/580190

FAOSTAT. 2014. *Crop production*. Food and Agriculture Organization of the United Nations Statistics Division, Rome.

Flowers, T., Yeo, A. 1995. Breeding for salinity resistance in crop plants: where next? *Functional Plant Biology*, no. 22 (6):875–884. https://www.publish.csiro.au/fp/PP9950875

Foolad, M.R. 2007. Genome mapping and molecular breeding of tomato. *International Journal of Plant Genomics,* no. 2007: 52. https://www.ncbi.nlm.nih.gov/pmc/articles/PMC2267253/

Fortmeier, R., Schubert, S. 1995. Salt tolerance of maize (*Zea mays* L.): The role of sodium exclusion. *Plant, Cell & Environment,* no. 18 (9):1041–1047. https://onlinelibrary.wiley.com/doi/abs/10.1111/j.1365-3040.1995.tb00615.x

Hauser, F., Horie, T. 2010. A conserved primary salt tolerance mechanism mediated by HKT transporters: a mechanism for sodium exclusion and maintenance of high K^+/Na^+ ratio in leaves during salinity stress. *Plant, Cell and Environment,* no. 33 (4):552–565. https://onlinelibrary.wiley.com/doi/full/10.1111/j.1365-3040.2009.02056.x

Hoagland, D.R., Arnon, D.I. 1950. The water-culture method for growing plants without soil. *Circular California Agricultural Experiment Station,* no. 347 (2nd edition). University of California, College of Agriculture, Berkeley, California. USA.

Hochmuth, G.J., Hochmuth, R.C. 2001. *Nutrient solution formulation for hydroponic (perlite, rockwool, NFT) tomatoes in Florida.* HS796 *Univ Fla Coop Ext Serv,* Gainesville.

Klepper, B., Browning, V.D., Taylor, H.M. 1971. Stem diameter in relation to plant water status. *Plant Physiology,* no. 48 (6):683–685. http://www.plantphysiol.org/content/48/6/683

Louws, F.J., Rivard, C.L., Kubota, C. 2010. Grafting fruiting vegetables to manage soilborne pathogens, foliar pathogens, arthropods and weeds. *Scientia Horticulturae,* no. 127 (2):127–146.

Mousa, M.A., Al-Qurashi, A.D., Bakhashwain, A.A. 2013. Response of tomato genotypes at early growing stages to irrigation water salinity. *Journal of Food, Agriculture & Environment,* no. 11 (2):501–507. https://www.sciencedirect.com/science/article/abs/pii/S0304423810004462

Navarrete, M., Jeannequin, B., Sebillotte, M. 1997. Vigour of greenhouse tomato plants (*Lycopersicon esculentum* Mill.): Analysis of the criteria used by growers and search for objective criteria. *Journal of Horticultural Science,* no. 72 (5):821–829. https://www.tandfonline.com/doi/abs/10.1080/14620316.1997.11515576

Negrão, S., Schmöckel, S., Tester, M. 2017. Evaluating physiological responses of plants to salinity stress. *Annals of Botany,* no. 119: 1–11. https://www.ncbi.nlm.nih.gov/pmc/articles/PMC5218372/

Oztekin, G.B., Tuzel, Y. 2011. Comparative salinity responses among tomato genotypes and rootstocks. *Pakistan Journal of Botany,* no. 43 (6):2665–2672.

Rao, E.S., Kadirvel, P., Symonds, R.C., Ebert, A.W. 2013. Relationship between survival and yield related traits in *Solanum pimpinellifolium* under salt stress. *Euphytica,* no. 190 (2):215–228. https://link.springer.com/article/10.1007/s10681-012-0801-2

Reimann, C. 1992. Sodium exclusion by Chenopodium species. *Journal of Experimental Botany,* no. 43 (4):503–510. https://academic.oup.com/jxb/article-abstract/43/4/503/610792?redirectedFrom=fulltext

Rush, D.W., Epstein, E. 1976. Genotypic responses to salinity differences between salt-sensitive and salt-tolerant genotypes of the tomato. *Plant Physiology,* no. 57 (2):162–166. https://www.jstor.org/stable/4264309?seq=1

Rush, D.W., Epstein, E. 1981. Comparative studies on the sodium, potassium, and chloride relations of a wild halophytic and a domestic salt-sensitive tomato species. *Plant Physiology,* no. 68 (6):1308–1313. https://www.ncbi.nlm.nih.gov/pmc/articles/PMC426093/

Rzepka-Plevnes, D., Kulpa, D., Smolik, M., Glowka, M. 2007. Somaclonal variation in tomato *L. pennelli* and *L. peruvianum* f. glandulosum characterized in respect to salt tolerance. *International Journal of Food, Agriculture and Environment,* no. 5 (2):194–201.

Santa-Cruz, A., Martinez-Rodriguez, M.M., Perez-Alfocea, F., Romero-Aranda, R., Bolarin, M.C. 2002. The rootstock effect on the tomato salinity response depends on the shoot genotype. *Plant Science,* no. 162 (5):825–831. http://dx.doi.org/10.1016/S0168-9452(02)00030-4

Schachtman, D., Munns, R., Whitecross, M. 1991. Variation in sodium exclusion and salt tolerance in *Triticum tauschii*. *Crop Science*, no. 31 (4):992–997. https://acsess.onlinelibrary.wiley.com/doi/abs/10.2135/cropsci1991.0011183X003100040030x

Schwarz, D., Rouphael, Y., Colla, G., Venema, J.H. 2010. Grafting as a tool to improve tolerance of vegetables to abiotic stresses: Thermal stress, water stress and organic pollutants. *Scientia Horticulturae*, no. 127 (2):162–171. https://www.sciencedirect.com/science/article/abs/pii/S0304423810004243

Valipour, M. 2014. Drainage, waterlogging, and salinity. *Archives of Agronomy and Soil Science*, no. 60 (12):1625–1640. https://www.tandfonline.com/doi/abs/10.1080/03650340.2014.905676

Wahb-Allah, M.A., Alsadon, A.A., Ibrahim, A.A. 2011. Drought tolerance of several tomato genotypes under greenhouse conditions. *World Applied Sciences Journal*, no. 15 (7):933–940. http://citeseerx.ist.psu.edu/viewdoc/download?doi=10.1.1.389.6993&rep=rep1&type=pdf

Walker, R. 1986. Sodium exclusion and potassium-sodium selectivity in salt-treated trifoliate orange (*Poncirus trifoliata*) and Cleopatra mandarin (*Citrus reticulata*) plants. *Functional Plant Biology*, no. 13 (2):293–303. https://www.publish.csiro.au/fp/PP9860293

Wichelns, D., Qadir, M. 2015. Achieving sustainable irrigation requires effective management of salts, soil salinity, and shallow groundwater. *Agricultural Water Management*, no. 157:31–38. https://www.sciencedirect.com/science/article/abs/pii/S0378377414002558

Zijlstra, S., Groot, S., Jansen, J. 1994. Genotypic variation of rootstocks for growth and production in cucumber; possibilities for improving the root system by plant breeding. *Scientia Horticulturae*, no. 56 (3):185–196. https://www.sciencedirect.com/science/article/abs/pii/0304423894900019

27 Root Architecture and Productivity of Three Grass Species under Salt Stress

Liping Wang, Junjie Yi, and Theo Elzenga

CONTENTS

27.1 INTRODUCTION

Soil salinity severely limits the productivity and quality of crops and, therefore, poses a threat to food security throughout the world. Among the various abiotic stress factors, soil salinization is one of the most serious land degradation threats to modern agriculture. Increasing and more extensive salinization of arable land, combined with the projected growth of the human population, necessitates using more salt-tolerant crop plants.

The main functions of plant roots are anchoring the plant and the uptake of nutrients and water. Root system architecture (RSA) alludes to the spatial configuration of the root system and combines several structural features such as the length of primary roots and their spread, the number, angle and length of lateral roots and responds to external environmental conditions such as water availability, nutrition and ion concentrations in the soil (Khan et al., 2016). Grasses (Poaceae) have seminal roots, which consist of primary axes, formed during embryogenesis and already present in the un-germinated caryopsis, and the laterals which emerge from these with time (Leszek, 2012). Water and nutrient capture depends largely on laterals of seminal roots and rhiszosphere properties (Ahmed et al., 2018; Carminati et al., 2017; Boudiar et al., 2020). Under salt stress, the RSA of grasses can be reshaped in

DOI: 10.1201/9781003112327-27

various ways. Salt stress has been shown to block primary root meristem division and inhibit cell elongation, causing inhibition and alterations to cell morphology, in combination with changing root length and the number of primary, lateral and seminal roots (Potters et al., 2007).

As RSA is highly variable and responds plastically to environmental conditions, it is considered a new target for breeding efforts. An efficient RSA improves plant growth and health by improving nutrient and water uptake under stress conditions, including salinity (Li et al., 2018; Berg et al., 2018). Salt-marsh grasses offer natural examples of plants with high salt tolerance and understanding their properties and strategies could yield critical parameters to be used in screening for crop plant varieties with potentially high salt tolerance. Native salt-marsh species have adapted to frequent flooding with sea water and are considered to be potentially highly salt tolerant (Rouger and Jump, 2015). *Puccinellia maritima,* or common salt-marsh grass, is one of the dominant grass species on salt marshes in Western Europe and is an important food source for migrating geese in early spring (Fokkema et al., 2016). *Festuca rubra* is another dominant species on saltmarshes but can also be found on non-saline soils (Gray and Scott 1977; Rouger et al., 2014).

In this study, we compared these two salt marsh grass species to *Lolium perenne*, a perennial pasture grass, widely used as fodder for cows and sheep, that can potentially also be used for biofuel purposes (Jauhar, 1993). To identify the relevant traits for salt-stress tolerance, we measured the development of the root systems and productivity of the three species under both control and saline conditions.

27.2 MATERIALS AND METHODS

Caryopses of grasses were germinated in 5-liter containers filled with vermiculite moistened with tap water. After germination, uniform seedlings with one leaf were selected and transferred to plastic pots filled with vermiculite and placed in a culture room (temperature 22/18°C, light/dark: 14 h/8h, 150 μmol m^{-2} s^{-1}). The seedlings were watered with tap water every day till the third leaf reached 3 cm. After that, the seedlings were watered with a 25% Hoagland solution for 4 days. At day 18 after germination, the salt treatment started. The 25% Hoagland solution was supplemented with NaCl. To avoid salt shock, the salinity was applied gradually with stepwise increases of 25 mM NaCl, every 2–3 days, until the final concentrations of 50, 100, 150 and 200 mM were reached. Plant samples were taken at 3, 6, 9 and 12 days after the final NaCl concentration was reached. Each pot contained three plants, and for each treatment three replicates were used.

When sampling, whole plants were carefully removed from the pots and rinsed under running tap water to remove any adhering vermiculite, avoiding any damage to the roots. Subsequently, the roots were stained in a solution of 0.5 g/L Neutral Red for 10 min and rinsed 5 times in distilled water. The roots were spread out to reduce crossing roots as much as possible on an Epson 10000 flatbed scanner controlled by WinRhizo, Arabido 2009 software. The images of the scanner were cropped to size and color inverted with Paint and then analyzed with RootNav 2.0 software, making distinctions between primary roots, seminal roots and lateral roots (Figure 27.1 for a schematic representation of the root architecture). Since there is no option for

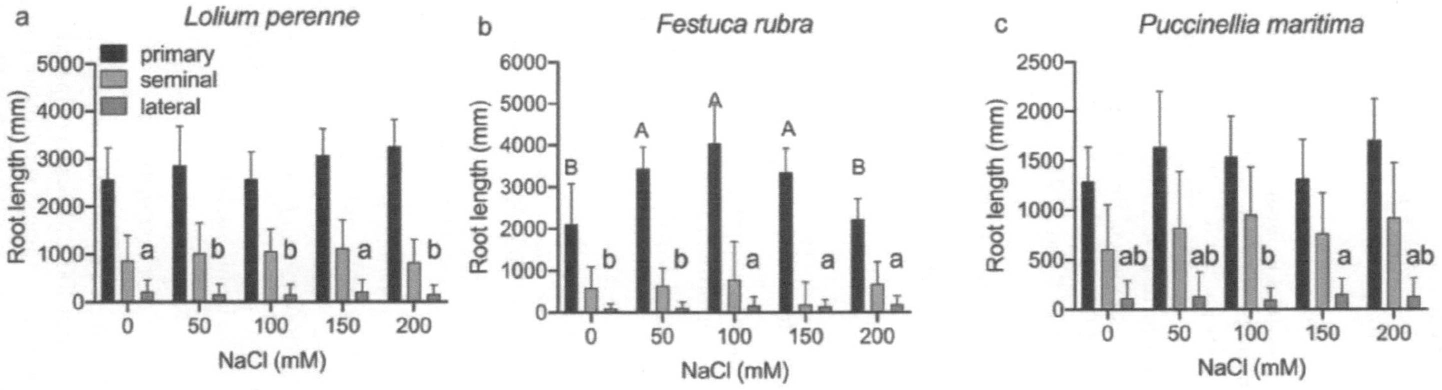

FIGURE 27.1 Effect of salinity on root length after 9-day of exposure to NaCl. Samples were taken at 3-day, 6-day, 9-day and 12-day after start of the salt treatment. Five salt concentrations were used: 0mM (control) 50mM, 100mM, 150mM, 200mM NaCl (n= 9).

seminal roots in this software, the primary roots were marked as 1st primary and the other seminal roots were marked as sequential lower order primaries. The statistical analysis ANOVA using the Tukey test was performed using Prism (version 8 for MacOs, Graphpad).

The fresh weight of roots and shoots were measured directly after scanning of the root system. Three replicate seedlings in one pot were weighed together to increase accuracy since individual seedlings were still small.

For determination of the RSA parameters, the 12-day old plants of *L. perenne* were too large to be analyzed by RootNav. The analysis of *Lolium* plants could, therefore, only be followed up to day 9. To determine the effect of salinity on growth rate and RSA, root and shoot weight and its ratio denoted as R/Sh, the relative ratios of results under 100 or 200 mM NaCl and control treatments were calculated according to formulas 27.1 and 27.2

$$\left(\text{value}_{100\text{mM}} - \text{value}_{0\text{mM}}\right) / \text{value}_{0\text{mM}} \times 100\% \tag{27.1}$$

$$\left(\text{value}_{200\text{mM}} - \text{value}_{0\text{mM}}\right) / \text{value}_{0\text{mM}} \times 100\%. \tag{27.2}$$

27.3 RESULTS

27.3.1 Effect of Salinity on Root Length

Salinity affected the primary root length (PRL) differently in the three grass species studied (Figure 27.2). The PRL of *Lolium perenne* and *Puccinellia maritima* was increased when exposed to NaCl, but *Festuca rubra* showed a significant increase ($P = 0.05$) at intermediate salt concentrations, a stimulation that was no longer apparent at 200 mM NaCl.

The seminal root length (SRL) of the three grass species also showed an increase with salinity, but this difference was not significant. Compared with the PRL, the SRL of *L. perenne* was much shorter. The SRL ranged from 6% to 45% of the PRL in *L. perenne* and *F. rubra* and was not affected significantly by salinity. In contrast, the SRL of *P. maritima* was very low at 0 mM NaCl (4%) but increased strongly (by 50–62%) when exposed to salinity.

The effect of salinity on lateral root length (LRL) differed significantly between the three species. The LRL of *L. perenne* was highest in the control plants and was reduced significantly at the higher salinity levels. In contrast, the LRL of *F. rubra* was highest under high salinity (100, 150 and 200 mM NaCl) and was lower at 0 and 50 mM NaCl. In *P. maritima* the LRL was relatively high under all salinity conditions but was not strongly affected by high salinity levels.

27.3.2 Effect of Salinity on Root Architecture

Root architecture was strongly affected by salinity stress as shown in Table 27.1. The total SRL increased in *L. perenne* at 100 mM NaCl compared to control, but the contribution of the lateral roots decreased (lateral root total length by –32% and –36% at 100 and 200 mM NaCl, respectively). *F. rubra* showed the most positive increases

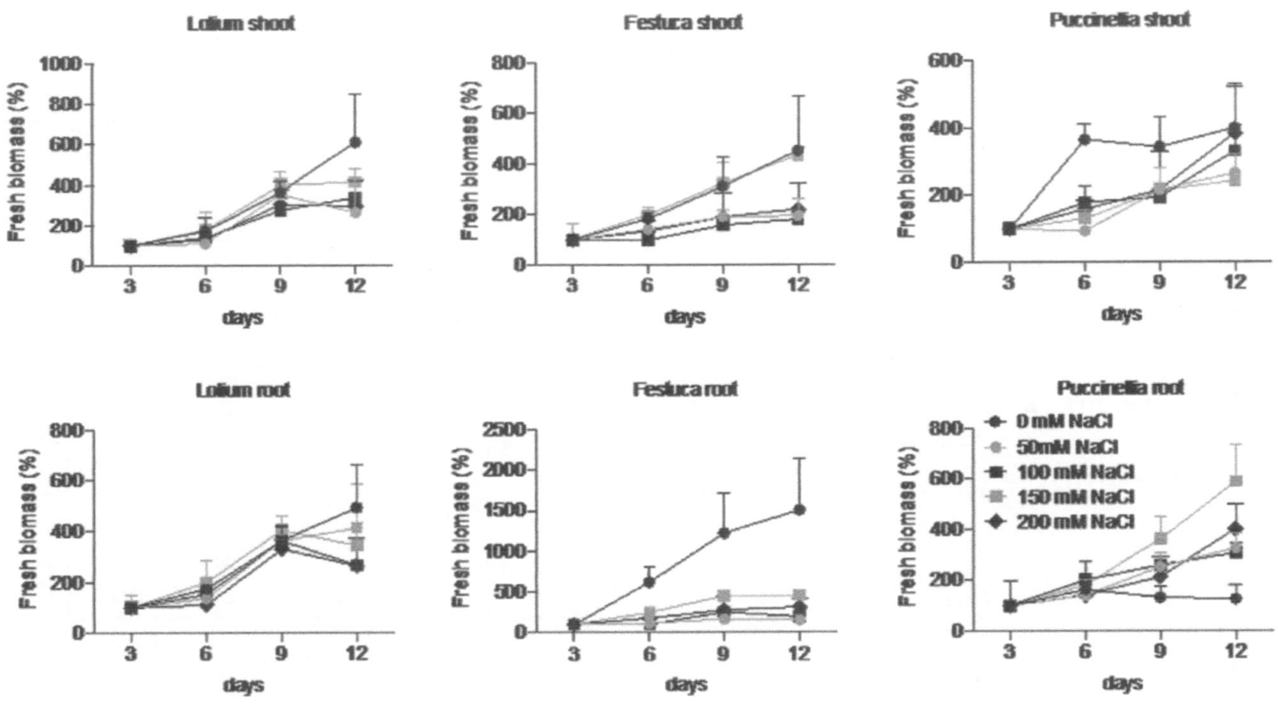

(a)

FIGURE 27.2 Effect of salinity on root and shoot productivity of three grasses. Uniform germinated seedlings with one leaf of *F. rubra*, *P. maritima*, and *L. perenne* were transferred to pots with vermiculite and watered with 25% Hoagland solution. Salt treatments started ~17-18 days after germination. Salt was gradually added to plants to avoid the salinity shock. End concentration NaCl were 0, 50, 100, 150, 200 mM, samples were taken at 3, 6, 9, 12 days after start of the salinity treatment (n=3). The top 6 panels indicate the increase in biomass in shoot and root in % (biomass at day 3 = 100%). The bottom 6 panels indicate the calculated relative growth rate (RGR) based on the data in the top panels. *(Continued)*

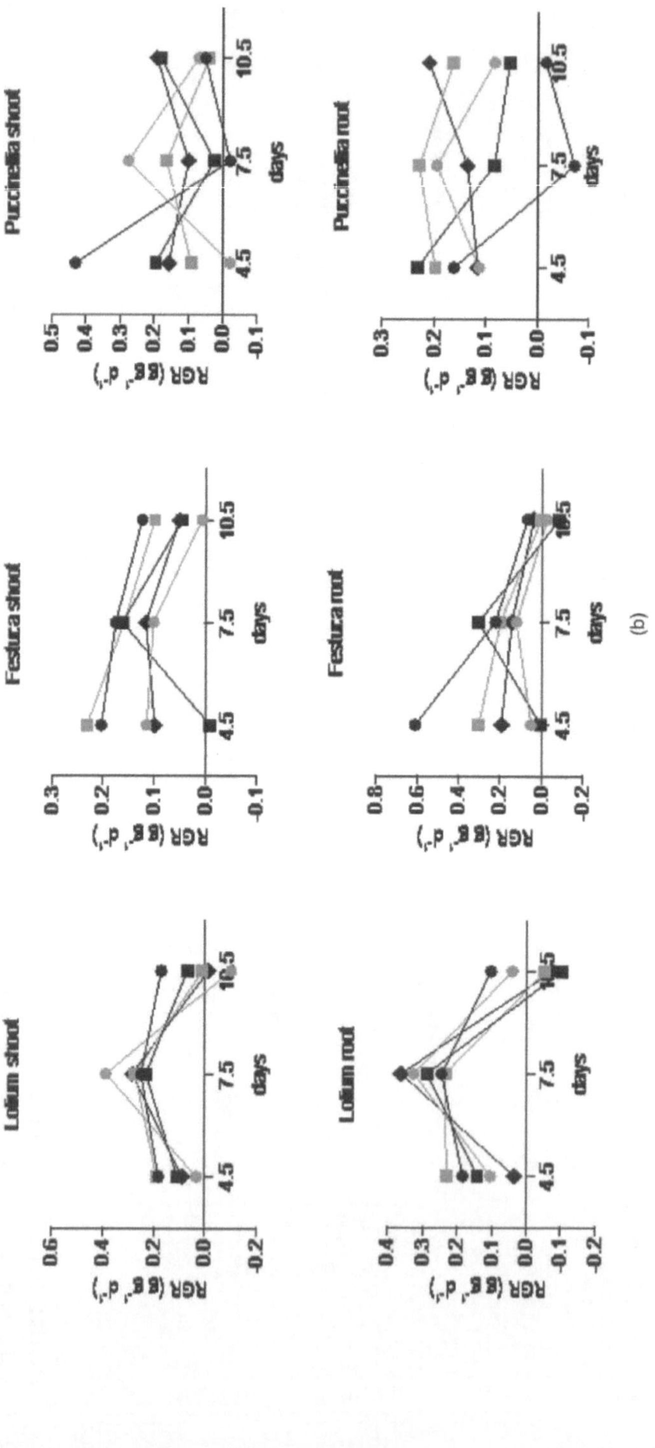

FIGURE 27.2 (Continued)

TABLE 27.1
Percentage NaCl-Induced Change in Root Architecture Parameters

	NaCl (mM)	Primary Length	Seminal Number	Seminal Mean Length	Seminal Total Length	Lateral Number	Lateral Mean Length	Lateral Total Length
L. perenne	100[1]	0	1	1	37	–6	–19	–32
F. rubra	100	86	–3	11	8	134	92	279
P. maritima	100	16	34	1	126	–6	–20	–12
L. perenne	200[2]	27	–5	–5	9	–13	–14	–36
F. rubra	200	6	–13	15	0	46	75	128
P. maritime	200	29	11	11	86	25	1	60

[1] Comparison of 100 mM NaCl to control.
[2] Comparison of 200 mM NaCl to control.

among three grasses. Apart from the seminal root number (a moderate decrease of –3% and –13% at 100 and 200 mM NaCl, respectively) all the other 6 parameters of root architecture increased at both 100 and 200 mM NaCl, the highest being the lateral root total length by 279%. However, for primary root, lateral root and seminal root (except for number of seminal root), an inhibition by 200 mM compared to 100 mM NaCl was apparent.

Each treatment has 9 replicates. Indicated is the percentage change after 9 days of exposure to NaCl in the nutrient solution compared to the control treatment.

Contrary to *F. rubra*, with *P. maritima* the highest stimulation was observed at 200 mM: both primary and seminal total root length of *P. maritima* increased at both 100 and 200 mM relative to the control. The seminal total length increased by 126% at 100 mM NaCl. Although the primary and seminal roots showed an increase at 100 mM, the lateral roots showed a slight decrease. This trend continues at 200 mM: the 7 root traits consistently increased in *P. maritima* especially with the seminal total length (87%), while the lateral length was not affected (1 to 2% in lateral mean length).

27.3.3 Effect of Salinity on Shoot and Root Productivity

Figure 27.2 shows the development of root and shoot fresh weight relative to the fresh weight present at day 3. Comparing the overall pattern of biomass increase for the three species, we observed very distinct differences. Although in all species growth was highest in the control (with the interesting exception of *P. maritima*, where prolonged exposure did show more or equal growth in the presence of NaCl), comparing the immediate (day 6) and post-acclimation (day 12) does yield some interesting differences. In *L. perenne*, the exposure to salt did not result in a significant inhibition of growth. Up to day 9, there was identical growth in both shoot and root in all treatments. However, at day 12 the high NaCl treatments did lead to such a strong effect that the root biomass showed a negative trend. In *F. rubra*, exposure to salinity

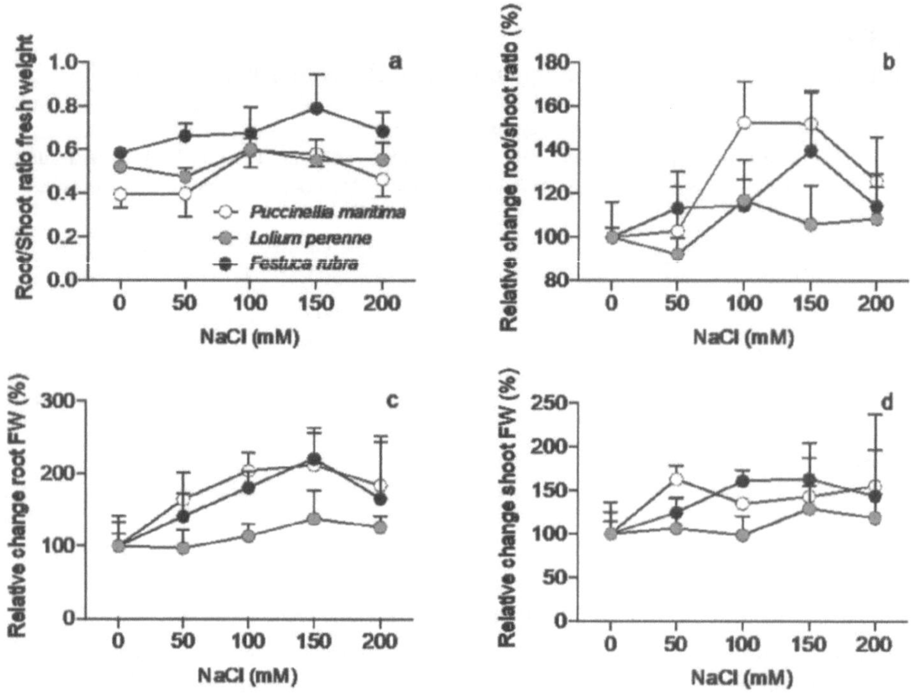

FIGURE 27.3 Effect of salinity on root shoot ratio after 9 days of growth. See legend to Figure 9.2 for details. Fresh Root shoot (R/Sh) ratio, salt to control Fresh R/Sh ratio and salt to control fresh root and fresh shoot were calculated based on fresh root and shoot weight. (a) Fresh Root Shoot weight ratio. (b- d) The R/Sh, Root weight and Shoot divided by the control, respectively (n=9).

induced an immediate growth reduction in both root and shoot and this growth reduction was present at a constant level throughout the whole 12 days period. A third pattern is observed in *P. maritima* where NaCl did have an instant effect leading to growth inhibition at day 6. However, this effect was transient, and at day 12 the growth in the presence of NaCl was as vigorous as under control conditions. In all three species, the observed patterns were most manifest in root biomass, but were also visible in the shoot development.

27.3.4 Effect of Salinity on Root to Shoot Ratio

The fresh weight data at day 9 were used to analyze the effect of increasing salinity on the root the shoot ratio (R/Sh). In Figure 27.4a, the ratios are directly plotted and although already differences the effect of salinity can be discerned, the difference became more apparent when we calculated the relative (compared to control) salinity-induced change in R/Sh (Figure 27.4b). The relative R/Sh of *L. perenne* did not significantly change, whereas in both *F. rubra* and *P. maritima* the R/Sh increased with salinity, only to drop again at the highest NaCl concentration. This increase in

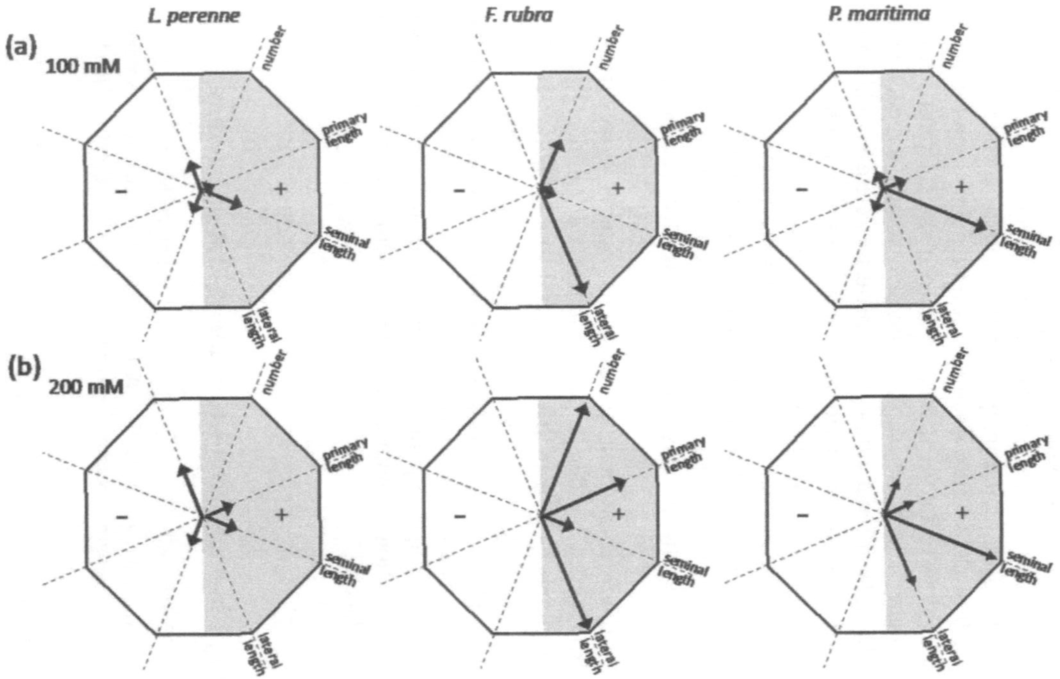

FIGURE 27.4 The effect of 100 and 200 mM NaCl on the following root architecture parameters: root number, primary root length, seminal root length and lateral root length. An arrow pointing to the white (left hand) half indicates a decrease in the indicated parameter, an arrow pointing to the grey (right hand) half indicates an increase in the parameter value. (a) The effect of 100 mM NaCl compared to the control conditions, (b) the effect of 200 mM NaCl. The changes are relative to the control (no salt) values and calculated by:

$$(\text{index}_{100\text{mM}} - \text{index}_{0\text{mM}}) / \text{index}_{0\text{mM}} * 100\% \text{ for } 100 \text{ mM compared to control and}$$

$$(\text{index}_{200\text{mM}} - \text{index}_{0\text{mM}}) / \text{index}_{0\text{mM}} * 100\% \text{ for } 200 \text{ mM compared to control.}$$

R/Sh in the two salt marsh species was caused by a relative increase in the root fresh weight with salinity, while the shoot biomass was not affected. In *L. perenne*, NaCl did not induce a similar increase in root biomass, resulting in a constant R/Sh.

27.4 DISCUSSION

Salinity had a significant effect on RSA in all three species in this study. However, the way the root system architected was affected differed between species. The patterns are summarized in Table 27.2. In the comparison between the species, we have mainly taken data of day 9 of the salinity treatment, a time point when the detrimental effects of high salinity were not yet apparent in *L. perenne*, the most salinity sensitive species. At the higher salinity levels, the *L. perenne* showed a reduction in biomass (negative growth) at day 12 and although the patterns were even more pronounced, it is likely that from that day on physiological processes were not within the normal operational range and would have introduced an unwanted bias.

Most of the results we obtained confirm that *L. perenne* is a moderately salt-sensitive species. In short-term experiments, the plant behaves as an excluder, maintaining growth when exposed, probably by excluding Na^+ from entering the plant. This strategy fails at longer exposure periods as is evident from the reduction in root biomass at day 12.

F. rubra and *P. maritima* exhibit radically different patterns, both different from *L. perenne*, but also from each other. At moderate levels of salinity, *F. rubra* increases strongly in the number and length of its lateral roots (Figure 27.2 and Tables 27.1, 27.2), reduces its shoot growth (Figure 27.3) and increases R/Sh ratio. All these adjustments to the growth pattern indicate an acclimation to a more restricted uptake of water and nutrients (Koevoets et al. 2016). Effectively, the *F. rubra* invests in the root system, increasing its uptake potential, while simultaneously decreasing the demand for nutrients (reduction of shoot growth) and water (smaller shoot leads to less evaporating leaf surface).

TABLE 27.2
Salinity-Induced Changes in Root Architecture after 9-Day of Exposure

	100mM NaCl[1]				200mM NaCl2			
	Primary Length	Seminal Length	Lateral Length	Lateral Number	Primary Length	Seminal Length	Lateral Length	Lateral Number
L. perenne	+	+	− −	− −	++	+	− −	−
F. rubra	+++	++	+++++	+ +++	+	+	++++	++
P. maritima	++	++++	− −	−	++	+++	+++	++

The relative percentage were calculated by Salt effect: (salt-control)/control*100%

0 ~10%: +; 10%~50%: ++; >50%: +++; >100%: ++++; >200%: +++++

0~−10%: −; −10%~−50%: − −, <−50%: − − −; <−100%: − − − −; <−200%: − − − − −.

[1] Comparison of 100 mM NaCl to control $(index_{100mM} - index_{0mM}) / index_{0mM} * 100\%$.

[2] Comparison of 200 mM NaCl to control $(index_{200mM} - index_{0mM}) / index_{0mM} * 100\%$.

P. maritima also increases its effective root system, but does so by increasing its primary, seminal and lateral root system. Remarkable is the stepwise increase: increasing primary and seminal roots at moderate salinity and mainly lateral root at the highest NaCl concentration. The overall growth of *P. maritima* (both shoot and root growth) is not significantly affected after an acclimation period of 3 to 6 days, making *P. maritima* in our study the most salinity tolerant species.

Although both *F. rubra* and *P. maritima* were effective in coping with salinity, the differences do confirm the relatively higher salinity tolerance of *P. maritima*, true to its ecological niche on the lower salt marsh. This comparative study clearly shows that grasses that are (highly) adapted to saline conditions are capable of restructuring their architecture by specifically investing in the development of lateral and seminal roots, increasing their capacity to take up water and nutrients. In breeding programs, these traits could be exploited, with the caveat that these traits could come at a cost, as they are in our examples associated with (temporary) reduced above ground productivity.

ACKNOWLEDGMENTS

Acknowledgments are due to Ms. Zhang Yan Pudussery for help in isolating uniform seedling for this experiment, and to Dr. Jan Henk Venema for introducing instrument and software of root scanning. Thanks to the USDA-ARS for providing the seeds of P. maritima. This mention of trade names or commercial products is solely for the purpose of specific information and does not imply recommendation or endorsement by the U.S. Department of Agriculture Research Service. Thanks for the support from the China Scholarship Council (CSC) and the University of Groningen. Author Contributions Wang, Liping and Theo Elzenga conceptualized and designed the experiments; Wang, L performed evaluation of root system archtecture and productivity parameters; Wang, L, Yi, J and Theo Elzenga analyzed the data; The manuscript was written jointly with contributions from all the authors. All authors have read and approved the manuscript.

REFERENCES

Ahmed, M.A., Zarebanadkouki, M., Meunier, F., Javaux, M., Kaestner, A., and Carminati, A. (2018) Root type matters: Measurement of water uptake by seminal, crown, and lateral roots in maize. Journal of Experimental Botany 69, 1199–1206. https://doi.org/10.1093/jxb/erx439

Berg, T. van den and Tusscher, K.H. (2018) Lateral root priming synergystically arises from root growth and auxin transport dynamics. bioRxiv 361709. https://doi.org/10.1101/361709

Boudiar, R., Casas, A.M., Gioia, T., Fiorani, F., Nagel, K.A., and Igartua, E. (2020) Effects of Low Water Availability on Root Placement and Shoot Development in Landraces and Modern Barley Cultivars. Agronomy 10, 134. https://doi.org/10.3390/agronomy10010134

Carminati, A., Benard, P., Ahmed, M.A. et al. (2017) Liquid bridges at the root-soil interface. Plant Soil 417, 1–15. https://doi.org/10.1007/s11104-017-3227-8

Fokkema,W., de Boer, W., van der Jeugd, H.P., Dokter, A., Nolet,B.A., De Kok, L.J., Elzenga, J.T.M., and Olff, H. (2016) The nature of plant adaptations to salinity stress has trophic consequences. Oikos 125, 804–811. doi: 10.1111/oik.02757

Gray, A.J. and Scott, R. (1977) *Puccinellia maritima (Huds.) Parl.: (Poa maritima Huds.; Glyceria maritima* (Huds.) Wahlb.). Journal of Ecology 65, 699–716.

Jauhar P. P. (1993). Cytogenetics of the Festuca-Lolium complex: Relevance to breeding. Berlin: Springer. pp. 1–8

Khan, M.A., Gemenet, D.C., and Villordon, A. (2016) Root system architecture and abiotic stress tolerance: Current knowledge in root and tuber crops. Frontiers in Plant Science 7, 1584. https://doi.org/10.3389/fpls.2016.01584

Koevoets, I., Venema, J.H., Elzenga, J.T.M., and Testerink, C. (2016) Roots withstanding their environment: Exploiting root system architecture responses to abiotic stress to improve crop tolerance. Frontiers in Plant Science 7(112), 1335. doi: 10.3389/fpls.2016.01335

Leszek, P. and Vincent, D. (2012) Maize (*Zea mays*)Gross Anatomy Ontology - seminal root system – Classes NCBO BioPortal. http://bioportal.bioontology.org/ontologies/ZEA

Li, P., Zhang, Y., Yin, S., Zhu, P., Pan, T., Xu, Y., Wang, J., Hao, D., Fang, H., Xu, C., and Yang, Z. (2018) QTL-by-environment interaction in the response of maize root and shoot traits to different water regimes. Frontiers in Plant Science 9, 229. doi:10.3389/fpls.2018.00229

Potters, G., Pasternak, T., Guisez, Y., and Palme, K. (2007) Stress-induced morphogenic responses: growing out of trouble? Trends in Plant Science, 12, 98–105. doi: 10.1016/j.tplants.2007.01.004

Rouger, R., Vallejo-Marin, M., and Jump, A.S. (2014) Development and cross-species amplification of microsatellite loci for *Puccinellia maritima*, an important engineer saltmarsh species. Genetics and Molecular Research 13, 3426–3431. doi: 10.4238/2014.April.30.3

Rouger, R. and Jump, A.S. (2015) Fine-scale spatial genetic structure across a strong environmental gradient in the saltmarsh plant *Puccinellia maritima*. Evolutionary Ecology 29, 609–623. https://doi.org/10.1007/s10682-015-9767-6

Schnepf, A., Leitner, D., Landl, M., Lobet, G., Hieu, Mai, T. H., Morandage, S., Sheng, C., Zörner, M., Vanderborght, J., and Vereecken, H. (2018) CRootBox: A structural–functional modelling framework for root systems. Annals of Botany 121, 1033–1053. doi:10.1093/aob/mcx221

Sparks, E.E. and Benfey, P.N. (2017) The contribution of root systems to plant nutrient acquisition. M.A. Hossain, T. Kamiya, D.J. Burritt, L.S.P. Tran, T. Fujiwara (Eds.), Eds.), In: Plant Macronutrient Use Efficiency, Academic Press, London, UK, pp. 83–92. doi:10.1016/B978-0-12-811308-0.00005-3

28 Quinoa, a Promising Halophyte with Modified Planting Date, and Minimum Water and Pesticide Requirements for Fars Province, Iran

Rezvan Talebnejad, Ali Reza Sepaskhah, and Maryam Bahrami

CONTENTS

28.1 INTRODUCTION

Quinoa (*Chenopodium quinoa* Willd.) is an Andean crop domesticated in southern Peru and Bolivia close to Titicaca Lake. Quinoa is adapted to a wide range of marginal agricultural soils, including those with high salinity and those prone to drought. Recently, some research has primarily addressed salt and drought tolerance in quinoa (Jacobsen et al. 2003; Trognitz 2003; Talebnejad and Sepaskhah 2018). Its salt tolerance is the result of osmotic adjustment, sodium exclusion, and xylem loading and potassium retention (Adolf et al. 2013; Razzaghi et al. 2015). In addition, the presence of a thick plant cuticle, sunken stomata and calcium oxalate crystals in leaf

DOI: 10.1201/9781003112327-28

vesicles all help to decrease leaf transpiration, increasing the tolerance of quinoa to drought stress (Jacobsen et al. 2009; Azurita-Silva et al. 2015; Issa Ali et al. 2019). Quinoa is a highly nutritious gluten-free crop, having a balanced composition of essential amino-acids sometimes scarce in legumes and cereals (Repo-Carrasco et al. 2003); it is also rich in Ca, Fe, and Mg, and has a high content of vitamins A, B2, and E (Ruales and Nair 1992; Adolf et al. 2013; Nowak et al. 2016). Väkeväinen et al. (2020) investigated the viability of two quinoa varieties, Pasankalla, and Rosada de Huancayo, for the development of fermented spoonable vegan products and reported that quinoa has a significant potential to be used for probiotic products. Moreover, quinoa has shown advantages for those who suffer from diabetes, dyslipidemia, obesity, or celiac disease with promising effects on health (Ceyhun et al. 2019). Romano et al. (2020) demonstrated that quinoa powders extracted from the spry-drying process are able to retain important nutritional components including lipids, antioxidants, and proteins, which suggests that a broad diversity of food products could be based on quinoa. This super grain is therefore a promising halophyte in agricultural production, which may cope with growing food demand in semi-arid areas faced with scarce water resources and soil and water salinization.

Quinoa is well adapted to grow under unfavorable soil and climatic conditions, and is rapidly gaining interest throughout the world, even in non-native regions such as Iran, because of its high nutritional value and high resistance to adverse impacts of climate change. Quinoa has different response mechanisms to endure the lack of water, including physiological strategies that operate at the levels of antioxidant defense, cell membrane stabilization, plant growth regulation, stomatal conductance, and osmotic adjustment (Hinojosa et al. 2018).

Improving water productivity (WP) in agricultural farm management can occur by increasing the economic yield of crops or decreasing the amount of irrigation water required to produce the crop. While farmers are interested in the economic output of irrigation systems, environmental policy makers are more concerned about fresh water consumption in agricultural production and other environmental impacts. The stable management of crop production requires a trade-off between economic and environmental objectives. Deficit irrigation is one of the common methods used to deal with water scarcity and limitations in available water resources: its aim is to maximize the amount of yield per unit of water consumed and is achieved by providing crops with less than the full potential evapotranspiration amount (English and Raja 1996). The full meeting of the water demands of the plant may not be an efficient management approach due to the scarcity of water resources. Acceptable yields may be possible by reducing the supply of irrigation water at different growth stages (Geerts et al. 2008). The use of deficit irrigation to achieve optimal crop yields in regions with seasonal drought has attracted the attention of many researchers (Garcia et al. 2003; Geerts et al. 2006; Kaya et al. 2015). Field experiments and modelling on quinoa yield by Geerts et al. (2009) in Bolivia showed that deficit irrigation could significantly increase the WP of quinoa for seed yield; the threshold at which dry matter production began to decrease was at approximately 55% of full irrigation. In addition, in an experiment conducted in southern Italy, Riccardi et al. (2014) found that applying irrigation water at 25% of full irrigation lead to the maximum WP of quinoa (1.12 kg/m^3). Greenhouse studies in Iran showed that quinoa grown with saline groundwater at 0.8 m depth and application of irrigation at 30% of full requirement (i.e. 70% deficit) had a decrease in seed yield of only 36% compared with that obtained in full irrigation, while there was a 12% increase in quinoa WP (Talebnejad and Sepaskhah 2015a; 2016).

Planting date is one of the most critical challenges in non-native crop cultivation. Air temperature and day length associated with different crop planting dates have been identified as critical factors in gaining economic yields on farm (Hinojosa et al. 2018). High temperature during the flowering and seed set stages can significantly decrease yields and is one of the major barriers to the global extension of quinoa cultivation (Pulvento et al. 2010; Hirich et al. 2014; Walters et al. 2016; Yang et al. 2016; Hinojosa et al. 2018). The optimum planting date for quinoa in the Ontario region (Canada) has been reported to be between May and June, while plantings in July have a 50% decrease in yield due to the plants being still immature at first frost occurrence (Nurse et al. 2016). (These researchers have also stated that the recommended spacing of the planting rows should be greater than 75 cm to improve the weeding process.) In experiments in the south Mediterranean region (Egypt), researchers concluded that plant growth is higher when planting occurs at the beginning of the winter season (i.e. the last quarter of December) compared to sowing in the second quarter of December or in January. In this work, the best sowing condition was found at a relative humidity of 68.8% and with 977 hours of sunshine. Risi and Galwey (1991) analyzed the performance of two cultivars, Baer and Blanca de Junin at Cambridge, England. Their results indicated that Baer sown in March had higher yields than Blanca de Junin, implying that Baer has a higher adaptability to temperate latitudes.

Low temperature and frost during the vegetative and flowering stages of quinoa are important for effective crop management (Jacobsen et al. 2005; Rosa et al. 2009). Bois et al. (2006) surveyed the effects of various ambient temperatures on ten different quinoa cultivars; they found that temperatures below 2°C delayed germination, but thermal sensitivity was independent of the geographic origin of the cultivars. Freezing temperatures of –6°C for nearly 4 h had detrimental effects on all cultivars, but –3°C had no harmful influence.

In recent years, quinoa has been introduced to leading farmers in Fars province, Iran as an alternative to high water consumption crops such as rice or maize (Talebnejad and Sepaskhah 2018). Local farmers plant quinoa in spring (March and April); however, plantings at this time have a high requirement for irrigation water. A farm-level experiment was conducted at the Drought Research Center of Shiraz University, Fars province, Iran (29° 56′ N, 52°02′ E, 1810 m above sea level) to assess the possibility of planting a Danish-bred quinoa cultivar (Titicaca, no. 5206) in September instead of the usual planting date of March. It was hypothesized that this alternative planting date would reduce crop water consumption.

28.2 MATERIALS AND METHODS

The experimental field was plowed at the beginning of the cultivation season, and triple superphosphate at a rate of 50 kg/ha was mixed into the soil at plowing. Urea was applied at 250 kg/ha at the vegetative and flowering stages of crop development.

TABLE 28.1
Mean Maximum and Minimum Monthly Temperatures during the Year of the Trial

	Mean Monthly Temperature °C	
Month	**Maximum**	**Minimum**
March	15	1
April	23	4
May	24	8
June	33	13
July	35	15
August	36	14
September	32	10
October	28	6
November	19	-0.5

The site was leveled, and 36 plots of 2 m × 2 m area were established; these were bunded with ridges ~30 cm high. Quinoa seed (Titicaca, no. 5206) was planted at 1–2 cm depth; each plot had 6 rows (spaced 33 cm apart) and the distance between each seed within rows was 15 cm. The climatic condition of the study area was semi-arid, and the soil had a silty clay loam texture. Mean maximum and minimum daily temperatures during the two growing seasons is shown in Table 28.1.

After 2–3 weeks, the plants were thinned by hand to a stand density of ~20 plants/m^2. Plots were hand weeded every two weeks during the experiment.

The amount of water needed for irrigation was determined using the modified Penman-Monteith equation based on the daily potential evapotranspiration and crop coefficient as follows (Razzaghi and Sepaskhah 2012).

$$ET_o = \frac{0.408\Delta(R_n - G) + \gamma\left[\frac{900}{T+273}\right]U_2(e_s - e_a)}{\Delta + \gamma(1 + 0.34U_2)} \tag{28.1}$$

where ET_o is the daily reference evapotranspiratyion (mm/day), T is the daily mean temperature (°C), e_s is the mean daily saturated vapor pressure (kPa), e_a is the mean daily actual vapor pressure (kPa), G is the sensible heat flux to soil (MJ/m^{2}d), R_n is the daily net radiation flux (MJ/m^{2}d), γ is the psychometric constant (kPa/°C), and Δ is the slope of the saturation vapor pressure curve at the mean temperature (kPa/°C).

Meteorological data were collected from the weather station located at the School of Agriculture, Shiraz University. Using the modified Penman-Monteith equation in combination and with the results of our previous study (Talebnejad and Sepaskhah 2015b), the crop coefficients of quinoa had been previously evaluated as 0.58, 1.2, and 0.8, respectively for initial, mid, and final growth stages. Therefore, the crop's standard evapotranspiration was achieved by the following equation.

$$ET_C = K_c \times ET_o \tag{28.2}$$

where ET_c is the daily standard crop evapotranspiration (mm/d), and K_c is the crop coefficient.

The experiment had a factorial complete randomized block design with three replications. There was three irrigation treatments (100%, 75%, and 50% of full irrigation – FI) and two planting dates – 3 March 2018 (spring) and 1 September 2018 (autumn). Surface irrigation was applied every 7-day. The amount of water for full irrigation was 792 mm for the spring planting (March), and 465 mm for the autumn planting (September). Total rainfall was 104 mm for the autumn planting, while there was no rainfall for the spring planting. Harvest dates for the spring and autumn planting were 3 July and 29 November 2018, respectively.

At harvest, panicles were separated from shoots. Achenes were separated from the panicles, and their covers were robed from the seeds. The seeds were dried in the open air for 48–72 h to determine seed yield. Plant shoots were dried in an oven at 65°C for 48–72 h to determine shoot dry matter. The WP of the seed yield was calculated by dividing the seed yield by amount of irrigation water applied. Also, the WP of the total dry matter (TDM) was calculated by dividing the TDM by the mount of applied irrigation water.

Leaf gas exchange parameters were measured with an LCi analyzer (Li-Cor Inc, Nebraska, USA) at 13:00 pm before irrigation at bud formation. Measured parameters included the net rates of photosynthesis and transpiration (A_n and T_r), stomatal conductance (g_s). The ratio A_n/g_s was also calculated as an index of intrinsic water use efficiency (IWUE).

The effects of deficit irrigation and planting date on quinoa yield were evaluated using analysis of variance, and the means were compared using Duncan's multiple range test. All data were normally distributed so no transformation of data was required.

28.3 RESULTS AND DISCUSSION

28.3.1 Seed Yield

Table 28.2 compares the effects the two planting dates, three irrigation regimes on seed yield, TDM, WP of the seed yield, and WP of the TDM. The average yield of quinoa seed was 4.6 Mg/ha for the spring (March) planting, but this increased by 10% to 5.06 Mg/ha for the autumn (September) planting. There was, no significant difference in yield between the FI and 0.75FI irrigation regimes, but the yields with 0.50 FI were significantly lower with both the spring and autumn plantings. Deficit irrigation (50% FI) decreased the seed yield by 17% in spring and by 14% in autumn. The lowest yield was 3.8 Mg/ha for the spring planting with 0.50FI. Our findings show that with higher rainfall and lower evapotranspiration, the autumn planting of quinoa used less irrigation water with no significant drop in seed yield compared with that obtained in the spring planting.

28.3.2 Dry Matter Yield

TDM was higher for the crop planted in autumn than for the crop planted in spring (Table 28.2). The highest TDM (13.6 Mg/ha) was for the autumn planting date with

TABLE 28.2
Seed Yield (SY), Total Dry Matter (TDM), WP of Seed Yield (WP_{SY}), and WP of Total Dry Matter (WP_{TDM}) with the Different Planting Date and Irrigation Regimes

	Planting Date	
Irrigation Regime	**Spring**	**Autumn**
SY (Mg/ha)		
FI	4.60 ab*	5.06 a
0.75 FI	4.46 ab	4.67 a
0.50 FI	3.80 c	4.36 ab
TDM (Mg/ha)		
FI	10.25 c	13.6 a
0.75 FI	9.02 cd	12.02 b
0.50 FI	8.35 d	10.42 c
WP_{SY} (kg/m^3)		
FI	0.64 d	1.09 c
0.75 FI	0.74 d	1.48 b
0.50 FI	0.81 d	2.78 a
WP_{TDM} (kg/m^3)		
FI	1.43 d	2.93 c
0.75 FI	1.51 d	3.82 b
0.50 FI	1.72 d	6.64 a

* Means followed by the same letters in each trait were not different at the 5% level of significance

full irrigation, while the lowest TDM (8.35 Mg/ha) was for the spring planting with 0.50 FI. Results also revealed that the difference between the values in the autumn season is higher than those obtained in the spring planting. Decreasing irrigation water from full irrigation to 50% FI caused a 23% decrease in TDM for the autumn planting, and an 18% decrease in TDM for the spring planting.

28.3.3 Water Productivity

In the case of WP of seed yield, the most water efficient value was with the autumn planting and the 0.50 FI regime (2.78 kg/m^3 (Table 28.2), and the least efficient value was 0.64 kg/m^3 with the spring planting with FI. Data analysis revealed that with the autumn planting, increasing the water stress to mild (0.75FI) and high levels (0.50FI) significantly increased WP by 35% and 155%, respectively. By contrast, with the spring planting, the implementation of the mild and high water stress increased WP_{SY} by 16% and 26%, respectively, although these increases were not significantly different to the FI controls.

The WP of the TDM was also significantly affected by the different irrigation treatments with the autumn planting. Decreasing the irrigation to 75% and 50% of FI caused 30% and 127% increases in WP_{TDM}, respectively. By contrast, with the quinoa planted in spring there was no significant effect of decreasing irrigation on WP_{TDM}. The highest WP_{TDM} was 6.64 kg/m^3 for the autumn planting with 0.50FI; the lowest value was 1.43 kg/m^3 for the spring planting and full irrigation. Similar results were reported by Razzaghi et al. (2012).

In general, the deficit irrigation strategy increased WP_{seed} due to the use of less irrigation water. However, water stress in spring was more detrimental to seed yield as it decreased stomata conductance and restricted plant photosynthesis. Obviously, such a yield decrease is not desirable for farmers as it restricts their economic benefits. Therefore, quinoa autumn planting is recommended in order to secure farmers benefits. Improved varieties and plant breeding to develop water stress tolerant variety of quinoa would be a solution for this conflict.

28.3.4 Physiological Parameters

Table 28.3 summarizes the effects of the different planting dates and various irrigation regimes on the leaf gas exchange characteristics of quinoa.

28.3.4.1 Rate of Photosynthesis (A_n)

Averaged across all irrigation treatments, rates of photosynthesis were ~30% higher with the autumn planting than with the spring planting. Decreasing the application of irrigation water significantly decreased A_n for both spring and autumn plantings. These effects were greatest with the 0.50 FI treatment, which decreased A_n by 25% with the spring planting and by 17% with the autumn planting. The highest rate of photosynthesis (14.68 μmol/m^2/s) occurred in the autumn planting with full irrigation. The lowest rate of photosynthesis (9.08 μmol/m^2/s) occurred in the spring planting with 0.50 FI. The highest value of A_n was therefore ~60% greater than the lowest value.

28.3.4.2 Leaf Stomatal Conductance (g_s)

The changes in the photosynthetic rate caused by water stress were primarily due to stomatal closure. Averaged across all irrigation treatments g_s was ~35% higher for quinoa planted in autumn compared with the spring. Deficit irrigation decreased g_s with adverse effects being greatest at 50% FI; this decreased g_s by 62% with the spring planting and by 47% with the autumn planting. The highest g_s was 0.32 mol/m^2/s with the autumn planted full irrigation treatment and the lowest g_s was 0.10 mol/m^2/s with the spring planted 0.50 FI treatment. In general, g_s was more sensitive to water stress than A_n. The highest value of g_s was ~220% greater than the lowest value. Similar results were reported by Talebnejad and Sepaskhah (2016).

28.3.4.3 Transpiration Rate (T_r)

In general, T_r was ~9% higher with the spring than the autumn planting; this was likely to have occurred because of higher air temperature and lower relative humidity in spring than in autumn. Generally, transpiration has been shown to decrease in quinoa in arid and semi-arid climates under high water stress (Hinojosa et al. 2018).

TABLE 28.3
Photosynthesis Rate (A_n), Stomatal Conductance (g_s), Transpiration Rate (T_r), Intrinsic Water Use Efficiency (A_n/g_s), and Transpiration Efficiency (A_n/T_r) at Different Planting Date and Irrigation Regimes

Irrigation Regime	Planting Date	
	Spring	Autumn
A_n (µmol/m²/s)		
FI	12.05 b*	14.68 a
0.75 FI	9.91 c	13.20 ab
0.50 FI	9.08 c	12.20 b
g_s (mol/m²/s)		
FI	0.26 b	0.32 a
0.75 FI	0.18 cd	0.24 bc
0.50 FI	0.10 e	0.17 d
T_r (mmol/m²/s)		
FI	4.27 a	4.10 ab
0.75 FI	3.87 b	3.43 c
0.50 FI	3.12 cd	2.78 d
A_n/g_s (µmol/mol)		
FI	47.24 c	46.09 c
0.75 FI	55.06 c	56.52 c
0.50 FI	91.31 a	70.88 b
A_n/T_r (g/kg)		
FI	2.83 c	3.57 b
0.75 FI	2.50 c	3.84 b
0.50 FI	2.92 c	4.41 a

* Means followed by the same letters in each trait are not significantly different at 5% level of probability.

Our data were consistent with this view. With the 0.75 FI treatment, T_r was 9% lower in spring and 16% lower in autumn than with the equivalent full irrigation treatment. With the 50 FI treatment, the adverse effects were even stronger: T_r was 27% lower in spring and 32% lower in autumn than with the equivalent full irrigation treatment. The highest rate was 4.27 mmol/m²/s for the quinoa which planted in spring and treated with full irrigation. The lowest rate was 2.78 mmol/m²/s for the autumn planting and the 0.50 FI treatment.

Other factors calculated in Table 28.3 were the IWUE (evaluated as the ratio A_n/g_s; c.f. Rawson et al. 1977) and the transpiration efficiency (TE; defined as A_n/T_r). A decrease in irrigation from FI to 0.75 FI, had no significant effect on IWUE, however, decreasing irrigation from FI to 0.50 FI significantly increased IWUE with both the spring and autumn planting. The highest rise (93%) occurred with a 50% reduction in

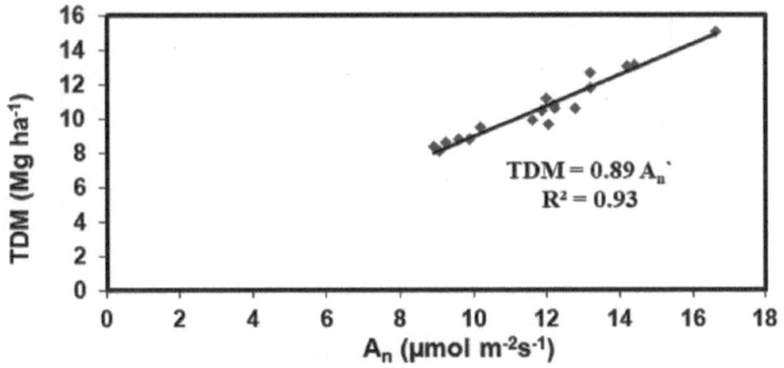

FIGURE 28.1 Relationship between total dry matter (TDM) and net photosynthetic rate (A_n).

irrigation in spring; by contrast, with the autumn planting scenario a 50% reduction in irrigation improved the IWUE by 54%. The transpiration efficiency was less sensitive than the IWUE to different water deficits, especially when the seeds were planted in spring. With the spring planting, there was no significant difference between any irrigation treatment; the value averaged across irrigation treatments was 2.75 g/kg. However, when the seeds were planted in autumn, the transpiration efficiency was higher than that obtained with the spring planting, and deficit irrigation increased the TE; the highest TE (4.41 g/kg) was with the irrigation 0.50 FI treatment.

Across all treatments, we also examined the relationships between growth and the physiological parameters measured. Figure 28.1 shows that there was a positive linear relationship between the TDM and the rate of photosynthesis (A_n). This behavior is in agreement with the similar results observed in wheat (Sikder et al. 2015), cotton (Brugnoli and Lauteri 1991) and saffron (Yarami and Sepaskhah 2015; Dastranj and Sepaskhah 2019).

Figure 28.2 shows that there was a positive linear relationship between A_n and g_s, which shows that water stress impacted on photosynthesis less than stomatal

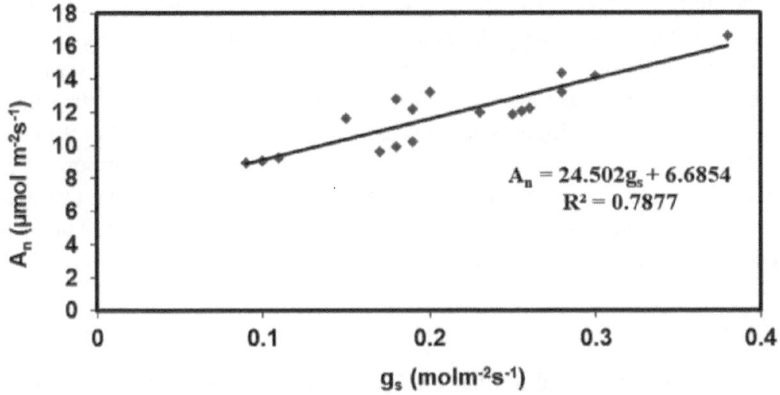

FIGURE 28.2 Relationship between net photosynthetic rate (A_n) and stomatal conductance (g_s).

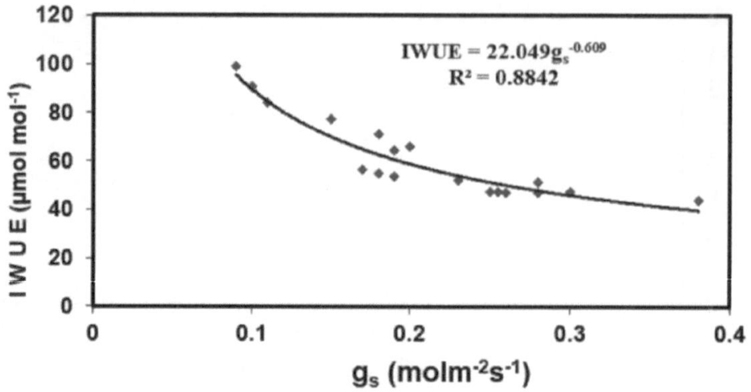

FIGURE 28.3 Relationship between intrinsic water use efficiency (IWUE) and stomatal conductance (g_s).

conductance and consequently decreased dry matter accumulation. This is in accordance with dry matter reduction under deficit irrigation treatments.

Figure 28.3 shows that there was a negative curvilinear relationship between intrinsic water use efficiency (IWUE) and stomatal conductance (g_s). This is consistent with similar trends reported in the literature (Razzaghi et al. 2015; Talebnejad and Sepaskhah 2016b; Mehrabi and Sepaskhah 2019); however, the data also suggest that an increase in IWUE through a reduction in g_s is more sensitive (higher slope in trendline equation) in semi-arid conditions as compared with that obtained in humid weather conditions as reported by Razzaghi et al. (2015).

28.4 CONCLUSION

The average quinoa seed yield was 4.6 Mg/ha for spring planting in March while it increased to 5.06 Mg/ha in autumn planting in September. Our results have shown that quinoa planted in autumn can complete its growth cycle, with significantly less irrigation water, and with no pesticide application, compared with quinoa planted in spring. The autumn (September) planting required 40% less irrigation water and had a 10% higher yield compared with the planting in spring (March) planting. The quinoa planted in early autumn flowered well, had a good seed set and avoided pest damage; by contrast, the spring planting required a pesticide application to control the pests that were promoted by the high temperatures occurring in June. It is therefore recommended that quinoa be planted in autumn in Fars Province, Iran. It is important to note that our research was conducted in an area with an arid climate with an average ratio of annual precipitation to average annual evaporation ratio (P/E) of 0.15. The results of our research may be generalized to similar climates for quinoa cultivation.

REFERENCES

Adolf, V.I., S.-E. Jacobsen, and S. Shabala. 2013. "Salt Tolerance Mechanisms in Quinoa (Chenopodium Quinoa Willd.)." *Environmental and Experimental Botany* 92: 43–54. https://doi.org/10.1016/j.envexpbot.2012.07.004.

Azurita-Silva, A., S.E. Jacobsen, F. Razzaghi, R. Alvarez Flores, K. Ruiz, A. Morales, and S. Herman. 2015. "Quinoa Drought Responses and Adaptation." In *In State of the Art Report of Quinoa in the World in 2013*, 157–171. Rome: FAO and CIRAD.

Bois, J.F., T. Winkel, J.P. Lhomme, J.P. Raffaillac, and A. Rocheteau. 2006. "Response of Some Andean Cultivars of Quinoa (Chenopodium Quinoa Willd.) to Temperature: Effects on Germination, Phenology, Growth and Freezing." *European Journal of Agronomy* 25 (4): 299–308. https://doi.org/10.1016/j.eja.2006.06.007.

Brugnoli, E. and M. Lauteri. 1991. "Effects of Salinity on Stomatal Conductance, Photosynthetic Capacity, and Carbon Isotope Discrimination of Salt-Tolerant (Gossypium Hirsutum L.) and Salt-Sensitive (Phaseolus Vulgaris L.) C3 Non-Halophytes." *Plant Physiology* 95 (2): 628 LP–635. https://doi.org/10.1104/pp.95.2.628.

Ceyhun S., Aybuke and N. Sanlier. 2019. "A New Generation Plant for the Conventional Cuisine: Quinoa (Chenopodium Quinoa Willd.)." *Trends in Food Science & Technology* 86: 51–58. https://doi.org/10.1016/j.tifs.2019.02.039.

Dastranj, M. and A.R. Sepaskhah. 2019. "Response of Saffron (Crocus Sativus L.) to Irrigation Water Salinity, Irrigation Regime and Planting Method: Physiological Growth and Gas Exchange." *Scientia Horticulturae* 257: 108714. https://doi.org/10.1016/j.scienta.2019.108714.

English, M. and S.N. Raja. 1996. "Perspectives on Deficit Irrigation." *Agricultural Water Management* 32 (1): 1–14. https://doi.org/10.1016/S0378-3774(96)01255-3.

Garcia, M., D. Raes, and S.E. Jacobsen. 2003. "Evapotranspiration Analysis and Irrigation Requirements of Quinoa (Chenopodium Quinoa) in the Bolivian Highlands." *Agricultural Water Management* 60 (2): 119–134. https://doi.org/10.1016/S0378-3774(02)00162-2.

Geerts, S., D. Raes, M. Garcia, C. Taboada, R. Miranda, J. Cusicanqui, T. Mhizha, and J. Vacher. 2009. "Modeling the Potential for Closing Quinoa Yield Gaps under Varying Water Availability in the Bolivian Altiplano." *Agricultural Water Management* 96 (11): 1652–1658. https://doi.org/10.1016/j.agwat.2009.06.020.

Geerts, S., D. Raes, M. Garcia, C. Del Castillo, and W. Buytaert. 2006. "Agro-Climatic Suitability Mapping for Crop Production in the Bolivian Altiplano: A Case Study for Quinoa." *Agricultural and Forest Meteorology* 139 (3–4): 399–412. https://doi.org/10.1016/j.agrformet.2006.08.018.

Geerts, S., D. Raes, M. Garcia, O. Condori, J. Mamani, R. Miranda, J. Cusicanqui, C. Taboada, E. Yucra, and J. Vacher. 2008. "Could Deficit Irrigation Be a Sustainable Practice for Quinoa (Chenopodium Quinoa Willd.) in the Southern Bolivian Altiplano?" *Agricultural Water Management* 95 (8): 909–917. https://doi.org/10.1016/j.agwat.2008.02.012.

Hinojosa, L., J.A. González, F.H. Barrios-Masias, F. Fuentes, and K.M. Murphy. 2018. "Quinoa Abiotic Stress Responses: A Review." *Plants*. https://doi.org/10.3390/plants7040106.

Hirich, A., R. Choukr-Allah, and S.-E. Jacobsen. 2014. "Quinoa in Morocco – Effect of Sowing Dates on Development and Yield." *Journal of Agronomy and Crop Science* 200 (5): 371–77. https://doi.org/10.1111/jac.12071.

Issa A., O. R. Fghire, F. Anaya, O. Benlhabib, and S. Wahbi. 2019. "Physiological and Morphological Responses of Two Quinoa Cultivars (Chenopodium Quinoa Willd.) to Drought Stress." *Gesunde Pflanzen* 71 (2): 123–33. https://doi.org/10.1007/s10343-019-00460-y.

Jacobsen, S.-E., A Mujica, and C.R. Jensen. 2003. "The Resistance of Quinoa (Chenopodium QuinoaWilld.) to Adverse Abiotic Factors." *Food Reviews International* 19 (1–2): 99–109. https://doi.org/10.1081/FRI-120018872.

Jacobsen, S.E., C. Monteros, J.L. Christiansen, L.A. Bravo, L.J. Corcuera, and A. Mujica. 2005. "Plant Responses of Quinoa (Chenopodium Quinoa Willd.) to Frost at Various

Phenological Stages." *European Journal of Agronomy* 22 (2): 131–139. https://doi.org/10.1016/j.eja.2004.01.003.

Jacobsen, S.-E., F. Liu, and C.R. Jensen. 2009. "Does Root-Sourced ABA Play a Role for Regulation of Stomata under Drought in Quinoa (Chenopodium Quinoa Willd.)." *Scientia Horticulturae* 122 (2): 281–287. https://doi.org/10.1016/j.scienta.2009.05.019.

Kaya, Ç.I., A. Yazar, and S.M. Sezen. 2015. "SALTMED Model Performance on Simulation of Soil Moisture and Crop Yield for Quinoa Irrigated Using Different Irrigation Systems, Irrigation Strategies and Water Qualities in Turkey." *Agriculture and Agricultural Science Procedia* 4: 108–118. https://doi.org/10.1016/j.aaspro.2015.03.013.

Mehrabi, F. and A.R. Sepaskhah. 2019. "Partial Root Zone Drying Irrigation, Planting Methods and Nitrogen Fertilization Influence on Physiologic and Agronomic Parameters of Winter Wheat." *Agricultural Water Management* 223: 105688. https://doi.org/10.1016/j.agwat.2019.105688.

Nowak, V., J. Du, and U.R. Charrondière. 2016. "Assessment of the Nutritional Composition of Quinoa (Chenopodium Quinoa Willd.)." *Food Chemistry* 193: 47–54. https://doi.org/10.1016/j.foodchem.2015.02.111.

Nurse, R.E., K. Obeid, and E.R. Page. 2016. "Optimal Planting Date, Row Width, and Critical Weed-Free Period for Grain Amaranth and Quinoa Grown in Ontario, Canada." *Canadian Journal of Plant Science* 96 (3): 360–366. https://doi.org/10.1139/cjps-2015-0160.

Pulvento, C., M. Riccardi, A. Lavini, R. D'Andria, G. Iafelice, and E. Marconi. 2010. "Field Trial Evaluation of Two Chenopodium Quinoa Genotypes Grown Under Rain-Fed Conditions in a Typical Mediterranean Environment in South Italy." *Journal of Agronomy and Crop Science* 196 (6): 407–411. https://doi.org/10.1111/j.1439-037X.2010.00431.x.

Rawson, H.M., J.E. Begg, and R.G. Woodward. 1977. "The Effect of Atmospheric Humidity on Photosynthesis, Transpiration and Water Use Efficiency of Leaves of Several Plant Species." *Planta* 134 (1): 5–10. https://doi.org/10.1007/BF00390086.

Razzaghi, F., S.-E. Jacobsen, C.R. Jensen, and M.N. Andersen. 2015. "Ionic and Photosynthetic Homeostasis in Quinoa Challenged by Salinity and Drought – Mechanisms of Tolerance." *Functional Plant Biology* 42 (2): 136–148.

Razzaghi, F., F. Plauborg, S.E. Jacobsen, C.R. Jensen, and M.N. Andersen. 2012. "Effect of Nitrogen and Water Availability of Three Soil Types on Yield, Radiation Use Efficiency and Evapotranspiration in Field-Grown Quinoa." *Agricultural Water Management* 109: 20–29. https://doi.org/10.1016/j.agwat.2012.02.002.

Razzaghi, F. and A. R. Sepaskhah. 2012. "Calibration and Validation of Four Common ET0 Estimation Equations by Lysimeter Data in a Semi-Arid Environment." *Archives of Agronomy and Soil Science* 58 (3): 303–319. https://doi.org/10.1080/03650340.2010.518957.

Repo-Carrasco, R., C. Espinoza, and S.-E. Jacobsen. 2003. "Nutritional Value and Use of the Andean Crops Quinoa (Chenopodium Quinoa) and Kaniva (Chenopodium Pallidicaule)." *Food Reviews International* 19 (1–2): 179–189. https://doi.org/10.1081/FRI-120018884.

Riccardi, M., C. Pulvento, A. Lavini, R. d'Andria, and S. E. Jacobsen. 2014. "Growth and Ionic Content of Quinoa under Saline Irrigation." *Journal of Agronomy and Crop Science* 200 (4): 246–260. https://doi.org/10.1111/jac.12061.

Risi, J. and N.W. Galwey. 1991. "Effects of Sowing Date and Sowing Rate on Plant Development and Grain Yield of Quinoa (Chenopodium Quinoa) in a Temperate Environment." *The Journal of Agricultural Science* 117 (3): 325–332. https://doi.org/DOI: 10.1017/S002185960006706X.

Romano, N., M. Micaela Ureta, M. Guerrero-Sánchez, and A. Gómez-Zavaglia. 2020. "Nutritional and Technological Properties of a Quinoa (Chenopodium Quinoa Willd.) Spray-Dried Powdered Extract." *Food Research International* 129: 108884. https://doi.org/10.1016/j.foodres.2019.108884.

Rosa, M., M. Hilal, J.A. González, and F.E. Prado. 2009. "Low-Temperature Effect on Enzyme Activities Involved in Sucrose–Starch Partitioning in Salt-Stressed and Salt-Acclimated Cotyledons of Quinoa (Chenopodium Quinoa Willd.) Seedlings." *Plant Physiology and Biochemistry* 47 (4): 300–307. https://doi.org/10.1016/j.plaphy.2008.12.001.

Ruales, J. and B.M. Nair. 1992. "Nutritional Quality of the Protein in Quinoa (Chenopodium Quinoa, Willd) Seeds." *Plant Foods for Human Nutrition* 42 (1): 1–11. https://doi.org/10.1007/BF02196067.

Sikder, S., J. Foulkes, H. West, J. De Silva, O. Gaju, A. Greenland, and P. Howell. 2015. "Evaluation of Photosynthetic Potential of Wheat Genotypes under Drought Condition." *Photosynthetica* 53 (1): 47–54. https://doi.org/10.1007/s11099-015-0082-9.

Talebnejad, R. and A.R. Sepaskhah. 2015a. "Effect of Deficit Irrigation and Different Saline Groundwater Depths on Yield and Water Productivity of Quinoa." *Agricultural Water Management* 159: 225–238. https://doi.org/10.1016/j.agwat.2015.06.005.

Talebnejad, R. and A.R. Sepaskhah. 2015b. "Effect of Different Saline Groundwater Depths and Irrigation Water Salinities on Yield and Water Use of Quinoa in Lysimeter." *Agricultural Water Management* 148: 177–188. https://doi.org/10.1016/j.agwat.2014.10.005.

Talebnejad, R. and A.R. Sepaskhah. 2016. "Modification of Transient State Analytical Model under Different Saline Groundwater Depths, Irrigation Water Salinities and Deficit Irrigation for Quinoa." *International Journal of Plant Production* 10 (3): 365–390. https://doi.org/10.22069/ijpp.2016.2903.

Talebnejad, R. and A.R. Sepaskhah. 2018. "Quinoa: A New Crop for Plant Diversification under Water and Salinity Stress Conditions in Iran." *Acta Horticulturae* 1190: 101–106. https://doi.org/10.17660/ActaHortic.2018.1190.17.

Talebnejad, R. and A.R. Sepaskhah. 2016. "Physiological Characteristics, Gas Exchange, and Plant Ion Relations of Quinoa to Different Saline Groundwater Depths and Water Salinity." *Archives of Agronomy and Soil Science* 62 (10): 1347–1367. https://doi.org/10.1080/03650340.2016.1144925.

Trognitz, B.R. 2003. "Prospects of Breeding Quinoa for Tolerance to Abiotic Stress." *Food Reviews International* 19 (1–2): 129–37. https://doi.org/10.1081/FRI-120018879.

Väkeväinen, K., F. Ludena-Urquizo, E. Korkala, A. Lapveteläinen, S. Peräniemi, A. von Wright, and C. Plumed-Ferrer. 2020. "Potential of Quinoa in the Development of Fermented Spoonable Vegan Products." *LWT* 120: 108912. https://doi.org/10.1016/j.lwt.2019.108912.

Walters, H., L. Carpenter-Boggs, K. Desta, L. Yan, J. Matanguihan, and K. Murphy. 2016. "Effect of Irrigation, Intercrop, and Cultivar on Agronomic and Nutritional Characteristics of Quinoa." *Agroecology and Sustainable Food Systems* 40 (8): 783–803. https://doi.org/10.1080/21683565.2016.1177805.

Yang, A., S.S. Akhtar, M. Amjad, S. Iqbal, and S.-E. Jacobsen. 2016. "Growth and Physiological Responses of Quinoa to Drought and Temperature Stress." *Journal of Agronomy and Crop Science* 202 (6): 445–453. https://doi.org/10.1111/jac.12167.

Yarami, N. and A.R. Sepaskhah. 2015. "Physiological Growth and Gas Exchange Response of Saffron (Crocus Sativus L.) to Irrigation Water Salinity, Manure Application and Planting Method." *Agricultural Water Management* 154: 43–51. https://doi.org/10.1016/j.agwat.2015.03.003.

29 Response of Quinoa to High Salinity under Arid Conditions

Mohammad Shahid and Sumitha Thushar

CONTENTS

29.1 INTRODUCTION

Quinoa (*Chenopodium quinoa* Willd.) is a pseudo-cereal crop, which is herbaceous and matures within 3-6 months. It belongs to the Amaranthaceae family that also includes prominent crops like beet and spinach. The cultivation of the plant started around 7,000 years ago (Kolata, 2009) after domestication of wild populations of *Chenopodium quinoa* (Pickersgill, 2007) in the Andes mountain range of South America. Quinoa is a highly nutritive crop as its grains contain all 9 essential amino acids, which few other crops possess. It has a low Glycine Index (GI) making it suitable for diabetics, but is rich in fiber, protein (14–20%), B vitamins and important minerals like iron, magnesium, potassium and manganese (Vaughn and Geissler, 2009). It also contains good carbohydrates and important fatty acids, making it a good food for human consumption (Koziol, 1992; Ranhotra et al., 1993; Repo-Carrasco et al., 2003). It is an ideal food for health-conscious people as well as for children in the third world countries where wholesome diets may not be available.

DOI: 10.1201/9781003112327-29

Though the typical quinoa grain yield is 0.4–0.9 t/ha, yields of up to 2–3.5 t/ha have been possible with improved agricultural practices (Jacobsen et al., 1994). High yields have been obtained from different parts of the world including Denmark (Jacobsen et al., 1994), France (L'Avenir Agricole, 2015), Pakistan (Iqbal et al., 2018) and India (Singh, 2018). Quinoa can also be used as a fodder crop. Average quinoa forage yields of around 10 t DM/ha have been recorded (Taváres et al., 1995).

Quinoa grows nicely in well-drained sandy soils of low nutritional status and with a pH range of 6.0–8.5. It is a facultative halophytic crop (Jacobsen et al., 2005; Koyro et al., 2008; Adolf et al., 2013) which makes it suited to marginal lands with salinity problems. Field experiments at low and high salinity have shown that certain quinoa cultivars can give good yields at high salinity (Iqbal et al., 2019). Salinity is increasing in different parts of the world and has affected about 7% of arable land (Panta et al., 2014), which is responsible for a substantial decline in agricultural production in those areas. For cultivation in salt-affected agricultural lands, crops tolerant to salinity are needed.

Salt-tolerant plants have three different strategies to cope with high salinity: tolerance against osmotic stress, exclusion of ions (Na^+ or Cl^-) from tissues, and tolerance to accumulated ions (Munns and Tester, 2008). Considerable variation in salinity tolerance exists amongst food crops. For example, within the major cereals, rice (*Oryza sativa* L.) has lowest tolerance, while barley (*Hordeum vulgare* L.) is the most salt-tolerant crop (Munns and Tester, 2008). The difference in salt-tolerance among dicotyledonous plants is greater. Forty-four percent of species in the Amaranthaceae are salt-tolerant in nature, which is the highest percentage among plant families (Flowers et al., 1986).

In saline conditions, Gómez-Pando et al. (2010) observed a sizeable difference amongst 200 quinoa genotypes for germination rate. Highly saline water (0.4 M NaCl) decreased the germination of quinoa seed by ~60% and decreased the fresh weight of germinating seeds by 60%, but it did not affect the dry weight of germinating seeds (Prado et al., 2000). NaCl reduces the fructose and glucose but increases the sucrose contents in quinoa seedlings. The increase in sucrose may be an indication of osmotic adaptation and/or a reduction in metabolic activities under highly saline conditions (Prado et al., 2000). Many quinoa accessions can tolerate salinities at the seedling stage of up to 150 mM NaCl (Ruiz-Carrasco et al., 2011) while some can resist salinities of more than seawater (Jacobsen et al., 2003).

There is no correlation between the salinity-tolerance of quinoa at the seed germination stage and during subsequent plant growth (Adolf et al., 2012). The genotypes that perform poorly at germination may fare well at later stages and vice versa. In a study by Hariadi et al. (2011), best quinoa plant growth was observed at 10–20 dS/m, while a decrease in yield was noted at 15 dS/m and above during other experiments (Jacobsen et al., 2001; 2003). In another study, the threshold salinity (above which yield was decreased) for a quinoa variety was found to be 11 dS/m (Wilson et al., 2002).

One of the main causes of decreased growth due to salinity is ion excess of sodium (Na^+), particularly in stems and leaves of plants. Na^+ exclusion is a key factor in maintaining the normal metabolism in plant cells. The ion exclusion is facilitated by a Na^+/H^+ exchanger located at the plasma membrane (Blumwald et al., 2000), which is encoded by the salt overly sensitive (SOS1) gene (Hasegawa et al., 2000; Qiu et al., 2002; Mullan et al., 2007). The accumulation of sodium ions in the cytosol is prevented by a tonoplast Na^+/H^+ exchanger NHX which shifts Na^+ into the vacuole (Apse et al., 1999; Shabala and Mackay, 2011). In quinoa, two of the SOS1 genes (cqSOS1A and cqSOS1B) have been found with homologies to the genes of other halophytic species (Maughan et al., 2009). Under non-saline conditions, the expression of cqSOS1A and cqSOS1B was about 4 times stronger in roots than in leaves. On the other hand, exposure to salinity (45 dS/m) led to over expression of these genes in leaves compared to root tissues (Maughan et al., 2009).

Various studies have been carried out at the International Center for Biosaline Agriculture (ICBA), Dubai, UAE to examine the salt-tolerance and adaptability of quinoa to the region. In one investigation, several quinoa genotypes were assessed for yield-related characteristics to examine their performance in hot environment and sandy soils (Rao and Shahid, 2012). In another study, different cultivars of the crop were grown in different regions of the world to explore their adaptation and tolerance against salinity (Choukr-Allah et al., 2016). The present field experiment was carried out at high and low salinities, to find the salt-tolerance of 11 accessions of quinoa, at ICBA during the cropping season of 2018–19. This trial also included 5 quinoa varieties that had been developed by ICBA.

Djulis (*Chenopodium formosanum* Koidz.) like quinoa is a pseudo-cereal crop native to Taiwan, and is also known as Taiwanese quinoa. It is mostly grown as a leafy vegetable in its native land, but its seeds are highly nutritious as they are rich in protein and dietary fiber, and contain essential amino acids. The seeds also contain antioxidants and pigments making it a healthy food (Tsai et al., 2011). Studies have shown that its seed helps in protecting the skin against the harmful effects of ultraviolet radiation (Hong et al., 2016)

29.2 MATERIALS AND METHODS

The trials were carried out at the field research facilities of the ICBA Dubai, United Arab Emirates (N 25° 05.847; E 055° 23.464), which is situated around 23 km from the Arabian Gulf, between October 2018 and April 2019.

The soils of the ICBA research area are relatively alkaline (pH 8.2), porous (45% porosity), calcareous (55% $CaCO_4$) and have a sandy texture (98% fine sand, 1% silt and 1% clay). With a saturation percentage of 26, the soil has extremely high drainage capacity. The saturated extract of the soil has an electrical conductivity (EC_e) of 1.2 dS/m. In accordance with soil classifications based on American Soil Taxonomy (Soil Survey Staff, 2010), the soil is Typic Torripsamments, hyperthermic and carbonatic (Shahid et al., 2009).

Before planting, compost was added to the sites selected for the field experiments at the rate of 40 tonnes per hectare (t ha^{-1}). Urea (46-0-0) was applied at the rate of 40 kg ha^{-1} after 4 weeks of seed germination, while NPK (20-20-20) was added at the rate of 30 kg ha^{-1} after 4 weeks of urea application. Both the synthetic fertilizers were applied using a fertigation method.

Eleven quinoa cultivars and one djulis variety were selected for the field experiment (Table 29.1). Each set of the accessions was sown using a randomized complete

TABLE 29.1
Eleven Different Accessions of Quinoa (1-11) and One Djulis Variety (12) Selected for the Yield Trials

S.N.	Accessions	S.N.	Accessions
1	ICBA-Q1*	7	NSL-84449
2	ICBA-Q2*	8	Ames-13215
3	ICBA-Q3*	9	Puno
4	ICBA-Q4*	10	Titicaca
5	ICBA-Q5*	11	NSL-106399
6	Ames-13757	12	Zhang Li 3†

* Quinoa varieties developed at International Center for Biosaline Agriculture (ICBA).
† Zhang Li 3 is a djulis variety developed in Taiwan.

block design (RCBD) in three replications. There were two water salinity (EC_w) treatments: low salinity (0.3 dS/m) and high salinity (15 dS/m). The high salinity treatment was started 2 weeks after planting.

The sowing was done in the 3rd week of October. Each plot size was 1 × 2 m. The seeds were sown manually by dibbling 3–4 seeds into the soil to a depth of 1–2 cm close to the dripper. The plant to plant and row to row distance was 25 cm. Once sowing was completed, the field was covered with acryl sheet to stop birds eating the planted seeds. After germination, only one seedling was kept at each spot. The field was covered with net at flowering to stop birds from eating seeds from the panicles (Figure 29.1).

Data on 9 agronomic and morphological traits (Table 29.2) were collected to determine the effect of salinity on the quinoa and djulis plants. For the traits, "days to flowering" and "days to maturity", the data were taken when half of the plants had

FIGURE 29.1 Quinoa yield trial at ICBA covered with nets to protect the seed from birds.

TABLE 29.2
Different Agronomic and Morphological Characteristics Studied in the Field Experiment

S.N.	Traits
1	Days to flowering
2	Days to maturity
3	Plant height
4	Number of primary branches per plant
5	Number of panicles per plant
6	Length of main panicle
7	Plant dry weight m^{-1}
8	Seed weight m^{-1}
9	Thousand seed weight

flowered and matured, respectively. All other characteristics were studied after the plants had reached maturity. Before taking the plant dry weight, the plants were kept in a dryer at 40°C for 48 hours.

A drip irrigation system was used for the experiment with drippers at 25 cm distance, which was part of the Supervisory Control and Data Acquisition system. Irrigation was twice a day for 5 minutes each time. Water was released from each dripper at a rate of 4 L/h. The plots were irrigated daily for both the saline and non-saline treatments. The chemical properties of the irrigation waters are given in Table 29.3.

Data on temperature, relative humidity and precipitation at the experimental field were taken from the weather station at ICBA (Table 29.4).

29.3 OBJECTIVES

The field experiment had the following objectives:

- To understand the effect of salinity on different morphological and agronomic traits of quinoa under field conditions
- To see how saltwater affects yield and yield related characteristics of the different quinoa cultivars developed at ICBA and other parts of the world
- To select quinoa accessions suitable for cultivation in marginal lands with salinity problems around the world

TABLE 29.3
Chemical Properties of the Irrigation Waters Used in the Field Experiment

Water Treatment	EC (dS/m)	pH	Soluble Ions (meq/L)						
			Cl	CO_3	HCO_3	SO_4	Na	K	Ca + Mg
low salinity	0.30	7.87	3.0	0.00	1.32	0.0	2.0	0.06	1.6
high salinity	15.0	7.43	13.4	0.12	3.52	62.4	109	2.28	73.6

TABLE 29.4
Temperature, Relative Humidity and Precipitation at the Experimental Field during the Cropping Season

Year	Month	Temperature Mean (°C)	Maximum Temperature Mean (°C)	Minimum Temperature Mean (°C)	Maximum Relative Humidity Mean (%)	Minimum Relative Humidity Mean (%)	Precipitation (mm)
2018	October	28.8	36.9	21.7	74.9	20	0
2018	November	23.0	30.1	16.3	75.5	26.2	0.2
2018	December	18.3	25.8	11.4	80.0	29.8	0.3
2019	January	16.8	24.4	10.0	86.2	28.3	25.2
2019	February	21.5	29.1	14.9	77.3	22.1	19.8
2019	March	20.2	28.0	13.3	76.7	19.2	0
2019	April	27.2	34.8	20.4	71.9	17.7	0
2019	May	32.6	40.5	24.7	57.8	13.9	0

29.4 RESULTS AND DISCUSSION

29.4.1 Days to Flowering

Quinoa genotypes have shown variation in emergence of the inflorescence and flowering according to their place of origin (Bhargava et al., 2007; Curti et al., 2016; Sosa-Zuniga et al., 2017). Our experiment showed similar variation. The cultivars Titicaca and NSL106399 had the earliest flowering (after 49 days). Zhang Li 3 flowered after 99 days making it the latest in this category (Figure 29.2). On average, high salinity hastened days to flowering by 3 days. The djulis variety Zhang Li 3 flowered 9 days earlier due to high salinity. While working on two quinoa varieties at

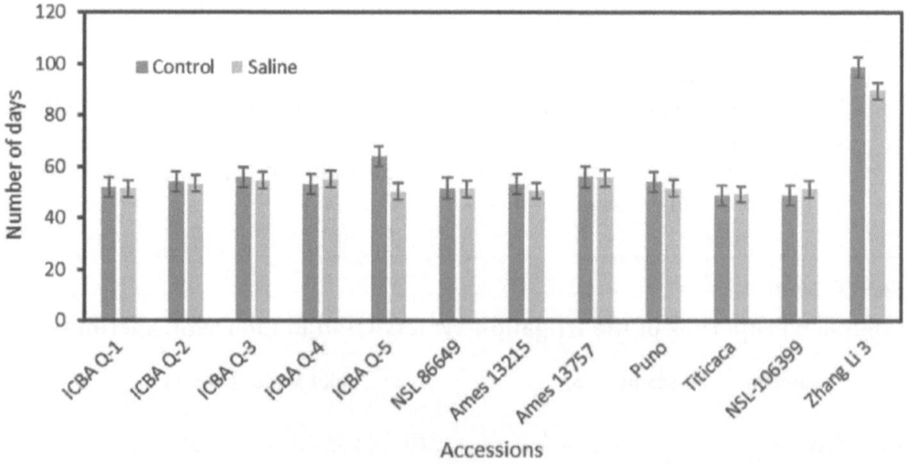

FIGURE 29.2 Days to flowering of quinoa and djulis accessions at low and high salinities.

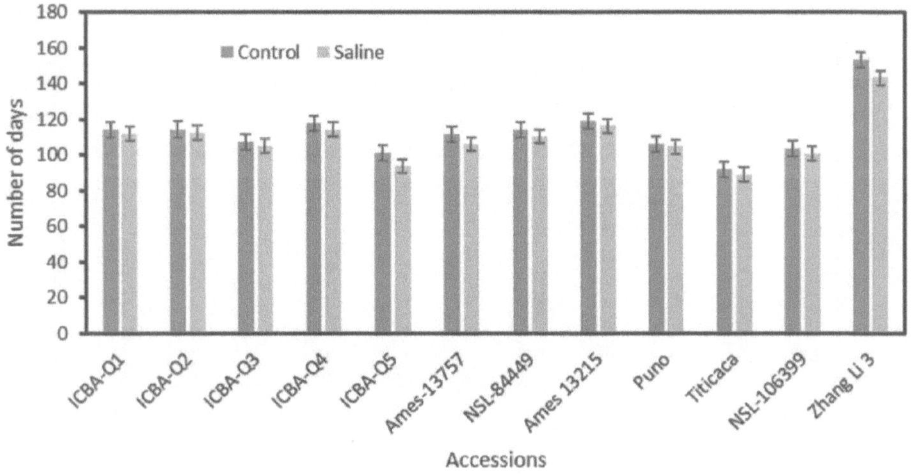

FIGURE 29.3 Days to maturity of quinoa and djulis accessions at low and high salinities.

4 different salinities, Algosaibi et al. (2017) found that high salinity hastened flowering by 4 to 8 days. The statistical analysis of the data shows a significant difference among the studied accessions.

29.4.2 Days to Maturity

Early maturity is a desirable trait for crops as it saves time and resources, especially, the precious irrigation water of farmers. Titicaca matured 92 days after planting making it the earliest of the 11 quinoa varieties studied (Figure 29.3). ICBA Q-5 (101 days) was the second earliest maturing variety. The djulis variety Zhang Li 3 took the longest period of time, maturing after 145 days. For quinoa growers who prefer early maturing varieties, Titicaca and ICBA would be ideal.

In control plots, the median maturity time for the quinoa varieties was 113 days, while irrigation with saline water decreased the maturity period by 5 days. Salinity reduced the days to maturity for all the planted accessions (Figure 29.3). Overall, a notable variation was observed for days to maturity amongst the cultivars using analysis of statistical data. Similar results were obtained by Algosaibi et al. (2017), working on quinoa genotypes under low and high salinities; in this work, high salinity treatment decreased the time to maturity by 27–35 days.

29.4.3 Plant Height

The quinoa plant can be used as a leafy vegetable (El-Naggar et al., 2018) as well as an alternative fodder for farm animals (van Schooten and Pinxterhuis, 2003). Quinoa leaves contains more protein than amaranth, spinach and moringa (Pathan et al., 2019), which makes it excellent vegetable and fodder. Tall plants will be better for both vegetable and fodder purposes as greater biomass can be expected for these.

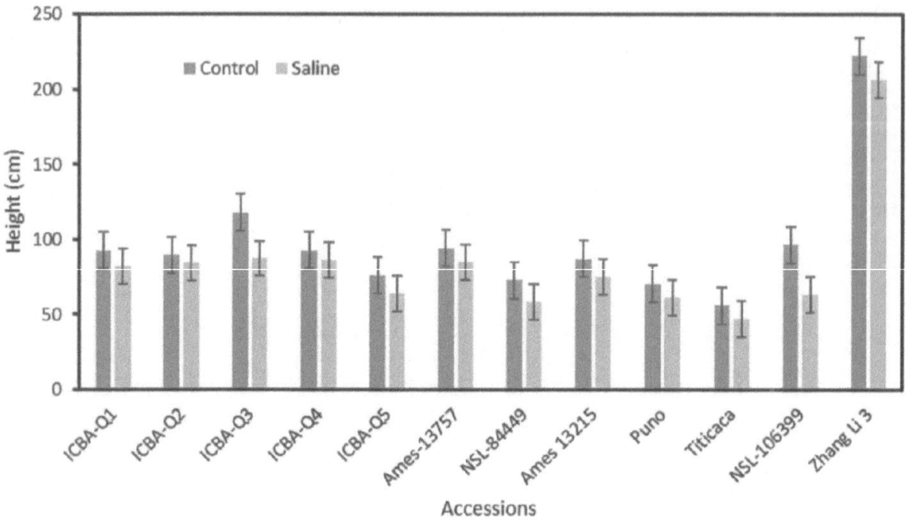

FIGURE 29.4 Plant heights of quinoa and djulis accessions at low and high salinities.

In our experiment, the greatest plant height was with the djulis variety Zhang Li 3 (~220 cm) followed by ICBA Q-3 (117 cm), which had almost half of the height of the former accession. The shortest accession amongst the 12 accessions was Titicaca (~56 cm tall) (Figure 29.4).

On average, use of saline water decreased plant height by ~10 cm. The most severely affected accession was NSL 106399, which lost ~34% of its height due to high salinity; the least affected cultivar was ICBA Q-2, with only a 6% decrease (Figure 29.4). Negative effects of salinity on quinoa plant height has also been observed by Long (2016), Koyro et al. (2008) and Algosaibi et al. (2017). Highly saline water decreases the plant capability to absorb water from soil which effects its growth negatively.

29.4.4 Number of Primary Branches per Plant

Statistical analysis of the data revealed a significant difference in the number of primary branches per plant in the 12 genotypes (Figure 29.5). The lowest number of primary branches was in Zhang Li 3 (1.1); the second lowest was in NSL 106399 (2.1). The highest number of branches was in NSL 84449 (9.4) and Ames-13215 (6.3).

In response to high salinity, quinoa reduced the number of primary branches; the average decrease was 9%. The greatest reduction in branches was with ICBA Q-1 (25% decrease), while Titicaca had no difference due to high salinity. Interestingly, Zhang Li 3 and Ames-13215 had an increase in number of branches due to irrigation with high saline water (Figure 29.5). Long (2016) also noted negative effects of high salinity on the number or branches in quinoa; their decline in the number of branches due to salinity was about 26%.

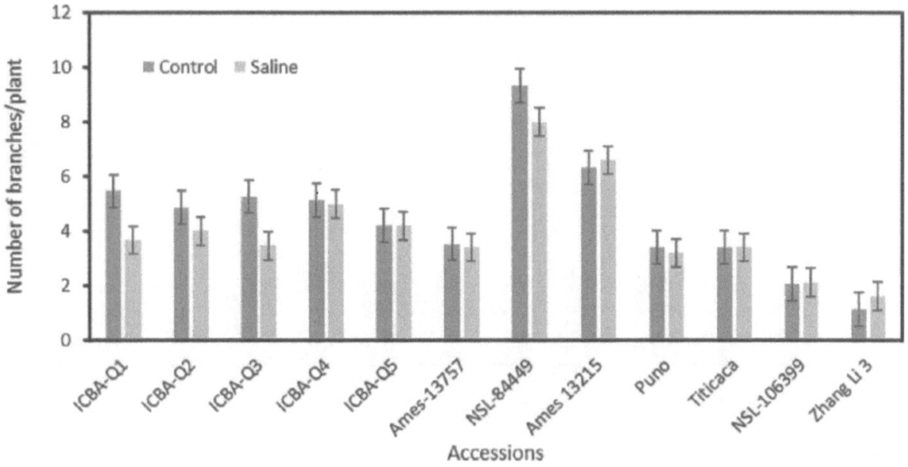

FIGURE 29.5 Number of branches per plant of quinoa and djulis accessions at low and high salinities.

29.4.5 Number of Panicles per Plant

The number of panicles per plant is related to grain yield. A larger number of panicles per plant may therefore indicate the prospect of better seed production in quinoa. The highest number of panicles per plant in our study were with NSL-84449 (9.3) and Ames-13215 (6.2). Djulis cultivar Zhang Li 3 had the lowest number of panicles (1.1) among the 12 studied quinoa cultivars (Figure 29.6).

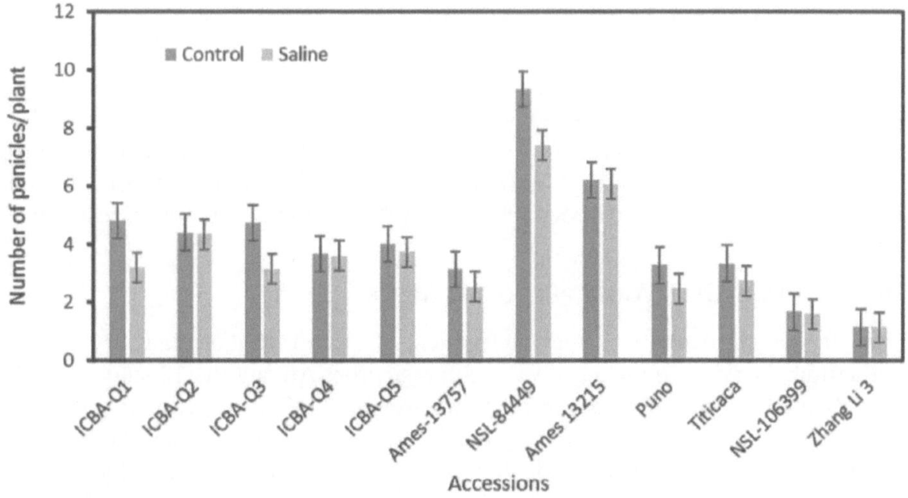

FIGURE 29.6 Number of panicles per plant of quinoa and djulis accessions at low and high salinities.

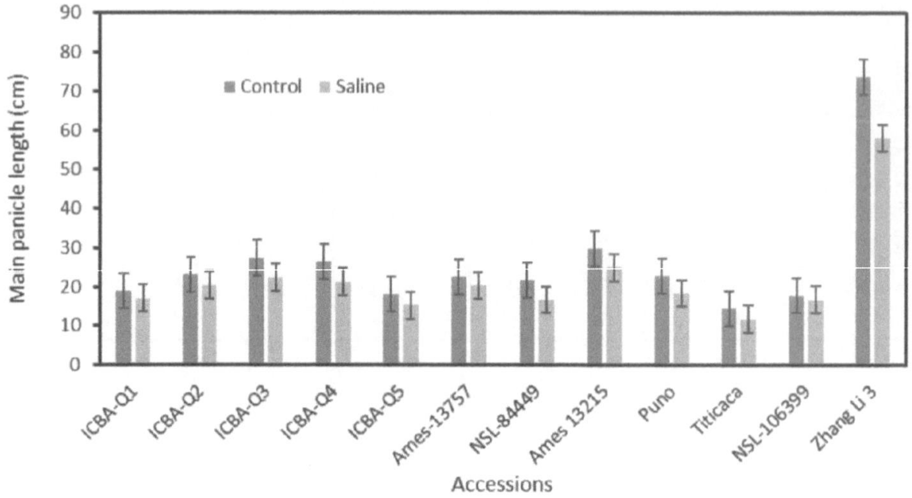

FIGURE 29.7 Panicle length of quinoa and djulis accessions at low and high salinities.

Irrigation with saline water decreased the panicle number in the quinoa plants by an average of 15%. The most affected accessions were ICBA Q-1 and ICBA Q-3, which showed a 34% reduction in panicle number, while no effect of salinity was recorded with Zhang Li 3 (Figure 29.6).

29.4.6 Length of Main Panicle

Panicle size is directly related to the amount of seed it contains. Larger panicles contain more grain than smaller panicles. Zhang Li 3 had the longest panicle (73 cm) followed by Ames-13215 (30 cm). The shortest panicle was in variety Titicaca (14 cm) (Figure 29.7).

High salinity treatment reduced the panicle length in quinoa on average by 17%. The negative effect was most obvious in djulis variety Zhang Li 3 which had a 21% decrease in panicle length (Figure 29.7). The same phenomenon was also observed by Long (2016) and Algosaibi et al. (2017) in quinoa, where reductions of 17% and 30% in panicle length was recoded respectively due to high saline water treatment.

29.4.7 Plant Dry Weight per Square Meter

Plant dry weight determines the fodder yield of a crop. Higher plant dry weight is a desirable trait for quinoa grown as a forage crop. Among the 12 studied cultivars, Zhang Li 3 had the heaviest plant dry weight (1440 g m^{-2}), which was more than double that of Ames-13757 the second-best performing accession (668 g m^{-2}). Because of its high vegetative production, Zhang Li 3 could be used as a fodder quinoa variety (Figure 29.8).

Overall, there was a 18% decrease in plant dry weight due to high salinity amongst the 12 accessions. NSL-84449 with a 7% reduction was the least affected accession,

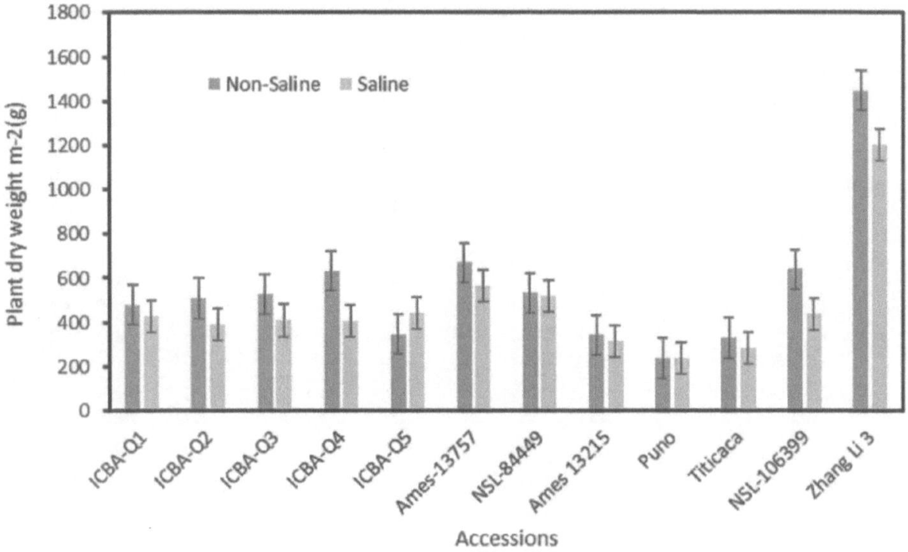

FIGURE 29.8 Plant dry weight of quinoa and djulis accessions at low and high salinities.

and Titicaca with a 44% reduction was the most affected accession (Figure 29.8). Negative effects of salinity on plant dry matter (15–25% decreases depending on the salinities of the irrigation waters used) were also noted by Long (2016). Algosaibi et al. (2017) observed similar salinity impacts on quinoa plant dry weight.

29.4.8 Seed Weight per Square Meter

Seed yield is the most important trait for the majority of the crops. For quinoa it defines the value of a variety as high yielding varieties are preferred for cultivation. For this key trait, the accessions Ames-13757 and NSL 106399 had highest yields (217 and 208 g seed m^{-2}, respectively). Ames-13215 had the lowest grain yield (Figure 29.9).

On average, the decline in seed yield in accessions because of high salinity was 25%. The highest decrease was with Puno (50%) and the lowest decrease was with NSL 84449 (Figure 29.9). Experiments conducted in other places have also shown negative impacts of salinity on seed yield (Long, 2016; Algosaibi et al., 2017; Iqbal et al., 2019).

29.4.9 Thousand Seed Weight

One thousand seed weight is an important trait for measuring seed quality; it also plays an important role in germination, seedling development and plant performance (Afshar et al., 2011).

Our study showed variation in this seed trait. Both ICBA Q-2 and ICBA Q-3 had a thousand seed weight of 3.6 g, which was the highest amongst the studied cultivars. The djulis variety Zhang Li 3 had the lowest thousand seed weight (1.1.g) (Figure 29.10).

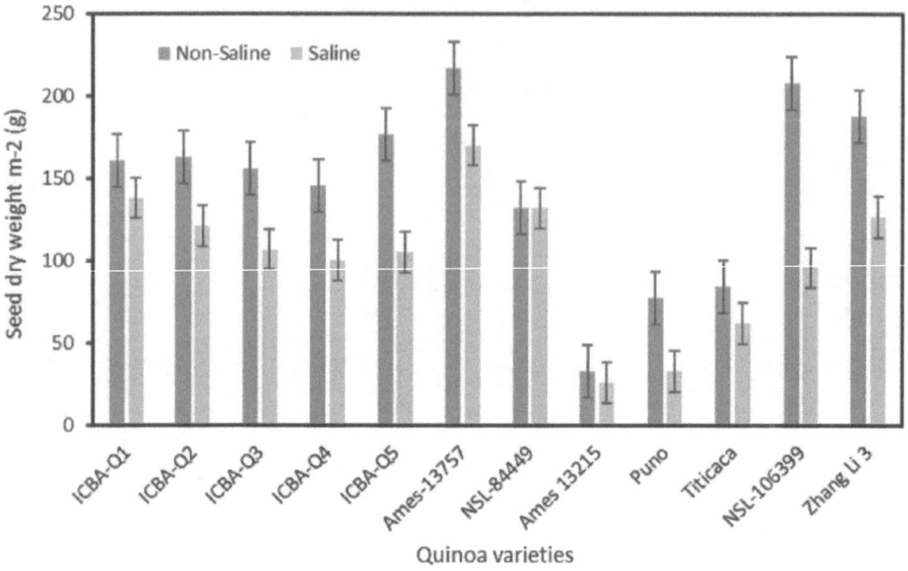

FIGURE 29.9 Seed yield of quinoa and djulis accessions at low and high salinities.

On average, for all accessions salinity decreased thousand seed weight by 7%. Zhang Li 3 didn't show any change in seed weight due to salinity, while other cultivars showed some decline (Figure 29.10). Long (2016) found a sizeable effect of high water-salinities, which decreased the thousand seed weight by 30–40%. While Algosaibi et al. (2017) observed smaller impact of salinity on this seed characteristic.

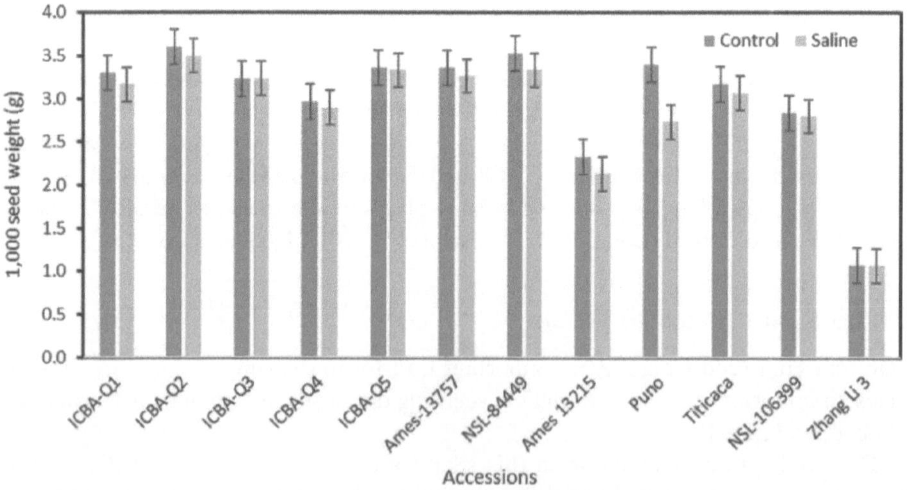

FIGURE 29.10 Thousand seed weight of quinoa and djulis accessions at low and high salinities.

29.5 CONCLUSION

Our results have indicated that at high salinity (EC_w 15 dS/m), the number of days to flowering and maturity decreased for all the quinoa accessions and the djulis cultivar. The salinity accelerated both flowering and maturity in the studied crops. For these traits, diversity exist in different quinoa cultivars. It was also obvious that salinity had negative effects on plant height, the number of primary branches, the number of panicles, plant dry weight, grain yield and thousand seed weight. Various quinoa accessions reacted differently against high salinity. Based on these results, the identified salt-tolerant quinoa cultivars can be introduced in the marginal lands of different countries with salt-effected large tracts of agricultural lands. High yielding salt-tolerant quinoa varieties will help in improving life in the rural areas of many poor countries, where salinity is affecting the production of other crops. Our study also showed that djulis (*Chenopodium formosanum*) is a less salttolerant crop in comparison with quinoa (*C. quinoa*).

REFERENCES

Adolf, V.I., S. Shabala, M.N. Andersen, F. Razzagh and S.-E. Jacobsen. 2012. Varietal differences of quinoa's tolerance to saline conditions. *Plant and Soil* 357 (1–2): 117–129.

Adolf, V. I., S. -E. Jacobsen and S. Shabala. 2013. Salt tolerance mechanisms in quinoa (*Chenopodium quinoa* Willd.). *Environmental and Experimental Botany* 92: 43–54. https://doi.org/10.1016/j.envexpbot.2012.07.004.

Afshar, H., M. Eftekhari, M. Faraji, A. G. Ebadi and A. Ghanbarimalidareh. 2011. Studying the effect of 1000 grain weight on the sprouting of different species of Salvia L. growing Iran. *Journal of Medicinal Plants Research* 5(16): 3991–3993.

Algosaibi, A. M., M. M. El-Garawany, A. E Badran, and A. M. Almadini. 2017. Effect of irrigation water salinity on the growth of quinoa plant seedlings. *Journal of Agricultural Science* 7(8): 205–214.

Apse, M. P., Aharon, G. S., Snedden, W. A., and Blumwald, E. 1999. Salt tolerance conferred by overexpression of a vacuolar Na+/H+ antiport in Arabidopsis. *Science* 285: 1256–1258.

Bhargava A., S. Shukla, S. Rajan and D. Ohri. 2007. Genetic diversity for morphological and quality traits in quinoa (*Chenopodium quinoa* Willd.) germplasm. *Genetic Resources and Crop Evolution* 54(1): 167–173.

Blumwald, E., G. S. Aharon and M. P. Apse. 2000. Sodium transport in plant cells. *Biochimica et Biophysica Acta-Biomembranes* 1465: 140–151.

Choukr-Allah, R., N. K. Rao, A. Hirich, M. Shahid, A. Alshankiti, K. Toderich, S. Gill and K. Butt. 2016. Quinoa for marginal environments: Towards future food and nutritional security in MENA and Central Asia regions. *Frontiers in Plant Science* 7: 346. doi: 10.3389/fpls.2016.00346.

Curti R. N., A. J. De la Vega, A. J. Andrade, S. J. Bramadi, and H. D. Bertero. 2016 Adaptive responses of quinoa to diverse agro ecological environments along an altitudinal gradient in North West Argentina. *Field Crops Research* 186: 10–18.

El-Naggar, A. M., S. A. Hussin, E. H. Abd El-Samad and S. S. Eisa. 2018. Quinoa as a new leafy vegetable crop in Egypt. *Arab University Journal of Agricultural Sciences* 26(2): 745–752.

Flowers, T. J., M. A. Hajibagheri and N. J. W. Clipson. 1986. Halophytes. *Quarterly Review of Biology* 61: 313–337.

Gómez-Pando, L. R., R. Álvarez-Castro and E. de la Barra. 2010. Effect of salt stress on Peruvian germplasm of *Chenopodium quinoa* Willd: A promising crop. *Journal of Agronomy and Crop Science* 196, 391–396.

Hariadi, Y., K. Marandon, Y. Tian, S-E. Jacobsen and S. Shabala. 2011. Ionic and osmotic relations in quinoa (Chenopodium quinoa Willd.) plant grown at various salinity levels. *Journal of Experimental Botany* 62 (1):185–193.

Hasegawa, P. M., R. A. Bressan, J. K. Zhu and H. J. Bohnert. 2000. Plant cellular and molecular responses to high salinity. *Annual Review of Plant Biology* 51: 463–499.

Hong, Y. H., Huang, Y. L., Liu, Y. C., and Tsai, P. J. 2016. Djulis (*Chenopodium formosanum* Koidz.) water extract and its bioactive components ameliorate dermal damage in UVB-irradiated skin models. *BioMed Research International* 2016: 7368797. http://dx.doi.org/10.1155/2016/7368797

Iqbal S., S. M. A. Basra, I. Afzal, A. Wahid, M. S. Saddiq, M. B. Hafeez and S.-E. Jacobsen. 2018. Yield potential and salt tolerance of quinoa on salt- degraded soils of Pakistan. *Crop Science*, 205: 13–21. doi: 10.1111/jac.12290

Jacobsen, S. E., I. Jørgensen, and O. Stølen. 1994. Cultivation of quinoa (*Chenopodium quinoa*) under temperate climatic conditions in Denmark. *Journal of Agriculture Science*, 122 (1): 47–52.

Jacobsen, S-E., H. Quispe, H. and A. Mujica. 2001. Quinoa: An alternative crop for saline soils in the Andes, Scientists and Farmer-Partners in Research for the 21st Century. CIP Program Report 1999–2000, pp. 403–408.

Jacobsen, S.-E., Mujica, A. and Jensen, C. R. 2003. The resistance of quinoa (*Chenopodium quinoa* Willd.) to adverse abiotic factors. *Food Reviews International* 19 (1 and 2): 99–109.

Jacobsen, S. E., C. Monteros, J. L. Christiansen, L. A. Bravo, L. J. Corcuera, and A. Mujica. 2005. Plant responses of quinoa (*Chenopodium quinoa* Willd.) to frost at various phonological stages. *European Journal of Agronomy* 22(2): 131–139.

Kolata, A. L. 2009. Quinoa: Production, Consumption and Social Value in Historical Context, Department of Anthropology, The University of Chicago. Chicago, Ill. USA.

Koyro, H. W., S. S. Eisa, and H. Lieth. 2008. Salt tolerance of *Chenopodium quinoa* Willd., grains of the Andes: Influence of salinity on biomass production, yield, composition of reserves in the seeds, water and solute relations. *Tasks for Vegetation Sciences* 43: 133–145.

Koziol, M. J., 1992. Chemical composition and nutritional evaluation of quinoa (*Chenopodium quinoa* Willd.). *The Journal of Food Composition and Analysis* 5: 35–68

L'Avenir Agricole. 2015. Quinoa, une culture à haute valeur ajoutée.

Long, N. V. 2016. Effects of salinity stress on growth and yield of quinoa (*Chenopodium quinoa* Willd.) at flower initiation stages. *Vietnam Journal of Agriculture Science* 14: 321–327.

Maughan, P. J., T. B. Turner, C. E. Coleman, D. B. Elzinga, E. N. Jellen, J. A. Morales, J. A. Udall, D. J. Fairbanks, and A. Bonifacio. 2009. Characterization of Salt Overly Sensitive 1 (SOS1) gene homoeologs in quinoa (*Chenopodium quinoa* Willd.). *Genome* 52: 647–657.

Mullan, D. J., T. D. Colmer and M. G. Francki. 2007. Arabidopsis-rice-wheat gene orthologues for Na+ transport and transcript analysis in wheat – *L. elongatum* aneuploids under salt stress. *Molecular Genetics and Genomics* 277: 199–212.

Munns, R. and M. Tester. 2008. Mechanisms of salinity tolerance. *Annual Review of Plant Biology*, 59: 651–681. https://doi.org/10.1146/annure

Panta, S., T. Flowers, P. Lane, R. Doyle, G. Haros and S. Shabala. 2014. Halophyte agriculture: Success stories. *Environmental and Experimental Botany* 107: 71–83. https://doi.org/10.1016/j.envexpbot.2014.05.006

Pathan S., F. Eivazi, B. Valliyodan, K. Paul, G. Ndunguru and K. Clark. 2019. Nutritional composition of the green leaves of quinoa (*Chenopodium quinoa* Willd.). *Journal of Food Research* 8 (6): 55–65.

Pickersgill, B. 2007. Domestication of plants in the Americas: Insights from Mendelian and molecular genetics. *Annals of Botany* 100 (5): 925–940.

Prado, F. E., C. Boero, M. Gallardo, and J. A. González. 2000. Effect of NaCl on germination, growth, and soluble sugar content in Chenopodium quinoa Willd. seeds. *Botanical Bulletin of Academia Sinica* 41: 27–34.

Qiu, Q. S., Y. Guo, M. A. Dietrich, K. S. Schumaker, and J. K. Zhu. 2002. Regulation of SOS1, a plasma membrane Na+/H+ exchanger in Arabidopsis thaliana, by SOS2 and SOS3. *Proceedings of the National Academy of Sciences of the United States of America* 99 (12): 8436–8441. http://dx.doi.org/10.1073/pnas.122224699

Ranhotra, G. S. J. A. Gelroth, B. K. Glaser, K. J. Lorenz, K. J. and D. L. Johnson. 1993. Composition and protein nutritional quality of quinoa. *Cereal Chemistry* 70 (3): 303–305.

Rao, N. K. and M. Shahid. 2012. Quinoa – A promising new crop for the Arabian Peninsula. *American-Eurasian Journal of Agricultural & Environmental Sciences* 12: 1350–1355.

Repo-Carrasco, R., C. Espinoza and S.-E. Jacobsen. 2003. Nutritional value and use of the Andean crops quinoa (*Chenopodium quinoa*) and Ka~niwa (*Chenopodium pallidicaule*). *Food Reviews International* 19 (1 and 2): 179–189.

Ruiz-Carrasco, K., F. Antognoni, A. K. Coulibaly, S. Lizardi, A. Covarrubias, E. A. Martínez, M. A. Molina-Montenegro, S. Biondi and A. Zurita-Silva. 2011. Variation in salinity tolerance of four lowland genotypes of quinoa (*Chenopodium quinoa* Willd.) as assessed by growth, physiological traits, and sodium transporter gene expression. *Plant Physiology and Biochemistry* 49, 1333–1341.

Shabala, S. and A. Mackay. 2011. Ion transport in halophytes. *Advances in Botanical Research* 57: 151–187.

Singh, D., 2018. *Quinoa (Chenopodium Quinoa Willd.)*. New Delhi, India: Scientific Publishers.

Shahid S. A., A. J. Dakheel, K. S. Mufti and G. Shabbir. 2009. Automated in-situ soil salinity logging in irrigated agriculture. *European Journal of Scientific Research* 26(2): 288–297.

Soil Survey Staff. 2010. *Keys to soil taxonomy*, 11th edn. Washington, DC: USDA-NRCS, US. Government Printing Office.

Sosa-Zuniga, V., V. Brito, F. Fuentes and U. Steinfort. 2017. Phenological growth stages of quinoa (*Chenopodium quinoa*) based on the BBCH scale. *Annals of Applied Biology* 171: 117–124. https://doi.org/10.1111/aab.12358

Taváres, O. B., M. G. D. Martínez, R. J. L. Ontiveros, and M. A. Orozco. 1995. Forage evaluation of 18 varieties of quinoa (*Chenopodium quinoa* Willd.) in Montecillo, Mexico. *Rev. Fac. Agron. (LUZ)*, 12 (1): 71–79.

Tsai, P.-J., Y.-S. Chen, C.-H. Sheu and C.-Y. Chen. 2011. Effect of nano grinding on the pigment and bioactivity of djulis (*Chenopodium formosanum* Koidz.). *Journal of Agricultural and Food Chemistry* 59(5):1814–1820.

Van Schooten H. A. and J. B. Pinxterhuis. 2003. Quinoa as an alternative forage crop in organic dairy farming. Optimal forage systems for animal production and the environment grassland. *Science in Europe* 8: 445–448.

Vaughn, J. G. and C. A. Geissler (2009). *The new Oxford book of food plants*. Oxford, UK: Oxford University Press.

Wilson, C., J. J. Read, and E. Abo-Kassem. 2002. Effect of mixed-salt salinity on growth and ion relations of a quinoa and a wheat variety. *Journal of Plant Nutrition* 25 (12): 2689–2704.

30 The Potential of Edible Halophytes as New Crops in Saline Agriculture

The Ice Plant (Mesembryanthemum crystallinum L.) Case Study

Giulia Atzori

CONTENTS

30.1 INTRODUCTION

A growing population will result in an increased food global demand, with a greater consumption of processed food, meat, dairy and fish, all products known to add pressure to the food supply system (Godfray et al. 2010). The trend in world hunger

DOI: 10.1201/9781003112327-30

characterized by a steady decline in the last decades, reverted in 2015, with today more than 820 million people chronically hungry. Such a situation restricts the achievement of the Zero Hunger target by 2030 (FAO IFAD UNICEF WFP and WHO 2019). Also, about 2 billion people in the world experience moderate or severe food insecurity, with the lack of regular access to nutritious and sufficient food leading to a greater risk of malnutrition and poor health (FAO IFAD UNICEF WFP and WHO 2019). Global climate change represents a further threat, especially in marginal and already-stressed agricultural ecosystems, including areas affected by salinity (Cheeseman 2016). In these regions, the world's major crops are not adequate to supply the calories, proteins, fats and nutrients people need: new crops are needed, specifically appropriate to such particular ecological conditions (Cheeseman 2016).

Globally, the irrigation of conventional crops accounts for about 70% of total freshwater (FAO 2011). Such a percentage is already high for areas where freshwater is not limited but becomes impracticable where this resource is scarce. Sustainable agriculture in saline environments requires improved crops and efficient water use (Jez et al. 2016): with respect to this, the domestication of edible species that have naturally adapted to saline environments (Cheeseman 2015), namely halophytes, is an interesting approach to consider (Atzori et al. 2019; Ventura et al. 2015; Rozema and Schat 2013; Rozema and Flowers 2008; Rozema et al. 2013; Glenn et al. 1998). Halophytes can be defined as salt-tolerant plants capable of growth and reproduction at soil salinities greater than 200 mM NaCl, roughly corresponding to ~40% of salinity of seawater (Flowers and Colmer 2008). This group of plants is estimated to comprise 5,000–6,000 species (Glenn et al. 1999), an important number of which are edible species already consumed in many world regions, mainly as wild and not (yet) as cultivated crops. The interest in these species is timely, as their domestication could allow for the exploitation of more available brackish water and seawater sources for sustainable food production in salt-rich environments where conventional crops are proving inadequate; the growth of these plants could also benefit from the macro- and microelements which are important components of these water sources (Rozema and Flowers 2008).

The exploitation of endemic halophytes has the objective of developing local or regional food crops to feed people most at risk for food insecurity because of soil salinity or groundwater salinization (Cheeseman 2015). Since nutrition is an urgent issue in world areas affected by salinity, the development of new crops starting from wild, salt-tolerant relatives of conventional major crops (such as rice, wheat and barley), as opposed to using genetic resources to improve existing crop varieties, represents a valid option (Cheeseman 2015). As another even quicker opportunity, there is a large number of endemic salt-resistant species already used as food that have received very little attention in the scientific literature (Ventura et al. 2015): one "famous" example of species that started as a marginal indigenous crop and then experienced a rapid expansion and acceptance at a global level is quinoa (*Chenopodium quinoa* Wild.), which interestingly is highly salt tolerant (Smil 2001). Following the example of quinoa, the use of other species could face a similar expansion.

Regarding crops' nutritive qualities, the effect of salinity on the production of secondary metabolites has been richly studied with regard to plant salt tolerance, even if such compounds have rarely been considered as quality parameters for healthy food production and commercial purposes (Ventura et al. 2015). Halophytes production of secondary metabolites in response to salt stress is well known (Flowers and Muscolo 2015): such metabolites, or compatible solutes, seem to have different functions, among which a role in the prevention of oxygen radical production or in the scavenging of reactive oxygen species (Hasegawa et al. 2000). The secondary metabolites include simple and complex sugars, amino acids, polyols and antioxidants, which could potentially be utilized in functional food. Following the definition of Buhmann and Papenbrock (2013), defining functional food as having disease-preventing and/or health-promoting benefits, a saline environment could then potentially enhance the quality of products.

In addition to high value nutritional components, halophytes can also accumulate undesired factors including oxalates, nitrates and salts (Ventura et al. 2015). Yet, agrotechnical practices can be applied in order to decrease their content: examples are represented by the reduced use of NO_3^- fertilization in favor of NH_4^+ to decrease the oxalate content in *Portulaca oleracea* (Palaniswamy et al. 2002) or by adjusting iron fertilization to decrease nitrate accumulation in *Aster tripolium* (Ventura et al. 2013). Also cooking methods can provide a way to decrease the content of such undesired factors (Caparrotta et al. 2019). Nonetheless, species-specific investigations are required because of the different species' responses, e.g. significantly decreasing nitrates in the halophyte *Tetragonia tetragonioides* with increasing salinity (Atzori et al. 2020) as opposed to their increased accumulation in *Aster tripolium* with increasing salinity (Ventura et al. 2013). Also sodium accumulation is species-specific; for example in *Mesambryanthemum crystallinum* adult leaves accumulate more Na^+ than young leaves (representing the edible part of the species): such a strategy prevents the edible leaf product from having a too high sodium content (Atzori et al. 2017).

One last issue that the development of halophyte-based crops could address is represented by soil remediation. Phytodesalination is defined as a species aptitude to remove salts from soils by accumulating this in the tissues (Rabhi et al. 2015). A number of halophyte species are characterized by an enhanced ability to take up sodium. Examples of phytodesalinating halophytes are *Mesambryanthemum crystallinum* (Loconsole et al. 2019; Tembo-Phiri 2019; Cassaniti and Romano 2011; Hasanuzzaman et al. 2014), *Tetragonia tetragonioides* (Hasanuzzaman et al. 2014; Bekmirzaev et al. 2011; Neves et al. 2007, 2008; Atzori et al. 2020), *Salsola soda* and *Portulaca oleracea* (Graifenberg et al. 2003; Karakas et al. 2017; Bekmirzaev et al. 2011). Interestingly, phytodesalination is the only existing process in terms of sodium removal that occurs under non-leaching conditions (Rabhi et al. 2015), thus having an important potential value in water-scarce areas.

Even if in the last decades many results indicating the potential of halophytes as possible new crops have been published, scientific documentation of large-scale experiments is still limited and no cultivation protocols have been optimized for such crops (Ventura et al. 2015). Research is, in fact, still needed to ensure the lasting sustainability of saline agriculture, since adequate cultivation systems are of importance. Coastal sandy soils seem an ecologically safe choice for large-scale halophyte production without the risk of salt contamination that could occur on fertile soils. Similarly, also underground freshwater contamination must be avoided. As a different option to consider, cropping systems as soil-less methods would prevent soil contamination and alleviate environmental concerns as the irrigation water would be reused until its depletion (Atzori et al. 2016; 2019). The belief in the importance of halophytes as potential sources of food in saline environments is widespread, particularly because they do not compete with the requirements of conventional food crops in terms of water and soil. In fact, research is currently directed both to the determination of the salt tolerance of halophytes and also to the improvement of their agricultural traits such as yield, palatability, chemical composition, use of mechanical harvesting, testing of market potential and, finally, securing farmers' income (Ventura et al. 2015).

30.2 ICE PLANT (*MESEMBRYANTHEMUM CRYSTALLINUM L.*)

Common ice plant (*Mesembryanthemum crystallinum* L.) is an annual prostrate succulent member of the Aizoaceae family, Caryophyllales (Figure 30.1). It is native to southern and eastern Africa and has now been introduced into western Australia, around the Mediterranean, the coasts of the western United States, Mexico, Chile, and the Caribbean (Adams et al. 1998). This species is already consumed as a vegetable crop in India, Australia, New Zealand and in some countries in Europe (Agarie et al. 2009), e.g. in Germany (Herppich et al. 2008) and the Netherlands.

M. crystallinum is typically distributed on coastal sand dunes and saline areas. It is tolerant to low temperatures and salt accumulation in the top soil and grows in well-drained sandy and loamy soils, even if nutritionally poor and saline, with an ambient temperature range from 12 to 30°C (Loconsole et al. 2019). It has also been

FIGURE 30.1 Ice plant during the flowering phase.

FIGURE 30.2 Examples of epidermal bladder cells.

known as a traditional medicine, characterized by demulcent and diuretic effects (Bouftira et al. 2012) and by naturally occurring superoxide dismutase (SOD) and related anti-oxidant molecules, which have a role in the protection of the skin against radiation exposure (Bouftira et al. 2008). The species is also characterized by antiseptic properties (Ksouri et al. 2008).

30.2.1 Physiology and Morphology

Ice plant has a developmentally programmed switch which enables it to move from C3 to Crassulacean acid metabolism (CAM) photosynthesis: this switch can be accelerated by salinity and drought stresses (Adams et al. 1998) and is connected to the transition from the juvenile to adult phase. The ability to switch to CAM allows plants to accumulate CO_2 during the night and then use it during the day, increasing both water use efficiency and carbon fixing ability (Loconsole et al. 2019).

Morphologically, the above-ground part of the ice plant is covered by epidermal bladder cells (shown in Figures 30.1 and 30.2), giving to the plant a shiny and gleaming appearance from which derives the common name of ice plant (Bohnert and Cushman 2000; Loconsole et al. 2019). The bladder cells are modified unicellular trichomes, ranging from 1 to 3 mm in diameter (Vivrette and Muller 1977), filled with a water solution and functioning as peripheral salinity and water reservoirs providing protection from short-term high salinity or water deficit stress (Agarie et al. 2007; Luttge et al. 1978).

Five stages of development have been described in terms of plant morphology and physiology (Adams et al. 1998):

Phase 1: *Germination:* Only the cotyledons are present; C3 photosynthesis; CAM has not yet been induced.

Phase 2: *Juvenile:* Seven leaf pairs develop on the primary axis; there are no side shoots and no flowers; C3 photosynthesis; CAM has not yet been induced; visible epidermal bladder cells.

Phase 3: *Adult:* Secondary leaves are on side shoots; senescence of primary leaves; no flowers; CAM becomes gradually inducible; visible epidermal bladder cells.

Phase 4: *Flowering:* Flowers occur at the terminus of the primary and side axes; CAM photosynthesis; epidermal bladder cells increase in number and size.

Phase 5: *Seed-formation:* Seed capsules are visible; water uptake ceases; epidermal bladder cells are prominent.

30.2.2 Salt Tolerance

Thanks to its considerable resistance to salt and drought stress (Bloom 1979; Vivrette and Muller 1977), the ice plant was studied from the 1980s onwards as a model species (Bohnert et al. 1988). A number of laboratory experiments aiming at elucidating the physiological and molecular mechanisms behind its stress resistance have been published since then including Agarie et al. (2007), Barker et al. (2004), Cosentino et al. (2010), Kore-eda et al. (2004), Oh et al. (2015), Sanada et al. (1995), Thomas et al. (1992), Thomas and Bohnert (1993) and Winter and Holtum (2007).

Mesembryanthemum is a sodium includer: after salt stress, sodium accumulates in a gradient from roots (about 70 mM) to the growing shoot apices, reaching a concentration of 1 M in the epidermal bladder cells. At the cellular level sodium is effectively partitioned to the vacuoles especially in the epidermal bladder cells (Bohnert and Cushman 2000). Also, salinity induces the accumulation of osmolytes, methylated inositols (ononitol and pinitol) and proline, with the objective of balancing the sodium accumulation in the vacuoles, where sodium may exceed 1 M concentration (Adams et al. 1998). Although the epidermal bladder cells are formed in the plant's juvenile phase, they remain pressed to the surface in unstressed plants, whereas under salinity conditions their volume increases dramatically (Adams et al. 1998).

30.3 FIELD EXPERIMENT

A field experiment was conducted in 2015 to test the growth of common ice plant under conditions of increasing soil salinity (complete results published in Atzori et al., 2017). The twofold aim of the experiment was:

1. To evaluate the crop potential of *M. crystallinum*, by determining the effects of a range of irrigation salinities on the growth and productive performance of the plant in an agricultural setting. A full screening of morphological and physiological characteristics was conducted to investigate adaptations to salinity.
2. To assess the effects of different salinity levels on the accumulation of mineral elements (especially sodium and calcium) related to physiological adaptation and nutritive value of the crop.

30.3.1 Materials and Methods

30.3.1.1 Research Location, Irrigation and Soil Salinity

M. crystallinum was grown in an experimental field on Texel island (53.012837°N, 4.755306°E), the Netherlands, from May to August 2015. The experimental field was divided into 21 plots (8 × 20 m each) with seven salt concentrations randomly distributed and replicated three times: a view of the experimental field is shown in Figure 30.3.

Drip lines, shown in Figure 30.4, were located at 40 cm intervals and provided plots with irrigation characterized by different salinity levels. The irrigation water

FIGURE 30.3 Particular of the field experiment.

FIGURE 30.4 Particular of the drip irrigated field.

was a mixture of fresh water collected from a rainwater basin and natural seawater (electrical conductivity ECw 35 dSm^{-1}) from a nearby ditch fed from the Waddensea. Using a custom built proportional-integral-derivative controller, fresh and saline waters were mixed with an automatic accuracy check of salinity levels in the irrigation water. Drainage pipes, located 60 cm below the surface with 5 m spacing between any two pipes, assured the rapid drainage of irrigation water and aeration of the soil.

The seven salinity treatments used in the experiment (each repeated in three plots) had electrical conductivities (EC_w) values of 2 dSm^{-1} (control), 4, 8, 12, 16, 20 and 35 dSm^{-1}. The soil salinity was monitored by means of samplers capable of collecting soil pore water; these were collected in all plots on three occasions during the experiment.

30.3.1.2 Plant Material, Samplings and Growth Measurements

Seeds of *M. crystallinum* were sown on the 14 April 2015 and young seedlings were transferred into the experimental fields at the rate of 30 plants per plot after one month. Three sampling events were performed during the experiment:

1. *T0 (time zero sampling):* Six untreated juvenile plants were harvested 5 weeks after germination. Shoot fresh weight and dry weight data were collected ($n = 6$). Dry material was used to assess the Na^+ and Ca^{2+} concentration in juvenile plants;
2. *T1 (at potential commercial maturity, as young fully expanded leaves were ready to be harvested):* 16 July 2015. Harvests were made of 3 plants per plot, 9 per treatment). Shoot fresh weight and the fresh weight of 3 young

fully expanded leaves per plant were recorded. Leaf area (LA), specific leaf area (SLA) and leaf succulence were determined on young fully expanded leaves (n = 9). Young fully expanded leaves (n = 9) were also analyzed for concentrations of carotenoids. Dried biomass was then determined and samples from young fully expanded leaves were used for measuring the Na^+ and Ca^{2+} concentration;

3. *T2 (at the end of the crop cycle):* 11 August 2015. Harvests were made of 3 plants per plot, 9 per treatment) with the same measurements performed in the T1 sampling.

30.3.2 Results and Discussion

30.3.2.1 Seawater Irrigation Extended the Growing Season

Figure 30.5 shows the growth of the ice plant shoots. There were no significant effects of salinity at T1, whereas at T2 saline conditions (especially EC_w 20 dSm^{-1}) led to higher biomass accumulation, both in terms of fresh and dry weight, compared to freshwater irrigation (control).

Similarly, for the FW and DW of young fully expanded leaves (yfel, 3 per plant), saline conditions also led to significant differences. At T2, salinity caused an increased weight of yfel (corresponding to the edible part of the tested species) compared to the control (Figure 30.6). Moreover, while total shoot DW significantly increased compared to control only at the 20 dSm^{-1} treatment, the youngest fully expanded leaves increased in DW with the majority of salinity treatments (i.e. 8, 12, 20 and 35 dSm^{-1}).

The better performance of plant biomass at higher salinity levels suggest growth stimulation by increased salinity, acting similarly to other halophytes. For many dicotyledonous halophytes, optimal growth occurs in fact at concentrations of

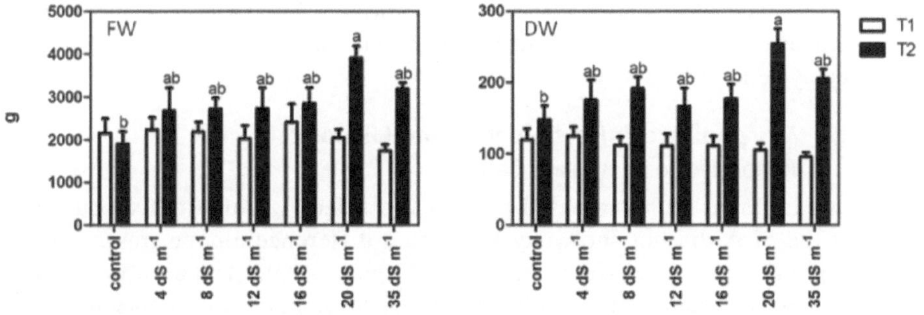

FIGURE 30.5 Fresh weight (FW) and dry weight (DW) of the ice plant shoot, collected at T1 and T2. Values are means ± s.e. (n = 9) expressed in grams per plant. Different letters indicate significant differences among treatments at $P < 0.01$ in FW and at $P < 0.05$ in DW plot (Tukey's Test). (Reprinted with permission from Elsevier from Atzori et al. "Effects of Increased Seawater Salinity Irrigation on Growth and Quality of the Edible Halophyte *Mesembryanthemum crystallinum L.* under Field Conditions" *Agricultural Water Management* 187 (2017): 37–46.)

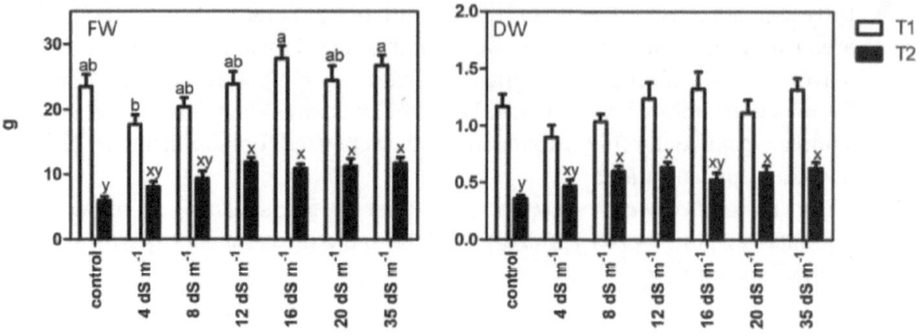

FIGURE 30.6 FW and DW of 3 young fully expanded leaves per plant, collected at T1 and T2. Values are means ± s.e. ($n = 9$) expressed in grams. Different letters indicate significant differences among treatments at the same harvest event at $P < 0.01$ (Tukey's Test). (Reprinted with permission from Elsevier from Atzori et al. "Effects of Increased Seawater Salinity Irrigation on Growth and Quality of the Edible Halophyte *Mesembryanthemum crystallinum* L. under Field Conditions" *Agricultural Water Management* 187(2017): 37–46.)

50–250 mM NaCl (corresponding to EC_w values of ~5 and ~25 dSm^{-1}) in the root medium (Flowers and Colmer 2008). In our experiment, seawater irrigation was related to good growth performance of the ice plant, indicating its potential for saline agriculture. Salinity conditions also led to an extension of the growing season: control plants started to senesce and proceed to the seed production phase in August, whilst in plants treated at high salinity senescence was delayed by about one month. Likewise, Adams et al. (1998) reported that salinity slows down the plant's developmental physiology. Such a slow-down in development is particularly important for our testing of ice plant as a potential salt-tolerant crop in saline agriculture: young leaves, the species edible part, are picked leaving the plant undamaged and can extend the vegetative stage enabling an extra month of harvest compared to plants grown under non-saline conditions. Moreover, the number of young leaves produced and the grams of potential harvest were increased by the extended cycle length.

30.3.2.2 Morphological, Physiological and Osmotic Response to Increased Salinity

As shown in Table 30.1, increased salinity did affect LA and leaf succulence, whereas the SLA did not show any significant difference among treatments at any sampling event. Over time, the main morphological change observed was the increased leaf succulence at higher salinity levels, possibly representing an essential part of the ice plant salt tolerance. In fact, halophytes balance their growth rate with their requirement for the salt needed for osmotic adjustments (Flowers and Yeo 1986): the increase in leaf succulence plays, in fact, a major role in the osmotic adjustment to a low external water potential induced by salinity (Flowers and Colmer 2008). Moreover, an increased leaf succulence, thus an enhanced water content per unit area, translates to an augmented carbon assimilation capacity per unit area, assuring plants growth despite a possibly relatively

TABLE 30.1
Morphological Adaptations of 3 Young Fully Expanded Leaves per Plant at T1 and T2

Treatment	Young Fully Expanded Leaves, T1 LA (cm^2)	SLA ($cm^2\ g^{-1}$)	Leaf Succulence ($g\ cm^{-2}$)	Young Fully Expanded Leaves, T2 LA (cm^2)	SLA ($cm^2\ g^{-1}$)	Leaf Succulence ($g\ cm^{-2}$)
Control	120.67 ± 10.47	110.31 ± 12.91	0.20 ± 0.01^{b}	25.38 ± 2.46^{b}	70.12 ± 4.14	0.24 ± 0.02^{c}
4dS m^{-1}	86.76 ± 6.96	101.42 ± 7.90	0.20 ± 0.01^{ab}	30.75 ± 3.42^{ab}	66.51 ± 4.98	0.27 ± 0.01^{bc}
8dS m^{-1}	91.58 ± 5.91	88.68 ± 3.69	0.22 ± 0.01^{ab}	38.83 ± 3.66^{a}	63.86 ± 3.53	0.24 ± 0.02^{bc}
12dS m^{-1}	102.87 ± 7.93	87.83 ± 7.70	0.23 ± 0.01^{ab}	37.92 ± 2.27^{a}	60.86 ± 3.56	0.31 ± 0.01^{ab}
16dS m^{-1}	116.55 ± 9.20	93.59 ± 9.35	0.24 ± 0.01^{a}	36.48 ± 2.30^{ab}	72.24 ± 4.07	0.30 ± 0.01^{ab}
20dS m^{-1}	110.66 ± 8.31	105.68 ± 11.88	0.22 ± 0.01^{ab}	37.44 ± 3.34^{ab}	66.52 ± 5.31	0.30 ± 0.01^{abc}
35dS m^{-1}	113.64 ± 8.08	87.09 ± 5.23	0.24 ± 0.01^{ab}	34.81 ± 2.24^{ab}	56.82 ± 3.07	0.33 ± 0.01^{a}

Note: Values are means ± s.e. (n=9). Different letters in the same column indicate significant differences at $P<0.05$ (Tukey's Test).
Data obtained with permission from Elsevier from Atzori et al. "Effects of Increased Seawater Salinity Irrigation on Growth and Quality of the Edible Halophyte *Mesembryanthemum crystallinum* L. under Field Conditions" *Agricultural Water Management* 187(2017): 37–46.

low SLA (de Vos et al. 2013). Indeed, in dicotyledonous halophytes, the increase in leaf succulence is often connected to a SLA decrease (Rozema et al. 2015; de Vos et al. 2013; de Vos et al. 2010; Geissler et al. 2009; Ayala and O'Leary 1995), a morphological adaptation associated with the plants need to limit transpiration (Flowers and Flowers 2005). Nevertheless, no significant decrease of SLA occurred in treated plants compared to the control, confirming that none of the treatments did effectively stress the plant, but increased its physiological activity and yield. In fact, in this experiment, LA of salt-treated leaves did rise compared to the control (even if significantly only at the intermediate salinity treatments). It can be suggested that the ice plant leaf area was not reduced by salinity because another feature helped in regulating the leaf ion concentration: the epidermal bladder cells, which are filled with a water solution and function as peripheral salinity and water reservoirs (Agarie et al. 2007; Luttge et al. 1978).

Beyond morphology adjustments, at a physiological level, the concentration of carotenoids, reported in Figure 30.7, augmented in time at increased salinity, whereas the control remained stable. Plants with the higher concentration of carotenoids at both harvests were those of the 20 dS m^{-1} treatment.

30.3.2.3 Nutritive Quality of the Edible Leaves

Figure 30.8 shows the concentration of sodium and calcium in the edible leaves. The concentrations of both sodium and calcium was significantly higher in every treatment compared to the control, at both sampling events.

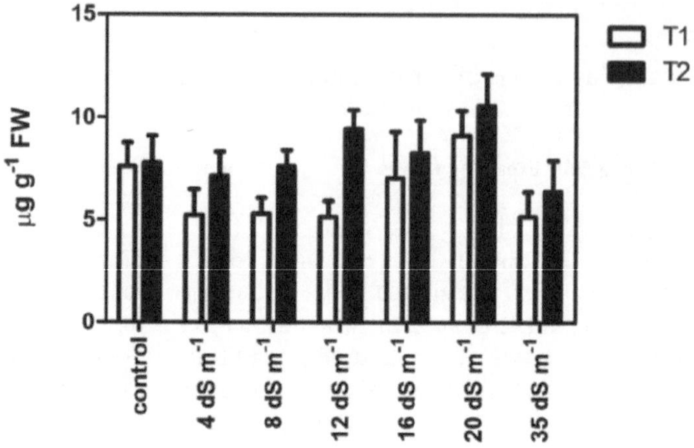

FIGURE 30.7 Concentration of carotenoids in young fully expanded leaves at T1 and T2. Values are means ± s.e. ($n = 9$) expressed in microgram per gram of fresh weight. No significant differences were assessed among treatments at $P<0.05$ (Tukey's Test). (Reprinted with permission from Elsevier from Atzori et al. "Effects of Increased Seawater Salinity Irrigation on Growth and Quality of the Edible Halophyte *Mesembryanthemum crystallinum* L. under Field Conditions" *Agricultural Water Management* 187(2017): 37–46.)

The sodium concentration results were expected because of the sodium includer strategy of *Mesembryanthemum*, with an increasing sodium gradient from roots to shoot apices (Bohnert and Cushman 2000). Interestingly, adult leaves accumulated more Na^+ than young leaves: this strategy prevents the edible leaf product from having too high sodium content, which could otherwise have negative effects for human health. The increased presence of sodium might also suggest that the ice plant leaves could be used as potential salt substitute.

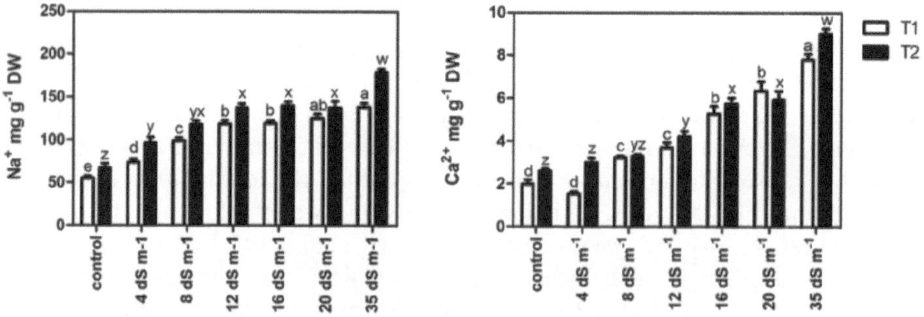

FIGURE 30.8 Concentration of Na^+ and Ca^{2+} in young fully expanded leaves at T1 and T2. Values are means ± s.e. ($n = 9$) expressed in milligram per gram of dry weight. Different letters indicate a significant difference among treatments at the same harvest event at $P < 0.0001$ (Tukey's Test). (Reprinted with permission from Elsevier from Atzori et al. "Effects of Increased Seawater Salinity Irrigation on Growth and Quality of the Edible Halophyte *Mesembryanthemum crystallinum* L. under Field Conditions" *Agricultural Water Management* 187(2017): 37–46.)

The increased calcium concentration with increasing salinity is in accord with the results reported by a number of authors (Agarie et al. 2007; Adams et al. 1998; Yang et al. 2007) and may be connected with the salt tolerance of the ice plant; experimental evidence has correlated increased Ca^{2+} with salt adaptation (Parida and Das 2005). Calcium is believed to protect the structure and function of membranes under salt stress (Yan et al. 1995), and its concentration increase under salinity stress may ameliorate the inhibitory effect on growth (Epstein 1972). From the point of view of the quality of food, the significant increase of Ca^{2+} may point to an interesting nutritional improvement achievable under salinity conditions, as calcium is among the main mineral elements lacking in the diet of over two-thirds of the world's population (White and Broadley 2009). Encouragingly, a strong correlation also exists between the ability of many plants to accumulate Ca^{2+} and Mg^{2+} (White and Broadley 2009). Furthermore, species from families within the Caryophyllales tend to accumulate high concentrations of Mg^{2+} and Zn^{2+} in leaves (White and Broadley 2009; Broadley et al. 2004; White 2001). Investigations on possible patterns between salinity and other mineral elements with an important role in human diet (i.e. Cu^{2+}, Fe^{2+}, Mg^{2+}, Zn^{2+}) could add important information to the ice plant nutritional enrichment opportunity in saline environments.

Finally, the carotenoids – another nutritive goal – rose between T1 and T2 in all salt-treated plants, while no increase was found in the control. Also in this feature, the ice plant seems to distinguish itself positively compared to some other halophyte plants in which the carotenoid concentration may decrease with increasing salinity (Redondo-Gomez et al. 2010; Aghaleh et al. 2009; Qiu et al. 2003).

30.3.2.4 Prospective for Saline Agriculture

As none of the tested salt concentrations resulted in biomass loss, it seems possible to cultivate *M. crystallinum* for saline agriculture, at salinities up to EC_w values of 20–35 dSm^{-1}. Perhaps even higher salinity levels are possible since we did not identify a threshold for substantial biomass reduction, although the highest biomass production was suggested to occur at an EC_w of about 20 dSm^{-1}. The already appreciated taste of saline agriculture vegetables in different countries (Rozema and Schat 2013) and of the ice plant, in particular, helped by its gleaming appearance (Agarie et al. 2009; Herppich et al. 2008), also encourage this possibility.

Many results indicating the potential of halophytes as possible new crops have been published in the last decades. Yet, scientific documentation of large-scale experiments is still limited, especially addressing cultivation protocols (Ventura et al. 2015). Since adequate cultivation systems are of major importance, research is still needed to ensure the lasting sustainability of saline agriculture.

REFERENCES

Adams, Patricia, Den E. Nelson, Shigehiro Yamada, Wendy Chmara, Richard G. Jensen, Hans J. Bohnert, and Howard Griffiths. 1998. "Tansley Review No. 97 Growth and Development of *Mesembryanthemum crystallinum* (Aizoaceae)." *New Phytologist* 138: 171–190. https://doi.org/10.1046/j.1469-8137.1998.00111.x.

Agarie, Sakae, Akiko Kawaguchi, Akiko Kodera, Haruki Sunagawa, Hide Kojima, Akihiro Nose, and Teruhisa Nakahara. 2009. "Potential of the Common Ice Plant,

Mesembryanthemum crystallinum as a New High-Functional Food as Evaluated by Polyol Accumulation." *Plant Production Science* 12 (1): 37–46. https://doi.org/10.1626/pps.12.37.

Agarie, Sakae, Toshifumi Shimoda, Yumi Shimizu, Kathleen Baumann, Haruki Sunagawa, Ayumu Kondo, Osamu Ueno, Teruhisa Nakahara, Akihiro Nose, and John C Cushman. 2007. "Salt Tolerance, Salt Accumulation, and Ionic Homeostasis in an Epidermal Bladder-Cell-Less Mutant of the Common Ice Plant *Mesembryanthemum crystallinum*." *Journal of Experimental Botany* 58 (8): 1957–1967. https://doi.org/10.1093/jxb/erm057.

Aghaleh, M., V. Niknam, H. Ebrahimzadeh, and K. Razavi. 2009. "Salt Stress Effects on Growth, Pigments, Proteins and Lipid Peroxidation in *Salicornia persica and S. europaea*." *Biologia Plantarum* 53 (2): 243–248. https://doi.org/10.1007/s10535-009-0046-7.

Atzori, Giulia, Werther Guidi Nissim, Stefania Caparrotta, Elisa Masi, Elisa Azzarello, Camilla Pandolfi, Pamela Vignolini, Cristina Gonnelli, and Stefano Mancuso. 2016. "Potential and Constraints of Different Seawater and Freshwater Blends as Growing Media for Three Vegetable Crops." *Agricultural Water Management* 176: 255–262. https://doi.org/10.1016/j.agwat.2016.06.016.

Atzori, Giulia, Stefano Mancuso, and Elisa Masi. 2019. "Seawater Potential Use in Soilless Culture: A Review." *Scientia Horticulturae* 249: 199–207. https://doi.org/10.1016/j.scienta.2019.01.035.

Atzori, Giulia, Werther Guidi Nissim, Tania Macchiavelli, Federico Vita, Elisa Azzarello, Camilla Pandolfi, Elisa Masi, and Stefano Mancuso. 2020. "*Tetragonia tetragonioides* (Pallas) Kuntz. as Promising Salt-Tolerant Crop in a Saline Agriculture Context." *Agricultural Water Management* 240: 106261. https://doi.org/10.1016/j.agwat.2020.106261.

Atzori, Giulia, Arjen C. de Vos, Marc van Rijsselberghe, Pamela Vignolini, Jelte Rozema, Stefano Mancuso, and Peter M. van Bodegom. 2017. "Effects of Increased Seawater Salinity Irrigation on Growth and Quality of the Edible Halophyte *Mesembryanthemum crystallinum* L. under Field Conditions." *Agricultural Water Management* 187: 37–46. https://doi.org/10.1016/j.agwat.2017.03.020.

Ayala, Felix, and James W. O'Leary. 1995. "Growth and Physiology of *Salicornia bigelovi* Torr. at Suboptimal Salinity." *International Journal of Plant Sciences* 156 (2): 197–205.

Barker, David H, Jeff Marszalek, Jeff F Zimpfer, and William W Adams I I I. 2004. "Changes in Photosynthetic Pigment Composition and Absorbed Energy Allocation during Salt Stress and CAM Induction in *Mesembryanthemum crystallinum*." *Functional Plant Biology* 31: 781–787.

Bekmirzaev, G, J Beltrao, M A Neves, and C Costa. 2011. "Climatical Changes Effects on the Potential Capacity of Salt Removing Species." *International Journal of Geology* 5 (3): 79–85.

Bloom, Arnold J. 1979. "Salt Requirement for Crassulacean Acid Metabolism in the Annual Succulent, *Mesembryanthemum crystallinum*." *Plant Physiology* 36: 749–753.

Bohnert, Hans I, James A Ostrem, John C Cushman, Christine B Michalowski, Jutta Rickers, Gabriele Meyer, E. Jay DeRocher, et al. 1988. "*Mesembryanthemum crystallinum*, a Higher Plant Model for the Study of Environmentally Induced Changes in Gene Expression." *Plant Molecular Biology Reporter* 6 (1): 10–28.

Bohnert, Hans J., and John C. Cushman. 2000. "The Ice Plant Cometh: Lessons in Abiotic Stress Tolerance." *Journal of Plant Growth Regulation* 19: 334–346. https://doi.org/10.1007/s003440000033.

Bouftira, I, C Abdelly, and S Sfar. 2008. "Characterization of Cosmetic Cream with *Mesembryanthemum crystallinum* Plant Extract: Influence of Formulation Composition on Physical Stability and Anti-Oxidant Activity." *International Journal of Cosmetic Science* 30: 443–452.

Bouftira, Ibtissem, Hizem Hela, Mahmoud Amor, Chedly Abdelly, and Sfar Souad. 2012. "Effect of *Mesembryanthemum crystallinum* Extract against DMH-Induced Colon Carcinogenesis in Experimental Animals." *International Journal of Research in Pharmaceutical and Biomedical Sciences* 3 (3): 1038–1043.

Broadley, Martin R, Helen C Bowen, Helen L Cotterill, John P Hammond, Mark C Meacham, Andrew Mead, and Philip J White. 2004. "Phylogenetic Variation in the Shoot Mineral Concentration of Angiosperms." *Journal of Experimental Botany* 55 (396): 321–336. https://doi.org/10.1093/jxb/erh002.

Buhmann, Anne, and Jutta Papenbrock. 2013. "An Economic Point of View of Secondary Compounds in Halophytes." *Functional Plant Biology* 40: 952–967.

Caparrotta, Stefania, Elisa Masi, Giulia Atzori, Ismael Diamanti, Elisa Azzarello, Stefano Mancuso, and Camilla Pandolfi. 2019. "Growing Spinach (*Spinacia oleracea*) with Different Seawater Concentrations: Effects on Fresh, Boiled and Steamed Leaves." *Scientia Horticulturae* 256: 1–7. https://doi.org/10.1016/j.scienta.2019.05.067.

Cassaniti, C and D Romano. 2011. "The Use of Halophytes for Mediterranean Landscaping." *European Journal of Plant Science and Biotechnology* 5 (Special Issue 2): 57–63.

Cheeseman, John M. 2016. "Food Security in the Face of Salinity, Drought, Climate Change, and Population Growth." In *Halophytes for Food Security in Dry Lands*, edited by M.A. Khan, M. Ozturk, B. Gul, and M.Z Ahmed, 111–24. Elsevier Inc. https://doi.org/10.1016/B978-0-12-801854-5.00007-8.

Cheeseman, John M. 2015. "The Evolution of Halophytes, Glycophytes and Crops, and Its Implications for Food Security under Saline Conditions." *New Phytologist*, 206 (1): 557–570.

Cosentino, Cristian, Elke Fischer-Schliebs, Adam Bertl, Gerhard Thiel, and Ulrike Homann. 2010. "Na +/H + Antiporters Are Differentially Regulated in Response to NaCl Stress in Leaves and Roots of *Mesembryanthemum crystallinum*." *New Phytologist* 186: 669–680. https://doi.org/10.1111/j.1469-8137.2010.03208.x.

Epstein, E. 1972. *Mineral Nutrition of Plants; Principles and Perspective*. New York: Wiley.

FAO. 2011. "The State of the World's Land and Water Resources for Food and Agriculture (SOLAW), Managing Systems at Risk." *Food and Agriculture Organization*. Rome and London: The Food and Agriculture Organization of the United Nations and Earthscan. https://doi.org/978-1-84971-326-9.

FAO IFAD UNICEF WFP and WHO. 2019. "The State of Food Security and Nutrition in the World 2019. Safeguarding against Economic Slowdowns and Downturns." Vol. 10. Rome. https://doi.org/10.26596/wn.201910395-97.

Flowers, T J and S A Flowers. 2005. "Why Does Salinity Pose Such a Difficult Problem for Plant Breeders ?" *Agricultural Water Management* 78: 15–24. https://doi.org/10.1016/j.agwat.2005.04.015.

Flowers, T J and A R Yeo. 1986. "Ion Relations of Plants under Drought and Salinity." *Australian Journal of Plant Physiology* 13: 75–91.

Flowers, Timothy J., and Timothy D. Colmer. 2008. "Salinity Tolerance in Halophytes." *New Phytologist* 179: 945–963. https://doi.org/10.1111/j.1469-8137.2008.02531.x.

Flowers, Timothy J, and Adele Muscolo. 2015. "Introduction to the Special Issue: Halophytes in a Changing World." *AoB PLANTS*, 7: 1–5. https://doi.org/10.1093/aobpla/plv020.

Geissler, Nicole, Sayed Hussin, and Hans-Werner Koyro. 2009. "Elevated Atmospheric CO 2 Concentration Ameliorates Effects of NaCl Salinity on Photosynthesis and Leaf Structure of *Aster tripolium* L." *Journal of Experimental Botany* 60 (1): 137–251. https://doi.org/10.1093/jxb/ern271.

Glenn, Edward P., J. Jed Brown, and Eduardo Blumwald. 1999. "Salt Tolerance and Crop Potential of Halophytes." *Critical Reviews in Plant Sciences* 18 (2): 227–255. https://doi.org/10.1016/S0735-2689(99)00388-3.

Glenn, Edward P., J. Jed Brown, and James W. O'Leary. 1998. "Irrigating Crops with Seawater." *Scientific American* 279 (2): 76–81. https://doi.org/10.1038/scientificamerican0898-76.

Godfray, H. Charles J., John R. Beddington, Ian R. Crute, Lawrence Haddad, David Lawrence, James F. Muir, Jules Pretty, Sherman Robinson, Sandy M. Thomas, and Camilla Toulmin. 2010. "Food Security: The Challenge of Feeding 9 Billion People." *Science* 327: 812–818. https://doi.org/10.1016/j.geoforum.2018.02.030.

Graifenberg, A, L Botrini, L Giustiniani, F Filippi, and M Curadi. 2003. "Tomato Growing in Saline Conditions with Biodesalinating Plants: *Salsola soda* L., and *Portulaca oleracea* L. " *Acta Horticulturae* 609: 301–305.

Hasanuzzaman, Mirza, Kamrun Nahar, Md Mahabub Alam, Prasanta C. Bhowmik, Md Amzad Hossain, Motior M. Rahman, Majeti Narasimha Vara Prasad, Munir Ozturk, and Masayuki Fujita. 2014. "Potential Use of Halophytes to Remediate Saline Soils." *BioMed Research International*, 2014: 589341. https://doi.org/10.1155/2014/589341.

Hasegawa, Paul M, Ray A Bressan, Jian-Kang Zhu, and Hans J. Bohnert. 2000. "Plant Cellular and Molecular Responses to High Salinity." *Annual Review of Plant Physiology and Plant Molecular Biology* 51: 463–499.

Herppich, W. B., S. Huyskens-Keil, and M. Schreiner. 2008. "Effects of Saline Irrigation on Growth, Physiology and Quality of *Mesembryanthemum crystallinum* L., a Rare Vegetable Crop." *Journal of Applied Botany and Food Quality* 82: 47–54.

Jez, Joseph M, Soon Goo Lee, and Ashley M Sherp. 2016. "The next Green Movement: Plant Biology for the Environment and Sustainability." *Science* 353 (6305): 1241–1244.

Karakas, Sema, Mehmet Ali Cullu, and Murat Dikilitas. 2017. "Comparison of Two Halophyte Species (*Salsola soda* and *Portulaca oleracea*) for Salt Removal Potential under Different Soil Salinity Conditions." *Turkish Journal of Agriculture and Forestry* 41: 183–190. https://doi.org/10.3906/tar-1611-82.

Kore-eda, Shin, Mary Ann Cushman, Inna Akselrod, Davina Bufford, Monica Fredrickson, Elizabeth Clark, and John C Cushman. 2004. "Transcript Profiling of Salinity Stress Responses by Large-Scale Expressed Sequence Tag Analysis in *Mesembryanthemum crystallinum*." *Gene* 341: 83–92. https://doi.org/10.1016/j.gene.2004.06.037.

Ksouri, Riadh, Wided Megdiche, Hanen Falleh, Nejla Trabelsi, Mondher Boulaaba, Abderrazak Smaoui, and Chedly Abdelly. 2008. "Influence of Biological, Environmental and Technical Factors on Phenolic Content and Antioxidant Activities of Tunisian Halophytes." *Comptes Rendus Biologies* 331: 865–873. https://doi.org/10.1016/j.crvi.2008.07.024.

Loconsole, Danilo, Bernardo Murillo-Amador, Giuseppe Cristiano, and Barbara De Lucia. 2019. "Halophyte Common Ice Plants: A Future Solution to Arable Land Salinization." *Sustainability* 11 (6076): 1–16.

Luttge, U., E. Fischer, and E. Steudle. 1978. "Membrane Potentials and Salt Distribution in Epidermal Bladders and Photosynthetic Tissue of *Mesembryanthemum crystallinum* L." *Plant Cell and Environment* 1 (2): 121–129. https://doi.org/10.1111/j.1365-3040.1978.tb00753.x.

Neves, M. A., M. G. Miguel, C. Marques, T. Panagopoulos, and J. Beltrao. 2007. "*Tetragonia tetragonioides* – A Potential Salt Removing Species. Response to the Combined Effects of Salts and Calcium." In Proc of the 3rd IASME/WSEAS IntConf on Energy, Environment, Ecosystems and Sustainable Development, 60–64.

Neves, M A, M G Miguel, C Marques, T Panagopoulos, and J Beltrao. 2008. "The Combined Effects of Salts and Calcium on Growth and Mineral Accumulation of *Tetragonia tetragonioides* – A Salt Removing Species." *WSEAS Transactions on Environment and Development* 4 (1): 1–5.

Oh, Dong-ha, Bronwyn J Barkla, Rosario Vera-Estrella, Omar Pantoja, Sang-yeol Lee, Hans J Bohnert, and Maheshi Dassanayake. 2015. "Cell Type-Specific Responses to Salinity – the Epidermal Bladder Cell Transcriptome of *Mesembryanthemum crystallinum*." *New Phytologist* 207: 627–644.

Palaniswamy, Usha R, Bernard B Bible, and Richard J Mc Avoy. 2002. "Effect of Nitrate: Ammonium Nitrogen Ratio on Oxalate Levels of Purslane." In Trends in New Crops and New Uses, edited by ASHS, Janick J, 453–455. Alexandria.

Parida, Asish Kumar, and Anath Bandhu Das. 2005. "Salt Tolerance and Salinity Effects on Plants: A Review." *Ecotoxicology and Environmental Safety* 60 (3): 324–349. https://doi.org/10.1016/j.ecoenv.2004.06.010.

Qiu, Nianwei, Qingtao Lu, and Congming Lu. 2003. "Photosynthesis, Photosystem II Efficiency and the Xanthophyll Cycle in the Salt-Adapted Halophyte *Atriplex centralasiatica*." *New Phytologist* 159: 479–486. https://doi.org/10.1046/j.1469-8137.2003.00825.x.

Rabhi, Mokded, Abdallah Atia, Chedly Abdelly, and Abderrazak Smaoui. 2015. "New Parameters for a Better Evaluation of Vegetative Bioremediation, Leaching, and Phytodesalination." *Journal of Theoretical Biology* 383: 7–11. https://doi.org/10.1016/j.jtbi.2015.07.027.

Redondo-Gomez, S., E. Mateos-Naranjo, M. E. Figueroa, and A. J. Davy. 2010. "Salt Stimulation of Growth and Photosynthesis in an Extreme Halophyte, *Arthrocnemum macrostachyum*." *Plant Biology* 12: 79–87. https://doi.org/10.1111/j.1438-8677.2009.00207.x.

Rozema, J., D. Cornelisse, Y. Zhang, H. Li, B. Bruning, D. Katschnig, R. Broekman, B. Ji, and P. van Bodegom. 2015. "Comparing Salt Tolerance of Beet Cultivars and Their Halophytic Ancestor: Consequences of Domestication and Breeding Programmes." *AoB PLANTS* 7 (0): plu083. https://doi.org/10.1093/aobpla/plu083.

Rozema, Jelte and Timothy Flowers. 2008. "Crops for a Salinized World." *Science* 322 (5907): 1478–1480. https://doi.org/10.1126/science.1168572.

Rozema, Jelte, Adele Muscolo, and Tim Flowers. 2013. "Sustainable Cultivation and Exploitation of Halophyte Crops in a Salinising World." *Environmental and Experimental Botany* 92: 1–3. https://doi.org/10.1016/j.envexpbot.2013.02.001.

Rozema, Jelte, and Henk Schat. 2013. "Salt Tolerance of Halophytes, Research Questions Reviewed in the Perspective of Saline Agriculture." *Environmental and Experimental Botany* 92: 83–95. https://doi.org/10.1016/j.envexpbot.2012.08.004.

Sanada, Yukuka, Hiroko Ueda, Kazuhiro Kuribayashi, Toshiya Andoh, Fumio Hayashi, Naoto Tamai, and Keishiro Wada. 1995. "Novel Light-Dark Change of Proline Levels in Halophyte (*Mesembryanthemum crystallinum* L.) and Glycophytes (*Hordeum vulgare* L. and *Triticum aestivum* L.) Leaves and Roots under Salt Stress." *Plant Cell Physiology* 36 (6): 965–970.

Smil, Vaclav. 2001. *Feeding the World: A Challenge for the Twenty-First Century*. Cambridge, Massachussets, London, England: The MIT Press.

Tembo-Phiri, Chimwemwe. 2019. "Edible Fynbos Plants: A Soil Types and Irrigation Regime Investigation on *Tetragonia decumbens* and *Mesembryanthemum crystallinum*." Stellenbosch University.

Thomas, J. C. and H. J. Bohnert. 1993. "Salt Stress Perception and Plant Growth Regulators in the Halophyte *Mesembryanthemum crystallinum*." *Plant Physiology* 103 (4): 1299–1304. https://doi.org/103/4/1299 [pii].

Thomas, J.C., R.L. De Armond, and H.J. Bohnert. 1992. "Influence of NaCl on Growth, Proline, and Phosphoenolpyruvate Carboxylase Levels in *Mesembryanthemum crystallinum* Suspension Cultures." *Plant Physiology* 98 (2): 626–631. https://doi.org/10.1104/pp.98.2.626.

Ventura, Yvonne, Amram Eshel, Dov Pasternak, and Moshe Sagi. 2015. "The Development of Halophyte-Based Agriculture: Past and Present." *Annals of Botany* 115: 529–540. https://doi.org/10.1093/aob/mcu173.

Ventura, Yvonne, Malika Myrzabayeva, Alikulov Zerekbay, Shabtai Cohen, Zion Shemer, and Moshe Sagi. 2013. "The Importance of Iron Supply during Repetitive Harvesting of *Aster tripolium*." *Functional Plant Biology* 40: 968–976.

Vivrette, Nancy J., and Cornelius H. Muller. 1977. "Mechanism of Invasion and Dominance of Coastal Grassland by *Mesembryanthemum crystallinum*." *Ecological Monographs* 47 (3): 301–318. https://doi.org/10.2307/1942519.

Vos, Arjen. C. de, Rob Broekman, Maartje P. Groot, and Jelte Rozema. 2010. "Ecophysiological Response of *Crambe maritima* to Airborne and Soil-Borne Salinity." *Annals of Botany* 105: 925–937. https://doi.org/10.1093/aob/mcq072.

Vos, Arjen C. de, Rob Broekman, Catia C. de Almeida Guerra, Marc van Rijsselberghe, and Jelte Rozema. 2013. "Developing and Testing New Halophyte Crops: A Case Study of Salt Tolerance of Two Species of the Brassicacea, *Diplotaxis tenuifoli* and *Cochlearia officinalis*." *Environmetal and Experimental Botany* 92: 154–164.

White, P J and M R Broadley. 2009. "Biofortification of Crops with Seven Mineral Elements Often Lacking in Human Diets–Iron, Zinc, Copper, Calcium, Magnesium, Selenium and Iodine." *New Phytol* 182 (1): 49–84. https://doi.org/10.1111/j.1469-8137.2008.02738.x.

White, Philip J. 2001. "The Pathways of Calcium Movement to the Xylem." *Journal of Experimental Botany* 52 (358): 891–899.

Winter, Klaus and Joseph A M Holtum. 2007. "Environment or Development ? Lifetime Net CO 2 Exchange and Control of the Expression of Crassulacean Acid Metabolism in *Mesembryanthemum crystallinum* 1." *Plant Physiology* 143: 98–107. https://doi.org/10.1104/pp.106.088922.

Yan, Bin, Dai Qiujie, Liu Xiaozhong, Huang Shaobai, Wang Zhixia, and Wang Zongli. 1995. "The Study on Increasing Salt Resistance of Rice by Calcium." *Acta Agronomica Sinica*. 21(6):685–690

Yang, Chunwu, Jianna Chong, Changyou Li, Changmin Kim, Decheng Shi, and Deli Wang. 2007. "Osmotic Adjustment and Ion Balance Traits of an Alkali Resistant Halophyte *Kochia sieversiana* during Adaptation to Salt and Alkali Conditions." *Plant Soil* 294: 263–276. https://doi.org/10.1007/s11104-007-9251-3.

31 Salicornia Species

Current Status and Future Potential

Tanmay Chaturvedi, Aslak H.C. Christiansen, Iwona Gołębiewska, and Mette H. Thomsen

CONTENTS

31.1 INTRODUCTION

The salinization of agricultural soils is an ever-increasing challenge that poses major constraints to agricultural productivity worldwide. An estimated 7% of the world's total land area is salt-affected to some degree; this occurs in all climatic zones, but especially in the arid and semi-arid regions of the world (Wicke et al. 2011). The majority of these salt-affected soils have arisen naturally, through the release of soluble salts from weathering processes of parental material, with sodium chloride being the most abundant salt (Szabolsc 1989). Other natural processes include the deposition of oceanic salt by wind and rain and the intrusion of seawater in low lying coastal areas (Munns and Tester 2008). Besides the natural processes resulting in salinization over time, anthropogenic factors, such as land clearing of natural deep-rooted vegetation and irrigation, has resulted in the salinization of agricultural land. Irrigation and land clearing can lead to a rise in the water table and thereby an increased concentration of soluble salts in the root zone, through the processes of evaporation and capillary rise (Barrett-Lennard 2002). The extent and the severity of salinity will likely increase with climate change, resulting in an increased risk of flooding, rising seawater levels, and changes in the global precipitation pattern

DOI: 10.1201/9781003112327-31

(Hossain 2010; IPCC 2018). In addition, the growing global population, expected to reach 9.7 billion in 2050, will require an unprecedented growth in food production and will thereby put further stress on already scarce land and water resources (United Nations 2019). Hence, there is an urgent need to find ways of utilizing saline land and saline water resources to maintain agricultural productivity and meet the growing global demand for food, water, and energy.

Most of the agricultural crops grown around the world are salt-sensitive glycophytes that suffer significant yield reductions when exposed to even mildly saline conditions, due to osmotic, and eventually, also ionic stress linked to the accumulation of salts in the leaf tissue (Munns and Tester 2008). Despite the high genetic diversity within crop species and advances in molecular genetics, little success has been made in developing new salt-tolerant varieties, as salinity tolerance is a multigenetic trait (Flowers et al. 2010; Ismail and Horie 2017). However, great potential lies in cultivating naturally salt-tolerant plants that can achieve comparatively larger biomass yields than conventional crops under saline conditions (Flowers and Colmer 2008; Flowers and Muscolo 2015). Halophytes can be defined as plants that can survive and reproduce in environments with a salinity level of 200 mM or above and constitute about 2% of the world's flora (Flowers and Colmer 2008; Bennett et al. 2013). Until now, the main interest in halophytes has been to gain insight into the physiological and molecular salt-tolerance mechanisms employed by these species, neglecting the crop potential of halophytes (Katschnig et al. 2013; Yvonne et al. 2013). It has been advocated that the most direct way of achieving salt-tolerant crops is through the domestication of these plants i.e. by enrolling potential halophytic crop plants in conventional plant breeding programs (Hodges et al. 1993; Zerai et al. 2010; Rozema and Flowers 2015). The species that have gained the most interest as crop species are *Atriplex* spp. (for forage), *Distichlis* spp. (for grain), and *Salicornia* spp. (as oilseed) (Glenn et al. 2013; Ventura et al. 2015). However, *Salicornia* has gained special commercial attention because of its other uses; it is currently being grown in Europe and North America as a vegetable, with attempts being made at large-scale commercial cultivation with seawater irrigation for production of oilseed and as an animal feed (Abdal 2009; Bailis and Yu 2012; Gunning 2016). However, throughout domestication process, there have been a range of agronomic challenges that have had to be tackled regarding its use as a seed crop, such as unsynchronized flowering, small seed size, seed shattering, and seed recoveries of <75% when harvesting (Glenn et al 1998; Zerai et al. 2010; Glenn et al. 2013). Breeding efforts have been deployed to improve these undesirable traits and new *Salicornia* varieties (such as SOS 10) have been developed (Glenn et al 1998; 2013; Zerai et al. 2010). Beside the production of seeds, alternative ways of utilizing other parts of *Salicornia* such as biomass and valuable components have been explored, and these will be elaborated in the following sections.

31.2 SPECIES AND GEOGRAPHICAL LOCATIONS

The genus *Salicornia* belongs to the family Amaranthaceae (previously Chenopodiaceae); it is an annual, succulent plant characterized by leafless stems and branches, with sessile flowers often arranged in 3-flower cymes per bract, and aggregated in dense terminal spike-like thyrses (Kadereit et al. 2007). The reproductive biology seems to be dominated by inbreeding in the diploid species, although out-crossing does occur particularly in the tetraploid species such as *S. bigelovii* (Noble et al. 1992). Most species have an erect or prostate growth habit, vary in height (10–60 cm), degree of branching (dependent on the environmental and climatic conditions) and have a preference for non-shaded sites (Davy et al. 2001). *Salicornia* spp. are widely distributed in the temperate, boreal, and subtropical parts of the northern hemisphere and can be found growing in and around coastal and inland salt marshes, salt pans, salt lakes, and mudflats. The environments that *Salicornia* inhabits are often affected by diurnal and seasonal fluctuations in the duration of submergence, waterlogging, and salinity levels. A high level of physiological plasticity has, therefore, been found in *Salicornia* spp. resulting in a broad phenotypic variation between populations under differing environmental conditions (Rozema et al. 1987). To cope with the stressful edaphic factors found in salt marsh environments, with salinities reaching twice the concentration of seawater (1 M NaCl), *Salicornia* spp. have developed extreme salt tolerance (Flowers et al. 1986; Glenn et al. 1991; Ventura and Sagi 2013). This high salt-tolerance relies on the compartmentalization of salts in the vacuoles accompanied by the synthesis of compatible solutes, enabling osmotic adjustment, while at the same time avoiding the toxic effects of Na^+ and Cl^- in the cytosol (Munns and Tester 2008). Compatible solutes such as sucrose, proline, and glycine-betaine not only serve to maintain osmotic pressure but also act as osmoprotective compounds that maintain protein integrity and protect the cytosol from ion toxicity and free radicals (Slama et al. 2015).

The *Salicornia* genus includes 25–30 species, although no present agreement exists on the exact number of accepted species (Kadereit et al. 2007). The high degree of physiological plasticity together with an extremely reduced leaf and flower morphology, providing few diagnostic characters, has led to a complex taxonomy (Kadereit et al. 2007; 2012). The complexity of the taxonomic characterization has also led to the use of the names *Salicornia europaea L.* and *Salicornia herbacea L.* in a broad sense to include many different genotypes, with same species being given different names in different regions (Davy et al. 2001; Kadereit et al. 2012). Depending on the region, *Salicornia* is known by the common names: samphire, sea asparagus, pickled sea–weed, crow's foot green, hamcho, glasswort, or sea-beans (Feng et al. 2013). Analysis of ribosomal DNA polymorphism and ETS sequence data have confirmed genetically distinct forms; however, these techniques have been insufficient to resolve morphologically distinct species (Noble et al. 1992; Singh et al. 2014). To discriminate between species, seed and fruit characters have been recognized as potentially useful diagnostic traits (Rhee et al. 2009).

Despite the taxonomic difficulties arising from phenotypic plasticity and morphological parallelism, some recognized species have attracted more interest than others. *S. europaea* (common glasswort) is one of the most common species found in Europe, characterized by an erect growth habit, height of 10–30 cm, and a fairly rich degree of branching (Davy et al. 2001). This species is mostly recognized for its culinary uses and medicinal properties and can be found at local markets around Europe and North America (Gunning 2016). *Salicornia bigelovii Torr.* (dwarf glasswort) belongs to the North American tetraploid branch of *Salicornia* and can be distinguished from other species by its acute and sharply mucronate leaf and bract tips (Kadereit et al. 2007). It can be found growing in subtropical regions, with an erect growth habit (up to 50 cm tall) and has been one of the most sought-after species in the effort to cultivate halophytes with seawater in coastal desert regions (Glenn et al. 1991; 1998; 2013; Hodges et al. 1993; Ventura et al. 2015).

31.3 CHEMICAL CHARACTERIZATION

Each species of *Salicornia* differs in its composition. This difference might simply be due to the difficulty in characterizing species and is more likely to be the result of environmental conditions, due to the large phenotypic plasticity. The variations within species can be attributed to variation in harvesting time, method of cultivation, and the fraction of the plant, which has undergone the analysis. Table 31.1 compares the compositional analysis of *S. arabica, S. bigelovii*, and *S. herbacea*. While the extremely low moisture and high carbohydrate contents stand out for *S. arabica*, the remainder of the components seems to be in agreement with studies of other species.

The compositional analyses of leaves and stems (also known as seed spikes or pods as referred to in the study) of *S. bigelovii* were compared against each other (Cybulska et al. 2014b). Since the cultivation was carried out at four salinity levels (10, 20, 30, and 50 ppt of NaCl concentration) and three fertilization levels (1.0, 1.5, and 2.0 g N/m^2), the influence of these parameters was also investigated in the same study. Table 31.2 shows that salinity had an effect on the total extractives, glucan, and lignin contents. The fertilizer concentration did not have a significant influence on any of the components. Salinity has a significant impact on the total extractives, lignin content, and glucan content.

S. brachiata was studied in greater detail; the crude polysaccharide fraction of the biomass was treated with four different solvents, namely cold water (CW), hot water (HW), ammonium oxalate (OX), and aqueous sodium hydroxide (ALK). In addition, the biomass was separated into roots, stems, and tips, a distinction,

TABLE 31.1
Compositional Analysis of Salicornia spp. (g/100 gm)

	S.arabica	*S.bigelovii*	*S.bigelovii*	*S.herbacea*		
Component	**Plant Powder**	**Fresh Tips**	**Seed**	**Tips**	**Shoots**	**Root**
Moisture	7.39–8.45	87.06–89.78		90.9	73.9	66.2
Crude Protein	1.11–1.37	1.44–1.64	30–33	1.7	2.0	2.0
Crude Lipid	N/A	0.36–0.38	26–33	0.2	0.3	0.3
Crude Ash	17.3–20.02	3.99–4.73	5–7	4.7	6.1	6.2
Salt	N/A	N/A	N/A	3.3	3.9	2.8
Total Sugar	86.32–86.33	4.02–4.94	N/A	2.2	13.4	22.8
Uronic Acid	2.96–3.7	N/A	N/A	0.3	1.4	1.9
Sulfate	9.64	N/A	N/A	N/A		
Crude Fiber	N/A	0.7–0.96	5–7	N/A		
References	(Hammami et al. 2018)	(Lu et al. 2010)	(Glenn et al. 1991)	(Min et al. 2002)		

TABLE 31.2
The Influence of Different Plant Fractions, Salinity of Water Used for Irrigation, and Fertilizer Concentration on the Composition of Raw *S. bigelovii*

				Significance†		
Component [g/100 g TS]	**Shoots**	**Tips**	**Shoots + Tips**	**Fraction (Shoots/tips)**	**Salinity [ppt]**	**Fertilizer [g N/m²]**
Glucan	16.02–27.12	4.73–8.03	7.52–10.6	***	**	ns
Xylan	13.51–22.63	4.00–7.27	7.32–8.06	***	ns	ns
Arabinan	2.29–5.93	3.82–5.24	3.38–7.54	***	ns	ns
Klason Lignin	11.61–23.63	7.44–21.31	5.4–8.26	**	***	ns
Structural Ash	2.18–8.11	4.60–11.76	6.8	***	*	**
Total Extractives	25.82–44.04	54.04–66.97	50.13–57.23	***	***	ns
Water extractives ash-free			48.93–49.33	***	ns	ns
Ethanol extractives ash-free			2.0–2.22	***	***	ns

† *** = $P < 0.001$; ** = $P < 0.01$; * = $P < 0.05$; ns = not significant.
Source: Cybulska et al. 2014b

which has not been shown in Table 31.3, but can be found in the original study. The uronic acid and protein concentrations for *S. brachiata* are comparable to those in Table 31.1. The CW and HW fractions contained predominantly glucose, arabinose, and galactose, whereas the OX fractions of all the three parts were principally composed of arabinose, galactose, and rhamnose monosaccharides. Extraction of proteins was higher in the CW, HW, and OX extracts, and comparatively very low in the ALK extracts.

Hammani et al. (2018) measured the presence of various monosaccharides after the acid hydrolysis of *S. arabica*. Arabinose, galactose, ribose, xylose and glucose had the highest concentrations amongst all the monosaccharides detected in *S. arabica,* which is similar to the monosaccharides listed in Table 31.3 (Hammami et al. 2018).

The concentrations of minerals in *S. bigelovii* and *S. herbacea* are shown in Table 31.4. *Salicornia* has a high sequestration of salts and minerals, especially Na, K, Ca, and Mg. The deposition in shoot cells is caused by the fact that these minerals are transported in the transpiration stream; however, they are compartmentalized in vacuoles to avoid cytosolic toxicity, since the plant actively utilizes Na and Cl as osmolytes. These high levels are one of the reasons why *Salicornia* cannot be used as a staple food in the human diet in big proportions. However, the consumption of *Salicornia* in small amounts can provide a mineral supplementation, as well as due to phytochemicals – a supply of antioxidants (Cybulska et al. 2014a).

The total amino acids in *S. bigelovii* account for 10.8 g/kg of fresh weight (Lu et al. 2010). The various amino acids present in *S. bigelovii* and *S. herbacea* have been compared in Table 31.5. The amino acid concentrations vary depending upon the

TABLE 31.3
The Polysaccharide Fractions of *S. brachiata* Were Calculated Based on a w/w %

	Total Sugar	Rhamnose	Ribose	Glucose	Xylose	Arabinose	Mannose	Galactose	Uronic Acid	Protein
CW	45–56	2.9–6.14	1.19–2.26	6.98–39.66	1.94–3.93	24.58–34.29	5.11–10.05	17.69–51.99	0.49–1.34	5.50–10.0
HW	44–53	7.34–9.59	1.88–2.32	10.82–22.21	4.70–6.39	35.57–42.42	8.9–11.67	16.19–27.26	0.86–2.11	8.81–13.8
OX	39–54	10.0–28.0	1.37–2.05	9.07–10.13	1.96–5.11	35.94–39.82	5.67–8.91	17.02–27.34	1.31–1.85	6.13–7.94
ALK	52–58	0.7–12.87	2.93–3.85	0.20–5.23	4.62–86.16	9.77–53.58	0.5–3.44	2.65–17.48	0.57–1.24	1.75–4.56

Source: Sanandiya and Siddhanta 2014.
Abbreviations: CW = Coldwater extract; HW = hot water; OX = aqueous ammonium oxalate; ALK = aqueous sodium hydroxide.

TABLE 31.4
Analysis of Mineral Elements in *Salicornia* spp.

	S. bigelovii (mg/100g of Fresh Weight)	*S. herbacea* (mg/100g)		
Mineral Elements	**Whole Plant**	**Tip**	**Shoot**	**Root**
Ca	60–64	237.5	158.8	22.1
Cd	0.1			N/A
Cr	<0.1			N/A
Cu	7.7–10.5	3.1	1.1	2.1
Fe	1	31.5	66.2	84.8
K	168–184	650.1	740.1	741.1
Mg	112–124	46.5	54.0	52.5
Mn	N/A	7.2	3.9	3.0
Na	927–1069	1003.4	1218.1	1333.8
Ni	N/A	1.1	0.7	0.4
P	17–19	N/A		
Pb	0.1–0.3	N/A		
Zn	39.1–41.9	13.4	29.6	2.4
References	(Lu et al. 2010)	(Min et al. 2002)		

TABLE 31.5
Amino Acid Profile of *Salicornia* spp.

	S. bigelovii (mg/100g of Fresh Weight)	*S. herbacea* (mg/100g)		
Amino Acid	**Total Plant**	**Tip**	**Shoot**	**Roots**
Alanine	67–71	79.9	88.7	98.2
Arginine	66–70	77.0	36.1	57.0
Asparagine	114–118	137.1	140.2	165.5
Cysteine	3	-	-	11.1
Glutamic acid	160–166	144.8	160.5	182.3
Glycine	52–54	76.9	80.4	122.9
Histidine	25–27	34.0	79.3	54.4
Isoleucine	45–49	110.7	107.5	94.7
Leucine	93–95	115.5	98.1	128.4
Lysine	72–74	79.8	310.2	178.9
Methionine	9	23.2	52.2	23.3
Phenylalanine	54–56	73.2	63.3	67.7
Proline	73–93	88.8	18.4	86.8
Serine	66–70	67.5	72.7	94.2
Threonine	54–56	70.9	69.8	81.2
Taurine	-	7.6	21.4	37.7
Tyrosine	43–45	10.8	-	-
Valine	54–64	72.9	126.1	94.7
References	(Lu et al. 2010)	(Min et al. 2002)		

fraction of the plant being studied. Glycine is abundant in the roots but not in the stems and leaves. By contrast, proline is abundant in the leaves and roots, compared to the stems. Arginine, glycine, histidine, lysine, proline, and valine are other amino acids which have concentrations that vary with the part of the plant (leaf, stem, root) being sampled.

The fatty acid and oil contents of *Salicornia* seeds are shown in Table 31.6. Myristic, palmitic, stearic, and arachidonic acids are the main saturated fatty acids present in *Salicornia* biomass, but are a minor fraction of the total lipids present. Oleic acid is the only reported monounsaturated fatty acid. Linoleic (found in the highest concentration) and linolenic acid are the main polyunsaturated fatty acids. The presence of higher concentration of polyunsaturated fatty acids makes *Salicornia* a prospective biomass for advanced fuel production and an alternative source for essential fatty acids in the human diet.

The green succulent part of *Salicornia* is easy to process through simple juicing or cold pressing. Since this fraction of the biomass is high in moisture content, it is suitable for juicing. Twenty-five percent of the dry matter of the green fraction of *Salicornia* can be extracted as juice components, and 75% of the dry matter ends up in the fiber rich pulp (Alassali et al. 2017). The protein content of *S. europaea* is 2.3 mg/g of fresh weight (FW), which in other green herbaceous plants such as spinach and celery leaf are 2.6 mg/g of FW (Wang et al. 2007; Lu et al. 2010). The shoots of *S. herbacea* have been reported to contain less protein (1.9 mg/g FW) in comparison to the roots (2.2 mg/g FW). Based on the nitrogen content, the protein in the green fraction of *S. sinus-persica* was estimated to be 13% of the pulped biomass after juicing, on a dry matter basis (Islam and Adams 2000; Alassali et al. 2017). Of the dry matter content of the juice, 7.6% was comprised of proteins and 4.6% of lipids, while the remaining constituents were ash and sugars.

S. bigelovii has been reported to contain 569, 159, 58 mg/kg of fresh weight of total chlorophyll, β-carotene, and ascorbic acid, respectively (Lu et al. 2010). The rich presence of β-carotene and ascorbic acid, which are sources of vitamin A and C, respectively, emphasize the nutritional value of *Salicornia* and make a case for its regulated consumption in the human diet.

31.4 COMMERCIAL APPLICATIONS

Salicornia has value as a nutrient source for humans, feed for animals and fish, feedstock for biofuel production, in the pharmaceutical industry, in phytoremediation, and as biofilters for aquaculture. Amongst its various applications, we limit ourselves to discussing the most well-documented areas.

31.4.1 Food Products

The tender green tips of *Salicornia* have been used as food ingredients for salads, as garnishes, or cooked in the absence of salt in a similar manner to spinach. An Apulian traditional dish consists of boiled *Salicornia* cooked with extra virgin olive oil and garlic and often accompanied with fish or seafood (Loconsole et al. 2019). *Salicornia* contains various nutrients such as proteins, vitamins (e.g. C, B1) minerals,

TABLE 31.6
Fatty Acid Composition in Seeds of *Salicornia* spp.

Fatty Acid*	Common Name	*S. bigelovii* Oil (% of Lipids)	*S. bigelovii* Oil (% of Lipids)	*S. bigelovii* Seeds (% of Lipids)	*S. europaea* Seeds (% of Lipids)	*S. brachiata* Shoots (% of Lipids)	*S. ramosissima* Shoots (% of Lipids)
14:0	Myristic acid	-	0.18	-	-	-	0.29
16:0	Palmitic acid	7.7–8.7	8.50	7.0–8.50	6.0–7.8	23.7–27.9	21.59–22.69
18:0	Stearic acid	1.6–2.4	-	1.24–1.69	0.7–1.1	6.58–7.82	4.16–6.83
18:1 n-9	Oleic acid	12.0–13.3	19.99	12.33–16.83	21.0–22.6	3.04–9.2	2.21–5.73
18:2 n-6	Linoleic acid	73.0–75.2	63.40	74.66–79.49	69.8–72.4	25.36–26.04	19.04–21
18:3 n-3	Linolenic acid	2.4–2.7	1.34	1.5–2.3	-	28.18–29.94	38.97–40.23
20:00	Arachidonic acid	-	6.59	-	-	0.78–1.03	0.66–0.72
Total lipids (g/100g)		26–33	-	27.7–32.0	27–29	17.82	-
References		(Glenn et al. 1991)	(Attia et al. 1997)	(Anwar et al. 2002)	(Austenfeld 1986)	(Mishra et al. 2015; Patel et al. 2019)	(Maciel et al. 2018)

* Fatty acids are denoted by the number of carbons in the molecule, then the number of double bonds, followed by the position of the first double bond in relation to the methyl end.

polysaccharides, and bioactive compounds (Lu et al. 2010). This wide range of nutrients has propelled the increased use of *Salicornia* in human diets and puts this plant into a group of super foods alongside the likes of kale and quinoa (Rowney 2013). A quick search online can show hundreds of webpages recommending Salicornia-based recipes.

In one study, *S. herbacea* extract was mixed with milk and tested as yoghurt for human consumption in three varying concentrations, namely 0.25, 0.5, and 1% (w/v). 0.25% was considered the best suited for color, flavor, viscosity, sweetness, sourness, and overall palatability (Cho et al. 2008). The use of *S. herbacea* is reported to nutritionally improve the quality of traditional Korean soy sauce *meju* and *kanjang* (Kim et al. 2011). The Fe, K, and Mg contents of *makgeolli* (Korean wine) have also been reported to be higher in *Salicornia*-based wine than traditional wheat- and rice-based wine, thus suggesting the presence of *Salicornia* in the *nuruk* culture had enhanced yeast growth (Jeon et al. 2010). The bioactive profile coupled with a high protein content and the presence of chlorophyll, β-carotene, and ascorbic acid also allows for *Salicornia* to serve as a supplement for current types of fish feed (Lu et al. 2010). *Salicornia* is already being used by companies such as Phyto Corporation, Atecmar Coop, Koppert Cress, Radiant Inc, Chanel, and MAC in products like nutritional supplements, tea, chips, salt substitutes, toothpastes, skin care products, dairy products, animal feed, weight loss supplements, and superfood ingredients (Sung et al. 2009; Feng et al. 2013; Shin and Lee 2013; Karan et al. 2018). The high content of polyunsaturated fatty acids such as linoleic and linolenic acid (which cannot be produced in the human body) in *Salicornia* seeds make them a suitable source for essential fatty acids in human diet.

31.4.2 Feed Products

There are some indications that *Salicornia* spp. can be used in livestock production. When *Salicornia bigelovii* was used as a substitute for Rhodes grass (*Chloris gayana)* as a forage for Damascus kid goats, there was a two-fold increase in consumption, thus offering a low cost and readily available substitute for goat feed (Glenn et al. 1992). Both washed (to reduce salt concentrations) and unwashed *S. bigelovii* was fed to goats. High salt concentrations in the forage did not inhibit the forage consumption patterns of the animals, however the water intake of the goats did slightly increase (Glenn et al. 1992). The increase in consumption of water by animals is an important factor when considering *Salicornia* as a feed supplement for livestock in arid regions, where fresh water is a scarce resource. *S. bigelovii* seed cake has also been mixed in broiler diets as an alternative protein feed (Attia et al. 1997).

31.4.3 Phytochemicals

Phytochemicals are naturally occurring chemical compounds found in plants. These chemicals are classified on the basis of protective function, physical characteristics, and chemical characteristics (Meagher and Thomson 1999). Phytochemicals are not essential nutrients for human health; however, their role in boosting the human immune system to fight common diseases has been well documented (Taofiq et al. 2017). Halophytes have been considered useful in medicinal application due to the presence of a wide variety of secondary metabolites such as alkaloids, flavonoids, tannins, terpenoids, saponins, and coumarins (Bandaranayake 2002). Prior to describing the presence of phytochemicals in *Salicornia* spp. and their respective medicinal effects, it is worth mentioning how these various phytochemicals are categorized. Phytochemicals can be broadly categorized as primary and secondary metabolites. Primary metabolites include sugars, amino acids, and proteins. Secondary metabolites include alkaloids, phenols, terpenes, steroids, and saponins. Phenolics, which are a large portion of the secondary metabolites, can be further categorized into smaller groups such as flavonoids, tannins, and phenolic acids, based on the structure of the chemical compounds. Flavonoids are the largest and most studied group of plant phenols (Saxena et al. 2013). Phenolic acids are a diverse group that can be further subdivided into hydroxycinnamic acids and hydroxybenzoic acids (Taofiq et al. 2017). Caffeic, ferulic, p-coumaric, rosmarinic, chlorogenic, cinnamic, and sinapic acids are some of the most common hydroxycinnamic acids, which are of importance to *Salicornia* spp. Phenolic polymers, also known as tannins, can be further categorized into hydrolysable and condensed tannins.

S. europaea was reported to contain phenolic compounds, alkaloids, flavonoids, and saponins (Lellau and Liebezeit 2001). Flavonoids have been reported to be used in treating hypertension, scurvy, and cephalalgia, and possess anti-inflammatory abilities (Min et al. 2002; Lellau and Liebezeit 2003). Until now, nine flavonoids, four chromone compounds, four triterpenoid saponins, and one new triterpenoid saponin have been identified in *S. europaea* (Arakawa et al. 1982; Yin et al. 2012). The four triterpenoid saponins are oleanolic acid glucoside, chikusetsusaponin methyl ester, calenduloside E, and calenduloside E 6'-methyl ester, and the latest discovered, dihydroxyoleanenoic acid glucopyranosyl ester (Cybulska et al. 2014a). *S. europaea* ethanol extracts have been reported to contain high concentrations of quinic acid, rosmarinic, and p-coumaric acids and lower concentrations of hesperidin, rutin, malic, and rhamnetin acid (Zengin et al. 2018). While the presence of alkaloids in *Salicornia spp.* has been contradictory at times, it has been mentioned that the chemical diversity of each species is controlled by the environment and harvesting time, both of which influenced the results of the specific secondary metabolites analyses (Lellau and Liebezeit 2001). In addition, the location, date and even time of the day can influence the presence of secondary metabolites, thus making it difficult to accurately predict the exact concentrations of secondary metabolites in a particular species.

S. herbacea is commonly known as tungtungmadi in Korea and has been used in traditional medicine and as seasonal vegetables. *S. herbacea* has been used as folk medicine for treating diarrhea, nephropathy, and constipation (Rhee et al. 2009). Contemporary pharmacological studies have verified the antioxidative, anti-inflammatory, and immunomodularity capabilities of this halophyte. The manner in which *S. herbacea* extracts suppress inflammation suggests that they can be used in treating cancer, autoimmune diseases (e.g. rheumatoid arthritis), vascular diseases (e.g. atherosclerosis), and metabolic diseases (e.g. diabetes) (Rhee et al. 2009).

The antioxidative capacity of *S. herbacea* was tested in an ethyl acetate soluble fraction based on the scavenging activity of the 1,1-diphenyl-2-picrylhydrazyl free radical (Young et al. 2005; Wang et al. 2017). A chlorogenic acid is an ester of caffeic acid and quinic acid, both of which have individually been reported to possess antioxidative properties (Medina et al. 2007; Zengin et al. 2018). Tungtungmadic acid is a derivative of chlorogenic acid, which is chemically classified as 3-caffeoyl-4-dihydrocaffeoyl quinic acid (Young et al. 2005). Other bioactive compounds that have been reported to be extracted from *S. herbacea* include β-sitosterol, isorhamnetin-3-O-β-D-glucopyranoside, stigmasterol, uracil, quercetin 3-O-β-D-glucopyranoside, isoquercitrin 6''-O-methyloxalate, methyl 4-caffeoyl-3-dihydrocaffeoyl (salicornate), 3,5-dicaffeoylquinic acid, methyl 3,5-dicaffeoyl quinate, and 3,4-dicaffeoylquinic acid (Lee et al. 2004; Kim et al. 2011; Cho et al. 2016). Isorhamnetin-3-O-β-D-glucopyranoside has been identified to be a leading compound in treating diabetes and/or prevention of diabetes and its related complications (Lee et al. 2005). Betaine obtained in methanol extracts of *S. herbacea* has been claimed to lower the level of homocysteine in blood, and thereby providing protection against cardiovascular ailments (Rhee et al. 2009). *S. herbacea* extracts have been examined for their immunomodulatory abilities on monocyte/macrophage lineage cells (Im et al. 2006). Macrophages are unique cells in immune systems capable of a dual role, initiating immune responses and serving as effector cells.

S. brachiata is considered a traditional medicine for treating hepatitis and has been tested for its antiviral activity (Bandaranayake 2002). The presence of bioactive compounds, minerals, amino acids, polyphenols, proteins, reducing sugars, and pigments known for antioxidative properties such as betacyanin and betaxanthin has been reported (Escribano et al. 1998; Parida et al. 2018).

31.4.4 Biofilters for Aquaculture

Unfiltered effluents from aquaculture contain large amounts of non-utilized nutrients and organic substances that can cause hypertrophication (also known as eutrophication) and toxification of adjacent ecosystems. The uncertainty of control parameters, such as pH, temperature, and dissolved oxygen level in open ponds, adds further to the complexity of estimating a reliable recovery of fish from open pond systems. The cost of procuring clean and fresh water while maintaining habitat suitable for discharge of effluents has led to an increased interest in Recirculatory Aquaculture Systems (RAS) (Martins et al. 2010; Dalsgaard et al. 2013). As RAS seems to provide viable alternative to current fish culture practices, however, this too like any new technology has its shortcomings. The high upfront capital cost of RAS and operating cost for maintaining the availability of clean water, round the clock electricity, and availability of nutrients are some of the cost factors that require optimization. Additionally, the cost associated with adding denitrification filtration systems that convert ammonia excreted by fish in the effluents to nitrogen, have high operating costs. A wide range of plants have been studied to conceptually design combined aquaculture and hydroponic systems (Watten and Busch 1984; Turcios and Papenbrock 2014). In this chapter, we restrict ourselves to focus on saline aquaculture systems and how salt-tolerant plants can be combined to design a hybrid aquaponic system. Halophytes can serve as biofilters in cleaning the effluents from aquaculture systems (Glenn, et al. 1999; Buhmann and Papenbrock 2013a; Buhmann et al. 2015). The use of plants as filters will have two impacts on an aquaculture system. First, it will reduce the stress on the filtration systems resulting in cost saving. Second, if the halophytes can be utilized commercially as food, feed, or biomass to derive biochemical products, they can provide an alternative revenue stream for aquaculture systems. This secondary revenue stream can offset the dependence of aquaculture industries on fish markets and provide a new avenue to diversify income portfolios (Buhmann and Papenbrock 2013b). Fish grown for commercial purposes in controlled environments such as RAS, take a large fraction of the nitrogen and phosphorus supplied in the feed. Using halophytes as biofilters can serve as the link between aquaculture systems and hydroponic cultivation.

Salicornia spp. have been studied to grow in constructed wetlands with the aim of reducing the treatment of waste-streams emanating from aquacultures. *Salicornia* has been confirmed to take up 85% of the nitrogen and 73% of dissolved inorganic phosphorous from wastewater into plant tissues (Webb et al. 2012). While the majority of nitrogen removal in wetlands happens due to microbial processes, the increased uptake by *Salicornia* can be accredited to its resilient adaptability to changing environmental conditions. The total uptake of nitrogen and phosphorous by *Salicornia* is significantly impacted by the surface versus subsurface flow and level of nutrients present in the flow streams (Brown et al. 1999; Shpigel et al. 2013). The growth and visual quality of *Salicornia* grown in a RAS-hydroponic systems has been reported to be excellent with the halophytes retaining 9% of the N and 10% of the P introduced from fish feed (Waller et al. 2015). While using *Salicornia* as a biofilter for aquaculture has economic value, the hydroponic systems coupled with aquaculture can help close the nutrient cycle, bringing us a step closer to circular production systems.

31.4.5 Fuels and Energy

Salicornia in early stages of growth is succulent and ideal for food consumption. As the plant matures, it loses moisture content, but remains high in protein and minerals. Once the plant dries out, the seeds can be separated from the straw fraction and utilized as a source for fuel, while the straw can serve as a source for extracting phytochemicals and carbohydrates. Thus, the same plant can be utilized for various purposes such as feed, fuels, type of fuel, and bioenergy.

The Sustainable Biofuels Research Consortium established in 2011, in Abu Dhabi, funded The Seawater Energy and Agriculture System (SEAS) project to investigate the possibility of integrating aquaculture, halophyte agriculture, and mangrove silviculture to produce sustainable biofuels. The SEAS project site was envisioned to consist of 10% aquaculture ponds, 70% *Salicornia* fields, and 20% mangrove wetlands (Warshay et al. 2017). *S. bigelovii* had been chosen for this project due to its high oil content in seeds (26–31%), of which 73–75% is linoleic acid (Glenn et al. 1991). *S. bigelovii* was studied to be hydro-processed into jet fuel or diesel or transesterified to produce fatty acid methyl esters (Warshay et al. 2011). The leftover fraction post oil seed processing (seed cake) is rich in proteins and could be used as animal fodder. The left over straw fraction of the biomass (which is not oilseeds) could be used as a feedstock via the Fischer Tropsch process for the production of jet fuel or diesel fuel through the cellulose to ethanol pathway or through gasification to produce syngas (Warshay et al. 2011). It was estimated that using a combination of one of these approaches, coupled with the utilization of the straw fraction would lead to 63–80 g of CO_2–equivalent reduction/passenger–km of greenhouse gas emission (68% reduction in GHG emissions) by substituting conventional fuels with biofuels from *S. bigelovii* (Warshay et al. 2011). However, the high concentration of salt in *Salicornia* posed a risk of corrosion of equipment and this also inhibited enzymatic hydrolysis and fermentation. The removal of salt needed to occur by washing the biomass with fresh water, which is a valuable resource in arid parts of the world. Thus, the removal of salt before processing the biomass for sugar recovery became a crucial step in determining the feasibility of biofuel production from *Salicornia*.

S. bigelovii biomass contains 5–16 g/100g of lignin and 16–55 g/100gm of carbohydrate in the total solids. Enzymatic hydrolysis at the optimized pretreatment temperature of 210°C resulted in 91% glucose recovery (Cybulska et al. 2014c). This corresponds to 100–111 kg ethanol/dry ton of *S. bigelovii*, while in comparison corn stover has an ethanol potential of 230 kg/dry ton (Kadam and McMillan 2003; Brown et al. 2014). However, it must be pointed out that *Salicornia* has a higher biomass yield per hectare (20 tons/ha) than corn stover (9.4 ton/ha) (Brown et al. 2018). In addition, *Salicornia* can be used for its oilseeds and the straw as protein rich feed, thus providing more product choices while utilizing less resources (arable land and fresh water for irrigation). Post pretreatment, *Saccharomyces cerevisiae* used in the simultaneous saccharification and fermentation of *S. bigelovii* has resulted in up to 98% ethanol yield (Bañuelos et al. 2018). While the cellulose sugars in the straw fraction can be utilized for ethanol production, the hemicellulose sugars can be used for biogas production (150 L methane/kg VS biomass) and the oilseed can be used for biodiesel or Bio-Synthetic Paraffin Kerosene production (Ashraf et al. 2016). The importance of moving from first generation (food-based biomass) to second generation (non-food based) biofuels has been well documented and discussed by Carriquiry et al. (2011). Species like *Salicornia* provide a new dimension in this regard, using non-arable land to grow crops that can be utilized for food and fuel (Marriott and Pourazadi 2017).

31.5 CONCLUSIONS AND FUTURE PERSPECTIVES

Salicornia spp. are widely distributed across the globe and are tolerant to saline water. With the discovery that these halophytes can be used as food, fuel, and in bio-products, more researchers undertook the task of classifying this genus and analyzing its constituents. The *Salicornia* genus includes up to 30 species, however, inbreeding, a high degree of physiological plasticity and few diagnostic characters, has led to an extremely challenging taxonomy, which has only begun to be better understood in the past two decades. *S. sinus-persica* was earlier misunderstood to be *S. europaea* until a taxonomic revision was published (Akhani 2008); no further studies of this species are available under this name. While *S. arabica, S. europaea,*

S. bigelovii, S. brachiata, S. ramosissima, and *S. herbacea* have all been studied to varying extents (for their proximate compositional analysis, polysaccharide fractions, carbohydrate fractions, mineral elements, amino acid profile, and fatty acid composition) the most comprehensive information is available for *S. bigelovii, S. europaea*, and *S. herbacea*.

Salicornia spp. are halophytes that show promise for the production of biomass for a range of applications, including:

1. Selective nutrients and proteins for food to be consumed by humans
2. Value added chemicals that can be incorporated into animal feeds to lower the cost of fodder and enhance its quality
3. Phytochemicals which have been used traditionally to treat diseases and can now be selectively extracted from plants to treat patients suffering from chronic illness
4. Nutrient uptake from soil, especially when used in combination with aquaculture systems, which tend to release large quantities of underutilized nutrients in their effluent streams
5. Biofuels derived from the oil-rich seeds

Early research with *Salicornia* spp. was riddled with difficulties in identifying the species and in taxonomical challenges due to phenotypic plasticity and morphological parallelism; some of these still persist to this day, continuing to cause difficulties in identifying species. However, with persistent global interest in the species and curiosity in exploiting its oil-rich seeds and phytochemical potential, considerable progress has been achieved. To harness the benefits of this crop, we need to build upon the existing knowledge base by formulating research studies that help bridge information gaps.

On the cultivation and physiology front, the optimization of growth conditions (such as effects of salinity, soil nutrients, and weather conditions) for *Salicornia* using waste effluents streams from aquacultural effluents need to be demonstrated on a pilot scale. In addition, a growth manual needs to be developed which includes details on sowing time and depth, fertilizer timing and requirement, and harvest time, dependent on which fraction of the biomass is being sought. Taxonomic studies of species need to be undertaken on a global scale to verify the presence of distinct species and get an overview of the genetic variation of the species.

Insights on the complex senescence processes of different *Salicornia* species will spur innovation in designing robust processing steps for extracting plant derivatives. These processes should be able to utilize the varying proteins, secondary metabolites and oil, at different stages of growth to yield the most valuable product for that growth stage. In the past, plants have been utilized commercially for food, fuels, and primary metabolites. Rarely has one plant been able to provide all three products, while also without the need for fresh water for irrigation and arable land for cultivation. In these aspects, *Salicornia* is a truly novel biomass. In an era, that recognizes the importance of reducing and recycling waste, optimizing resource utilization, and developing alternative uses of energy, the production of *Salicornia* could be a strong contender in building a sustainable model for a circular economy.

ACKNOWLEDGMENT

TC, AC, IC have contributed in writing and editing all sections of the manuscripts. MHT has reviewed and helped in conceptualizing this chapter and defining its scope.

REFERENCES

Abdal, M S. 2009. "Salicornia Production in Kuwait." *World Applied Sciences Journal* 6 (8): 1033–1038.

Akhani, H. 2008. "Taxonomic Revision of the Genus Salicornia L. (Chenopodiaceae) in Central and Southern Iran." *Pakistan Journal of Botany* 40 (4 SPEC. ISS.): 1635–1655.

Alassali, A., I. Cybulska, A.R. Galvan, and M.H. Thomsen. 2017. "Wet Fractionation of the Succulent Halophyte Salicornia Sinus-Persica, with the Aim of Low Input (Water Saving) Biorefining into Bioethanol." *Applied Microbiology and Biotechnology* 101 (4): 1769–1779. doi:10.1007/s00253-016-8049-8.

Anwar, F., M.I. Bhanger, M. Khalll A. Nasir, and S. Ismail. 2002. "Analytical Characterization of Salicornia Bigelovii Seed Oil Cultivated in Pakistan." *Journal of Agricultural and Food Chemistry* 50 (15): 4210–4214. doi:10.1021/jf0114132.

Arakawa, Y., Y.Z. Asada, and H. Ishida. 1982. "Structures of New Two Isoflavones and One Flavanone from Glasswort (Salicornia Europaea L.)." *Journal of the Faculty of Agriculture, Hokkaido University* 61 (1): 1–12.

Ashraf, M.T., C. Fang, T. Bochenski, I. Cybulska, and A. Alassali. 2016. "Estimation of Bioenergy Potential for Local Biomass in the United Arab Emirates." *Emirates Journal of Food and Agriculture* 28 (2): 99–106. doi:10.9755/ejfa.2015-04-060.

Attia, F. M., A. A. Alsobayel, M. S. Kriadees, M. Y. Al-Saiady, and M. S. Bayoumi. 1997. "Nutrient Composition and Feeding Value of Salicornia Bigelovii Torr Meal in Broiler Diets." *Animal Feed Science and Technology* 65 (1–4): 257–263. doi:10.1016/S0377-8401(96)01074-7.

Austenfeld, F.-A. 1986. "Nutrient Reserves of Salicornia Europaea Seeds." *Physiologia Plantarum* 68 (3): 446–450.

Bailis, R. and E. Yu. 2012. "Environmental and Social Implications of Integrated Seawater Agriculture Systems ProducingSalicornia Bigeloviifor Biofuel." *Biofuels* 3 (5): 555–574. doi:10.4155/bfs.12.50.

Bandaranayake, W. M. 2002. "Bioactivities, Bioactive Compounds and Chemical Constituents of Mangrove Plants." *Wetlands Ecology and Management* 10 (6): 421–452. doi:10.1023/A:1021397624349.

Bañuelos, J.A., I. Velázquez-Hernández, M. Guerra-Balcázar, and N. Arjona. 2018. "Production, Characterization and Evaluation of the Energetic Capability of Bioethanol from Salicornia Bigelovii as a Renewable Energy Source." *Renewable Energy* 123: 125–134. doi:10.1016/j.renene.2018.02.031.

Barrett-Lennard, E.G. 2002. "Restoration of Saline Land through Revegetation." *Agriculture and Water Management* 53 (2002): 213–226.

Bennett, T.H., T.J. Flowers, and L. Bromham. 2013. "Repeated Evolution of Salt-Tolerance in Grasses." *Biology Letters* 9 (2): 20130029

Brown, J.J., I. Cybulska, T. Chaturvedi, and M.H. Thomsen. 2014. "Halophytes for the Production of Liquid Biofuels." In *Sabkha Ecosystems: Volume IV: Cash Crop Halophyte and Biodiversity Conservation*, IV: 67–72. doi:10.1007/978-94-007-7411-7.

Brown, J.J., P. Das, and M. Al-Saidi. 2018. "Sustainable Agriculture in the Arabian/Persian Gulf Region Utilizing Marginal Water Resources: Making the Best of a Bad Situation." *Sustainability* 10 (5): 1–16. doi:10.3390/su10051364.

Brown, J.J., E.P. Glenn, K.M. Fitzsimmons, and S.E. Smith. 1999. "Halophytes for the Treatment of Saline Aquaculture Effluent." *Aquaculture* 175 (3–4): 255–268. doi:10.1016/S0044-8486(99)00084-8.

Buhmann, A.K., U. Waller, B. Wecker, and J. Papenbrock. 2015. "Optimization of Culturing Conditions and Selection of Species for the Use of Halophytes as Biofilter for Nutrient-Rich Saline Water." *Agricultural Water Management* 149: 102–114. doi:10.1016/j.agwat.2014.11.001.

Buhmann, A. and J. Papenbrock. 2013a. "Biofiltering of Aquaculture Effluents by Halophytic Plants: Basic Principles, Current Uses and Future Perspectives." *Environmental and Experimental Botany* 92: 122–133. doi:10.1016/j.envexpbot.2012.07.005.

Buhmann, A. and J. Papenbrock. 2013b. "An Economic Point of View of Secondary Compounds in Halophytes." *Functional Plant Biology* 40 (9): 952–967. doi:10.1071/FP12342.

Carriquiry, M.A., X. Du, and G.R. Timilsina. 2011. "Second Generation Biofuels: Economics and Policies." *Energy Policy* 39 (7): 4222–4234. doi:10.1016/J.ENPOL.2011.04.036.

Cho, J.Y., J.Y. Kim, Y.G. Lee, H.J. Lee, H.J. Shim, J.H. Lee, S.J. Kim, K.S. Ham, and J.H. Moon. 2016. "Four New Dicaffeoylquinic Acid Derivatives from Glasswort (Salicornia Herbacea L.) and Their Antioxidative Activity." *Molecules* 21 (8):1097. doi:10.3390/molecules21081097.

Cho, Y.S., S.I. Kim, and Y.S. Han. 2008. "Effect of Slander Glasswort Extract Yogurt on Quality during Storage." *Korean Journal of Food Science and Technology* 24 (2): 212–221.

Cybulska, I., G. Brudecki, A. Alassali, M. Thomsen, and J. Brown. 2014. "Phytochemical Composition of Some Common Coastal Halophytes of the United Arab Emirates." *Emirates Journal of Food and Agriculture* 26 (12): 1046. doi:10.9755/ejfa.v26i12.19104.

Cybulska, I., T. Chaturvedi, A. Alassali, G.P. Brudecki, J.J. Brown, S. Sgouridis, and M.H. Thomsen. 2014. "Characterization of the Chemical Composition of the Halophyte Salicornia Bigelovii under Cultivation." *Energy & Fuels* 28 (6): 3873–3883. doi:10.1021/ef500478b.

Cybulska, I., T. Chaturvedi, G.P. Brudecki, Z. Kádár, A.S. Meyer, R.M. Baldwin, and M.H. Thomsen. 2014. "Bioresource Technology Chemical Characterization and Hydrothermal Pretreatment of Salicornia Bigelovii Straw for Enhanced Enzymatic Hydrolysis and Bioethanol Potential." *Bioresource Technology* 153: 165–172. doi:10.1016/j.biortech.2013.11.071.

Dalsgaard, J., I. Lund, R. Thorarinsdottir, A. Drengstig, K. Arvonen, and P.B. Pedersen. 2013. "Farming Different Species in RAS in Nordic Countries: Current Status and Future Perspectives." *Aquacultural Engineering* 53: 2–13. doi:10.1016/j.aquaeng.2012.11.008.

Davy, A.J., G.F. Bishop, and C.S.B.B Costa. 2001. "Salicornia L. (Salicornia Pusilla J. Woods, S. Ramosissima J. Woods, S. Europaea L., S. Obscura P.W. Ball & Tutin, S. Nitens P.W. Ball & Tutin, S. Fragilis P.W. Ball & Tutin and S. Dolichostachya Moss)." *Journal of Ecology* 89 (4): 681–707. doi:10.1046/j.0022-0477.2001.00607.x.

Escribano, J., M.A. Pedreño, F. García-Carmona, and R. Muñoz. 1998. "Characterization of the Antiradical Activity of Betalains from Beta Vulgaris L. Roots." *Phytochemical Analysis* 9 (3): 124–127. doi:10.1002/(SICI)1099-1565(199805/06)9:3<124::AID-PCA401>3.0.CO;2-0.

Feng, L., B. Ji, and B. Su. 2013. "Economic Value and Exploiting Approaches of Sea Asparagus, a Seawater-Irrigated Vegetable." *Agricultural Sciences* 04 (09): 40–44. doi:10.4236/as.2013.49b007.

Flowers, T.J. and T.D. Colmer. 2008. "Salinity Tolerance in Halophytes." *New Phytology* 179 (4): 945–963. doi:10.1111/j.1469-8137.2008.02531.x.

Flowers, T.J., H.K. Galal, and L. Bromham. 2010. "Evolution of Halophytes, Multiple Origins of Salt Tolerance in Land Plants." *Functional Plant Biology* 37: 604–612.

Flowers, T.J., M.A. Hajibagheri, and N.J.W. Clipson. 1986. "Halophytes." *The Quarterly Review of Biology* 61: 313–337.

Flowers, T.J. and A. Muscolo. 2015. "Introduction to the Special Issue: Halophytes in a Changing World." *AoB PLANTS* 7. doi:10.1093/aobpla/plv020.

Glenn, E.P., J.J. Brown, and J.W. O′leary. 1998. "Irrigating Crops with Seawater." *Scientific American Inc.* 279 (2): 76–81.

Glenn, E.P., J.J. Brown, and E. Blumwald. 1999. Salt Tolerance and Crop Potential of Halophytes. *Critical Reviews in Plant Sciences* 18: 227–255. doi:10.1080/07352689991309207.

Glenn, E.P., W.E. Coates, J.J. Riley, R.O. Kuehl, and R.S. Swingle. 1992. "Salicornia Bigelovii Torr.: A Seawater-Irrigated Forage for Goats." *Animal Feed Science and Technology* 40 (1): 21–30. doi:10.1016/0377-8401(92)90109-J.

Glenn, E.P., T. Anday, R. Chaturvedi, R. Martinez-Garcia, S. Pearlstein, D. Soliz, S.G. Nelson, and R.S. Felger. 2013. "Three Halophytes for Saline-Water Agriculture: An Oilseed, a Forage and a Grain Crop." *Environmental and Experimental Botany* 92: 110–121. doi:10.1016/j.envexpbot.2012.05.002.

Glenn, E.P., J.W. O′leary, M.C. Watson, L.T. Thomson, R.O. Kuehl, J.W.O. Leary, M.C. Watson, T.L. Thompson, and R.O. Kuehl. 1991. "Salicornia Bigelovii Torr an Oilseed Halophyte for Seawater Irrigation." *Science* 251 (4997): 1065–1067.

Gunning, D. 2016. "*Cultivating Salicornia Europaea (Marsh Samphire).*" Dublin: Irish Sea Fisheries Board.

Hammami, N., A.B. Gara, K. Bargougui, H. Ayedi, F.B. Abdalleh, and K. Belghith. 2018. "Improved in Vitro Antioxidant and Antimicrobial Capacities of Polysaccharides Isolated from Salicornia Arabica." *International Journal of Biological Macromolecules* 120: 2123–2130. doi:10.1016/j.ijbiomac.2018.09.052.

Hodges, C.N., T.L. Thompson, J.J. Riley, and E.P. Glenn. 1993. "Reversing the Flow: Water and Nutrients from the Sea to the Land." *Ambio* 22: 483–490.

Hossain, M A. 2010. "Global Warming Induced Sea Level Rise on Soil, Land and Crop Production Loss in Bangladesh." In *19th World Congress of Soil Science, Soil Solutions for a Changing World, Brisbane. Http://Www. Ars. Usda. Gov/Services/Docs. Htm.*

Im, S.A., K. Kim, and C.K. Lee. 2006. "Immunomodulatory Activity of Polysaccharides Isolated from Salicornia Herbacea." *International Immunopharmacology* 6 (9): 1451–1458. doi:10.1016/j.intimp.2006.04.011.

IPCC. 2018. "Summary for Policymakers." In: Global Warming of 1.5°C. An IPCC Special Report on the Impacts of Global Warming of 1.5°C above Pre Industrial Levels and Related Global Greenhouse Gas Emission Pathways, in the Context of Strengthening the Global Response to. World Meteorological Organization.

Islam, M., and M.A. Adams. 2000. "Nutrient Distribution among Metabolic Fractions in 2 Atriplex Spp." *Journal of Range Management* 53 (1): 79–85. doi:10.2307/4003396.

Ismail, A.M, and T. Horie. 2017. "Genomics, Physiology, and Molecular Breeding Approaches for Improving Salt Tolerance." *Annual Review of Plant Biology* 68: 405–434. doi:10.1146/annurev-arplant-042916-040936.

Jeon, B.Y., H.N. Seo, A. Yun, I.H. Lee, and D.H. Park. 2010. "Effect of Glasswort (Salicornia Herbacea L.) on Nuruk-Making Process and Makgeolli Quality." *Food Science and Biotechnology* 19 (4): 999–1004. doi:10.1007/s10068-010-0140-9.

Kadam, K.L., and J.D. McMillan. 2003. "Availability of Corn Stover as a Sustainable Feedstock for Bioethanol Production." *Bioresource Technology* 88 (1): 17–25. doi:10.1016/S0960-8524(02)00269-9.

Kadereit, G., M. Piirainen, J. Lambinon, and A. Vanderpoorten. 2012. "Cryptic Taxa Should Have Names: Reflections in the Glasswort Genus Salicornia (Amaranthaceae)." *Taxon* 61 (6): 1227–1239.

Kadereit, G., P. Ball, S. Beer, L. Mucina, D. Sokoloff, P. Teege, A.E. Yaprak, and H. Freitag. 2007. "A Taxonomic Nightmare Comes True: Phylogeny and Biogeography of Glassworts (Salicornia L., Chenopodiaceae)." *Taxon* 56 (4): 1143–1170. doi:10.2307/25065909.

Karan, S., C. Turan, and M.K. Sangun. 2018. "Use of Glasswort (Saliconia Europaea) Plant as Raw Material in Cosmetics." In 2nd International Cosmetic Congress. Antalya, Turkey.

Katschnig, D., R. Broekman, and J. Rozema. 2013. "Salt Tolerance in the Halophyte Salicornia Dolichostachya Moss: Growth, Morphology and Physiology." *Environmental and Experimental Botany* 92: 32–42. doi:10.1016/j.envexpbot.2012.04.002.

Kim, J.Y., J.Y. Cho, Y.K. Ma, K.Y. Park, S.H. Lee, K.S. Ham, H.J. Lee, K.H. Park, and J.H. Moon. 2011. "Dicaffeoylquinic Acid Derivatives and Flavonoid Glucosides from Glasswort (Salicornia Herbacea L.) and Their Antioxidative Activity." *Food Chemistry* 125 (1): 55–62. doi:10.1016/j.foodchem.2010.08.035.

Kim, J.K., B.Y. Jeon, and D.H. Park. 2011. "Development of Kanjang (Traditional Korean Soy Sauce) Supplemented with Glasswort (Salicornia Herbacea L.)." *Journal of Food Science and Nutrition* 16 (2): 165–173. doi:10.3746/jfn.2011.16.2.165.

Lee, Y.S., S.L. Hye, H.S. Kuk, B.K. Kim, and S. Lee. 2004. "Constituents of the Halophyte Salicornia Herbacea." *Archives of Pharmacal Research* 27 (10): 1034–1036. doi:10.1007/BF02975427.

Lee, Y.S., S. Lee, H.S. Lee, B.K. Kim, K. Ohuchi, and K.H. Shin. 2005. "Inhibitory Effects of Isorhamnetin-3-O-β-D-Glucoside from Salicornia Herbacea on Rat Lens Aldose Reductase and Sorbitol Accumulation in Streptozotocin-Induced Diabetic Rat Tissues." *Biological and Pharmaceutical Bulletin* 28 (5): 916–918. doi:10.1248/bpb.28.916.

Lellau, T.F. and G. Liebezeit. 2001. "Alkaloids, Saponins and Phenolic Compounds in Salt Marsh Plants from the Lower Saxonian Wadden Sea." *Senckenbergiana Maritima* 31 (1): 1–9. doi:10.1007/BF03042831.

Lellau, T.F. and G. Liebezeit. 2003. "Activity of Ethanolic Extracts of Salt Marsh Plants from the Lower Saxonian Wadden Sea Coast against Microorganisms." *Senckenbergiana Maritima* 32 (1–2): 177–181. doi:10.1007/BF03043093.

Loconsole, D., G. Cristiano, and B. De Lucia. 2019. "Glassworts: From Wild Salt Marsh Species to Sustainable Edible Crops." *Agriculture (Switzerland)* 9 (1):14. doi:10.3390/agriculture9010014.

Lu, D., M. Zhang, S. Wang, J. Cai, X. Zhou, and C. Zhu. 2010. "Nutritional Characterization and Changes in Quality of Salicornia Bigelovii Torr. during Storage." *LWT - Food Science and Technology* 43 (3): 519–524. doi:10.1016/j.lwt.2009.09.021.

Maciel, E., A. Lillebø, P. Domingues, E. da Costa, R. Calado, M. Rosário, and M. Domingues. 2018. "Polar Lipidome Profiling of Salicornia Ramosissima and Halimione Portulacoides and the Relevance of Lipidomics for the Valorization of Halophytes." *Phytochemistry* 153: 94–101. doi:10.1016/j.phytochem.2018.05.015.

Marriott, L. and E. Pourazadi. 2017. "Industrialisation of Saline Cultivation for Second-Generation Biofuels: Progress and Challenges." *Environmental Technology Reviews* 6 (1): 15–25. doi:10.1080/21622515.2016.1272643.

Martins, C.I.M., E.H. Eding, M.C.J. Verdegem, L.T.N. Heinsbroek, O. Schneider, J.P. Blancheton, E. Roque d'Orbcastel, and J.A.J. Verreth. 2010. "New Developments in Recirculating Aquaculture Systems in Europe: A Perspective on Environmental Sustainability." *Aquacultural Engineering* 43 (3): 83–93. doi:10.1016/j.aquaeng.2010.09.002.

Meagher, E. and C. Thomson. 1999. *Vitamin and Mineral Therapy in Medical Nutrition and Disease.* ed. G. Morrison and L. Hark, Malden, MA: Blackwell Science Inc.

Medina, I., J.M. Gallardo, M.J. González, S. Lois, and N. Hedges. 2007. "Effect of Molecular Structure of Phenolic Families as Hydroxycinnamic Acids and Catechins on Their Antioxidant Effectiveness in Minced Fish Muscle." *Journal of Agricultural and Food Chemistry* 55 (10): 3889–3895. doi:10.1021/jf063498i.

Min, J.-G., D.-S. Lee, T.-J. Kim, J.-H. Park, T.-Y. Cho, and D.-I. Park. 2002. "Chemical Composition of Salicornia Herbacea L." *Preventive Nutrition and Food Science* 7: 105–107. doi:10.3746/jfn.2002.7.1.105.

Mishra, A., M.K. Patel, and B. Jha. 2015. "Non-Targeted Metabolomics and Scavenging Activity of Reactive Oxygen Species Reveal the Potential of Salicornia Brachiata as a Functional Food." *Journal of Functional Foods* 13: 21–31. doi:10.1016/j.jff.2014.12.027.

Munns, R. and M. Tester. 2008. "Mechanisms of Salinity Tolerance." *Annual Review of Plant Biology* 59: 651–681. doi:10.1146/annurev.arplant.59.032607.092911.

Nations, United. 2019. *World Population Prospects 2019: Highlights.*

Noble, S.M., A.J. Davy, and R.P. Oliver. 1992. "Ribosomal DNA Variation and Population Differentiation in Salicornia L." *New Phytologist* 122: 553–565.

Parida, A.K., A. Kumari, A. Panda, J. Rangani, and P.K. Agarwal. 2018. "Photosynthetic Pigments, Betalains, Proteins, Sugars, and Minerals during Salicornia Brachiata Senescence." *Biologia Plantarum* 62 (2): 343–352. doi:10.1007/s10535-017-0764-1.

Patel, M.K., S. Pandey, H.R. Brahmbhatt, A. Mishra, and B. Jha. 2019. "Lipid Content and Fatty Acid Profile of Selected Halophytic Plants Reveal a Promising Source of Renewable Energy." *Biomass and Bioenergy* 124: 25–32. doi:10.1016/j.biombioe.2019.03.007.

Rhee, M.H., H.J. Park, and J.Y. Cho. 2009. "Salicornia Herbacea: Botanical, Chemical and Pharmacological Review of Halophyte Marsh Plant." *Journal of Medicinal Plants Research* 3 (8): 548–555.

Rowney, J.. 2013. "Samphire: The Next Superfood ?" *Weekend Notes.* https://www.weekendnotes.com/samphire-the-next-superfood/.

Rozema, J. and T. Flowers. 2015. "Crops for a Salinized World." *Science* 322 (5907): 1478–1480.

Rozema, J., J.C. van der List, H. Schat, J. van Diggelen, and R.A. Broekman. 1987. "Ecophysilogical Response of Salicornia Dolichostachya and Salicornia Brachysrachya to Seawater Inundation." *Vegetation between Land and Sea*, 11:180–186.

Sanandiya, N.D. and A. K. Siddhanta. 2014. "Chemical Studies on the Polysaccharides of Salicornia Brachiata." *Carbohydrate Polymers* 112: 300–307. doi:10.1016/j.carbpol.2014.05.072.

Saxena, M., J. Saxena, R. Nema, D. Singh, and A. Gupta. 2013. "Phytochemistry of Medicinal Plants." *Journal of Pharmacognosy and Phytochemistry Phytochemistry* 1 (6): 168–182. doi:10.1007/978-1-4614-3912-7_4.

Shin, M.G. and G.H. Lee. 2013. "Spherical Granule Production from Micronized Saltwort (Salicornia Herbacea) Powder as Salt Substitute." *Preventive Nutrition and Food Science* 18 (1): 60–66. doi:10.3746/pnf.2013.18.1.060.

Shpigel, M., D. Ben-Ezra, L. Shauli, M. Sagi, Y. Ventura, T. Samocha, and J. J. Lee. 2013. "Constructed Wetland with Salicornia as a Biofilter for Mariculture Effluents." *Aquaculture* 412–413: 52–63. doi:10.1016/j.aquaculture.2013.06.038.

Singh, D., A.K. Buhmann, T.J. Flowers, C.E. Seal, and J. Papenbrock. 2014. "Salicornia as a Crop Plant in Temperate Regions: Selection of Genetically Characterized Ecotypes and Optimization of Their Cultivation Conditions." *AoB PLANTS* 6. doi:10.1093/aobpla/plu071.

Slama, I., C. Abdelly, A. Bouchereau, T. Flowers, and A. Savoure. 2015. "Diversity, Distribution and Roles of Osmoprotective Compounds Accumulated in Halophytes under Abiotic Stress." *Annals of Botany* 115 (3): 433–447. doi:10.1093/aob/mcu239.

Sung, J.H., S.H. Park, D.H. Seo, J.H. Lee, S.W. Hong, and S.S. Hong. 2009. "Antioxidative and Skin-Whitening Effect of an Aqueous Extract of Salicornia Herbace." *Bioscience, Biotechnology and Biochemistry* 73 (3): 552–556. doi:10.1271/bbb.80601.

Szabolsc, I. 1989. *Salt-Affected Soils.* Boca Raton, FL: *CRC Press Inc.*

Taofiq, O., A.M. González-Paramás, M.F. Barreiro, I.C.F.R. Ferreira, and D.J. McPhee. 2017. "Hydroxycinnamic Acids and Their Derivatives: Cosmeceutical Significance, Challenges and Future Perspectives, a Review." *Molecules* 22 (281). doi:10.3390/molecules22020281.

Turcios, A.E. and J. Papenbrock. 2014. "Sustainable Treatment of Aquaculture Effluents-What Can We Learn from the Past for the Future?" *Sustainability (Switzerland)* 6 (2): 836–856. doi:10.3390/su6020836.

United Nations, Department of Economic and Social Affairs, Population Division (2019). World Population Prospects 2019: Highlights (ST/ESA/SER.A/423).

Ventura, Y., A. Eshel, D. Pasternak, and M. Sagi. 2015. "The Development of Halophyte-Based Agriculture: Past and Present." *Annals of Botany* 115 (3): 529–540. doi:10.1093/aob/mcu173.

Ventura, Y. and M. Sagi. 2013. "Halophyte Crop Cultivation: The Case for Salicornia and Sarcocornia." *Environmental and Experimental Botany* 92: 144–153. doi:10.1016/j.envexpbot.2012.07.010.

Waller, U., A.K. Buhmann, A. Ernst, V. Hanke, A. Kulakowski, B. Wecker, J. Orellana, and J. Papenbrock. 2015. "Integrated Multi-Trophic Aquaculture in a Zero-Exchange Recirculation Aquaculture System for Marine Fish and Hydroponic Halophyte Production." *Aquaculture International* 23 (6): 1473–1489. doi:10.1007/s10499-015-9898-3.

Wang, H., Z. Xu, X. Li, J. Sun, D. Yao, H. Jiang, T. Zhou, et al. 2017. "Extraction, Preliminary Characterization and Antioxidant Properties of Polysaccharides from the Testa of Salicornia Herbacea." *Carbohydrate Polymers* 176: 99–106. doi:10.1016/j.carbpol.2017.07.047.

Wang, X., X. Li, X. Deng, H. Han, W. Shi, and Y. Li. 2007. "A Protein Extraction Method Compatible with Proteomic Analysis for the Euhalophyte Salicornia Europaea." *Electrophoresis* 28 (21): 3976–3987. doi:10.1002/elps.200600805.

Warshay, B., J.J. Brown, and S. Sgouridis. 2017. "Life Cycle Assessment of Integrated Seawater Agriculture in the Arabian (Persian) Gulf as a Potential Food and Aviation Biofuel Resource." *International Journal of Life Cycle Assessment* 22 (7): 1017–1032. doi:10.1007/s11367-016-1215-5.

Warshay, B., J. Pan, and S. Sgouridis. 2011. "Aviation Industry's Quest for a Sustainable Fuel: Considerations of Scale and Modal Opportunity Carbon Benefit." *Biofuels* 2 (1): 33–58. doi:10.4155/bfs.10.70.

Watten, B.J. and R.L. Busch. 1984. "Tropical Production of Tilapia (Sarotherodon Aurea) and Tomatoes (Lycopersicon Esculentum) in a Small-Scale Recirculating Water System." *Aquaculture* 41 (3): 271–283.

Webb, J.M., R. Quintã, S. Papadimitriou, L. Norman, M. Rigby, D.N. Thomas, and L.L. Vay. 2012. "Halophyte Filter Beds for Treatment of Saline Wastewater from Aquaculture." *Water Research* 46 (16): 5102–5114. doi:10.1016/j.watres.2012.06.034.

Wicke, B., E. Smeets, V. Dornburg, B. Vashev, T. Gaiser, W. Turkenburg, and A. Faaij. 2011. "The Global Technical and Economic Potential of Bioenergy from Salt-Affected Soils." *Energy & Environmental Science* 4 (8): 2669-2681. doi:10.1039/c1ee01029h.

Yin, M., X. Wang, M. Wang, Y. Chen, Y. Dong, Y. Zhao, and X. Feng. 2012. "A New Triterpenoid Saponin and Other Saponins from Salicornia Europaea." *Chemistry of Natural Compounds* 48 (2): 258–261. doi:10.1007/s10600-012-0216-2.

Young, C.C., K.C. Hyo, Y.Y. Jae, Y.K. Ji, H.H. Eun, H.K. Yung, and G.J. Hye. 2005. "Tungtungmadic Acid, a Novel Antioxidant, from Salicornia Herbacea." *Archives of Pharmacal Research* 28 (10): 1122–1126.

Yvonne, V, Sagi, M. 2013. "Halophyte Crop Cultivation: The Case for Salicornia and Sarcocornia." *Environmental and Experimental Botany* 92: 144–153. doi:10.1016/j.envexpbot.2012.07.010

Zengin, G., Z. Aumeeruddy-Elalfi, A. Mollica, M.A. Yilmaz, and M.F. Mahomoodally. 2018. "In Vitro and in Silico Perspectives on Biological and Phytochemical Profile of Three Halophyte Species—A Source of Innovative Phytopharmaceuticals from Nature." *Phytomedicine* 38: 35–44. doi:10.1016/J.PHYMED.2017.10.017.
Zerai, D.B., E.P. Glenn, R. Chatervedi, Z. Lu, A.N. Mamood, S.G. Nelson, and D.T. Ray. 2010. "Potential for the Improvement of Salicornia Bigelovii through Selective Breeding." *Ecological Engineering* 36 (5): 730–739. doi:10.1016/j.ecoleng.2010.01.002.

32 Plant Growth-Promoting Bacteria as an Alternative Strategy for the Amelioration of Salt-Stress Effects in Plants

Živko Jovanović and Svetlana Radović

CONTENTS

32.1 PLANT GROWTH-PROMOTING BACTERIA AND SALT TOLERANCE IN PLANTS

Plant growth-promoting bacteria (PGPB) represent a group of rhizospheric and/or endophytic bacteria that colonize interior or exterior of roots. Although most of these belong to the genera and *Bacillus* and *Pseudomonas*, they may also belong to the *Microbacterium*, *Pantoea*, *Rhizobium*, *Burkholderia*, *Methylobacterium*, *Azospirillum*, *Poenibacillus*, *Micorcoccus* and *Variovorax*. All of them could improve tolerance to host plants during abiotic stresses (Akram et al. 2016; Shahid et al. 2018; Abbas et al. 2019). Those bacteria could improve plant growth under normal conditions, as well as enhance tolerance against salinity stress by various mechanisms. Those mechanisms could be divided into two groups – direct and indirect mechanisms. Direct mechanisms represent improving the bioavailability of different mineral nutrients, such as iron (for siderophore production) and phosphorus

DOI: 10.1201/9781003112327-32

(phosphate solubilization), nitrogen fixation, production of different phytohormones (indole-acetic acids, auxins, ethylen). Among the indirect mechanisms are the activation of antioxidative defense, production of exopolysaccharides (EPS) and compatible osmolytes and the biosynthesis of volatile organic compounds (VOC) (Yang et al. 2009; Numan et al. 2018; Abbas et al. 2019). PGPB are able to activate "memory" defence in plants and induce response against salinity through microbe-associated molecular patterns. Such mechanisms, providing tolerance in stress conditions, are known as "induced systemic tolerance – IST" (Abbas et al. 2019). Interestingly, PGPB able to activate IST could be salt-tolerant, also, but this is not obligatory.

32.1.1 Mechanisms of PGPB-Mediated Salt-Stress Tolerance

PGPB are able to use a plethora of mechanisms, which directly or indirectly ameliorate salt-stress effects in plants (Hashem et al. 2016; Egamberdieva et al. 2019). All these mechanisms act synergistically in order to cope with emerging stress, activating different protection mechanisms, such as antioxidative systems and osmoprotectants.

32.1.1.1 Synthesis of Plant Growth Regulators

Many PGPB have the ability to release phytohormones, such as gibberelic acids (GA), auxins (indole-3-acetic acid – IAA, dominantly), cytokinins (CK) and abscisic acid (ABA). Apart from these hormones, some PGPB strains could modulate plant hormone status by releasing metabolites as well as the enzymes, such as 1-aminocyclopropane-1-carboxylase (ACC) deaminase. Together these can play essential roles in ameliorating salt-stress effects in plants. Acting as specific regulators, these molecules are able to modulate plant metabolism and morphology.

32.1.1.1.1 Auxins

Many strains of PGPB can produce auxins, such as indole-3-butyric acid or indole-3-acetic acid (IAA), or their precursors (Martinez-Morales et al. 2003; Spaepen et al. 2007; Numan et al. 2018). IAA is the most important and the most investigated auxin, acting as a key player in many processes, like cell division, differentiation and extension, but also in gravitropic and phototropic responses (Korasick et al. 2013) and circadian rhythms (Vinterhalter et al. 2015). This hormone can be produced in the tryptophan-based pathway via the formation of indole-3-pyruvic acid or indole-3-acetamide (Vaishnav et al. 2016). It is interesting that plant root cells secrete tryptophan, produced in the phenylpropanoid pathway from chorismate, and secreted amino acid is taken up by microorganisms situated in the rhizosphere. Production of IAA by PGPB is one of the most common and widely studied bacterial signaling molecules involved in plant-microbe interaction. There is a lot of experimental data confirming the importance of IAA-producing PGPB in minimizing the deleterious effects of salinity on crop plants. Generally, plants inoculated with IAA-producing PGPB grow more main roots and laterals, due to the effects of the hormone on apical meristems. These changes increase plant access to soil nutrients (Manulis et al. 1994; Nicolas et al. 2004). In addition, inoculated plants have more leaves, as IAA is also the main auxin promoting shoot growth (Malhotra and Srivastava 2009). All these effects of IAA-producing PRGP on plants are considered as adaptive responses to salinity (Vaishnav et al. 2016). Experiments on *Phaseolus vulgaris* exposed to 50 mM NaCl showed that application of *Azospirillum brasiliense* could alleviate the negative effects of stress, by encouraging higher branching of roots and activating the production of flavonoids (Dardanelli et al. 2008). Yao et al. (2010) showed that *Pseudomonas putida* was able to modulate production of IAA in plant tissues, increasing growth parameters in cotton plants. Some PGPB, like salt-tolerant *Streptomyces* isolates that produce IAA, were able to improve the root system of wheat plants exposed to salt stress (Sadeghi et al. 2012). *Bacillus amyloliqefaciens* SQR 9 was able to enhance salt-stress tolerance in maize seedlings exposed to 100 mM NaCl. Maize seedlings inoculated with that bacteria strain showed not only increased total chlorophyll and soluble sugar contents, but also enhanced glutathione content. In addition, SQR9 inoculation activates peroxidase (POD) and catalase (CAT) and reduced the ABA level induced by salinity. As a consequence of plant-microbe interaction, upregulation of many of the genes (*RBCS, RBCL, NCED, HKT1, NHX1, NHX2 and NHX3*) involved in photosynthesis and metabolism has been observed (Chen et al. 2016; Ilangumaran and Smith 2017). In Arabidopsis and tomato (*Solanum lycopersicum*) plants, inoculation with *Enterobacter* sp. EJ01 (isolated from a halophyte) improved the tolerance of plants to salt stress (200 mM NaCl). In these plants, inoculation with EJ01 activated the expression of genes involved in dehydration, such as DRE-binding proteins DREB2b, LEA genes RAB18 and a gene involved in biosynthesis of proline – P5CS1 (Ilangumaran and Smith 2017). More recent experiments have shown that the salt-tolerant PGPB *Leclercia adecarboxylata* regulates sugar biosynthesis, the production of different organic acids and chlorophyll fluorescence in tomato under salt stress (Kang et al. 2019). Those effects are also accomplished via IAA.

32.1.1.1.2 Abscisic Acid

Although it has been proposed that many PGPB produce ABA *in vitro*, there are relatively few studies about the role of ABA in plant-microbe interactions. In addition, it is not clear whether ABA, produced by bacteria, can modulate ABA status in plants exposed to enhanced salinity (Dodd et al. 2010; Ilangumaran and Smith 2017). Some reports have shown that PGPB inhibit ABA production in salt-stressed plants, while others claim that bacteria enhance the accumulation of ABA, providing better conditions for plant survival under salt stress (Vaishnav et al. 2016). However, the experimental data show that PGPB modulate not only ABA biosynthesis, but also ABA-mediated signaling pathways. In wheat, inoculation with *Dietzia natronolimnaea* STR1 induced salinity tolerance in plants exposed to 150 mM NaCl, via modulation of an ABA-signaling cascade. Inoculation upregulates *TaABARE* (an ABA-responsive gene) and *TaOPR1* (the 12-oxophytodienoate reductase gene), which lead to the activation of TaMYB and TaWRKY, and finally upregulated the expression of the stress-related gene *TaST* (salt stress-induced). Inoculation of cucumber (*Cucumis sativus*) with *Burkholderia cepacia* SE4, *Promicromonospora* sp. SE188 and *Acinetobacter calcoaceticus* SE370 provided better performance (higher biomass) of plants grown at 120 mM NaCl. It was interesting that inoculated plants had a lower ABA level compared with control (non-inoculated) plants, as well as an increased water potential and decreased electrolyte leakage. In addition, inoculation increased salycilic acid and giberellin content (Kang et al. 2014). Inoculation of cotton (*Gossypium hirsutum*) seeds with *Pseudomonas putida* Rs-198 increased plant biomass and reduced ABA level. Non-inoculated plants showed a higher accumulation of ABA in leaves (Yao et al., 2010). Barnawal et al. (2017) observed that, in wheat plants, inoculation with *B. subtilis* LDR2 and *Arthrobacter protophormiae* SA3 increased ABA content under salt-stress conditions (100 mM NaCl). That effect was accomplished by upregulation of the *TaCTR1* (Serine/Threonine protein kinase – ethylene responsive) and *TaDRE2* (drought-responsive element) genes.

32.1.1.1.3 Gibberellin

There are a few data about the gibberellin-producing PGPB, especially about their role in acquiring salt-stress tolerance in plants. It is well known that some plants, inoculated with different *Rhizobium* strains (GA7 producers), have longer roots (Numan et al. 2018). Cassán et al. (2014) showed that inoculation of wheat and soybean with GA-producing *A. brasilense* enhanced plant growth.

32.1.1.1.4 Cytokinins

CK are a group of purine-derivates acting as plant hormones: they regulate many processes, like cell division, differentiation of root callus, shoot formation, chloroplast maturation and stomatal conductance in higher plants (Cassán et al. 2014). It is well known that plants maintain their totipotent stem cells in root and shoot meristems due to the action of CK (Howell et al. 2003; Leibfried et al. 2005). Although many experimental data have shown that increased growth of inoculated plants is correlated with cytokinin-producing PGPB (Nieto and Frankenberger 1991; Arkhipova et al. 2005), the role of bacterial CK in salt-stress tolerance is unknown. Egamberdieva (2019) showed that some *Pseudomans* strains (*P. aurantiaca* TSAU22, *P. extremorientalis* TSAU6 and *P. extremorientalis* SAU20) could increase plant growth as well as break salinity (100 mM NaCl) induced seed dormancy.

32.1.1.1.5 Ethylene

Ethylene, the only gaseous plant hormone, is known as a stress hormone. As stress increases the ethylene content in plants, the transcription of auxin response factors is inhibited leading to a perturbation in plant growth. Plants use the Yang cycle for ethylene biosynthesis, in which ACC, as the precursor, is converted into ethylene by the ACC oxidase enzyme (Yoon and Kieber 2013). Salinity stress induces the accumulation of ethylene, as well as its precursor ACC, in the leaves, which results in decreased photosynthesis and foliar senescence (Ghanem et al. 2008). Many PGPB are able to secrete ACC deaminase, an enzyme that restricts ethylene biosynthesis in plants. ACC deaminase converts ACC to ammonia and α-ketobutyrate and thereby decreases ethylene levels in plants. Many experimental data show that ACC deaminase-producing PGPB could provide tolerance to plants in saline soil. Thus, cucumber plants are able to grow in saline soil due to the stimulation effects of *Stenotrophomonas rhizophila* (Egamberdieva et al. 2011). Nadeem et al. (2009) reported that *Pseudomonas fluorescens* and *Enterobacter* spp. could improve the yield of maize grown in a salt-affected soil. In addition, tomato seedlings, inoculated with *Pseudomonas putida* UW4, were able to grow in saline conditions (90 mM NaCl). Interestingly, these plants showed increased shoot growth after 6 weeks in saline conditions. In that case, the bacteria used for inoculation, although producing ACC deaminase, were able to up regulate the expression of *TocGTPase* gene. That gene coding is part of the chloroplast protein import apparatus, facilitating the import of proteins involved in stress response (Yan et al. 2014). Peanut tolerance to saline condition has been improved by inoculation with *Brachybacterium saurashtrense* (JG-06), *Brevibacterium casei* (JG-08) and *Haererohalobacter* (JG-11), ACC-producing PGPB (Shukla et al. 2012). Apart from its role in decreasing ethylene level in plants, preventing them from senescence, it has been found that ACC deaminase producing PGPB strains could have other effects on plants, such as the production of pigments under drought and salt stress, the biosynthesis of compatible solutes and the stabilization of membranes (Tiwari et al. 2018). In pea plants (*Pisum sativum* cv. Alderman), inoculated with *Variovorax paradoxus* 5C-2, under salt-stress conditions – 70 and 130 mM NaCl, there was an increased K^+/Na^+ ratio in shoots. This is very important for pea survival under saline conditions, as experimental data have shown that pea varieties have different sensitivity to salinity (Miljuš-Djukić et al. 2013). These strains could also have the effects on nodule formation in legume crops (Ahmad et al. 2011; Egamberdieva et al. 2019). PGPB that produce both ACC deaminase and auxin (IAA) have a great potential for plant protection from different stresses. Accumulation of IAA activates the enzyme ACC synthase, increasing the ACC level and subsequently ethylene level. The excess ACC could be destroyed by ACC-deaminase producing PGPB, allowing the promotion of plant growth under stress conditions mediated by IAA (Glick 2012; Ilangumaran and Smith 2017).

32.1.1.2 Volatile Organic Compounds

VOC play one of the crucial roles in plant-microorganism interactions. Due to high vapor pressure, these compounds can enter the atmosphere as vapors. VOC-producing PGPB are able to modulate many signaling pathways in plants, resulting in the promotion of growth and the activation of induced systemic resistance – ISR (Ryu et al. 2004; Zhang et al. 2008; Numan et al. 2018). Using a microarray, Zhang et al. (2007) showed that the VOC of *B. subtilis* GB03 in *A. thaliana* seedlings altered the expression of more than 600 genes. Under saline conditions, VOCs were able to upregulate the expression of HKT1 – the high affinity K^+ transporter, thus lowering accumulation of Na^+ in plant tissues (Zhang et al. 2008). *Pseudomonas simiae* AU, a VOC producing PGPB, was able to induce salt tolerance in soybean (*Glycine max*) exposed to high salinity (100 mM NaCl) by decreasing the accumulation of sodium cations in roots and increasing the accumulation of proline. Analysis of proteins revealed upregulation of different vegetative storage proteins involved in Na^+ homeostasis and the RuBisCO large subunit (Vaishnav et al. 2016). Ledger et al. (2016) showed that VOCs produced by *Paraburkholderia phytofirmanas* PsJN induced salinity tolerance *in vitro* and in the soil. Also, all growth parameters, such as fresh weight, length of primary roots and rosette area were higher in Arabidopsis plants treated with VOCs comparing with non-inoculated plants (Ledger et al. 2016).

32.1.1.3 Exopolysaccharides

EPS or surface polysaccharides secreted by bacteria are responsible for bacterial attachment to root surfaces, soil particles as well as to other bacteria. Some of PGPB secrete EPS (Tewari and Arora 2014; Khan and Bano 2019). It is important that EPS could act as a barrier around plant roots, thus supporting plant growth under saline conditions (Vaishnav et al. 2016). Inoculation of chickpea (*Cicer arietinum* var. CM-98) with EPS-producing PGPB *Halomonas variabilis* HT1 and *Planococcus rifietoensis* RT4 resulted in increased plant growth at 200 mM NaCl (Qurashi and Sabri 2012). Yang et al. (2016) inoculated quinoa (*Chenopodium quinoa*) seeds with *Enterobacter* sp. MN17 and *Bacillus* sp. MN54 and showed plant survival under saline irrigation conditions (400 mM NaCl). Recently, it has been shown that salt-tolerant EPS producing *B. subtilis* subsp. *inaquosorum* and *Marinobacter lypoliticus* SM19 reduced drought and salinity stress effects in wheat (Atouei et al. 2019). In *A. thaliana*, EPS producing *Pseudomonas* sp. was able to upregulate the *LOX2* gene which encodes a lipoxygenase. This enzyme is a component of the jasmonic acid biosynthesis pathway (Chu et al. 2019).

Apart from exopolysaccharides, many PGPB produce lipo-chitooligosaccharides. These molecules are secreted by rhizobia as Nod-factors (NFs) and their secretion is induced by flavonoids present in root exudates. NFs, thus, initiate nodule formation (Ilangumaran and Smith 2017). Miransari and Smith (2009) reported that inoculation of soybean plants with *Bradyrhizobium japonicum* 532C enhanced nodulation and growth under mild salinity stress (36 mM and 61 mM NaCl).

32.1.1.4 Osmoprotectants

Salt stress is basically a two-component stress and plant growth and development are first affected by the osmotic impacts of salt stress. The accumulation of salt ions leads to a decrease in the osmotic balance in plants. Plants accumulate compatible osmolytes to improve plant-water relations and maintain cell structures protected from osmotic shock. Many PGPB also use this mechanism for protection against osmotic stress (Ilangumaran and Smith 2017). Choudhary et al. (2012) reported that many PGPB, such as *Burkholderia*, *Arthobacter*, *Bacillus* and *Pseudomonas*, are able to enhance proline levels in plants under abiotic stress. Some endophytic bacteria, producing ACC deaminase and IAA, are able to increase proline content in sweet pepper (*Capsicum annum*). Maize plants (*Zea mays*) inoculated with *Rhizobium* and *Pseudomonas* accumulated more proline comparing with control (non-inoculated) plants (Bano and Fatima 2009). According to the experimental data, it is evident that the accumulation of proline under abiotic stress is a result of the activation of the pyrroline-5-carboxylate synthase (*P5CS*) gene; thus, bacterial treatment activates the expression of *P5CS* gene (Kumari et al. 2015). Some PGPB also produce polyamines, such as spermidine (*Bacillus megaterium* BOFC15) and soluble sugars (trehalose). Many of PGPB have genes for the biosynthesis of trehalose (Qin et al. 2018; Shim et al. 2019).

32.1.1.5 Antioxidant Enzymes

Many reports have shown that inoculation of plants with PGPB reduced the oxidative stress in plants and its deleterious effects (Manaf and Zayed 2015; Islam et al. 2016).

PGPB strains are able to activate and even increase the concentration of antioxidative enzymes. Under saline conditions, the activity of soluble POD, superoxide dismutase, CAT and glutathione reductase has been increased by PGPB (Jha and Subramanian 2013; Sen and Chandrasekhar 2015; Ansari et al. 2019). Plant inoculation with *Azospirillum lipoferum* FKI induced an increasing level of antioxidant gene transcripts (El-Esawi et al. 2019). Kohler et al. (2009) reported that lettuce plants inoculated with PGPB showed increased activity of CAT under salinity stress. In Jatropha plants exposed to salt stress, Patel and Saraf (2013) observed an increased level of APX and CAT activity in leaves. Although many reports have claimed that plants inoculated with PGPB showed increased activities of antioxidative enzymes, the exact mechanism for enzyme activation remains unknown.

32.1.1.6 Nutrient Acquisition

Salinity can affect plant growth and productivity due to a limitation of nutrients in the soil. As salinity changes the pH of soil, some nutrients become less available for plants. PGPB are able to solubilize different nutrients making them available for plants. PGPB exert these effects by different mechanisms, such as siderophore production, increased phosphorous solubilization, organic and inorganic phosphate solubilization and non-symbiotic nitrogen fixation.

Siderophores are agents able to chelate iron and represent an important trait for the promotion of plant growth. Siderophores produced by different PGPB have high affinity for iron, thus preventing the proliferation of phytopathogens (Numan et al. 2018). Siderophores are secreted in the rhizosphere, so plant roots can uptake iron from them via two mechanisms – by chelate degradation or by direct uptake (Rajkumar et al. 2010). In saline soil, the availability of Fe^{3+} ions can be reduced due to changes in pH (Thomine and Lanquar 2011); plant inoculation with siderophore-producing PGPB can, therefore, help plants to overcome the nutrient stress caused by high salinity.

Phosphorus is a crucial macronutrient for plants. In organic form, it is present as inositol phosphates, phosphoesters, phosphodiesters and phosphotriesters (Sindhu et al. 2010). A large part of the soil phosphoros pool is immobilized and unavailable for plants (Numan et al. 2018). PGPB can play an important role in the transformation of phosphoros, making it available for plants. Phosphate solubilizing bacteria belong to the genera *Bacillus*, *Pseudomonas*, *Brevibacterium*, *Serratia*, *Xanthomonas*, *Corynebacterium* and *Alcaligenes* (Sindhu et al. 2010). These bacteria are able to hydrolize unavailable forms of phosphorus into more available forms. As salinity can cause the precipitation of available phosphorus, PGPB could solubilize precipitated forms, providing nutrients to plants under salinity stress.

PGPB have the ability to provide other nutrients to plants, such as nitrogen, potassium, sulfur and zinc (Vaishnav et al. 2016). Some PGPB can enhance the nodulation process, especially in saline soil, which is important because *Rhizobium* are often not effective for the induction of nodulation under saline conditions (Ahmad et al. 2011).

32.2 FUTURE PERSPECTIVES

In order to successfully use PGPB strains in the region affected by increased salinity the best way is to use halotolerant bacteria for plant and/or seed inoculation. These PGPB should be isolated from salt-affected soils, which make them able to colonize roots under salinity (Paul and Nair 2008; Vaishnav et al. 2016). Ramadoss et al. (2013) demonstrated that five PGPB halotolerant bacteria could ameliorate salt stress in wheat plants. In these experiments, they used 320 mM NaCl for inducing salt stress and observed that inoculation was able to increase root length by up to 71.7%. Apart from the use of halotolerant bacteria, it is speculated that is possible to use PGPB which are not halotolerant (tested in laboratory conditions with growing medium supplemented with NaCl) for the enhancement of plant growth under saline conditions. In that case it should be the use of bacterial cultures for seed inoculation, so called "seed priming". In this way, PGPB could activate different mechanisms in plants, including ISR, making them more tolerant to salinity. Some plant seeds, inoculated with non-halotolerant PGPB, were able to germinate faster under salt stress, compared with non-inoculated control seeds.

The commercialization of PGPB strains is a complex process, which includes many stages – from isolation, screening and testing in laboratory conditions, up to testing in the field, estimating their efficacy and making adequate formulations. It is necessary also to test the viability of formula and industrial production, etc. (Bhattacharyya and Jha 2012; Vaishnav et al. 2016).

Many of the mechanisms of osmo-adaptation of PGPB still need to be elucidated; establishing these could have a big impact on the process of improvement of productivity in agriculturally important plants grown in saline agro-ecosystem (Paul 2013). Further work is also required to investigate the regulatory networks which provide salt (and other) stress tolerance to bacteria. Also, plant-microbe interaction, particularly under saline condition, should be deeper investigated. Although many mechanisms are well documented, the possible interaction among them is still controversial. A deeper analysis of the bacterial communities associated with plant roots, as well as examination of the dynamic changes of microbiomes under saline condition could provide valuable data.

It is very important that the research findings obtained under laboratory conditions be extended to field trials and in different geographical regions. Also important is that this research should be conducted on different plants, even on different cultivars, which is a major challenge. The major issue for the end-users is using developed bioinoculant formula on many different plants. Having in mind that PGPB are diverse, in respect to their plant-growth promoting effects, the recommendation is to use no single PGPB strains, but several strains or consortia. Using diverse microbes in consortia formulations is potentially a promising strategy for the alleviation not only of salinity (and other abiotic) stress in plants, but also biotic stress – phytopathogens (Egamberdieva et al. 2019).

PGPB used for enhancing the plant productivity in saline agro-ecosystem have the potential to help in achieving food security and improving global food production. Moreover, they can change the quality of soil and help in combating the adverse effects of climate change (Arora 2019).

32.3 CONCLUSION

Salt stress is still a major problem for agriculture, limiting the yield of many agriculturally important crops. Although plant-growth promoting bacteria, including halotolerant types, represent a promising tool for the alleviation of salt-stress effects on plants, a lot is yet to be explored, at the biochemical as well as at the molecular levels. Apart from helping plants, PGPB in saline agro-ecosystem have the potential to improve soil qualities, like soil fertility. These microorganisms, used as biocontrol agents, as a part of alternative green biotechnologies, should be extensively used in the future not only in saline soils, but also in other marginal soil types. Many issues, including problems with the formulation and production of final products, as well as potential legal and social impacts, should be further analyzed.

REFERENCES

Abbas R, Rasul, Aslam K, Baber M, Shahid M, Mubeen F, Naqqash, T. 2019. Halotolerant PGPR: A hope for cultivation of saline soils. *Journal of King Saud University Science* 31: 1195–1201.

Ahmad M, Zahir ZA, Asghar HN, Asghar M. 2011. Inducing salt tolerance in mung bean through coinoculation with rhizobia and plant-growth-promoting rhizobacteria containing 1-aminocyclopropane-1-carboxylate deaminase. *Canadian Journal of Microbiology* 57: 578–589.

Akram MS, Shahid M, Tariq M, Azeem M, Javed MT, Saleem S, Riaz S. 2016. Deciphering *Staphylococcus sciuri* SAT-17 Mediated Anti-oxidative Defense Mechanisms and Growth Modulations in Salt Stressed Maize (*Zea mays* L.). *Frontiers in Microbiology* 7:867.

Ansari FA, Ahmad I, Pichtel J. 2019. Growth stimulation and alleviation of salinity stress to wheat by the biofilm forming *Bacillus pumilus* strain FAB10. *Applied Soil Ecology* 3: 45–54.

Arkhipova TN, Veselov SU, Melentiev AI, Martynenko EV, Kudoyarova GR. 2005. Ability of bacterium *Bacillus subtilis* to produce cytokinins and to influence the growth and endogenous hormone content of lettuce plants. *Plant Soil* 272: 201–209.

Arora NK. 2019. Impact of climate change on agriculture production and its sustainable solutions. *Environmental Sustainability* 2:95–96.

Atouei MT, Pourbabaee AA, Shorafa M. 2019. Alleviation of salinity stress on some growth parameters of wheat by exopolysaccharide-producing bacteria. *Iranian Journal of Science and Technology Transaction A* 43: 2725–2733.

Bano A, Fatima M. 2009. Salt tolerance in *Zea mays* (L.) following inoculation with Rhizobium and Pseudomonas. *Biology and Fertility of Soils* 45: 405–413.

Barnawal D, Bharti N, Maji D, Chanotiya CS, Kalra A. 2014. ACC deaminase-containing *Arthrobacter protophormiae* induces NaCl stress tolerance through reduced ACC oxidase activity and ethylene production resulting in improved nodulation and mycorrhization in *Pisum sativum. Journal of Plant Physiology* 171:884–894.

Bhattacharyya PN, Jha DK. 2012. Plant growth-promoting rhizobacteria (PGPR): Emergence in agriculture. *World Journal of Microbiololgy and Biotechnology* 28: 1327–1350.

Cassán F, Vanderleyden J, Spaepen S. 2014. Physiological and agronomical aspects of phytohormone production by model plant-growth-promoting rhizobacteria (PGPR) belonging to the genus Azospirillum. *Journal of Plant Growth Regulation* 33:440–459.

Chen L, Liu Y, Wu G, Veronican Njeri K, Shen Q, Zhang N. et al. 2016. Induced maize salt tolerance by rhizosphere inoculation of *Bacillus amyloliquefaciens* SQR9. *Physiologia Plantarum* 158: 34–44.

Choudhary D. 2012. Microbial rescue to plant under habitat-imposed abiotic and biotic stresses. *Applied Microbiology and Biotechnology* 96:1137–1155.

Chu TN, Tran BTH, Van Bui L, Hoang MTT. 2019. Plant growth-promoting rhizobacterium *Pseudomonas* PS01 induces salt tolerance in Arabidopsis thaliana. *BMC Research Notes* 12:11.

Dardanelli MS, Fernández de Córdoba FJ, Espuny MR, Rodríguez-Carvajal MA, Soría-Díaz ME, Gil-Serrano AM, Okon Y, Megías M. 2008. Effect of *Azospirillum brasilense* coinoculated with *Rhizobium* on *Phaseolus vulgaris* flavonoids and Nod factor production under salt stress. *Soil Biology and Biochemistry* 40:2713–2721.

Dodd IC, Zinovkina NY, Safronova V I, Belimov AA. 2010. Rhizobacterial mediation of plant hormone status. *Annals of Applied Biology* 157: 361–379.

Egamberdieva D, Wirth S, Bellingrath-Kimura SD, Mishra J, Arora NK. 2019. Salt- tolerant plant growth promoting rhizobacteria for enhancing crop productivity of saline soils. *Frontiers in Microbiology* 10:2791.

Egamberdieva, D, Kucharova, Z, Davranov, K, Berg, G, Makarova, N, Azarova T, Chebotar V, Tikhonovich I, Kamilova F, Validov S, Lugtenberg B. 2011. Bacteria able to control foot and root rot and to promote growth of cucumber in salinated soils. *Biology and Fertility of Soils* 47: 197–205.

El-Esawi MA, Alaraidh IA, Alsahli AA, Alamri SA, Ali HM, Alayafi AA. 2018. *Bacillus firmus* (SW5) augments salt tolerance in soybean (*Glycine max* L.) by modulating root system architecture, antioxidant defense systems and stress-responsive genes expression. *Plant Physiology and Biochemistry* 132: 375–384.

Ghanem ME, Albacete A, Martínez-Andújar C, Acosta M, Romero-Aranda R, Dodd IC, Lutts S, Pérez-Alfocea F. 2008. Hormonal changes during salinity-induced leaf senescence in tomato (*Solanum lycopersicum* L.) *Journal of Experimental Botany* 59:3039–3050.

Glick B. 2012. Plant growth-promoting bacteria: Mechanisms and applications. *Scientifica*. 2012:1–15.

Hashem A, Abd_Allah EF, Alqarawi AA, Al-Huqail AA, Wirth S, Egamberdieva D. 2016. The interaction between arbuscular mycorrhizal fungi and endophytic bacteria enhances plant growth of *Acacia gerrardii* under salt stress. *Frontiers in Microbiology* 7:1089.

Howell SH, Lall S, Che P. 2003. Cytokinins and shoot development. *Trends in Plant Science* 8: 453–459.

Ilangumaran G, Smith DL. 2017. Plant growth promoting rhizobacteria in amelioration of Salinity stress: a systems biology perspective. *Frontiers in Plant Science* 8:1768.

Islam F, Yasmeen T, Arif MS, Ali S, Ali B, Hameed S et al. 2016. Plant growth promoting bacteria confer salt tolerance in *Vigna radiata* by up-regulating antioxidant defense and biological soil fertility. *Plant Growth Regulation* 80: 23–36.

Jha Y, Subramanian RB. 2013. Paddy plants inoculated with PGPR show better growth physiology and nutrient content under saline condition. *Chilean Journal of Agricultural Research* 73: 213–219.

Kang SM, Radhakrishnan R, Khan AL, Kim MJ, Park JM, Kim BR, Shin DH, Lee IJ.2014. Gibberellin secreting rhizobacterium, *Pseudomonas putida* H- 2- 3 modulates the hormonal and stress physiology of soybean to improve the plant growth under saline and drought conditions. *Plant Physiology and Biochemistry* 84: 115–124.

Kang SM, Shahzad R, Bilal S, Khan A L, Park YG, Lee K E et al. 2019. Indole-3-acetic-acid and ACC deaminase producing *Leclercia adecarboxylata* MO1 improves *Solanum lycopersicum* L. growth and salinity stress tolerance by endogenous secondary metabolites regulation. *BMC Microbiology* 19:80.

Khan N, Bano A. 2019. Exopolysaccharide producing rhizobacteria and their impact on growth and drought tolerance of wheat grown under rainfed conditions. *PLoS One* 14: e0222302.

Kohler J, Hernandez JA, Caravaca F, Roldan A. 2009. Induction of antioxidant enzymes is involved in the greater effectiveness of a PGPR versus AM fungi with respect to increasing the tolerance of lettuce to severe salt stress. *Environmental and Experimental Botany* 65: 245–252.

Korasick DA, Enders TA, Strader LC. 2013. Auxin biosynthesis and storage forms. *Journal of Experimental Botany* 64: 2541–2555.

Kumari S, Vaishnav A, Jain V, Choudhary DK. 2015. Bacterial-mediated induction of systemic tolerance to salinity with expression of stress alleviating enzymes in soybean (*Glycine max* L. Merrill). *Journal of Plant Growth Regulation* 34: 558–573.

Ledger T, Rojas S, Timmermann T, et al. 2016. Volatile-mediated effects predominate in *Paraburkholderia phytofirmans* growth promotion and salt stress tolerance of *Arabidopsis thaliana*. *Frontiers in Microbiology* 7:1838.

Leibfried A, To JPC, Busch W, Stehling S, Kehle A, Demar M, Kieber JJ, Lohmann JU. 2005. WUSCHEL controls meristem function by direct regulation of cytokinin-inducible response regulators. *Nature* 438: 1172–1175.

Malhotra M, Srivastava S. 2009. Stress-responsive indole-3-acetic acid biosynthesis by *Azospirillum brasilense* SM and its ability to modulate plant growth. *European Journal of Soil Biology* 4573–4580.

Manaf HH, Zayed MS. 2015. Productivity of cowpea as affected by salt stress in presence of endomycorrhizae and *Pseudomonas fluorescens*. *Annals of Agricultural Sciences* 60: 219–226.

Manulis S, Shafri H, Epstein E, Lichter A, Barash I. 1994. Biosynthesis of Indole 3-acetic acid via the indole 3-acetamide pathway in *Streptomyces* spp. *Microbiology* 140: 1045–1050.

Martínez-Morales LJ, Soto-Urzúa L, Baca BE, Sánchez-Ahédo, JA. 2003. Indole-3-butyric acid (IBA) production in culture medium by wild strain *Azospirillum brasilense*. *FEMS Microbiology Letters*, 228: 167–173.

Miljuš-Djukić J, Stanisavljević N, Radovic S, Mikic A, Maksimovic V, Jovanović Ž. 2013. Differential response of three contrasting pea (*Pisum arvense*, *P. sativum* and *P. fulvum*) species to salt stress: assessment of variation in antioxidative defence and miRNA expression. *Australian Journal of Crop Science* 7:2145–2153.

Miransari M, Smith D. 2009 Alleviating salt stress on soybean (*Glycine max* (L.) Merr.)-*Bradyrhizobium japonicum* symbiosis, using signal molecule genistein. *European Journal of Soil Biology* 45:146–152.

Nadeem SM, Zahir ZA, Naved M, Arshad M. 2009. Rhizobacteria containing ACC-deaminase confer salt tolerance in maize grown on salt-affected fields. *Canadian Journal of Microbiology* 55: 1302–1309.

Nicolás L, Ines J, Acosta M, Sánchez-Bravo J. 2004. Role of basipetal auxin transport and lateral auxin movement in rooting and growth of etiolated lupin hypocotyls. *Physiologia plantarum* 121: 294–304.

Nieto K, Frankenberger W.1991. Influence of adenine, isopentyl alcohol and Azotobacter chroococcum on the vegetative growth of *Zea mays*. *Plant and Soil* 135: 213–221.

Numan M, Bashir S, Khan Y, Mumtaz R, Shinwari ZK, Khan AL, Khan A, AL-Harrasi A. 2018. Plant growth promoting bacteria as an alternative strategy for salt tolerance in plants: A review. *Microbiology Research* 209: 21–32.

Patel T, Saraf M. 2017. Biosynthesis of phytohormones from novel rhizobacterial isolates and their in vitro plant growth-promoting efficacy. *Journal of Plant Interactions* 12: 480–487.

Paul D. 2013. Osmotic stress adaptations in rhizobacteria. *Journal of Basic Microbiology* 53: 101–110.

Paul D, Nair S. 2008. Stress adaptations in a plant growth promoting rhizobacterium (PGPR) with increasing salinity in the coastal agricultural soils. *Journal of Basic Microbiology* 48:378–384.

Qin S, Feng WW, Zhang YJ, Wang TT, Xiong YW, Xing K. 2018. Diversity of bacterial microbiota of coastal halophyte *Limonium sinense* and amelioration of salinity stress damage by symbiotic plant growth-promoting actinobacterium *Glutamicibacter halophytocola* KLBMP 5180. *Applied and Environmental Microbiology* 84: e1533–e1518.

Qurashi AW, Sabri, AN. 2012. Bacterial exopolysaccharide and biofilm formation stimulate chickpea growth and soil aggregation under salt stress. *Brazilian Journal of Microbiology* 43: 1183–1191.

Rajkumar M, Ae N, Prasad MNV, Freitas H. 2010. Potential of siderophore-producing bacteria for improving heavy metal phytoextraction. *Trends in Biotechnology* 28; 142–149.

Ramadoss D, Lakkineni VK, Bose P, Ali S, Annapurna K. 2013. Mitigation of salt stress in wheat seedlings by halotolerant bacteria isolated from saline habitats. *Springer Plus*, 2: 1–7.

Ryu CM, Farag MA, Hu CH, Reddy MS, Kloepper JW, Paré PW. 2004. Bacterial volatiles induce systemic resistance in Arabidopsis. *Plant Physiology* 134:1017–1026.

Sadeghi A, Karimi E, Dahaji PA, Javid MG, Dalvand Y, Askari H. 2012. Plant growth promoting activity of an auxin and siderophore producing isolate of Streptomyces under saline soil conditions. *World Journal of Microbiology and Biotechnology* 28: 1503–1509.

Sen S, Chandrasekhar CN. 2015. Effect of PGPR on enzymatic activities of rice (*Oryza sativa* L.) under salt stress. *Asian Journal of Plant Science and Research* 5: 44–48.

Shahid M, Akram MS, Khan MA, Zubair M, Shah SM, Ismail M, Shabir G, Basheer S, Aslam K, Tariq M. 2018. A phytobeneficial strain *Planomicrobium* sp. MSSA-10 triggered oxidative stress responsive mechanisms and regulated the growth of pea plants under induced saline environment. *Journal of Applied Microbiology*, 124: 1566–1579.

Shim JS, Seo JS, Kim Y, Koo Do Choi Y et al. 2019. Heterologous expression of bacterial trehalose biosynthetic genes enhances trehalose accumulation in potato plants without adverse growth effects. *Plant Biotechnology Reports* 13: 409–418.

Shukla PS, Agarwal PK, Jha B. 2012. Improved salinity tolerance of (*Arachis hypogaea* L.) by the interaction of halotolerant plant-growth-promoting rhizobacteria. *Journal of Plant Growth Regulation* 31: 195–206.

Sindhu SS, Dua S, Verma MK, Khandelwal A. 2010. Growth Promotion of Legumes by Inoculation of Rhizosphere Bacteria. In *Microbes for legume improvement*, ed. Khan MS, Musarrat J, Zaidi A. Vienna: Springer.

Spaepen S, Vanderleyden J, Remans R. 2007. Indole-3-acetic acid in microbial and microorganism-plant signalling. *FEMS Microbiology Reviews* 31: 425–448.

Tewari S, Arora NK. 2014. Multifunctional exopolysaccharides from *Pseudomonas aeruginosa* PF23 involved in plant growth stimulation, biocontrol and stress amelioration in sunflower under saline conditions. *Current Microbiology* 69: 484–494.

Thomine S, Lanquar V. 2011. Iron Transport and Signaling in Plants. In *Transporters and pumps in plant signaling, signaling and communication in plants 7M*, ed. Geisler and K. Venema, 99–131, Berlin Heidelberg: Springer-Verlag.

Tiwari G, Duraivadivel P, Sharma S, Hariprasad P. 2018. 1-Aminocyclopropane-1-carboxylic acid deaminase producing beneficial rhizobacteria ameliorate the biomass characters of *Panicum maximum* Jacq. by mitigating drought and salt stress. *Scientific Reports* 8:17513.

Vaishnav A, Varma A, Tuteja N, Choudhary DK. 2016. PGPR-mediated amelioration of crops under salt stress. In *Plant microbe interaction: an approach to sustainable agriculture*, ed.D. Choudhari, A.Varna and N.Tuteja, 205–226, Singapore: Springer.

Vinterhalter D, Vinterhalter B, Miljuš-Djukić J, Jovanović Ž, Orbović V. 2015. (Erratum) Daily changes in the competence for photo- and gravitropic response by potato plantlets. *Journal of Plant Growth Regulation* 34:440–450.

Yan J, Campbell JH, Glick BR, Smith MD, Liang Y. 2014. Molecular characterization and expression analysis of chloroplast protein import components in tomato (*Solanum lycopersicum*). *PLoS One* 9(4): e95088.

Yang A, Akhtar SS, Iqbal S, Amjad M, Naveed M, Zahir ZA et al. 2016. Enhancing salt tolerance in quinoa by halotolerant bacterial inoculation. *Functional Plant Biology* 43: 632–642.

Yang J, Kloepper JW, Ryu CM. 2009. Rhizosphere bacteria help plants tolerate abiotic stress. *Trends Plant Sci.* 14: 1–4.

Yao L, Wu Z, Zheng Y, Kaleem I, Li C. 2010. Growth promotion and protection against salt stress by *Pseudomonas putida* Rs-198 on cotton. *European Journal of Soil Biology* 46: 49–54.

Yoon GM, Kieber JJ. 2013. 14-3-3 regulates 1-aminocyclopropane-1-carboxylate synthase protein turnover in Arabidopsis. *Plant Cell* 25:1016–1028

Zhang H, Kim MS, Krishnamachari V, Payton P, Sun Y, Grimson M, et al. 2007. Rhizobacterial volatile emissions regulate auxin homeostasis and cell expansion in Arabidopsis. *Planta* 226: 839–851.

Zhang H, Kim MS, Sun, Dowd SE, Shi H, Paré P W. 2008. Soil bacteria confer plant salt tolerance by tissue-specific regulation of the sodium transporter HKT1. *Molecular Plant-Microbe Interaction* 21: 737–744.

33 Tolerance to Environmental Stresses

Do Fungal Endophytes Mediate Plasticity in Solanum Dulcamara?

Sasirekha Munikumar, Karaba N. Nataraja, and Theo Elzenga

CONTENTS

33.1 INTRODUCTION

As the area of arable land for the optimal cultivation of food crops is decreasing due to inadequate irrigation management practices, land degradation, sea water level rise and global climate change, there is growing demand for developing stress-tolerant crops,

DOI: 10.1201/9781003112327-33

specifically crops that can withstand saline conditions. Various methods to develop salt-tolerant crops have been used, these include conventional breeding, transgenics, use of plant growth promoting (endophytic) bacteria, mycorrhizal or/and endophytic fungi.

Salt-tolerant crops could be developed by five possible ways: a) exploitation of existing variation present in crops; b) use of interspecific hybridization to improve the salt tolerance of existing crops; c) creating variation by using traditional breeding and genetic modification; d) use of halophytes as alternative crops; and e) breeding for stable yield instead of tolerance (Flowers and Yeo, 1989). While introduction of genes from wild salt-tolerant species has been explored for many plant species such as cereals, this did not lead to the development of many other salt-tolerant crops. As a wide range of halophytes is available, this approach is still being considered for generating salt-tolerant crops (Flowers, 2004).

33.1.1 Conventional Breeding Approach

Conventional breeding approaches have been explored to develop salt-tolerant varieties by integrating traits from wild parent species into superior crop varieties. The most notable results of this approach are some salt-tolerant cereals. Since, salinity tolerance is a complex and multigenic trait, this approach has had very limited success, as no commercial varieties have become available so far. The introduction of a gene decreasing Na+ uptake by durum wheat is a notable exception (Munns *et al.*, 2012). Identification and validation of important traits contributing to tolerance would be very crucial for trait introgression through conventional breeding. Conventional breeding has several disadvantages as the process is labor-intensive and time-consuming and depends on germplasm collections (Hanin *et al.*, 2016).

33.1.2 Transgenic Approach

Genetic-modification techniques have been extensively used to introduce very specific genetic changes into commercial varieties, thereby maintaining the elite traits found in commercial crops. Expression of foreign genes improved the level of salinity tolerance in various plant species (Srivastava *et al.*, 2016). Physiological traits targeted by genetic modification include the plant cell metabolic pathway, for instance, to increase the accumulation level of osmoprotectants when plants are exposed to saline conditions (Carillo *et al.*, 2011), to quench reactive oxygen species (ROS) produced during salinity-induced oxidative stress (Roy *et al.*, 2014), to reduce lipid peroxidation of membranes (Alvarez Viveros *et al.*, 2013) and to protect protein structural integrity and their functioning (Jha, 2019). As Na+ uptake and compartmentalization into the vacuole is a critical mechanism for plant survival under salinity stress, most work on developing salt-tolerant crops has been concentrated on ion transporter genes and the genes controlling their activity. Overexpressing genes encoding membrane transporters improved salinity tolerance in many crops (Gupta and Huang, 2014). Also, for the genetic-modification approach, the complex nature of salt-stress tolerance and our limited understanding of the processes or traits involved has been an obstacle for the successful introduction of tolerant crop varieties (Shrivastava and Kumar, 2015).

33.1.3 Plasticity Induced by Endophytes Increasing Ecological Amplitude

33.1.3.1 Ecological Amplitude

The range of conditions a species can survive in is known as the ecological amplitude of that species. Often this range is defined by the limitations imposed by abiotic conditions affecting plant functioning, such as temperature, light and precipitation. The species belonging to a particular ecosystem will have adaptations and responses that have been shaped by the prevailing environmental factors for that ecosystem (Ogbemudia *et al.,* 2018). All land plants possess regulatory mechanisms to respond to suboptimal environmental conditions, such as salinity, high and low temperature and drought. The ability to effectively tolerate stress conditions can differ strongly between species and is associated with the genetic traits a species has (Smallwood *et al.,* 1999). However, inherent plasticity and the association with microbial symbionts can also confer stress tolerance and a wider ecological amplitude.

33.1.3.2 Phenotypic Plasticity

The phenotype of a plant is the result of the interactions between its genotype and environment. Phenotypic plasticity is defined as the variation in phenotypic expression of a genotype in response to different environmental conditions, where it enhances the fitness and reproduction of an individual under those conditions. Plasticity may increase the chances of a population's persistence through time by allowing certain reversible, environment-induced changes in morphology and metabolism, and, thus, increase the geographical range of a species. Most plant species are extremely plastic by nature. In plants, many phenotypic characteristics of a species are strongly influenced by different environmental factors, much more so than in most animal species (Sultan, 1987).

The semi-woody vine, Bittersweet nightshade (*Solanum dulcamara*), has a widespread occurrence in ecologically contrasting habitats, ranging from wetlands to coastal dunes (Dawood *et al.,* 2014). Environmental conditions have a strong influence on many phenotypic characteristics of *S. dulcamara*. It survives under contrasting environments, such as flood-prone and dry habitats, by producing adventitious roots from preformed dormant primordia, and having a high root to shoot ratio, respectively. The plant is being extensively investigated as a model system for adaptive mechanisms, such as phenotypic plasticity and local adaptation, to cope with the contrasting environmental factors (Zhang *et al.,* 2016).

33.1.3.3 Symbiont-Induced Stress Tolerance

To various degrees, plants form symbioses with endosymbionts, which means that they are inhabited internally by diverse microbial communities. These microorganisms with endophytic lifestyles can play crucial roles in plant development, growth, fitness, and diversification (Hardoim *et al.,* 2015). Plants provide a unique ecological niche for diverse communities of symbiotic microbes which, in turn, have various benefits for the plant, such as the promotion of growth, nutrient, water use and photosynthetic efficiency, and adaptation to biotic and abiotic stresses.

Symbiotic fungi may contribute to plant adaptation to environmental stresses (Clay and Holah, 1999; Redman *et al.,* 2002) and thus are crucial to the overall development of plant communities (Petrini, 1986; Bacon and Hill, 1996; Rodriguez and Redman, 1997). There are two major groups of fungal symbionts known to be associated with plants, mycorrhizal fungi which reside only in roots but extend out into the rhizosphere, and endophytic fungi which reside entirely within the plant tissues and are associated with leaves, stems and roots (Rodriguez *et al.*, 2004). Symbiotic association with mycorrhizal or endophytic fungi can provide fitness benefits to several crop species (Redman *et al.*, 2002; Waqas *et al.*, 2012; Hamayun *et al.*, 2017).

Endophytes isolated from grasses confer stress tolerance to genetically distant plants, such as tomato and rice. Although, the evolutionary divergence of these plants occurred already approximately 400 million years ago, still they depend on their symbionts (Rodriguez and Redman, 2008; Redman *et al.*, 2011). Fungal endophytes, the species that colonize the whole plant are known to confer stress tolerance through habitat adapted symbiosis (HAS) mechanism where plants are proposed to associate with particular endophytes that increase tolerance or resistance to the predominant biotic or abiotic stresses of their habitats. Rodriguez et al. (2008) defined this phenomenon by endophytic fungi as an intergenomic epigenetic mechanism for plant adaptation and survival in high-stress environments, indicating that the increased stress resistance and effect of epigenetic changes are induced by the symbiont. Utilizing endophytes to mitigate climate change impacts has been suggested as a potential strategy for expanding agricultural production on marginal lands (Rodriguez *et al.*, 2008).

33.1.3.4 S. Dulcamara – Microbiome – Existing Gap

Although *S. dulcamara* with its extremely wide ecological amplitude has been adopted as a model system for acclimation to changing environmental conditions, its associated microbiome and the interaction between plant and microbiome are hardly studied. *S. dulcamara* and its endophyte community were, therefore, chosen in this study as a model for utilization of endophytic services as an adaptive strategy of plants to survive under various abiotic stress conditions. If *S. dulcamara* plants indeed depend on their associated endophytic fungal community to maintain plant fitness under abiotic stress conditions, comparing the endophytic community found in plants from different habitats will provide a much sought-after test of the 'habitat adapted symbiosis' hypothesis.

33.1.3.5 Fungal Endophytes

Endophytes are a huge and diverse group of fungi that colonize healthy plant tissues (Wilson, 1995; Pawłowska *et al.*, 2014). They reside in the intercellular spaces beneath the epidermis (Petrini, 1986) and are an enigmatic component of every terrestrial plant community, presenting us with intriguing evolutionary questions, relating to the origin and nature of this symbiosis and to the widespread dependence on symbionts for resistance to quite common stresses. Four major classes of endophytes are being distinguished: 1) the clavicipitaceous endophytes, 2) the non-clavicipitaceous endophytes that colonize the whole plant, 3) the non-clavicipitaceous endophytes that colonize shoots and 4) the non-clavicipitaceous endophytes that colonize roots (Rodriguez *et al.*, 2009). All land plants are known to be colonized by class 3 endophytes which are transmitted horizontally (Davis *et al.,* 2003). The symbiosis of the fungi found as endophytes could be mutualistic, providing fitness benefits, be commensalistic, or be dormant pathogenic, residing in the plants without any positive or negative effects (Lewis, 1985) until the conditions are favorable for disease development (Dodd, 1980). Endophytes may act as pathogens or mutualists by indirectly affecting the host health (Fravel *et al.,* 2003). Endophyte lineages might have evolved from pathogens over time as commensals and as a result now lack the traits that confer pathogenicity.

33.1.3.6 Salinity – As a Case Study

Soil salinization is slowly developing into a major global threat. An increasing percentage of arable land is affected by salinization, due to a rise in sea level and irrigation practices (Sanders, 2020). It is estimated that 1125 million hectares of land are affected by salt, of which, 76 million hectares are affected by human-induced salinization (Hossain, 2019).

In general, salinity limits the productivity in several, economically important crop plants. Munns et al. (1995) proposed a two-phase model to describe the osmotic and ionic effects of salinity on plants. High salinity induces rapid osmotic stress and there is also a delayed ionic stress, especially due to a high Na^+ concentration (Gupta and Huang, 2014). Osmotic stress affects various physiological processes, such as a reduction in photosynthetic activity and the production of ROS, by decreasing the water absorption by roots (Munns and Tester, 2008). Ionic stress is caused by an increased accumulation of ions, which affects the metabolic homeostasis (Greenway and Munns, 1980; Zhu, 2001). The osmotic stress effect is characterized by a rapid inhibition of growth rate of young leaves, while the ionic effects typically result in an increased rate of senescence in older leaves (Sekmen *et al.,* 2013). Our knowledge on how a plant senses salt stress is meager. The rate at which the plant senses toxicity differs greatly between salt-tolerant and -sensitive plants. During the first, osmotically induced phase, growth is inhibited in both tolerant and sensitive plants. During the second, ionic toxicity phase, the photosynthetic capacity is reduced and older leaves start senescing in salt-sensitive plants.

Plants experience a wide range of constraints under salinity stress. These constraints include accumulation of sodium ions in the rhizosphere, reduced water availability due to hyperosmotic stress induced by salinity, a rapid accumulation of ROS in plant tissues and alterations in cytosolic Ca^{2+} and K^+ homeostasis. Every constraint could be sensed by either plasma membrane bound or cytosolic sensors, triggering a cascade of events that ultimately leads to acclimation to the stress condition (Shabala *et al.,* 2015).

33.1.4 Classification of Plants Based on Their Response to Salinity

Based on their tolerance or sensitivity, plants are often classified as halophytes or glycophytes. Halophytes are species that can withstand a salt concentration of about 200mM, using specialized anatomical or morphological adaptations or avoidance mechanisms (Flowers *et al.,* 2010). These mechanisms help plants to survive and maintain growth under unfavorable conditions. Examples of traits found in these specialized halophytic species are salt glands extruding salt on the leaves (for instance, in *Avicennia marina*, a mangrove species) and well developed Casparian strips with early vacuolation and suberization of the hypo- and endodermis in roots (for instance in *Suaeda maritima*, Seepweed) (Sekmen *et al.*, 2013).

In contrast, glycophytes, evolved in areas with low soil sodium levels, are highly sensitive to salinity and have to maintain low sodium levels in their aboveground tissues by restricting the upward movement of excess ions to the shoot (Cheeseman, 2015). In glycophytic plants, root and shoot growth are already inhibited when exposed to moderate levels of salinity (Greenway and Munns, 1980).

33.1.5 Salinity Tolerance Mechanism in Plants

Salt tolerance is the complex mechanism which is often controlled by inter-correlated metabolic pathways. It is achieved by detoxification, ion homeostasis and maintenance of growth. First, the detoxification is based on the production of antioxidants and the enzymes responsible for the inactivation of radical oxygen species, such as superoxide dismutase and catalase (CAT). This is followed by the synthesis of compatible osmolytes such as glycine betaine, proline or sugar alcohols to maintain osmotic balance. Ionic homeostasis at the cellular level is accomplished through compartmentalization of excess ions in the vacuole. Finally, the expression of regulatory genes, like transcription factors, help to maintain plant growth at a high salt concentration (McCord, 2000; Zhu, 2001; Flowers and Colmer, 2008; Flowers *et al.*, 2010). Salt-tolerance mechanisms involve the complex interplay between the several genes associated with biochemical and cellular pathways. The up-regulation of several miRNAs is essential for alleviating salt stress (Goswami *et al.*, 2017).

Many salt responsive genes have been identified in plants, and several signaling cascades involved in the response to salinity have been predicted and validated in several model plants. In Arabidopsis Na^+, homeostasis is controlled by the Salt Overly Sensitive pathway (Zhu, 2002; Jamil *et al.*, 2011; Zhang *et al.*, 2012). It is stated that 'salinity tolerance is multigenic and needs to be viewed in a wider context than just that of membrane transport of NaCl'(Sanders, 2020).

33.1.6 Endophytic Fungi-Mediated Salinity Stress Tolerance

Unlike halophytes, glycophytes does not produce any special anatomical structures such as salt glands, or well developed Casparian strips, or suberization of hypo- or endodermis as an adaptive strategy. In order to overcome the effects of salinity, glycophytes have to maintain low Na^+ levels in the cytosol at the cellular level and low shoot Na^+ concentrations at the whole plant level. Maintenance of high cytosolic K^+/Na^+ ratios in shoots is suggested to be crucial for salt tolerance in glycophytes (Horie *et al.*, 2012; Cheeseman, 2015). In addition to their inherent mechanisms to maintain low sodium levels, glycophytes may also depend on their endosymbionts under saline conditions.

Most of the domesticated crop plants are glycophytes. These might be improved for better adaptation to saline environments with the use of beneficial symbionts. The microbial approach is considered as one of the environmental-friendly approaches to tackle the salinity issue in sustainable agriculture (Shokat and Großkinsky, 2019).

Studies have shown that inoculation with endophytic fungi enhances salinity tolerance in plants (Rodriguez *et al.*, 2004; Waller *et al.*, 2005; Baltruschat *et al.*, 2008; Khan *et al.*, 2011; 2012; 2013). Endophytes isolated from plants growing in a saline environment have the potential to confer salinity tolerance to other plants via habitat adapted mechanism (Rodriguez *et al.*, 2004). The endophyte *Fusarium culmorum*, FcRed1, isolated from a dune grass, *Leymus mollis* on coastal beach habitats of San Juan Island Archipelago, WA, shown to confer salinity tolerance to dune grass and tomato (Rodriguez *et al.*, 2008). Plants adapted to harsh environmental conditions like sand dunes and salt marshes in Mediterranean ecosystems harbor huge biodiversity of fungal root endophytes. These fungi could protect against the environmental stresses and reduce the symptoms caused by root pathogens (Lopez-Llorca and Macia-Vicente, 2009). Commercial rice varieties have had increased fitness benefits against several abiotic stress when inoculated with endophytes via HAS mechanism (Redman *et al.*, 2011). Tomato plants colonized by the endophytic fungi *Alternaria* spp. and *Trichoderma harzianum* had better water use efficiency, higher biomass and photosynthetic efficiency than the control plants when exposed to salt or drought stress (Azad and Kaminskyj, 2016).

The plant root colonizing, basidiomycete fungus *Piriformospora indica,* isolated from plants growing in the Indian Thar desert, is being one of the best-studied root endophytes so far, to significantly promote growth (30% increase in fresh weight of Chinese cabbage at 200 mM NaCl, Khalid *et al.*, 2018) as well as to improve salinity tolerance in symbiotic association with a wide variety of plant species such as barley, tomato, rice, soybean (Verma *et al.*, 1998; Waller *et al.*, 2005; Baltruschat *et al.*, 2008). Endophytic fungi, *Ampelomyces* sp. or *P. chrysogenum* improved growth and drought and salinity stress tolerance in tomato (Morsy *et al.*, 2020).

The endophytic fungus *Penicillium minioluteum* which was isolated from the roots of field-grown soybean is reported to synthesize gibberellins (GAs) and promote the growth of mutant rice line Waito-C and improved salinity stress tolerance in Soybean (Khan *et al.*, 2011). Pretreatment of soybean seeds with an endophytic fungus, *Fusarium verticillioides*, proved to be an effective method to improve growth under salinity stress conditions (Radhakrishnan *et al.*, 2013). An endophyte, *Aspergillus flavus* CSH1, confers salinity tolerance in Soybean, by modulating its phytohormone levels and antioxidant system (Lubna *et al.*, 2018). A similar effect was found in cucumber plants by an endophytic fungus, *Paecilomyces formosus* LHL10. In addition to GAs, endophytes also produced indole acetic acid to improve growth (Khan *et al.*, 2012). Another fungal endophyte, *Penicillium janthinellum* LK5 isolated from roots of tomato plants, improved the growth of abscisic acid (ABA)-deficient mutant *sitiens* plants under salinity (Khan *et al.*, 2013). *P. indica* alleviated salt stress in salt-sensitive barley cultivar Ingrid through decreasing lipid peroxidation, metabolic heat efflux and fatty acid desaturation in the leaves. Also, a significant increase in the amount of ascorbic acid and other antioxidant enzymes was recorded in barley root under salinity stress. It states that antioxidants might play a role in both endophyte-mediated plant tolerance to salinity (Baltruschat *et al.*, 2008).

Tomato plants inoculated with *P. indica* had a higher level of photosynthetic pigments and osmoprotectants like proline and glycine betaine under different salinity gradient (Ghorbani *et al.*, 2018). Plants inoculated with endophytes had an enhanced expression NHX1 gene related to salt tolerance which, in turn, regulates iron homeostasis. Inoculated plants had better osmotic tolerance under salinity stress by maintaining a higher Na^+ level in leaves (Molina-Montenegro *et al.*, 2018).

Functional characterization revealed that PiHOG1 plays a significant role in the salinity response of *P. indica* as well as helps the host to overcome salinity stress. PiHOG1, a stress regulator MAP kinase from *Piriformospora indica*, confers salinity stress tolerance in rice plants. HOG1 MAP kinase plays a vital role in the osmo-adaptation pathway in different fungi as it required to restoring osmotic pressure by increasing glycerol accumulation (Jogawat *et al.*, 2016). Random overexpression of *P. indica* genes (cDNA library) in *Escherichia coli* identified six genes that were upregulated in response to salinity. These are genes encoding cyclophilin, stearoyl-CoA desaturase, thiamine pyrophosphate-binding domain-containing protein, BCL-2 associated athanogene 3-like protein, cytochrome P450 and 60S ribosomal protein genes upregulated in 400mM NaCl-treated *P. indica* compared to the control. These genes are involved in different cellular processes such as metabolism, biosynthetic processes, DNA repair, protein turnover, transport and salt tolerance (Gahlot *et al.*, 2015).

There are several reports suggesting a role of endophytic fungus in stress tolerance and growth promotion (Rodriguez *et al.*, 2004; Waqas *et al.*, 2012; Hamayun *et al.*, 2017; Dhanyalakshmi *et al.*, 2019). However, the mechanism by which both fungal and plant partners are interacting and how stress tolerance is conferred by endophytic fungi are poorly understood (Johnson *et al.*, 2018; Hyde *et al.*, 2019).

33.1.7 Possible Mechanism of Salt Tolerance by Endophytes

The mechanism by which endophytic fungi improve stress tolerance in plants is still intriguing. Several attempts have been made to unravel the stress tolerance mechanism using *Piriformospora indica* and other endophytic fungi as model systems (Table 33.1). It is reported that *P. indica* colonization elevated the antioxidant capacity of plants to cope with oxidative stress induced by salinity in several plants, such as barley (Waller *et al.*, 2005; Baltruschat *et al.*, 2008), *Medicago truncatula* (Li *et al.*, 2017) and tomato (Abdelaziz *et al.*, 2019). Colonized plants had 1.5- and 8-fold higher levels of the osmoprotectant proline, compared to non-colonized rice and *Medicago truncatula*, respectively (Jogawat *et al.*, 2013; Li *et al.*, 2017). In symbiotic plants, the level of photosynthetic pigment like chlorophyll was higher, whereas the anthocyanin was lower in Rice, Arabidopsis and Tomato, growth was higher and ion homeostasis was better maintained when exposed to salt (Jogawat *et al.*, 2013, Abdelaziz *et al.*, 2017; 2019). In Maize, *P. indica* colonization improved stomatal regulation and improved K+ transport into the shoots (Yun *et al.*, 2018). The endophytic fungus, *Porostereum spadiceum,* modulated the level of phytohormones in Soybean,

TABLE 33.1
List of Fungal Endophytes Conferring Salinity Tolerance in Various Crops

Fungal endophytes	Crops	Mechanisms	References
Alternaria spp. and *Trichoderma harzianum*	Tomato	Improved the water use efficiency, higher biomass and photosynthetic efficiency	Azad and Kaminskyj, 2016
Fusarium culmorum, FcRed1	Dune grass, Tomato	Habitat adapted symbiosis - Intergenomic epigenetic mechanism;	Rodriguez *et al.*, 2008
Fusarium sp.	Rice	Improved assimilation rate and chlorophyll stability index;	Sampangi-Ramaiah *et al.*, 2020
Paecilomyces formosus LHL10	Cucumber	Accumulated higher levels of phytohormone such as gibberellic acid and indole acetic acid	Khan *et al.*, 2012
Penicillium brevicompactum and P. chrysogenum	Tomato and lettuce	Improved ecophysiological performance; enhanced expression of NHX1 gene	Molina-Montenegro *et al.*, 2018
Penicillium janthinellum LK5	Tomato	Decreasing lipid peroxidation and significantly increasing antioxidant enzyme activities; higher amount of abscisic and reduced levels of jasmonic acid	Khan *et al.*, 2013
Penicillium minioluteum; *Fusarium verticillioides;* *Aspergillus flavus* CSH1,	Soybean	Production of phytohormones and activating defense mechanisms of the host;	Khan *et al.*, 2011; Radhakrishnan *et al.*, 2013; Lubna *et al.*, 2018
Piriformospora indica	Barley, *Medicago truncatula,* Tomato Maize Tomato	Antioxidant capacity; Higher levels of the osmoprotectant; upregulated expression of several defense related genes and transcription factors Improved stomatal regulation and the K+ transport into the shoots By improving plant water status and gas exchange characteristics	Waller *et al.*, 2005; Baltruschat *et al.*, 2008; Li *et al.*, 2017; Abdelaziz *et al.*, 2019 Yun *et al.*, 2018 Ghorbani *et al.*, 2018
Porostereum spadiceum	Soybean	Increasing the levels of GA and isoflavones, and reducing levels of ABA and JA	Hamayun *et al.*, 2017
Ampelomyces sp. or *P. chrysogenum*	Tomato	Unknown mechanism	Morsy *et al.*, 2020

increasing the levels of GA and isoflavones, and reducing levels of ABA and JA under salt stress (Hamayun *et al.*, 2017). In *M. truncatula*, it was shown that *P. indica* colonization upregulated expression of several defense-related genes and transcription factors (Li *et al.*, 2017). In summary, endophytic fungi help plants to cope with salt stress through the production of phytohormones and activating antioxidant defense mechanisms of the host (Khan *et al.*, 2011; 2012; 2013; Radhakrishnan *et al.*, 2013; Baltruschat *et al.*, 2008). It is also shown to improve the ecophysiological performance such as plant water status and gas exchange characteristics of the host under saline conditions (Ghorbani *et al.*, 2018; Molina-Montenegro *et al.*, 2018).

So far there are no indications that symbiotic fungi protect plants from the effect of abiotic stress through totally novel mechanisms, but rather through increased mobilization of inherent stress protection processes. Nevertheless, the information available on the mechanisms of the acclimation response conferred by endophytic fungi still cannot answer the questions of *why* plants do depend on the presence of EF for their full stress resistance response. These studies have provided a *proof of concept* that endophytic fungi could be harnessed to ameliorate plant growth under abiotic stresses and provide an additional and alternate approach toward management of these stresses in crop plants.

33.1.8 Criteria to Select Potential Endophytic Fungal Isolate

In the next paragraph, we examine the ecological significance of endophytic fungi-induced stress tolerance in *Solanaceae* plants and test the hypothesis (Figure 33.1)

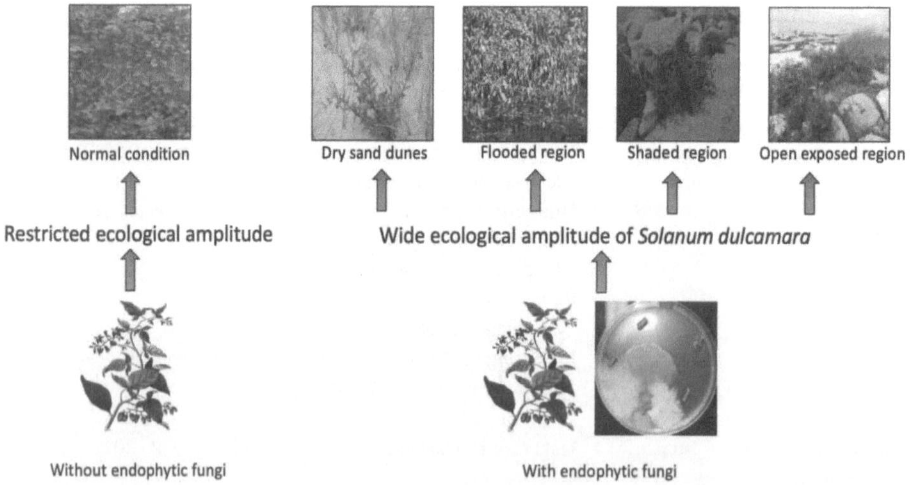

FIGURE 33.1 A hypothetical model showing endophytic fungi may increase ecological amplitude of *Solanum dulcamara*. When the plant has no fungal symbiont, it has a restricted ecological amplitude, whereas the plant harboring fungal symbiont has the potential to expand into a wider range of environmental conditions.

that endophytic fungi are instrumental in the enormous phenotypic plasticity of *Solanum dulcamara* under contrasting habitats. If this hypothesis holds true, a potential candidate should fulfill the following criteria.

A potential isolate should be able to: (1) colonize the host/selected non-host plants, (2) grow on high concentrations of salt/under simulated home stress conditions, and (3) improve stress tolerance under simulated stress conditions when inoculated to plants. To test our hypothesis, we sampled the dry coastal dunes and permanent wet regions in the island, Schiermonnikoog. Five plants were collected from each location and brought to the laboratory and processed same day. Figure 33.2 explains the methodology in brief.

33.2 RESULTS

The results showed that almost all plant parts were colonized by endophytic fungi (Figure 33.3). In total, 242 isolates were isolated from surface sterilized plant tissues, out of which 153 were identified based on Internal Transcribed Spacer sequencing. Colonization frequency was higher in leaves and roots, followed by stems. The identified fungi were divided into 21 genera from dry and 17 genera from wet habitats.

The endophytic fungal community differed between both habitat and plant parts (Figure 33.4). Based on the occurrence in the two habitats, the isolates were classified as common (occurring in both habitats) or specific (exclusively occurring in one of the habitats) (Figure 33.5). In both habitats, the genus *Alternaria* was dominant in leaves. From dry habitats, the isolate *Boeremia exigua* from stem and *Harzia velata* from roots were more abundant. From the wet habitat, the isolate unknown fungus_4 and *Colletotrichum coccodes* were recovered more frequently. In general, the genus, *Alternaria* accounted for highest proportion in dry and wet habitat (38% from the dry and 40% from the wet habitat).

33.2.1 In Vitro Screening for Potential Isolates

Based on their occurrence in common and specific habitats, twenty-one isolates were selected and tested for their salt-stress tolerance. Our assumption for this experiment is, if the isolate itself is able to tolerance high-stress condition, it could potentially also provide stress tolerance to plants upon inoculation and since most endophytic species have to infect every plant individually, it needs to be able to survive in saline soil to be effective. Almost all the isolates tested were grown up to 300 mM NaCl1 concentration and a few were grown at 600 and 1200 mM (Figure 33.6). The morphology of the endophytic fungi isolates changes with the salt treatment. The change in morphology is probably due to different metabolic reactions activated by different salt concentrations.

Based on *in vitro* screening, a few species were identified as salt tolerant. Recently, a method has been developed to test our isolates of interest using *P. indica,* as a positive control in Chinese cabbage under different water level. In brief, the experimental setup is as indicated in Figure 33.7. The results showed that plants inoculated with

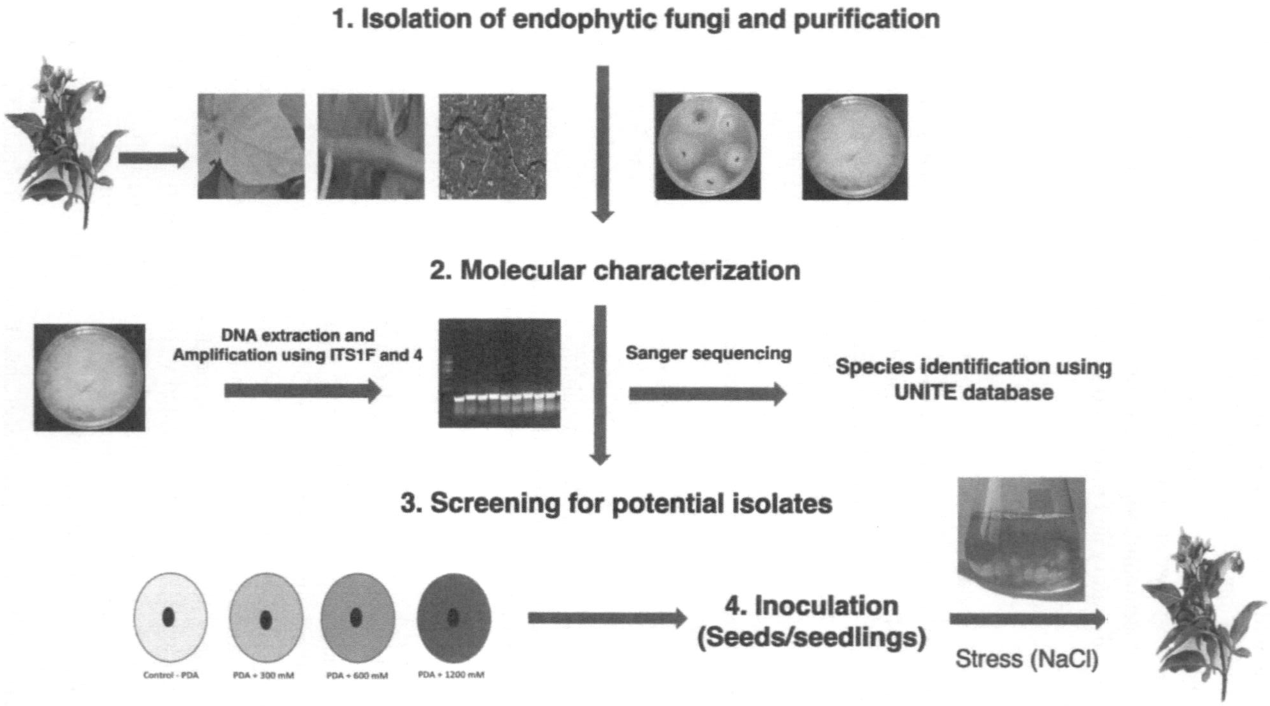

FIGURE 33.2 Overview of the approach used to identify the potential isolate. 1. Endophytic fungi were isolated from surface sterilized leaves, stem and root segments of *S. dulcamara* and cultured on potato dextrose agar media and purified into a single isolate. 2. DNA was extracted from each isolate and their ITS region were amplified using ITS1F and four primers. Samples were subjected to Sanger sequencing. The species were identified using UNITE database based on sequence homology. 3. The isolates were taxonomically classified and grouped into common and specific isolates based on their habitats. Screening was done to assess the salt-tolerance ability of selected fungal isolates, in which the fungal isolates were cultured under different salt concentrations and growth patterns were observed. 4. After selecting the potential candidates, plants were inoculated with those isolates and their ability to impart their abiotic stress tolerance ability to plants was assessed.

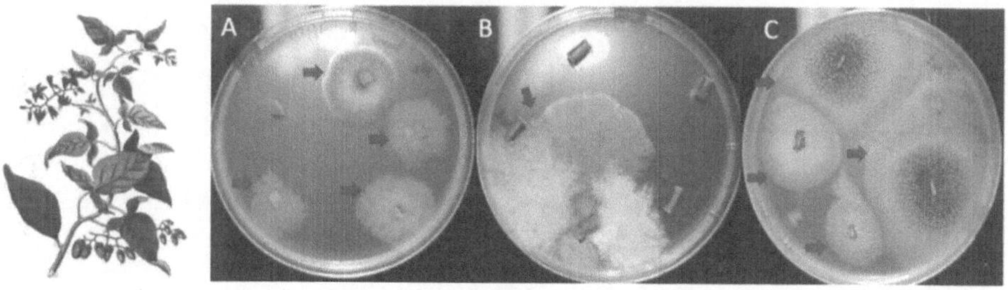

FIGURE 33.3 Isolation of endophytic fungi from *Solanum dulcamara*. Arrow heads indicate the emergence of endophytic fungi from the surface sterilized tissue segments. A-leaf, B-stem and C-root.

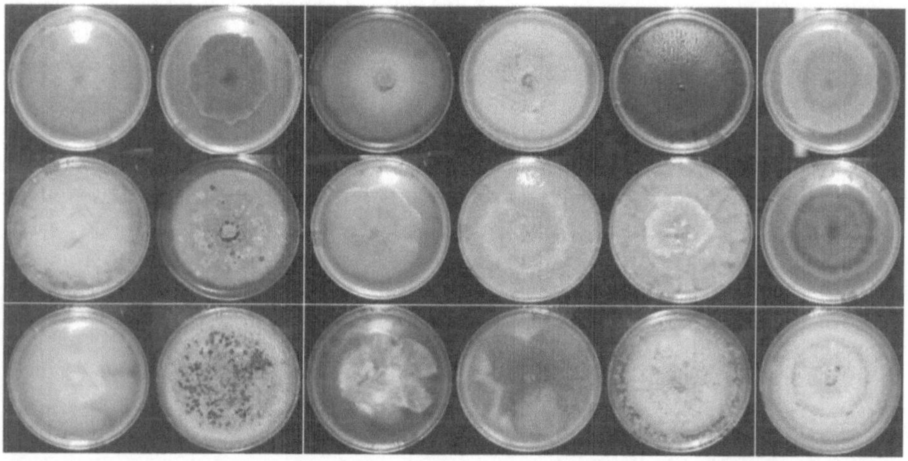

FIGURE 33.4 Diversity of endophytic fungi from *Solanum dulcamara* collected from contrasting habitats. From left to right: Row I: 1. *Gibberella avenacea* (W2); 2. *Alternaria arborescens* (W8); 3. *Epicoccum nigrum* (W36); 4. *Alternaria rosae* (W55II); 5. *Colletotrichum coccodes* (W46I); 6. *Alternaria tenuissima* (W12). Row II: 1. *Fusarium* sp. (W60); 2. *Epicoccum nigrum* (W74); 3. *Phomopsis velata* (D27); 4. W83; 5. *Diaporthe cotoneastri* (W118); 6. *Alternaria alternata* (D23). Row III: 1. *Neofabraea vagabunda* (D32); 2. *Boeremia exigua* (D74); 3. *Aureobasidium pullulans* (D107); 4. *Harzia velata* (D36I); 5. *Colletotrichum godetiae* (D83); 6. Pleosporaceae (D88). W and D indicate wet and dry habitats, respectively.

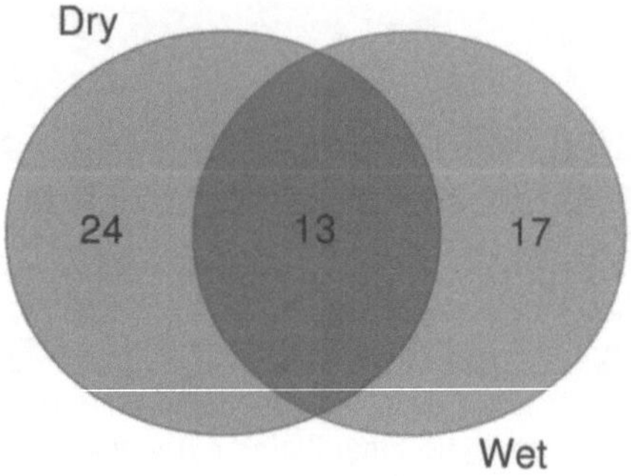

FIGURE 33.5 A Venn diagram represent the common (occurring in both habitats) and specific (exclusively occurring in one of the habitats) isolates recovered from *Solanum dulcamara* in contrasting habitats. In total, 24 and 17 isolates were specific to dry and wet habitat, respectively. 13 isolates were found in both dry and wet habitat.

P. indica accumulated higher photosynthates compared to the non-inoculated plants under different water levels. This was more evident in the biomass of the plants. This setup could be used further to test the abilities of selected isolates to confer salt/ drought stress tolerance.

As mentioned previously, the potential isolate has to fulfill three criteria. From the results, it is evident that the isolates fulfilled the first two criteria, that is, they are able to colonize the plant tissues and grow under high salt concentrations, as well as osmotic and drought stress conditions. The third criteria, the ability to impart stress tolerance upon inoculation with plants, is presently being tested in tomato a model for use in other Solanaceous crops.

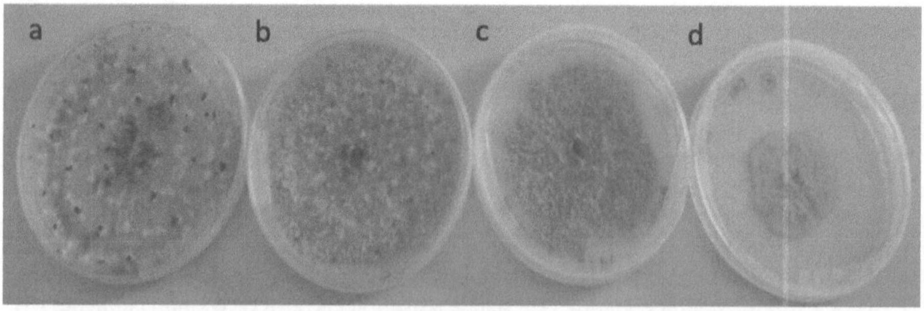

FIGURE 33.6 *In vitro* screening of endophytic fungus from *Solanum dulcamara* against salinity stress. Growth pattern of endophytic fungus, *Phomopsis velata* under different NaCl concentrations. (a) 0, (b) 300, (c) 600, and (d) 1200 mM NaCl. *P. velata* grew better under both control and 300 mM NaCl concentration. Under 600 and 1200 mM NaCl concentration, *P. velata* showed the mycelial growth inhibition of about 33 and 75%, respectively, compared to control.

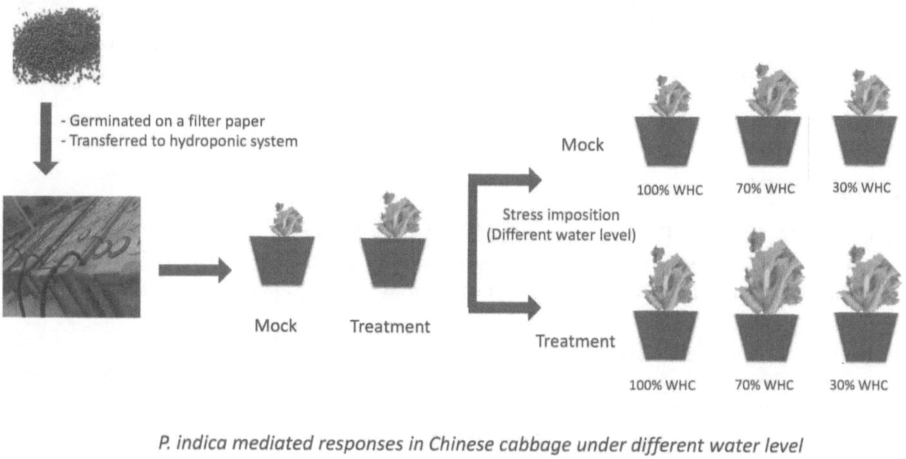

FIGURE 33.7 Experimental setup to test the potential isolate. *P. indica* mediated responses in Chinese cabbage under different water level. Seeds of Chinese cabbage were surface sterilized and germinated on a filter paper. After 4 days, the germinated seedlings were transferred to hydroponic system for better acclimatization. Uniform seedlings were transplanted onto the pots containing sterile sand. Before transplanting, one group of plants was treated with the 1% mycelial suspension and the other group of seedlings was mock treated with broth without fungus. Additionally, 2 ml of mycelial suspension/broth without fungus were pipetted out near the root zone. These seedlings were left for 4–6 days to establish a successful colonization of *P. indica*. Later, the plants were subjected to stress conditions; in this case, we imposed drought stress by maintaining certain water level (100, 70 and 30% water holding capacity-WHC). Once the required water level was achieved, these pots were maintained under stress for 10 days. Eco physiological measurements were done before and after imposing stress.

33.3 CONCLUSION

Recent studies have shown that symbioses with endophytes are of vital importance in the distribution of plant communities worldwide. These symbioses in many cases are responsible for adaptation of the plant to the biological and environmental stresses. Exploitation of such plant symbiotic organisms could be a potential approach leading to sustainable development of the agricultural production and ecosystem management. The novel endophytes identified from this study will be evaluated for their potential to improve crop productivity under suboptimal conditions such as salinity.

ACKNOWLEDGMENT

This research has been carried out in the Laboratory of Plant Physiology (GREEN) at the Groningen Institute for Evolutionary Life Sciences (GELIFES) according to the requirements of the Graduate School of Science and Engineering (Faculty of Science and Engineering, University of Groningen; Groningen, the Netherlands). This research was supported by an Adaptive Life Programme grant awarded to

Sasirekha Munikumar, J.T(Theo). M. Elzenga, J. D(Dick) van Elsas and Nataraja N. Karaba made possible by the Board of the University of Groningen, the Faculty of Science and Engineering and the Groningen Institute for Evolutionary Life Sciences (GELIFES).

REFERENCES

Abdelaziz, M. E., Kim, D., Ali, S., Fedoroff, N. V. and Al-Babili, S. (2017). The endophytic fungus *Piriformospora indica* enhances *Arabidopsis thaliana* growth and modulates Na^+/K^+ homeostasis under salt stress conditions. *Plant Science.* 263, 107–115. https://doi.org/10.1016/j.plantsci.2017.07.006.

Abdelaziz, M. E., Abdelsattar, M., Abdeldaym, E. A., Atia, M. A. M., Mahmoud, A. W. M., Saad, M. M. and Hirt, H. (2019). *Piriformospora indica* alters Na^+/K^+ homeostasis, antioxidant enzymes and LeNHX1 expression of greenhouse tomato grown under salt stress. *Scientia Horticulturae*, 256, 108532. https://doi.org/10.1016/j.scienta.2019.05.059.

Alvarez Viveros M. F., Inostroza-Blancheteau, C., Timmermann, T., González, M. and Arce-Johnson, P. (2013) Overexpression of *GlyI* and *GlyII* genes in transgenic tomato (*Solanum lycopersicum* Mill.) plants confers salt tolerance by decreasing oxidative stress. *Molecular Biology Report*, 40, 3281–3290.

Azad, K. and Kaminskyj, S. (2016). A fungal endophyte strategy for mitigating the effect of salt and drought stress on plant growth. *Symbiosis*, 1, 73–78.

Bacon, C. W. and Hill, N. S. (1996). Symptomless grass endophytes: Products of coevolutionary symbioses and their role in the ecological adaptations of grasses, in S.C. Redkin and L.M. Carris (eds.), *Endophytic Fungi in Grasses and Woody Plants*, St. Paul, MN, APS Press, pp. 155–178.

Baltruschat, H., Fodor, J., Harrach, B. D., Niemczyk, E., Barna, B., Gullner, G. and Skoczowski, A. (2008). Salt tolerance of barley induced by the root endophyte *Piriformospora indica* is associated with a strong increase in antioxidants. *New Phytologist*, 180(2), 501–510.

Carillo, P., Annunziata, M. G., Pontecorvo, G., Fuggi, A. and Woodrow, P. (2011). Salinity stress and salt tolerance, in A. Shanker and B. Venkateswarlu (eds.), Abiotic Stress in Plants - Mechanisms and Adaptations, IntechOpen, pp. 21–38. https://doi.org/10.5772/22331.

Cheeseman, J. M. (2015). The evolution of halophytes, glycophytes and crops, and its implications for food security under saline conditions. *New Phytologist*, 206(2), 557–570. https://doi.org/10.1111/nph.13217.

Clay, K. and Holah, J. (1999). Fungal endophyte symbiosis and plant diversity in successional fields. *Science*, 285, 1742–1744.

Davis, E. C., Franklin, J. B., Shaw, A. J. and Vilgalys, R. (2003). Endophytic *Xylaria* (Xylariaceae) among liverworts and angiosperms: Phylogenetics, distribution, and symbiosis. *American Journal of Botany*, 90(11): 1661–1667.

Dawood, T., Rieu, I., Wolters-Arts, M., Derksen, E.B., Mariani, C. and Visser, E. J. W. (2014). Rapid flooding induced adventitious root development from preformed primordia in *Solanum dulcamara. AoB Plants*, 6, plt058v1.

Dhanyalakshmi, K. H., Mounashree, D. C., Vidyashree, D. N., Earanna, N. and Nataraja, K. N. (2019). Options and opportunities for manipulation of drought traits using endophytes in crops. *Plant Physiology Reports*, 24(4): 555–562.

Dodd, J. (1980). The role of plant stresses in development of corn stalk rots. *Plant Disease*, 64, 533–537.

Flowers T.S. and Yeo A.R. (1989). Effects of salinity on plant growth and crop yields, in J.H. Cherry (eds.), *Environmental Stress in Plants.* NATO ASI Series (Series G:

Ecological Sciences), Berlin, Heidelberg, Springer, 19: 101–119. https://doi.org/10.1007/978-3-642-73163-1_11.

Flowers, T. J. (2004). Improving crop salt tolerance. *Journal of Experimental Botany*, 55(396), 307–319. https://doi.org/10.1093/jxb/erh003.

Flowers, T. J. and Colmer, T. D. (2008). Salinity tolerance in halophytes. *New Phytologist*, 179, 945–963.

Flowers, T. J., Galal, H. K. and Bromham, D. L. (2010). Evolution of halophytes: Multiple origins of salt tolerance in land plants. *Functional Plant Biology*, 37, 604–612.

Fravel, D., Olivain, C. and Alabouvette, C. (2003). *Fusarium oxysporum* and its biocontrol. *New Phytologist*, 157, 493–502.

Gahlot, S., Joshi, A., Singh, P., Tuteja, R., Dua, M., Jogawat, A., andTuteja, N. (2015). Isolation of genes conferring salt tolerance from *Piriformospora indica* by random overexpression in *Escherichia coli*. *World Journal of Microbiology & Biotechnology*, 31(8), 1195–1209. https://doi.org/10.1007/s11274-015-1867-5.

Ghorbani, A., Razavi, S. M., Ghasemi Omran, V. O. and Pirdashti, H. (2018). *Piriformospora indica* inoculation alleviates the adverse effect of NaCl stress on growth, gas exchange and chlorophyll fluorescence in tomato (*Solanum lycopersicum* L.). *Plant Biology*, 20(4), 729–736. https://doi.org/10.1111/plb.12717.

Goswami, K., Tripathi, A. and Sanan-Mishra, N. (2017). Comparative miRomics of salt-tolerant and salt-sensitive rice. *Journal of Integrative Bioinformatics*, 14(1), 1–18. doi: 10.1515/jib-2017-0002.

Greenway, H. and Munns, R. (1980). Mechanisms of salt tolerance in nonhalophytes. *Annual Review of Plant Physiology*, 31, 149–190.

Gupta, B. and Huang, B.. (2014). Mechanism of salinity tolerance in plants: Physiological, biochemical, and molecular characterization. *International Journal of Genomics*, 2014, 1–18. https://doi.org/10.1155/2014/701596.

Hamayun, M., Hussain, A., Khan, S. A., Kim, H. Y., Khan, A. L., Waqas, M., Irshad, M., Iqbal, A., Rehman, G., Jan, S. and Lee, I. J. (2017). Gibberellins producing endophytic fungus *Porostereum spadiceum* AGH786 rescues growth of salt affected soybean. *Frontiers in Microbiology*, 8, 686. doi: 10.3389/fmicb.2017.00686.

Hanin, M., Ebel, C., Ngom, M., Laplaze, L. and Masmoudi, K. (2016). New insights on plant salt tolerance mechanisms and their potential use for breeding. *Frontiers in Plant Science*, 7, 1–17. https://doi.org/10.3389/fpls.2016.01787.

Hardoim, P. R., van Overbeek, L. S., Berg, G., Pirttilä, A. M., Compant, S., Campisano, A., Döring, M.. and Sessitsch, A. (2015). The hidden world within plants: Ecological and evolutionary considerations for defining functioning of microbial endophytes. *Microbiology and Molecular Biology Reviews*, 79(3), 293–320.

Horie, T., Karahara, I. and Katsuhara, M. (2012). Salinity tolerance mechanisms in glycophytes: An overview with the central focus on rice plants, *Rice*, 5, 11.

Hossain, S. M. (2019). Present scenario of global salt affected soils, its management and importance of salinity research. *International Research Journal of Biological Sciences Perspective*, 1, 2663–5976.

Hyde, K. D., Xu, J., Rapior, S., Jeewon, R., Lumyon, S., Niego, A. G. T.…Nataraja, K. N., et al (2019). The amazing potential of fungi: 50 ways we can exploit fungi industrially. *Fungal Diversity* 97(1):1–136.

Jamil, A., Riaz, S., Ashraf, M. and Foolad, M. R. (2011). Gene expression profiling of plants under salt stress. *Critical Reviews in Plant Sciences*, 30, 435–458. doi:10.1080/07352689.2011.605739.

Jha, S. (2019). Transgenic approaches for enhancement of salinity stress tolerance in plants. in S. Singh, S. Upadhyay, A. Pandey, S. Kumar (eds.), *Molecular Approaches in Plant Biology and Environmental Challenges. Energy, Environment, and Sustainability*, Singapore, Springer, pp. 265–322. https://doi.org/10.1007/978-981-15-0690-1_14.

Jogawat, A., Saha, S., Bakshi, M., Dayaman, V., Kumar, M., Dua, M., … Johri, A. K. (2013). *Piriformospora indica* rescues growth diminution of rice seedlings during high salt stress. *Plant Signaling and Behavior*, 8(10): e26891. https://doi.org/10.4161/psb.26891.

Jogawat, A., Vadassery, J., Verma, N., Oelmüller, R., Dua, M., Nevo, E. and Johri, A. K. (2016). PiHOG1, a stress regulator MAP kinase from the root endophyte fungus *Piriformospora indica*, confers salinity stress tolerance in rice plants. *Scientific Reports*, 6(1), 36765. https://doi.org/10.1038/srep36765.

Johnson, J.M., Thürich, J., Petutschnig, E.K., Altschmied, L., Meichsner, D… Oelmüller, R. (2018). A poly(A) ribonuclease controls the cellotriose-based interaction between *Piriformospora indica* and its host Arabidopsis. *Plant Physiology*, 176: 2496–2514.

Khalid, M., Hassani, D., Liao, J., Xiong, X., Bilal, M. and Huang, D. (2018). An endosymbiont *Piriformospora indica* reduces adverse effects of salinity by regulating cation transporter genes, phytohormones, and antioxidants in *Brassica campestris* ssp. Chinensis. *Environmental and Experimental Botany*, 153, 89–99. https://doi.org/10.1016/j.envexpbot.2018.05.007.

Khan, A. L., Hamayun, M., Ahmad, N., Hussain, J., Kang, S. M., Kim, Y. H., … Lee, I. J. (2011). Salinity stress resistance offered by endophytic fungal interaction between *Penicillium minioluteum* LHL09 and glycine max. L. *Journal of Microbiology and Biotechnology*, 21(9), 893–902. https://doi.org/10.4014/jmb.1103.03012.

Khan, A.L., Hamayun, M., Kang, S. M., Kim, Y. H., Jung, H. Y., Lee, J. H. and Lee, I. J. (2012). Endophytic fungal association via gibberellins and indole acetic acid can improve plant growth under abiotic stress: An example of *Paecilomyces formosus* LHL10. *BMC Microbiology*, 12, 1–14. https://doi.org/10.1186/1471-2180-12-3.

Khan, A. L., Waqas, M., Khan, A. R., Hussain, J., Kang, S.-M., Gilani, S. A., … Lee, I.-J. (2013). Fungal endophyte *Penicillium janthinellum* LK5 improves growth of ABA-deficient tomato under salinity. *World Journal of Microbiology & Biotechnology*, 29(11), 2133–2144. https://doi.org/10.1007/s11274-013-1378.

Lewis, D. H. (1985). Symbiosis and mutualism: Crisp concepts and soggy semantics, in D.H. Boucher (ed.), *The Biology of Mutualism*, London, Croom Helm Ltd, pp. 29–39.

Li, L., Li, L., Wang, X., Zhu, P., Wu, H., Qi, S. (2017). Plant growth-promoting endophyte *Piriformospora indica* alleviates salinity stress in *Medicago truncatula*, *Plant Physiology and Biochemistry*, 119, 211–223.

Lopez-Llorca, L. V. and Macia-Vicente, J. G. (2009). In. Plant symbioses with fungal endophytes: perspectives on conservation and sustainable exploitation of Mediterranean ecosystems. I.S.S.N.: 0210-5004.

Lubna, Asaf, S., Hamayun, M., Latif, A., Waqas, M., Aaqil, M., Jan, R., … Hussain, A. (2018). Salt tolerance of *Glycine max*. L induced by endophytic fungus *Aspergillus flavus* CSH1, via regulating its endogenous hormones and antioxidative system. *Plant Physiology and Biochemistry*, 128, 13–23. https://doi.org/10.1016/j.plaphy.2018.05.007.

McCord, J. M. (2000). The evolution of free radicals and oxidative stress. *The American Journal of Medicine*, 108, 652–659.

Molina-Montenegro, M. A., Acuña-Rodríguez, I. S., Torres-Díaz, C., and Gundel, P. E. (2018). Root endophytes improve physiological performance and yield in crops under salt stress by up-regulating the foliar sodium concentration. *BioRxiv, preprint*, 1–24.

Morsy, M., Cleckler, B. and Armuelles-Millican, H. (2020). Fungal endophytes promote tomato growth and enhance drought and salt tolerance. *Plants*, 9, 877; https://doi.org/10.3390/plants9070877.

Munns, R., Schachtman, D. and Condon, A. (1995). The significance of a two-phase growth response to salinity in wheat and barley. *Functional Plant Biology*, 22(4), 561–569.

Munns, R. and Tester, M. (2008). Mechanisms of salinity tolerance. *Annual Review of Plant Biology*, 59, 651–681. PMID: 18444910.

Munns, R., James, R. and Xu, B. *et al.* (2012). Wheat grain yield on saline soils is improved by an ancestral Na^+ transporter gene. *Nature Biotechnology*, 30, 360–364. https://doi.org/10.1038/nbt.2120.

Ogbemudia, F., Ita, R. and Philips, T. (2018). Geospatial variability and ecological amplitudes of plants along nutrient gradients in Imo River wetland. *Asian Journal of Environment & Ecology*, 7(2), 1–10. https://doi.org/10.9734/ajee/2018/42881.

Pawłowska, J. and Wilk, M., Śliwińska-Wyrzychowska, A., *et al.* (2014). The diversity of endophytic fungi in the above-ground tissue of two *Lycopodium* species in Poland. *Symbiosis*, 63, 87–97.

Petrini, O. (1986). Taxonomy of endophytic fungi of aerial plant tissues', in N.J. Fokkema and J. van den Heuvel (eds.), *Microbiology of the Phyllosphere*, Cambridge, Cambridge University Press, pp. 175–187.

Radhakrishnan, R., Khan, A. L. and Lee, I. J. (2013). Endophytic fungal pre-treatments of seeds alleviate salinity stress effects in soybean plants. *Journal of Microbiology*, 51(6), 850–857. https://doi.org/10.1007/s12275-013-3168-8.

Redman, R. S., Sheehan, K. B., Stout, R. G., Rodriguez, R. J. and Henson, J. M. (2002). Thermotolerance conferred to plant host and fungal endophyte during mutualistic symbiosis. *Science*, 298, 1581.

Redman, R. S., Dunigan, D. D. and Rodriguez, R. J. (2011). Increased fitness of rice plants to abiotic stress via habitat adapted symbiosis: A strategy for mitigating impacts of climate change. *PLoS ONE*, 6(7), 1423–1428.

Rodriguez, R. J. and Redman, R. S. (1997). Fungal life-styles and ecosystem dynamics: biological aspects of plant pathogens, plant endophytes and saprophytes. *Advances in Botanical Research*, 24, 169–193.

Rodriguez, R. J., Redman, R. S. and Henson, J. M. (2004). The role of fungal symbioses in the adaptation of plants to high stress environments. *Mitigation and Adaptation Strategies for Global Change*, 9, 261–272.

Rodriguez, R. and Redman, R. (2008). More than 400 million years of evolution and some plants still can't make it on their own: plant stress tolerance via fungal symbiosis. *Journal of Experimental Botany*, 59(5), 1109–1114.

Rodriguez, R. J., Henson, J., Van, Volkenburgh, E., Hoy, M., Wright, L. and Beckwith, F. (2008). Stress tolerance in plants via habitat adapted symbiosis. *ISME Journal*, 2, 404–416.

Rodriguez, R. J., White Jr, J. F., Arnold, A. E. and Redman, R. S. (2009). Fungal endophytes: diversity and functional roles. *New Phytologist*, 182, 314–330. doi:10.1111/j.1469-8137.2009.02773.x.

Roy, S.J., Negrao, S. and Tester, M. (2014). Salt resistant crop plants. *Current Opinion in Biotechnology*. 26, 115–124.

Sampangi-Ramaiah, M. H., Jagadheesh, Dey, P., Jambagi, S., Vasantha Kumari, M. M., Oelmüller, R., … Uma Shaanker, R. (2020). An endophyte from salt-adapted Pokkali rice confers salt-tolerance to a salt-sensitive rice variety and targets a unique pattern of genes in its new host. *Scientific Reports*, 10(1), 1–14. https://doi.org/10.1038/s41598-020-59998-x

Sanders, D. (2020). The salinity challenge. *New Phytologist*, 225(3), 1047–1048. https://doi.org/10.1111/nph.16357.

Sekmen, A. H., Bor, M., Ozdemir, F. and Turkan, I. (2013). Current concepts about salinity and salinity tolerance in plants, in N. Tuteja and Gill, S. S. (ed.) *Climate Change and Plant Abiotic Stress Tolerance*, Hoboken, NJ, Wiley-Blackwell, pp. 163–188.

Shabala, S., Wu, H. H. and Bose, J. (2015). Salt stress sensing and early signalling events in plant roots: current knowledge and hypothesis. *Plant Science*, 241, 109–119.

Shokat, S. and Großkinsky, D. K. (2019). Tackling salinity in sustainable agriculture – What Developing Countries May Learn from Approaches of the Developed World, *Sustainability*, 11, 4558. doi:10.3390/su11174558.

Shrivastava, P. and Kumar, R. (2015). Soil salinity: A serious environmental issue and plant growth promoting bacteria as one of the tools for its alleviation. *Saudi Journal of Biological Sciences*, 22(2), 123–131. https://doi.org/10.1016/j.sjbs.2014.12.001

Smallwood, M. F., Calvert, C. M. and Bowles, D. J. (1999). *Plant Responses to Environmental Stress*, Oxford, BIOS Scientific Publishers Limited, pp. 224.

Srivastava, V. K., Raikwar S, Tuteja, R. and Tuteja, N. (2016). Ectopic expression of phloem motor protein pea forisome *PsSEO-F1* enhances salinity stress tolerance in tobacco. *Plant Cell Rep* 35(5):1021–1041.

Sultan, S. E. (1987). Evolutionary Implications of Phenotypic Plasticity in Plants, in M.K. Hecht, B. Wallace, G.T. Prance (eds.), *Evolutionary Biology*, Boston, MA, Springer, pp. 127–178. https://doi.org/10.1007/978-1-4615-6986-2_7.

Verma, S., Varma, A., Rexer, K. H., Hassel, A., Kost, G., Sarbhoy, A., Bisen, P., Bütehorn, B. and Franken, P. (1998). *Piriformospora indica*, gen. et sp. nov., a new root-colonizing fungus. *Mycologia*, 90 (5), 896–903. doi:10.2307/3761331.

Waller, F., Achatz, B., Baltruschat, H., Fodor, J., Becker, K., Fischer, M., … Kogel, K.-H. (2005). The endophytic fungus *Piriformospora indica* reprograms barley to salt-stress tolerance, disease resistance, and higher yield. *Proceedings of the National Academy of Sciences of the United States of America*, 102(38), 13386–13391. https://doi.org/10.1073/pnas.0504423102.

Waqas, M., Khan, A. L., Kamran, M., Hamayun, M., Kang, S. M., Kim, Y. H. and Lee, I. J. (2012). Endophytic fungi produce gibberellins and indoleacetic acid and promotes host-plant growth during stress, *Molecules*, 17, 10754–10773.

Wilson, D. (1995). Endophyte – The evolution of a term, and clarification of its use and definition. *Oikos*, 73, 274–276.

Yun, P., Xu, L., Wang, S. S., Shabala, L., Shabala, S. and Zhang, W. Y. (2018). *Piriformospora indica* improves salinity stress tolerance in *Zea mays* L. plants by regulating Na^+ and K^+ loading in root and allocating K^+ in shoot. *Plant Growth Regulation*, 86(2), 323–331. https://doi.org/10.1007/s10725-018-0431-3

Zhang, H., Han, B., Wang, T., Chen, S., Li, H., Zhang, Y. and Dai, S. (2012). Mechanisms of Plant Salt Response: Insights from Proteomics. *Journal of Proteome Research*, 11, 49–67.

Zhang, Q., Peters, J. L., Visser, E. J. W., de Kroon, H. and Huber, H. (2016). Hydrologically contrasting environments induce genetic but not phenotypic differentiation in *Solanum dulcamara*. *Journal of Ecology*, 104: 1649–1661. doi:10.1111/1365-2745.12648.

Zhu, J. K. (2001). Plant salt tolerance. *Trends in Plant Science*, 6: 66–71.

Zhu, J. K. (2002). Salt and drought stress signal transduction in plants. *Annual Review of Plant Biology*, 53, 247–273.

Index

T

U

W